ANNUAL REVIEW OF PLANT PHYSIOLOGY

ANNUAL REVIEW OF PLANT PHYSIOLOGY

LEONARD MACHLIS, *Editor*

University of California, Berkeley

WINSLOW R. BRIGGS, *Associate Editor*

Harvard University

RODERIC B. PARK, *Associate Editor*

University of California, Berkeley

VOLUME 23

1972

PUBLISHED BY

ANNUAL REVIEWS INC.

4139 EL CAMINO WAY

PALO ALTO, CALIFORNIA 94306, USA

ANNUAL REVIEWS INC.
PALO ALTO, CALIFORNIA, USA

Standard Book Number 8243–0623–6
Library of Congress Catalog Card Number: A51–1660

FOREIGN AGENCY
Maruzen Company, Limited
6 Tori-Nichome, Nihonbashi
Tokyo

PRINTED AND BOUND IN THE UNITED STATES OF AMERICA BY

GEORGE BANTA COMPANY, INC.

PREFACE

The Preface to last year's volume described how topics and authors are selected for the *Annual Review of Plant Physiology*. To complete the story, we will now follow a manuscript through the process that leads to publication.

When a manuscript arrives in the office of *Annual Reviews,* it goes first to the assistant editor. She edits the manuscript for spelling, grammar, sentence construction, and conformity to the style rules of *Annual Reviews.* Tables are checked for proper format, and illustrations are scaled to acceptable size and their legends edited.

The wealth of information contained in the bibliography requires special attention. There may be discrepancies in spelling of author names or variation in initials. Abbreviations of journal names are made to conform to the style established by *Chemical Abstracts,* and the references are checked for arrangement in letter-by-letter alphabetical order. Finally, the bibliography is checked against the text to be sure all references are cited, since names appearing in uncited references are not included in the Author Index at the back of the volume.

After these preliminaries, the manuscript is sent to the Editor or one of the Associate Editors, who do the substantive and scientific editing. This done, the manuscript is returned to the assistant editor, who puts it in order and sends it to the printer.

Approximately 4 to 6 weeks later, galley proofs are mailed to the author. The author corrects his copy and sends it to *Annual Reviews,* where it goes to another of the three editors to be read again. When all galleys have been returned to the assistant editor, she transfers the author and editor corrections to the press galley, arranges the articles in the order of their placement in the volume, and sends them to press for paging.

Within a month the page proof arrives. Again the chapters are divided among the Editor and Associate Editors, each one receiving the articles that he has not already seen either in manuscript or in galley. (Thus each article is read by all three editors in one of its forms.) Page proof receives careful attention, since the author does not see his work at this stage and this is the last opportunity to catch any errors that may have been overlooked. Meanwhile, another set of page proof has been sent to Mrs. Dorothy Read, who prepares the detailed Subject Index. A third set is used at the editorial office to prepare the Author Index, which lists every author cited in the volume. The Cumulative Index of Contributing Authors and the Cumulative Index of Chapter Titles from the past 10 volumes are prepared by the assistant editor. The preliminary pages, including the Preface, have already been set in type.

When all parts of the book are ready for publication, they are returned to press for final correction, printing, and binding. Early in June the production process is complete, and another volume of the *Annual Review of Plant Physiology* is ready for distribution.

The assistant editor, Jean Heavener, who prepared the preceding paragraphs, failed to mention several other matters she attends to—particularly letters, telegrams, and telephone calls related to overdue manuscripts, and the composing of letters to authors providing answers to a host of questions that invariably arise. I am sure it is obvious that the major work on some 20 to 24 manuscripts in each published book is done by the assistant editor. It is a tribute to the assistant editors of *Annual Reviews* that they direct the traffic in manuscripts, galleys, and page proofs and do the work described above not for just one book but for three each year.

Two changes in format are evident in this year's volume, both of which we hope will be useful to readers. Each chapter now includes on the title page an appropriate bibliographic citation, and each review now begins with a listing of Contents.

I wish now to thank the retiring member of the Editorial Committee, Martin Gibbs, for many important contributions during his five-year term of service. Dr. Gibbs' departure leaves a large vacancy because of his particularly broad background both in research areas and investigations, due in part to his position as Editor of *Plant Physiology*. I am pleased to welcome aboard the new member of the Committee, who this year is Dr. John B. Hanson.

My own resignation as Editor of the *Annual Review of Plant Physiology* becomes effective with the publication of this volume. I have been associated with the *Review* since its inception in 1950, serving as an Associate Editor through 1958 and since then as Editor. I have taken this action mainly because I felt I had played an influential role in the destiny of the *Review* long enough.

My decision was made difficult because of the stature and usefulness of the *Review*. This perhaps is most clearly evidenced by the almost unanimous acceptance each year of the invitations to prepare reviews by the most active investigators in the field and also by the willingness of already highly involved colleagues to serve on the Editorial Committee. The favorable reception accorded the *Review* by plant physiologists is a tribute to the work of a number of people. First are the Associate Editors. Originally there was just one—John Torrey. When he moved to Harvard, his place was taken by Winslow Briggs, then in the Department of Biological Sciences at Stanford and now at Harvard. In 1966 the *Review* acquired a second Associate Editor—Roderic Park, who was and is now my departmental colleague.

The extent to which the *Review* has successfully reflected, systematized, and made critical contributions to the development of plant physiology over the years has been the work of the Editorial Committees. How they function was described in detail in the Preface to Volume 22. The list of those involved is too long for me to acknowledge each one here, but the successive members have been thanked individually year by year.

In the first part of this Preface I described the important role of the assistant editor, presently Jean Heavener. To her and her predecessors I am most grateful for detailed work well done. Less directly involved in my own duties but nevertheless important for the completeness of the book have been the several individuals, acknowledged in the appropriate volumes, who have prepared the Subject Indexes.

My gratefulness to Murray Luck, who was Editor-in-Chief of *Annual Reviews* during most of my tenure, was expressed in an earlier Preface. Working with his successor, Robert Schutz, has been equally rewarding.

Now it is my pleasure to announce the changes in the editorial staff for next year. Winslow Briggs will be Editor, and his place as Associate Editor will be taken by my colleague at Berkeley, Russell Jones. To them and to Roderic Park, who continues as Associate Editor, my best wishes for the continued good heath of the *Review*.

LEONARD MACHLIS, *Editor*

CONTENTS

REPRINTS

The conspicuous number (7520 to 7538) aligned in the margin with the title of each review in this volume is a key for use in ordering reprints.

Reprints of most articles published in the *Annual Reviews of Biochemistry* and *Psychology* from 1961, and the *Annual Reviews of Microbiology* and *Physiology* from 1968, are now maintained in inventory. Beginning with July 1970, this reprint policy was extended to all other *Annual Reviews* volumes.

Available reprints are priced at the uniform rate of $1 each postpaid. Payment must accompany orders less than $10. The following discounts will be given for large orders: $5–9, 10%; $10–24, 20%; $25 or over, 30%. All remittances are to be made payable to Annual Reviews Inc. in U.S. dollars. California orders are subject to sales tax. One-day service is given on items in stock.

For orders of 100 or more, any *Annual Reviews* article will be specially printed and shipped within 6 weeks. Reprints which are out of stock may also be purchased from the Institute for Scientific Information, 325 Chestnut Street, Philadelphia, Pa. 19106. Direct inquiries to the reprint department, Annual Reviews Inc.

The sales of reprints of articles published in the *Reviews* has been expanded in the belief that reprints as individual copies, as sets covering stated topics, and in quantity for classroom use will have a special appeal to students and teachers.

Ann. Rev. Plant Physiol. 1972. 23:1–28

A CHEMIST AMONG PLANTS 7520

HUBERT BRADFORD VICKERY

Biochemist Emeritus, Department of Biochemistry, Connecticut Agricultural Experiment Station, New Haven, Conn.

CONTENTS

When one looks back upon more than 50 years of activity in chemistry, a natural question is how did this all start? I have a clear memory of the incident. Among the books given me by my father for Christmas when I was in the seventh grade in school in Yarmouth, Nova Scotia, was a copy of Steele's *Fourteen Weeks in Chemistry*. This was an elementary text at about high school level, but I read it with increasing interest, learned the symbols for the elements, and carried out such experiments as the preparation of carbon dioxide from baking soda and vinegar. During the subsequent 4 years in high school, I took every course that was offered in chemistry, physics, and botany, and by the time I was graduated at age 17, I knew that I wanted to become a chemist. There followed a year of teaching in a two-department country school, but I spent many evenings in the laboratory at the high school working through a text book on qualitative analysis.

In the fall of 1912, I entered Dalhousie University in Halifax, Nova Scotia, with credit for most of the courses required of freshmen. This was possible at that time because the fourth year of high school, then an appallingly rigorous course, was accepted by the colleges in Nova Scotia as the equivalent of the freshman year. But not the chemistry! Being young and very naive, I applied for entry into the honors course in chemistry. Professor MacKay questioned me thoroughly on what I had read and what laboratory experience

I had had, and then said that I might take the lectures and laboratory of the second-year course in chemistry, but that I must also do the laboratory work of the freshman year and write the final examinations of the first-year course. He indicated that if I was successful, they would consider admitting me into the honors course the following year. At Dalhousie, then as now, the willing horse got the heaviest load.

The honors course at Dalhousie allowed one to specialize in his chosen subject. He could substitute advanced courses in his field for the otherwise required courses for the Bachelor degree. In my case, this meant that I omitted Latin, history, economics, philosophy, and so forth, but took all of the lecture and laboratory courses offered in chemistry and as much of the advanced work in physics as could be crowded in. English, French, and German were of course required. They also had another interesting little trick. If one took a regular course and turned in a final examination paper that was practically perfect, he got a simple pass mark. To obtain anything better than this, he applied to the professor early in the year and was given a list of extra reading. Then at the final examination he was required either to write a second paper on this reading or to answer the first group of questions on the examination paper and then go on to a second group on the theory that if he was any good he could answer the regular questions very quickly and could spend the greater part of the 3-hour period on the really tough ones on the extra reading at the end. It was a demanding system. The attitude of the professors was, simply, here are the pearls, do as you like with them. If you successfully picked them up you got honors, otherwise you went home. Certainly, no one asked you embarrassing questions about why you failed the quiz last week.

In addition, to obtain honors in chemistry, one took all of the courses available, worked in the laboratory all day and every day, wrote the regular examinations, and then at the end of the senior year wrote a second set of four papers on everything one had been taught or had read about during the entire 4 years. There was also a requirement for a brief research, which in my case was a revision of one of the methods for the qualitative analysis of the alkaline earths. I was able to determine the limits within which each earth could be detected in the presence of excess of the others.

After graduation with honors in chemistry and chemical physics in the spring of 1915, there followed 2 years of teaching science in one of the high schools in Halifax, but the third year was interrupted on December 6th, 1917, by the explosion of a munitions ship in Halifax harbor which destroyed the entire north end of the city, killed more than 2000 people, and injured many thousands more. Luckily the school in which I taught was well over the brow of the hill above the point at which the ship exploded so that, although every window was broken and there was much structural damage, no one in the building was killed although most of us were cut by flying glass. If that morning I had gone up to my room instead of stopping to talk a few minutes with one of the other teachers, I should have been found later under the bricks of a chimney that had been overturned and had broken through the roof directly above my desk and chair.

It was many weeks before the school system of the city could be reorganized, and meanwhile I learned that the Imperial Oil Company had a position for a chemist at the refinery that had just been built across the harbor in the town of Dartmouth. However, 18 months as an analytical chemist in the oil industry convinced me that teaching offers a much more pleasant life, and I applied for and obtained the post of science teacher at the Provincial Normal School in Truro, Nova Scotia. Meanwhile I had completed the requirements for a Master of Science degree at Dalhousie by an extensive course of reading and a research project in which I determined the approximate position of the isochlor lines for the natural waters of the Province (1), a problem of some importance in sanitary surveys and one in which Professor MacKay was much interested.

My career as a teacher of prospective teachers was, however, a short one. In the early spring of 1920, I was summoned to Dalhousie to undertake the lectures and supervision of the laboratory classes for all of the courses in chemistry currently taught. Professor MacKay had died in the fall, and Professor Nickerson, who had attempted to carry on alone until further assistance could be found, had become ill and had collapsed in class. Looking back, I think that those two months when I tried to do the work of two men were about the toughest of my entire life. It was, of course, an impossible task, but the students were considerate, they also worked hard, and most of them passed their examinations successfully. And the university was not only kind but generous. They paid me well, and at the end rewarded me with an appointment to an 1851 Exhibition scholarship.

This scholarship, funded by the profits made from the great Exhibition of 1851 in London under the shrewd direction of Prince Albert, the husband of Queen Victoria, is not as well known as the later established Rhodes scholarship which was patterned after it. The Rhodes has no restrictions, but the 1851 is restricted to British citizens among whom it carries prestige equal to the Rhodes, and is further restricted to science. Today it is a moderately valuable scholarship, but in 1920 it was only 200 pounds, and the pound had dropped seriously in value just after the war. It was not much, but my wife and I decided that it was enough. Acting on the advice of Dr. Howard Bronson, the professor of physics at Dalhousie, a Yale man, the fall of 1920 found us in New Haven registered in the Yale Graduate School. During the next 2 years my wife earned the degree of PHT while I earned the PhD.

For the second time an element of pure chance directed the course of my scientific career. It happened that one of Dr. Thomas B. Osborne's assistants at the Connecticut Agricultural Experiment Station resigned early in 1921 while in the midst of a research problem in which Osborne was greatly interested. Chancing one day to meet Professor Treat B. Johnson, with whom I was studying at Yale, he asked if there was a young man in Johnson's group who would be willing to come out to the Station and do his research there.

Johnson had put me to work on a problem supported by government funds on the nitration of benzene, and it occurred to him that, come summer vacation, the man on this problem might be required to continue the work in one of the government military laboratories. Although security matters were not so gravely considered in those days as they were in later years, it was obvious that a Canadian citizen might not be welcome under the circumstances. Accordingly, a few days later, Johnson walked out to the Station laboratory with me and introduced me to Osborne, who asked if I would be willing to do my research under his direction. To a young and impecunious student, the chance to work in a large, well-equipped, and comfortable laboratory under a man of international reputation, with no fees or charges for reagents or apparatus, offered a choice to which there was only one answer. A few days later, after some intensive reading on fundamental protein chemistry, I began work in the laboratory where I have spent the rest of my professional life.

THE CONNECTICUT AGRICULTURAL EXPERIMENT STATION

Osborne was an extraordinary man. He was an acknowledged world authority of the time on the chemistry of proteins. A charming although sometimes crusty gentleman of independent wealth, for more than 30 years he had directed a laboratory which had demonstrated that the proteins of plant seeds are substances of definite and ascertainable chemical properties, that each seed protein differs in some respect from all others and is thus a specific substance, and together with Professor Lafayette B. Mendel of Yale, had, during the previous 10 years, shown that proteins may differ from each other in their nutritive effects. They had demonstrated that lysine and tryptophan are essential in the nutrition of rats, and had independently discovered vitamin A. Currently the work of the laboratory had to do with the occurrence of vitamin B in such food materials as fruits and leaf tissues as well as in yeast, and especially with the effect of carefully controlled diets of known composition on the phenomena of growth. Osborne himself was engaged in a study of the proteins of green leaves and in the fractionation of extracts from leaves in the hope of obtaining some evidence concerning the chemical nature of vitamin B. He set me to work on a study of the stability to acid hydrolysis of the form of nitrogen in proteins which is liberated as ammonia, the so-called amide nitrogen. At that time it was purely speculative that this nitrogen is combined in the protein as the amide groups of asparagine and glutamine.

The problem developed into a study of the rate at which the protein gliadin from wheat is hydrolyzed, with respect to both amide and peptide bonds, by boiling acid or alkali as a function of the concentration of the hydrolyzing reagent. A year later I was able to present to the authorities at Yale a dissertation which, to my intense relief, they accepted (2). The oral examination was more or less of an anticlimax. With the exception of Johnson himself, there was no one at that time on the chemical faculty at Yale who knew enough about protein chemistry to ask any but the most elementary questions, and I survived without too much damage.

Osborne then offered me an assistant's position in the laboratory, which at that time was mainly supported by annual grants from the Carnegie Institution of Washington. The salary was only a little more than I had been earning as a teacher in Nova Scotia, for Osborne was extremely careful with his budget, and he pointed out that under annual grants there was no assurance that the position would last for more than the current year. But the opportunity to continue a life of research in a laboratory directed by such an outstanding scientist was far too good to be missed, and I have never regretted my decision.

During the next 2 years, the laboratory was enlivened by the presence of A. C. Chibnall, who came to us from Schryver's laboratory at Imperial College in London. He had been busy for some time with efforts to prepare proteins from cabbage leaves after plasmolysis of the cells by treatment with ether water. Osborne and Wakeman's publication in 1920 of a paper describing a protein preparation from spinach leaves, which was the first on the proteins of leaves to be published for more than a century, stimulated him to apply for a postdoctoral fellowship which brought him to the only other laboratory in the world where work in this field was in progress. He arrived almost unannounced before Osborne got back from his prolonged summer vacation—which ended only after the partridge season opened—so I arranged for Chibnall to join Professor Mendel's group. When Osborne returned a few weeks later, I told him of this Englishman who wanted to join us. Osborne was very busy that morning, and my tale was greeted with a succession of grunts and finally, "Well, is he any good?" Assured on that point, he finally said, "All right, bring him out."

Chibnall succeeded in charming Osborne within the hour, and I was instructed to install him in the laboratory at once. As soon as Osborne saw his command of technique, his original approach, and his industry, he happily turned over all of the work on leaf proteins to him, and during the next 2 years Chibnall established methods for the isolation of preparations of leaf proteins that served him as well as others for the next decade or more to carry out distinguished research. Meanwhile we formed a friendship that, in spite of innumerable arguments, has lasted to the present day.

The Basic Substances in Alfalfa

Osborne invited me to share with him and Dr. Alfred J. Wakeman in the work on the search for evidence of the presence of vitamin B in fractions prepared from the juice of the alfalfa plant. At that time, the only methods available for the separation of nitrogenous organic substances involved the use of salts of heavy metals as precipitants. Silver, mercury, and lead salts, and phosphotungstic acid were the main reagents. Hydrogen sulfide was used to decompose the precipitated organic compounds, and barium hydroxide to decompose the phosphotungstic acid precipitates. All reagents must be removable by some simple technique. Wakeman had been using mercuric acetate and sodium carbonate, added alternately until an excess was present in a faintly alkaline solution, to precipitate amino nitrogen from extracts of the alfalfa plant, and nutrition tests for the presence of the vitamin in such fractions were in progress, a slow matter at best. It seemed to me that one must work on as large a scale as possible if one were to obtain chemical evidence for the identity of the substances in such fractions, for isolation of crystalline products and ultimate analysis were the only acceptable proofs of identity. The simple chromatographic methods on trace amounts of material used today were then many years in the future. The progress of the fractionation was to be followed by determinations of total, amino, and amide nitrogen.

Accordingly, I undertook the analysis of an extract prepared by grinding a large quantity of freshly cut alfalfa and expressing the juice with the hydraulic press. The juice was treated with an equal volume of alcohol which precipitated any protein present together with much inorganic material. The filtrate was then concentrated and "clarified," as it was called, by the addition of an excess of basic lead acetate, and lead was removed from the filtrate. Next was treatment with an excess of mercuric acetate and sodium carbonate, a procedure advocated years before by Neuberg and Kerb as a general method to precipitate amino acids. Removal of reagents, followed by precipitation of basic substances by phosphotungstic acid and treatment of both precipitate and filtrate with barium hydroxide to liberate the organic bases and remove the reagent, completed the process. This was done on solutions from both the mercuric acetate precipitate and the filtrate from it. One finally obtained fractions that were concentrated for fractional crystallization. The base fractions were treated with silver sulfate according to the Kossel method for the separation of the basic amino acids, and arginine and lysine were sought for before and after severe acid hydrolysis.

The outcome of this and several subsequent elaborate fractionations was most disappointing. After more than 3 years of study, Charles Leavenworth and I were able to account for only about 55 percent of the nitrogen and 30 percent of the solids in the alfalfa extract as identified substances weighed as well-characterized products in crystalline form. The main components were asparagine, a few amino acids separated by the Fischer esterification process and including serine, which was isolated from a plant for the first time, together with arginine, lysine, stachydrine, choline and adenine, and trace amounts of a few other basic substances (3). In 1958 in Canada, while studying stachydrine synthesis in alfalfa with the use of modern chromatographic methods, Marion found that homostachydrine, the betaine of pipecolic acid, is also present and is a contaminant of samples of stachydrine obtained by the classical procedure. He was kind enough to examine several of my preparations and, to my delight, found that one of them was in fact pure.

Many modifications of the procedure were made. For example, basic lead acetate was given up as a clarification reagent as the result of the work of a student, Carl Vinson, who joined us for awhile and isolated adenine, arginine, lysine, stachydrine, aspartic acid, and tyrosine, in small amounts to be sure, from a carefully washed basic lead acetate precipitate obtained from an extract from alfalfa. Obviously it is impossible to wash such voluminous precipitates adequately, since these substances were among the main components of the original extract, and only aspartic acid should form an insoluble lead compound. Inasmuch as barium hydroxide must be used frequently in later stages of the analysis, we subsequently clarified the extracts by adding an excess of barium hydroxide and alcohol at the start of the operation. This threw out a huge precipitate of barium salts of organic acids, mainly malic and malonic acid, together with barium compounds of much carbohydrate material. It also clarified the solutions as well as the lead reagent and precipitated little nitrogenous material other than any aspartic acid present.

The identification of adenine provided us with a puzzle for several months (4). While I was on vacation one summer, Leavenworth had observed that the substance we had isolated as a picrate, which crystallized in fine hair-like needles that decomposed at 298°, is precipitated from the extract by silver salts at acid reaction. This is a general property of purines, and we accordingly tried all known tests for purines on it. It did not give the murexide test, a fact which in the light of later knowledge should have been strong evidence that the substance was adenine, but any suspicions of this we entertained were suppressed by the descriptions of adenine picrate in the literature. This substance was alleged to separate in small rhombic crystals which decompose in the vicinity of 281°. One or two papers mentioned hair-like needles, but where this appearance was noted no decomposition point was given. Furthermore, in previous years Leavenworth had repeatedly isolated adenine of the behavior described in the literature from liver. Fortunately, Professor Mendel still had one of these old preparations. When heated under the conditions we were using in determing the decomposition point of our preparation from alfalfa, it decomposed at 290°. To settle the matter, we isolated adenine by the conventional process from liver, obtaining a preparation of the picrate that decomposed at 295°, converted this to the beautifully crystallized sulfate, and then back to the picrate. This time it crystallized as long pale yellow hair-like needles that decomposed sharply at 298° and was obviously identical with our preparations from alfalfa. This taught us a great deal. We no longer accepted descriptions in the literature at face value, and we learned that to assure ourselves of the purity of a preparation, it is necessary to crystallize the substance as at least two different derivatives or salts. Only in this way may major impurities be eliminated. The adenine isolated from alfalfa apparently separated as its picrate with little or no trace amounts of other purines as contaminants. It was too pure to be easily recognized!

The nutrition tests for the presence of vitamin B in the fractions from alfalfa were unsatisfactory, so I turned to a study of brewers yeast, an especially rich source of the vitamin. In the course of the fractionation I obtained a specimen of nicotinic acid, a substance which had previously been obtained from yeast by Funk and years later was recognized as one of the water-soluble B vitamins. In the 1920s, however, the observations that adenine is present in alfalfa and nicotinic acid in yeast were simply moderately interesting facts. The biochemical relationships of these substances were not recognized until many years later. In spite of many frustrations, however, we had learned a great deal about the composition of protein-free extracts from plants. Amino acids are present in appreciable amounts, asparagine often being the dominant one. They were also present in combinations which we took to be peptides since acid hydrolysis increased the amounts of amino nitrogen. The bases arginine and lysine occurred not only free but also in peptide combination. Furthermore purines, especially adenine, were present together with methylated bases, frequently in large amounts, the major one in alfalfa being stachydrine, in tobacco nicotine, in yeast choline. The greatest difficulty with the methods of analysis then available was, however, the interpretation of what had come to be called the basic nitrogen, that is, the total nitrogen precipitated by phosphotungstic acid. To be sure, much of this nitrogen did indeed belong to known basic substances, but considerable amounts could be shown to be present as peptides which contained monoamino acids. In addition, the barium phosphotungstate obtained when decomposing the alleged base precipitates contained much nitrogen which stubbornly resisted extraction even by boiling hydrochloric acid. The acid did indeed extract an assortment of monoamino acids from the precipitate, but never all of the nitrogen. Today we know that extracts from leaves may contain 50 or more soluble nitrogenous substances demonstrable by paper chromatography, but the identity of the peptide-like material that we encountered is still unknown, and in no case has anything approaching a quantitative accounting of the soluble nitrogen of extracts of leaves been obtained.

BASIC AMINO ACIDS

By 1927 it had become clear that the current methods for the analysis of extracts of plant tissues were hopelessly inadequate, and I accordingly turned to a study of the methods for the determination of the basic amino acids of proteins, a far simpler problem. Leavenworth and I improved the method for the separation of the silver compounds of histidine and arginine as a result of the observation that histidine silver is completely precipitated at pH 7.0 and that arginine silver does not begin to precipitate until the reaction becomes considerably more alkaline (5). Clearly the silver compounds form most completely at or near the isoelectric points of the respective bases. This led to a number of analyses of the bases yielded by several proteins with results which we believed were more nearly accurate than any previously published (6), and which were of material assistance to Professor Edwin Cohn at Harvard, with whom we collaborated in his attempts to account for the acid-binding capacity of proteins. We developed gravimetric methods to determine arginine and histidine, using flavianic acid for arginine and 3,4-dichlorobenzenesulfonic acid, a reagent suggested by Max Bergmann, for histidine, and learned to prepare both bases in quantity by fairly simple procedures. We also had the satisfaction of preparing lysine as a free base in crystalline form for the first time (7). A student, Charles A. Cook, who worked with us in 1931, prepared crystalline ornithine also for the first time.

TOBACCO LEAF CHEMISTRY

Meanwhile the work on the chemistry of leaf tissue also went on. At a conference in 1928, after Osborne had retired, and in which Chibnall, who had come over for a visit, joined Osborne and me, we discussed the direction in which the work should go. It was obvious that the complete analysis of the nitrogen compounds present in a leaf was impossible with available methods. Alfalfa was not a good choice of a leaf tissue to study since the plant has small leaves and is cut with a scythe; the tissue worked with was therefore largely stem. If we were to learn anything about the metabolism of the leaf we needed a large leaf from a plant easily grown in quantity in Connecticut, and especially one of known genetic relationships that would be available from year to year. The obvious choice was the tobacco plant, a major crop in the state. It is an amusing instance of my continued naiveté that I at no time considered how this decision would be regarded by the administration of the Station. When I told Director Slate of our plans, I was met with an enthusiastic response and the promise of greatly increased financial support. As a result, Dr. George W. Pucher joined us, and during the next 20 years the team of Vickery and Pucher, together with their associates, published no less than 74 papers and Station Bulletins mostly dealing with the chemistry of the tobacco plant.

Pucher was the most skilled and thorough laboratory worker that I have ever met. His approach to a problem was always original, difficulties were patiently surmounted, and in his work on the development of analytical methods, no possible source of interference was overlooked. When he had finished a project to his own satisfaction, he would prepare a voluminous report supported by table after table of data, and then, if I did not interfere, he would put notebooks and reports away in the vault and go on to the next project. Once he had found out how to do something, he lost all interest in the detail of how the results had been accomplished. It devolved upon me to prepare such reports for publication. In doing this, I recall no instance when I had failed to understand some point and questioned him that he could not find in his notebook some experiment or test that settled the matter at once.

From the start, Pucher was intrigued by the analytical problems presented by the tobacco plant. Nicotine is just volatile enough in hot alkaline solution to interfere with the determination of ammonia, a matter of fundamental importance in the study of amide metabolism. He solved that one, first by the use of Permutite, a cation exchange resin, to absorb the ammonia, filtering it and then distilling the ammonia from the Permutite (8). Later he devised equipment for distillation in vacuo with the use of alkaline buffers, a rapid, convenient, and accurate method now widely used (9). However, he soon settled down to the study of the organic acids of the plant. It was clear from the composition of the barium hydroxide and alcohol precipitates we were obtaining when clarifying leaf extracts that organic acids make up a substantial part of the organic solids present, sometimes even exceeding the amount of protein in the leaf. Furthermore, the tobacco technologists were interested in organic acids since the prized "graininess" of the surface of finished cigar wrapper tobacco was alleged to arise from crystals of salts of organic acids. In 1930, the only method known for the separation and determination of the common plant organic acids was the distillation of their esters in vacuo. Reliable specific methods were available for very few except for oxalic acid, although there was a polarimetric method to determine malic acid and a gravimetric pentabromoacetone method to determine citric acid, both subject to various uncertainties. Pucher greatly improved the technique of the ester distillation method to identify and determine the organic acids of the tobacco leaf (10). A few years later, his mastery of the method enabled him to identify the so-called "crassulacean malic acid" of the *Bryophyllum* leaf as isocitric acid (11), a fact which led in turn to an extensive study of the organic acid metabolism of this and related plants, as well as to the development of methods to prepare this almost unknown but very important acid in quantity and to improve its synthesis (12, 13). Both of these problems were carried further by me after Pucher's untimely death in 1947.

What was probably Pucher's most significant contribution was his development of a succession of methods to determine citric acid. When citric acid in solution in the correct concentration of sulfuric acid is treated with potassium bromide and potassium permanganate, it is converted quantitatively into pentabromoacetone. The earlier method described in the literature involved weighing this substance and, with proper corrections for solubility, gave excellent results. It occurred to Pucher that it should be possible to decompose this compound and titrate the bromide ion produced. Obviously this could be done satisfactorily on a smaller amount of citric acid than was necessary for the gravimetric method, and held out hope that a semi-micro method could be developed. The reagent chosen for the debromination was sodium sulfide, and when the appropriate conditions were found, and especially when the beautiful titration method of Sendroy was applied to determine the bromide, an excellent method accurate in the 1 to 16 mg range became available (14, 15). But Pucher did not stop there. He had noted that the debromination was accompanied by the appearance of a yellow to red color, depending on the amount of citric acid taken for the test, when the petroleum ether extract containing the pentabromoacetone was treated with sodium sulfide. This led to the development of an accurate colorimetric method to determine quantities of citric acid of the order of 1 mg or less (16), a method that subsequently has been modified by many workers, although without notable improvement in either accuracy or convenience, but which served essentially in its original form to enable Krebs to develop the tricarboxylic acid cycle hypothesis as the explanation of respiration in pigeon breast muscle, for which he later received a Nobel prize.

A valuable by-product of the study of the method to determine citric acid was the observation that any malic acid present in the mixture of organic acids subjected to oxidative bromination is converted quantitatively into a substance which is volatile with steam and which forms a very insoluble compound with dinitrophenylhydrazine. This product, when dissolved in hot pyridine and made alkaline, gave a strong blue color in every way suitable for the colorimetric estimation of malic acid (17). The method was undeniably tricky, many precautions were essential, and rigid adherence to the technique was necessary. However, in skilled and experienced hands it gave accurate recoveries of malic acid and served us for years until the development of a simple chromatographic method by Palmer in our laboratory in 1954. I shall never forget the delighted expression on Otto Folin's face when he visited us one day soon after Pucher had worked out the details of the method and demonstrated the magnificent blue color of the final solution to him. Folin did not understand the chemistry of the reaction, nor for that matter did we, but he fully appreciated the significance of that color. It was years later that the volatile oxidation product of malic acid was identified as glyoxal.

In our studies of the forms of nitrogen present in extracts from leaves, we had frequently encountered evidence for the presence of a substance which is rather easily decomposed with the formation of ammonia. H. E. Clark, who worked with us in the mid-thirties, found a great deal of this in the stems of tomato plants raised upon ammonium salts as the source of nitrogen. The behavior was that to be expected of glutamine, and Clark accordingly undertook the isolation of glutamine by the classical procedure of Schulze. He was rewarded with several grams of beautifully crystallized glutamine, then one of the rarest of the amino acids. This observation led to many further studies of amide metabolism as well as to the modernization of the Schulze procedure for preparing glutamine in quantity from beets. A micro method to determine it was worked out in collaboration with Chibnall, and, within a few years, glutamine (of various degrees of purity to be sure) became a regular item in chemical catalogs although an expensive one.

HISTORY OF CHEMISTRY

In 1927, Osborne was asked by the editors of *Physiological Reviews* to prepare a review of the literature dealing with the many hypotheses that had been advanced to account for the structure of proteins. Osborne himself was too busy in this his last year as head of the laboratory and asked me to write the paper. For months, I read the early and more recent literature and wrote my account of it. We had an excellent library and collection of reprints going back to the earliest days of the subject, and I rarely had to go down to the Yale library and then only for obscure papers. Each morning, when Mendel came out for his daily conference with Osborne on the nutrition work, I was asked to read what I had prepared. I was then subjected to the criticism of two of the keenest minds in American biochemistry. No infelicity of English expression escaped their attention, and I was showered with helpful suggestions for improvement and extension of what I had done. It was the most thorough course in technical writing I have ever experienced, as well as a revelation of what two masters of their own fields knew about their subjects. The outcome was a manuscript of which I was, I hope pardonably, proud

(18). Then the question of authorship came up. I maintained that since Osborne had been asked to write it and had shared in every paragraph, he should be the author. He replied that since I had written it, I was the author and he had had little to do with it. The debate grew quite heated, but I finally closed it with the question, "Are you ashamed to have your name on this paper?" Osborne thought for a moment and then said, "All right, but you are the senior author." For the second time in his life Osborne took the second position in authorship; the first was a paper with Chittenden in 1892!

In 1924, Mendel had asked me to give a course of ten lectures on protein chemistry in the department of Physiological Chemistry, an appointment which was continued annually until my retirement in 1963. Second-year graduate students were required to take it, and it served for a long time as a means to eliminate students who were unwilling or unable to take things seriously. Although this requirement was modified as time went on, the course was well attended. In the early years it was the only formal instruction in protein chemistry given at Yale. Later, as protein chemistry became more and more to be treated in the regular courses, I turned to fundamental amino acid chemistry as my subject with an occasional course on the history of protein chemistry. In 1930, I took occasion to work up lectures on the discovery of each of the protein amino acids. That fall, when visiting Professor C. L. A. Schmidt in Berkeley for a few days, I found that he had also worked up and delivered a similar course. We discussed the matter of preparing a joint paper for one of the review journals and sat down at once and divided up the list of amino acids for our individual treatment. Later we exchanged manuscripts as they were written, edited each other's material, and finally offered the whole to *Chemical Reviews*. To my surprise it was at once accepted, and in the printed form took 149 pages of the journal, nearly a whole issue (19). I suspect that that paper has repeatedly been the subject of seminar discussions in biochemical departments. The stock of reprints was exhausted in a few months, and I still get an occasional request for one, an evidence that the history of biochemistry is a matter of increasing consideration even to the present generation of students to most of whom a paper 10 years old is hopelessly outdated and of no further account.

Recently, at the request of the editors of *Advances in Protein Chemistry*, I have brought the material in that now 40-year-old review up to date. During this period at least 16 new amino acids have been detected in hydrolysates of various proteins, mainly of animal origin. Most of them are derivatives of the 20 amino acids that make up the list of commonly found products of the hydrolysis of proteins, and are the result of highly specific interactions of enzyme systems with already assembled polypeptide chains whereby various methylation or oxidation or halogenation reactions occur. This is an active field of study at the present time, and there is little doubt that the list of authentic protein amino acids will ultimately be at least twice as long as that recognized in 1931.

One of the most interesting bits of historical research I have been concerned with is the matter of the origin of the word protein. Since it first appeared in his papers, I had always assumed that the term was invented by Mulder in the late 1830s as a designation for the radical which he thought combines with phosphorus and sulfur in the so-called albuminous substances as they occur in nature. In 1948, Sir Harold Hartley gave an address in Stockholm on the occasion of the centenary of the death of Berzelius. He pointed out the great service Berzelius had made to chemistry in the matter of nomenclature, and in listing terms suggested by Berzelius mentioned the word protein. My friend Chibnall picked up this point in the published account of Sir Harold's lecture and wrote a concerned letter to me on the mistake I had made in a published paper on proteins. Correspondence, and later a delightful personal interview with Sir Harold, led to the preparation of a brief note on the origin of the word to be published simultaneously with one of his own (20). He had not appreciated the fact that Berzelius' suggestion had been totally overlooked by protein chemists.

It appears that Mulder was accustomed to send his analytical data to Berzelius for discussion—the correspondence between the two men runs to hundreds of pages in Soderbaum's magnificent edition of the Berzelius letters—and in a letter early in 1838 he set forth his idea that the albuminous materials in their native condition are compounds of "the animal substances lacking sulfur and phosphorus" with sulfur and phosphorus. This was an awkward expression and it seemed to Berzelius that a new technical term was required. In a letter to Mulder written on July 10, 1838, he suggested the word protein, derived from the Greek for "to be in the first place," as a suitable one on the grounds that these substances are clearly in the first place in human and animal nutrition. Mulder had not asked for a suggestion, and Berzelius obviously felt no strong proprietary rights in this one, for there are many words derived from the Greek *protos* meaning first. At all events, Mulder gladly accepted the suggestion and used the new word freely in a series of remarkable publications in which he expounded his ideas. Later, as a result of the long and bitter polemic with Liebig, who soon showed that Mulder's view was inadequate, the meaning of the word was expanded and later was universally adopted as a far better term than the German *Eiweiss* or English *albumin* with their implied restriction to the proteins of eggs.

REVISION OF THE NOMENCLATURE OF AMINO ACIDS

At a meeting of the editorial board of the *Journal of Biological Chemistry* in 1945, the current confusion in the nomenclature of the amino acids was discussed. For example, two papers on the metabolism of tryptophan had been published recently in the journal in one of which the amino acid was named $l(+)$-tryptophan, in the other $l(-)$-tryptophan although the same substance was meant. The difficulty arises from the fact that the specific rotation of six of the common protein amino acids is dextro if observed in acid solution but levo if observed in water. A further difficulty had arisen with the name of threonine. This had been called $d(-)$-threonine by W. C. Rose, its discoverer, but since its α-carbon atom has the configuration of the l-family (the β-carbon is d), it was frequently referred to in the literature as $l(-)$-threonine or even simply as l-threonine.

The suggestion made by H. B. Lewis and W. M. Clark, that the whole problem could be resolved if the small capital letters L and D were used as prefixes, was enthusiastically accepted. These prefixes were intended to designate the configurational family to which the α carbon belonged. I was asked to prepare a proposal to be made to the nomenclature committee of the American Chemical Society, and also to similar committees in Britain, for the complete reform of the nomenclature of amino acids. The outcome was a set of rules of nomenclature, developed after many revisions with the aid and criticism of committees in both this country and abroad, which was accepted as official by the American Chemical Society in 1947 and at once came into general use. The International Union of Pure and Applied Chemistry approved them in 1952 (21).

However, the rules as then worked out failed to deal satisfactorily with the difficult problem of naming the isomers of amino acids that have more than one center of asymmetry. Our committee had one version, our British colleagues another, and a number of insoluble problems had begun to accumulate. In 1961, I attempted to resolve this difficulty by appropriating the prefixes of carbohydrate nomenclature to the names of the more complex amino acids so as to define the configurational relationship. This general method of naming these substances had already been used by a few workers, but there was no consensus regarding precise details. Having had considerable experience with the behavior of committees on nomenclature, I approached the top committee, that of the International Union of Biochemistry, simultaneously with my approach to the national committees, and supported my suggestion with letters of approval from several leading amino acid and carbohydrate authorities. It worked; and in spite of my flagrant breach of protocol, the rule was officially approved within only 2 years (22). It serves as at least a temporary solution of the difficulty we had had 15 years before, and the new rule is now in common use.

BIOCHEMISTRY OF LEAVES

Pucher's sudden death in the fall of 1947 from a heart attack left the laboratory in an extremely awkward position. Wakeman had retired a few years before and had been succeeded by Marjorie Abrahams, who had become an expert in the methods to determine the organic acids. Pucher was in the midst of his studies of the organic acid metabolism of the *Bryophyllum* leaf and with the fundamental chemistry of isocitric acid. Leavenworth and I were busy with the basic amino acids and with glutamine, work in which Pucher had also shared, and there were two long reports with accompanying notebooks in the vault, one on the determination of starch in leaves (23), and a second on an elaborate experiment that Pucher had carried out at my suggestion in which a number of organic acids had been fed in culture solution to tobacco leaves (24). These had to be prepared for publication without the occasional conferences which had made the writing of previous papers a much simpler and far more pleasant task. Leavenworth's death on the first anniversary of Pucher's deprived me of a valued collaborator and friend with whom I had worked for more than 25 years, and it became clear that the program of the laboratory must be thoroughly revised. Hitherto I had spent most of my time on the basic amino acids and on amide metabolism in leaves. Pucher had worked with the organic acids and carbohydrates, and we worked together on many experiments in which detached tobacco leaves were studied during culture under various conditions and during the technical operation of curing.

In all of our culture work, we had depended upon the selection of a series of samples of leaves that we hoped would be initially identical with each other at the start of the treatment. One or two samples would be prepared for analysis at once to furnish a base line for the detection and measurement of the chemical changes which subsequently occurred during the treatment. We tried many methods. Random selection from a large pile of leaves, adjustment of the samples to initial equal weight, selection of leaves of the same size, and so forth. No method we had been able to devise gave a set of eight or ten samples in which the total nitrogen content, for example, had a coefficient of variation of less than about 5 percent. Thus our conclusions regarding any chemical changes that had occurred during the treatment were always subject to a sampling error of at least this magnitude, and we were constrained to accept as real only changes of the order of 10 percent or more. This was clearly not good enough.

STATISTICAL SAMPLING OF LEAVES

I enlisted the aid of Dr. Chester Bliss, the statistician on the station staff, took him out to the greenhouse, and asked him how one could pick a set of ten samples of initial identical composition from a row of large handsome tobacco plants. He examined the plants for a few minutes and then pointed out that there are three sources of variation to be considered. One arises from differences between individual plants, a second from the position of the leaf on the plant and a third, which he called a random component of variation, from such matters as slight differences in the composition of the soil in which each plant grew, slight differences in exposure to light, and such accidents as minor damage from insects and so forth. If one picked the leaves in such a way that each plant and each position of the leaf on the plant were equally represented in each sample, one should be able to minimize the variation between samples, and thereby reduce to an acceptable level the coefficient of variation of any component which was not altered by the treatment. In a few minutes, he worked out a method of picking the leaves, based on a systematic Latin square, and suggested that we try it.

Our response was to pick sets of samples from a group of *Bryophyllum calycinum* plants according to three systems. In one system reliance was placed on the assumption that opposite leaflets on the same pentafoliate leaf should be nearly identical in composition, in another we assumed that leaves of the same size should be identical, and in the third we picked the leaflets according to the systematized Latin square so that each plant and each leaflet position was equally represented in each sample. All samples were dried for analysis, and solids, ash, total nitrogen, protein nitrogen, and starch were determined. Bliss then subjected the data to an elaborate analysis which showed the great superiority of the statistical method (25). To obtain an accuracy equal to that given by this method, from four to five times as many samples would need to be taken by either of the other methods for each point examined. In other words, for equal accuracy the statistical method is about five times as efficient in terms of the analytical work required. Subsequently, we always obtained sets of samples of tobacco or *Bryophyllum* leaves which gave coefficients of variation of the nitrogen or ash content and sometimes even of the fresh weight of less than 2 percent. Since the analytical error for most components was itself of the order of 1 percent, it is clear that a considerable improvement in the reliability of our work had been made. There is a great deal of satisfaction, when one comes to plot the data from such a set of samples, in having the figures for some component that was not affected by the treatment yield a horizontal straight line. Admittedly, there are certain difficulties. To obtain samples for some of the experiments with *Bryophyllum* leaves, my whole crew had to turn up in the greenhouse well before sunrise and put in an hour or so of concentrated hard work. One's popularity on a bright summer morning is not improved by demands of this kind, but all later admitted, when the data were ready, that it had been worthwhile.

There were three lines of investigation that seemed important; the continued study of Crassulacean metabolism, an examination of the effects of culture of tobacco leaves in solutions of organic acids, and renewed study of the chemistry of isocitric acid, for our laboratory was probably the only one in the world where this hitherto extremely rare substance was available in substantial quantities.

Crassulacean Metabolism

Plants of the family Crassulaceae exhibit the phenomenon of diurnal variation of acidity to an extreme degree, for from 15 to 20 percent of the organic solids are often involved in the diurnal changes in composition that occur. During the night the concentration of organic acids increases at the expense of starch, while during the day it decreases again and starch accumulates. Malic acid, and to a lesser extent citric acid, are the chief acid reactants.

Because of an unfortunate systematic error in calculating the protein content of *Bryophyllum* leaves exposed to diurnal variation of light, a few experiments that Pucher had recently carried out seemed to show that protein varies in amount in a manner correlated with the variation in organic acid content. Luckily the error was found and corrected in an unhappy paper that we published shortly before Pucher died, but it was necessary to check a number of points and to use the new sampling method so as to improve our general accuracy. Accordingly, a long series of culture experiments was planned in which we hoped to confirm and follow up many of the earlier observations. One that caused us considerable trouble was an observation that starch may accumulate in small amounts in *Bryophyllum* leaves excised at daybreak and cultured in darkness, that is to say exposed to an artificially prolonged night. This was surprising as one normally assumes that starch is a product of photosynthesis. How could it accumulate in the dark? The outcome was that the starch in such leaves did indeed continue to diminish for the first few hours after the leaves were picked as malic acid continued to increase, but after the malic acid reached its maximum and began to diminish, as it invariably does in prolonged darkness, starch increased moderately for a few hours. Thus, if malic acid is at a high enough level, starch is formed independently of the illumination (26).

We also repeatedly confirmed the observation that isocitric acid in *Bryophyllum calycinum* leaves undergoes little if any change in amount during culture of the leaves either in light or darkness, this in spite of the later observation of Attila Klein, who worked with us in 1962 and made use of a specimen of isocitric acid labeled with ^{14}C that I had prepared. He found that isocitric does indeed play an important part in the general organic acid metabolism of the *Bryophyllum* leaf (27). To be sure, we had seen an occasional set of samples, particularly one which was unusually low in starch at daybreak, in which the isocitric acid was drawn into the metabolism for a short time after the leaves were maintained in culture in darkness, but as soon as the malic acid began to diminish, the drain upon the isocitric acid ceased. It became clear that much is still to be learned about the metabolism of isocitric acid in this species, the most important point probably being why and by what means it accumulates in Crassulacean plants in so phenomenal a way. And if, as seems likely, most of that present is segregated into a pool which is remote from the part which is exposed to the activity of the enzyme systems, especially the aconitase of the cells, how is this segregation accomplished?

We studied the effect of prolonged exposure of *Bryophyllum* leaves to both light (28) and darkness (29) and the capacity of the leaves to recover from stresses thereby brought about (30). Prolonged exposure to light appeared to damage the enzyme systems concerned with the synthesis of malic acid when the stressed leaves were placed in darkness. This effect was presumably associated with the small loss of protein that was observed. Other experiments dealt with the effect of temperature on the fundamental transformation of starch to malic acid which occurs in darkness, and with the effect of light upon the citric acid of leaves collected at sunrise (31). Citric acid behaves in a manner much like that of malic acid although on a lower scale of quantity, but the increases seen in the first few hours of exposure to light, when malic acid is still at a high level although diminishing in amount, led us to conclude that citric acid arises from malic acid by a series of reactions for which there are many analogies in the behavior of these two acids in the tobacco leaf. Thus the level attained by citric acid and the variation in quantity present are functions of the concentration of malic acid. The level attained by malic acid is in turn a function of that of the starch.

METABOLISM OF ORGANIC ACIDS IN TOBACCO LEAVES

Our interest in the organic acids of tobacco leaves, aside from the fact that they are major components of extracts of the tissue, was greatly stimulated by the observation made in the early 1930s that citric acid increases, sometimes to a phenomenal extent when the leaves are cured. This also occurred when leaves were cultured in water in darkness, and evidence gradually accumulated indicating that, far from being merely a buffer system provided to stabilize the hydrogen ion activity of the cells, the organic acids are extremely reactive metabolites in the cell system. The preliminary experiment in which Pucher had fed a wide assortment of acids to leaves cultured in darkness gave convincing evidence of this. Plans were accordingly laid to embark upon a study of the effects upon the organic acids of culture of tobacco leaves in solutions of a long series of organic acids. For this we needed improvements in our analytical methods, and Chester Hargreaves, who came to us as a postdoctoral fellow in 1950, was asked to examine the method for citric acid and to improve the method to determine isocitric acid based upon the use of aconitase from beef heart to transform it to citric acid. This he successfully accomplished (15). The whole problem of the analytical determination of the organic acids was placed upon a new footing, however, by the brilliant work of James K. Palmer, who joined us in 1954. He was asked to see what could be done by the chromatographic methods which were then coming into use in many fields. The outcome was a technical improvement of methods originally described by Busch, Hurlbert, and Potter in which the acids are separated on Dowex-1 by elution with a gradually increasing strength of formic acid (32). The apparatus was simple, the results were accurate, and the whole procedure was rapid. Moreover, by suitable modification of the reagents, the method could be adapted to the determination of a wide range of organic acids. But we did not discard Pucher's method for citric acid. On the contrary, it became even more useful. Citric acid, isocitric acid, and a third component that we ultimately identified as phosphoric acid are eluted together from Dowex-1 as a single peak on the plot of the titration of the fractions. Accordingly, these fractions were combined, citric acid and phosphoric acid were determined in the mixture, and when working with *Bryophyllum* leaves, Hargreaves' aconitase technique was used to determine isocitric acid. Later, when the beautiful enzymatic method of Grafflin and Ochoa to determine isocitric acid became available it was used exclusively.

Thus, with improved analytical methods and assisted in many of the more recent experiments by the availability of organic acids labeled with ^{14}C, we undertook to examine the behavior of most of the components of the tricarboxylic acid cycle as well as a number of other acids when fed to tobacco leaves cultured in darkness.

There are two approaches to such problems. With radioactive substrates, only very dilute solutions of the acid are necessary because of the great sensitivity of the analytical methods. Furthermore, small samples of leaf tissue, even single small disks, are adequate for many kinds of experiments. But if one is interested in the behavior of the carbohydrates, and the proteins, as well as in the loss or gain of organic material from respiration or photosynthesis, large samples are necessary and, to detect changes by ordinary analytical methods of which one can be certain, it is necessary to, as it were, flood the leaf tissue with substrate. Thus one uses 0.2 M solutions of the substrate and prolongs the period of treatment. To be sure, thoroughly abnormal conditions are established, but one obtains unequivocal evidence of the enzymatic behavior one is seeking to establish.

Perhaps a word about the fundamental technique of these experiments may be helpful. With the improvement in accuracy from the use of statistical sampling, it became possible to consider quantitative relationships between the amounts of the substrate taken up and the amounts of its metabolic products, and, as the work progressed and our confidence increased, we were sometimes in a position to suggest definite enzymatic reactions to account for the observations. The general method depends upon the assumption that tobacco leaves placed with their bases in a 0.2 M solution of the potassium salt of an organic acid will take up a quantity of the acid which can be determined in several ways. The increase in potassium, and the increase in the alkalinity of the ash usually gave closely agreeing results. With aspartic or glutamic acid, the increase in nitrogen also served, and with oxalic or phosphoric acid and, as we found, tartaric acid, the increase in the administered acid furnished an additional measure, since neither oxalic nor tartaric acid is metabolized detectably in the tobacco leaf but merely accumulates. The respiration of leaves cultured in darkness was measured by the loss of organic solids, the net photosynthesis in light by their increase. Most of the work was done upon leaves cultured in darkness. The sets of samples, usually ten in number and with tobacco each consisting of 20 leaves, were arranged in V-shaped troughs in an air-conditioned room at constant temperature and humidity. A sample was dried for analysis at the start, and samples were removed at intervals or at the termination of the experiment, usually 24 or 48 hours. After being dried and weighed, the leaves were equilibrated in the constant temperature room until they came to constant weight, and the tissue was then ground and preserved in closed bottles in the same room. The bottles were removed only for the short time required to weigh out samples for analysis. All data, obtained as percentages of this weight, were then calculated to the amount in grams or milliequivalents that would have been found if each of the samples of fresh leaves had weighed initially one kilogram. Thus, any difference between the amount of a component in the treated sample and the initial sample was a direct measure of the effect of the treatment. In all experiments, control samples were cultured in water and in 0.2 M potassium succinate to afford data for comparison with the effect of the acid under study.

The outstanding reaction seen in tobacco leaves cultured in water in the dark is the transformation of malic acid into citric acid. There was also a moderate loss of organic solids due to respiration and evidence for a little proteolysis. Both of these latter reactions were substantially increased by culture on salts of organic acids. As data accumulated, it became clear that two moles of malic acid are used for each mole of citric acid produced. This reaction was greatly promoted by culture on malate (33, 34), and also by succinate. It was reversed by culture on citrate (35). Succinic acid is normally present in tobacco leaves in only minute quantities and, when fed to the leaves, about 90 percent of that taken up disappears, being converted mainly into malic and citric acids.

Only about one-half of the fumaric acid taken up by tobacco leaves is converted into other acids, citric acid being the major product. In contrast, only about one-third of the maleic acid fed at pH 6 underwent chemical change. At pH 5, this fraction was one-fifth. There was a minor increase in citric acid, but the ultimate fate of most of the maleic acid that underwent change was not clearly evident (36). The major effect was an inhibition of several of the enzyme systems of the leaves, in particular the systems involved in the formation of citric acid and the utilization of malic acid, and also the proteolytic system. It could be inferred that the presence of active thiol groups in the structure of the enzymes involved is essential.

From culture solutions of glycolic acid adjusted to pH 5 and 6, nearly 90 percent of the anion taken up by the leaves disappeared, and glycolic acid is thus nearly as active a metabolite as succinic acid in this system (37). The major product is citric acid which accumulated to nearly the same extent as it does when citric acid itself is administered. In addition, reactions occurred which led to the accumulation of moderate amounts of succinic acid and of substantial amounts of oxalic acid, the first instance we have encountered in which an important effect upon oxalic acid occurred. Oxalic acid is usually regarded as an end product of organic acid metabolism, and the present evidence suggests that one of its precursors is glycolic acid. Malic acid behaved in a manner which suggested that it serves as an intermediate in the transformation of the glycolic acid to citric acid. It increased moderately during the first 24 hours of culture, but then diminished substantially.

Malonic acid, in contrast, underwent transformation to only a small extent. Traces of this acid are normally present in the tobacco leaf, and it is a substantial component of the organic acids in leaves of leguminous plants such as beans and alfalfa, but its position in the general metabolic scheme is by no means clear. However, when present in tobacco leaves at a concentration that approached 0.1M at the end of 48 hours of culture, it led to a marked accumulation of succinic acid, an evidence of the inhibition of succinic dehydrogenase (38, 39). The loss of malic acid was substantial, but there was little if any change in the amount of citric acid present until the second

24-hour period of the experiment: at pH 5 there was a small increase and at pH 6 this was greater. The evidence suggested that malonic acid contributed to some extent to the formation of citric acid. Interpretation of the general results in terms of the tricarboxylic cycle is manifestly difficult.

Several experiments in which (+)-tartaric acid was fed to tobacco leaves led to the then unexpected result that this substance is not at all metabolized; it merely accumulates and has no effect whatever upon the series of reactions whereby malic acid is converted into citric acid even when the total amount of tartaric acid in the system finally exceeded the total amount of malic acid present. This observation enabled us to look more closely into the validity of our assumption that the uptake of an organic acid at any reasonable pH of the culture solution can be accurately calculated from the increase in the alkalinity of the ash and the percentage of the acid neutralized at the pH of the solution as deduced from the dissociation constants. Sodium tartrate was fed to a series of samples in solutions in the range pH 2.3 to 6.2 in which the acid was neutralized from 23 to 99 percent. The uptake was calculated from the increase in the alkalinity of the ash and from the tartaric acid found. The mean result of eight observations was that the calculated uptake was 100.6 ± 8.3 percent of the amount of tartaric acid found in the leaves after 48 hours of culture, and the conclusion could be drawn that our calculations were valid within these limits (40, 41).

A further outcome of this experiment was the evidence for the substantially increased respiratory loss of organic solids when salts of organic acids are fed to tobacco leaves, presumably because of the increased demand for the energy required to convey the acid from the vascular system into the cells, and especially the evidence for a stimulation of decarboxylation reactions when the culture solution is adjusted to a pH reaction lower than that of an extract of the normal leaf. This can be interpreted as a protective response to the advent of acid from the culture solution; there is only minor loss of carboxyl groups when the reaction of the culture solution is at or above the normal reaction of extracts of the leaves.

The analogy in the behavior of (+)-tartaric acid and oxalic acid when fed to tobacco leaves raised the question, why is (+)-tartaric not metabolized? The literature indicates that in other than the grape vine, few higher plants are positively known to contain any tartaric acid whatever; it is not a common plant organic acid. However, in the leaves and fruit of the grape it is the dominant acid, and the situation is one that invites speculation. The demonstration in 1951 by Bijvoet and his associates that both asymmetric carbon atoms of naturally occurring (+)-tartaric acid are configurationally related to D-glyceraldehyde, and are thus of the same configuration as carbon atoms 2 and 3 of glucose, led us to suggest that tartaric acid arises from glucose by a sequence of oxidation reactions whereby carbon atoms 5 and 6 are removed and the endstanding carbon atoms of the remaining substance are oxidized to carboxyl groups. Apparently there is no enzyme system in the grape plant or in the tobacco leaf that attacks the resulting (+)-tartaric acid and it therefore accumulates. It is thus also an end product of organic acid metabolism although to be sure it is not commonly seen. The subsequent demonstration by Vennesland and her associates that *meso*-tartaric acid, in which one of the carbon atoms certainly belongs to the L-family, is metabolized in leaf tissues lent some color to this view and there is more recent evidence in its favor.

To obtain some light on the so-called dark fixation of carbon dioxide, tobacco leaves were cultured in the dark in 0.1 and 0.2 M solutions of potassium bicarbonate for 24 and for 48 hours, and the effects upon the organic acid metabolism were examined (42). Control samples were cultured in potassium fumarate, succinate, sulfate, and secondary phosphate. The data indicated that from one-fifth to one-quarter of the bicarbonate ion taken up entered into the organic acid metabolism, and that a stimulation of the formation of citric acid occurred of the same order as that observed when succinate or fumarate is fed. The molar relationship between the amounts of precursor used to that of the citric acid produced is essentially constant whether bicarbonate, succinate, fumarate, or L-malate is made available to the enzyme systems of the leaf, and it could be suggested that the bicarbonate ion is metabolized with the production of a four-carbon organic acid, possibly oxaloacetic acid, which was subsequently converted into citric acid.

The examination of the composition of the control sample cultured in secondary phosphate, the purpose of which had been merely to provide a control at an alkaline reaction not far from that of the bicarbonate solutions used, led to the unexpected observation that phosphoric acid is eluted by formic acid together with citric acid from the Dowex-1 column, and thus in turn to the identification of the puzzling unknown acid we had invariably found in this fraction. Even more interesting was the observation that citric acid accumulated to a considerably greater extent than it did in the water or potassium sulfate control. Analysis of the data made it quite clear that the alkaline solution of phosphate had taken up carbon dioxide from the air of the culture room, and that this culture solution was also, in effect, one of bicarbonate. However, more detailed study was obviously required, and an experiment was carried out in which the effect of phosphate solutions adjusted to reactions from pH 4.4 to 8.8 was examined (43). There was no marked effect upon the behavior of the malic and citric acids save in the sample cultured at pH 8.8. In this, citric acid accumulated to twice the amount found in the potassium sulfate control, confirming the previous observation. The only specific effect of the phosphate detected was the accumulation of minor amounts of water-soluble substances which contained organically bound phosphate.

The repeated observation that the major effect of the culture of tobacco leaves on solutions of the common organic acids is an increase of citric acid led to a test of the effect of pyruvic acid (44). Citric acid did indeed accumulate in substantial amounts, and there was only a small decrease in malic acid; pyruvic acid was clearly extensively involved in the transformation.

There was a substantial uptake, although little pyruvic acid survived the operations of drying the leaves and evaporation of the formic acid eluates from the column. Proof that pyruvic acid became involved in the reactions that occurred was obtained in a parallel test in which a single tobacco leaf was cultured in a solution that contained pyruvic acid labeled with radioactive carbon. After 48 hours, the citric acid present had approximately the same specific activity as the labeled pyruvic acid, and the trace amounts of succinic acid and fumaric acid in the leaf were also highly labeled. Malic acid was only moderately labeled, but a consideration of the large amount present from the start showed that about as much radioactive carbon was present in the malic acid at the end as had found its way into the citric acid. The evidence pointed to the presence in the leaves of a condensing enzyme system similar in function to that of Stern and Ochoa.

Analysis of the data suggested that citric acid is present in the tobacco leaf in two compartments or pools. One of these receives and stores newly formed citric acid, the other and probably much smaller pool is the scene of rapid enzymatic reactions which may well follow the plan of the tricarboxylic acid cycle. However, the evidence that this scheme of reactions is responsible for the entire respiration of the leaf is still far from convincing.

The availability of substantial amounts of the potassium salt of isocitric acid in our laboratory naturally led to a study of the metabolism in the tobacco leaf of this rare but extremely important substance. One of Pucher's last researches was the development of a method to prepare isocitric acid in quantity from the *Bryophyllum* plant. After much study, he had hit upon the dimethyl ester of isocitric lactone as a substance of suitable properties and had accumulated several hundred grams of this compound (13). When D.G. Wilson joined us for a 2-year postdoctoral tour in the mid-1950s, I suggested that he might carry out a preparation by Pucher's method and then go on to see if a chromatographic method could not be worked out, for we knew that isocitric lactone could be separated on the Dowex-1 column from both citric acid and isocitric acid which are eluted together. He succeeded in doing this, but had difficulty in obtaining a satisfactory preparation of the lactone. It occurred to him one day that monopotassium tartrate is an unusually insoluble salt, and that, since isocitric lactone is another dicarboxylic acid, its monopotassium salt might also be reasonably insoluble. The literature suggested that the first carboxyl group of isocitric lactone would be maximally neutralized at about pH 3.25, so he adjusted a concentrated solution of the lactone to this reaction and added alcohol. A beautifully crystallized salt promptly separated in nearly quantitative yield (45). Samples of this salt and the corresponding rubidium salt supplied to Dr. A. L. Patterson of the Institute for Cancer Research in Philadelphia enabled him to work out the crystal structure in detail and later to establish that the asymmetric α-carbon atom belongs to the D family in confirmation of the extraordinary work of Katsura and his associates in Japan in 1960.

When Wilson left us to return to Canada, there remained a collection of many solutions that represented ether extracts of the acids from *Bryophyllum* leaves, and it occurred to me to try to isolate his potassium salt directly from them. This was possible, although the yield was small and the material rather impure, but one day a small crop of crystals which were considerably more insoluble than the potassium salt of the lactone separated from an aqueous mother liquor. Analysis showed that it was the monopotassium salt of isocitric acid itself. The outcome of this observation was a method to prepare isocitric acid as its potassium salt which is simple enough to serve as an excellent student preparation, and within a few weeks we had hundreds of grams of it on hand (46). Until recently when it became available commercially, we were the source of supply of this unique compound to colleagues all over the world who were interested in the metabolism of isocitric acid.

When isocitric acid is fed to tobacco leaves, it behaves quite differently from its behavior in the leaves of Crassulacean plants (47). From a culture solution at pH 5, about two-thirds of the substantial amount taken up disappears as such, and about 40 percent of this quantity is apparently transformed into citric acid. Malic acid increased slightly, and in a parallel single leaf experiment in which synthetic isocitric acid labeled with ^{14}C in carbon atoms 3 and 4 was used, malic and succinic acids were extensively labeled as well as the citric acid. It could be concluded that, although a part of the citric acid may have arisen through the condensation reactions of the tricarboxylic acid cycle, much of it also arose through direct transformation by an aconitase.

An experiment in which L-glutamic acid was fed to tobacco leaves in darkness showed that this substance ranks in metabolic activity with succinic and glycolic acids in that about 90 percent of that taken up is transformed into other substances (48). The nitrogen appeared mostly as asparagine, together with an appreciable amount of aspartic acid and ammonia and a little glutamine amide nitrogen. More than four-fifths of the acquired nitrogen could be thus accounted for. Citric acid increased strikingly, and succinic acid increased to about one-half the extent that it did in the control sample cultured on succinate. The small increase in the so-called minor acids, which are eluted from the Dowex-1 column in advance of succinic acid, could be quantitatively accounted for by the glutamic and aspartic acids in this fraction. In the parallel single-leaf experiment in which radioactive glutamic acid was administered, both aspartic and succinic acids attained a specific activity equal to that of the glutamic acid supplied, and malic and citric acids were highly labeled, the glutamine somewhat less so. The trace of D-glutamic acid in the culture solution obviously was not attacked at all. The general picture suggested that three moles of glutamic acid yielded one mole each of citric acid, asparagine, and ammonia, and succinic acid is obviously an important intermediate in the reactions.

Among the many puzzles encountered in our studies of the chemical changes that occur when tobacco leaves are subjected to the commercial operation of curing was the apparent complete stability of asparagine under circumstances where all other amino acids save possibly glutamine are completely oxidized with the liberation of ammonia (49). Tobacco leaves to be used for cigar wrappers, after being picked from the plants as the leaves successively became technically ripe, are strung on cords attached at each end to wooden laths and hung in the curing shed where they remain for many weeks. The leaves soon turn yellow and then brown, most of the water they contain evaporates, and about one-half of the protein disappears. The outstanding chemical event is the formation of asparagine in an amount equivalent to about 70 percent of the nitrogen of the protein which disappeared with subsequent oxidation of the amino acids produced. Amino acids other than asparagine do not accumulate save in negligible traces. In an attempt to throw some light upon this behavior, tobacco leaves were cultured in darkness on aspartic acid adjusted to pH 5 and 6 and also upon asparagine (50). In a parallel single-leaf experiment, radioactive aspartate at pH 5.5 was used. About three-quarters of the aspartic acid taken up was metabolized, asparagine and ammonia being the major products together with a little glutamine. There was a small but detectable increase in protein during the first 24 hours, a most unusual observation. Citric acid increased in substantial amounts, but there was only a minor increase in malic acid. Succinic acid also increased appreciably, an evidence that it must have served as one of the important intermediates in the reactions that occurred. In the single-leaf experiment in which radioactive aspartic acid was fed, succinic acid attained the same specific activity as that of the L-aspartic acid in the aspartic acid taken up, while the small amount of aspartic acid present at the end had twice this specific activity, an evidence that the trace of radioactive D-aspartic acid in the substrate was not metabolized at all.

The leaves cultured on asparagine after 48 hours contained appreciably more asparagine than was taken up in spite of the fact that about one-quarter of the uptake had been metabolized. The main nitrogenous products were ammonia and glutamine, and the organic acids behaved in a manner very much like that in the leaves cultured on aspartic acid. The general picture was that of a system capable of metabolizing aspartic acid and asparagine efficiently in the presence of another system capable of synthesizing asparagine. After 48 hours of culture, the effects of the second system predominated, thus leading to a false impression of the unusual stability of asparagine in tobacco leaves whether cultured in solutions or subjected to the technical operation of curing. Aspartic acid and glutamic acids are obviously as readily metabolized as most of the other organic acids we have studied, but aspartic acid is unique in that tobacco leaves, and probably many other plant tissues, make use of the ammonia produced by the oxidation of amino acids for the formation of asparagine which accumulates. Observations of this accumulation led Prianishnikov many years ago to draw an analogy between the behavior of asparagine in plants and urea in the animal. He considered both substances to be produced as a response to the presence of ammonia. This old "detoxication" hypothesis is no longer insisted upon today, for there is much evidence that ammonia can accumulate in leaves of certain species, particularly rhubarb, without harm. However, it is a fact that the formation of asparagine is the most common outcome in any situation where plant tissues—whether seeds, shoots, or leaves—encounter conditions where the protein is hydrolyzed at a rate greater than that at which it is synthesized.

In looking back upon these busy years when we tried to throw a little light upon the behavior of organic acids in one narrowly restricted phase of plant metabolism, one sometimes wonders what it all amounted to. One conclusion is clear. At the beginning of our work very little was known about the organic acids of leaves save that they form a substantial part of the organic substances in this extraordinary tissue. Why they are there and what they do was mostly a matter for speculation. Today the organic acids are recognized to be the central metabolites of the systems involved in carbohydrate and protein chemistry, in the phenomena of photosynthesis and respiration, and in many more functions of the life process in plants.

The present-day appreciation of the significance of the organic acids is mainly the result of the work of biochemists who have concerned themselves with enzymes. Although there is a long history of the detection of this or that enzyme reaction in plant tissues, it was not until the fundamental techniques of protein fractionation at low temperatures began to be widely applied some 40 years ago that various systems of coordinated enzyme reactions were discovered and offered as explanations of the chemical events observed in living cells. Pucher and I were both trained as organic chemists with a bent for analytical chemistry. I was chiefly interested in nitrogenous substances, he in non-nitrogenous: we were not physiologists at all, but as we learned more about the composition of leaf tissues and how it changed as the result of treatment, the physiological, or rather the biochemical aspects of the observations impressed themselves more and more upon us. Back in the early thirties there were very few investigators who concerned themselves with the chemical composition of plants, in fact astonishingly little was known about it, and one of the greatest and most pleasant surprises of my professional life was the award in 1933 of the Hales prize by the American Society of Plant Physiologists. I had had no inkling that our work with leaves had been noticed by colleagues in this discipline. For many years we were chiefly busy with the development of analytical methods suitable for application to extracts of plants. Accuracy was the main consideration and revision of the techniques was continuous. Our concern was with substances and how they behaved. Thus I have been content to leave the development of detailed biochemical mechanisms to those whose special training equipped them to deal with this difficult field. In my view accurate measurement of the chemical changes that occur must precede attempts to account for them in terms of enzyme-catalyzed reactions.

The development of chromatographic methods in recent years has enormously broadened the field of analytical attack upon physiological problems and has diminished the scale upon which one may work to what an organic

chemist quite properly regards as traces of material. Yet I am still old-fash-
ioned enough to prefer a crop of crystals that one can weigh to the measurement of the location of a spot of color on a strip of filter paper. Thus as I
approach 80 years of age I am happy to turn over to my younger and more
broadly trained colleagues the responsibility for further progress in the study
of the organic acids of plants.

I cannot close, however, without a word of appreciation for the devoted
help of the members of my staff over the years. To Israel Zelitch, my successor as head of the laboratory, I am indebted for innumerable helpful discussions as well as for direct aid in many experiments. And to K. R. Hanson, J.
K. Palmer, A. Klein, A. N. Meiss, C. A. Hargreaves II, C. C. Levy, and
especially in the early years to G. W. Pucher, A. J. Wakeman, C. S. Leavenworth, Abraham White, H. E. Clark, and Emil Smith I owe an equal debt.
For more than 40 years Laurence Nolan gave practical assistance in preparing
the innumerable samples for analysis and in the invention of apparatus (51)
to simplify the work. Marjorie Abrahams and Katherine Clark have given
devoted and skilled technical assistance. I have unblushingly picked the
brains and used the ideas and abilities of all, and if there is any merit in what
we have done, by far the greater part of it belongs to these devoted colleagues.

LITERATURE CITED

1. Vickery, H. B. 1918. *Trans. Nova Scotia Inst. Sci.* 14:355
2. Vickery, H. B. 1922. *J. Biol. Chem.* 53:495
3. Vickery, H. B. 1924. *J. Biol. Chem.* 60:647; 61:117. Ibid 1925. 65: 81, 91, 657
4. Vickery, H. B., Leavenworth, C. S. 1925. *J. Biol. Chem.* 63:579
5. Vickery, H. B., Leavenworth, C. S. 1927. *J. Biol. Chem.* 72:403
6. Vickery, H. B., Leavenworth, C. S. 1928. *J. Biol. Chem.* 76:707; 79: 377
7. Vickery, H. B., Leavenworth, C. S. 1927. *J. Biol. Chem.* 75:115. Ibid 1928. 76:437, 701; 78:627
8. Vickery, H. B., Pucher, G. W. 1929. *J. Biol. Chem.* 83:1
9. Pucher, G. W., Vickery, H. B., Leavenworth, C. S. 1935. *Ind. Eng. Chem. Anal. Ed.* 7:152
10. Vickery, H. B., Pucher, G. W. 1931. *Conn. Agr. Exp. Sta. Bull.* 323
11. Pucher, G. W. 1942. *J. Biol. Chem.* 145:511
12. Pucher, G. W., Vickery, H. B. 1942. *J. Biol. Chem.* 145:525. 1946. 163:169
13. Pucher, G. W., Abrahams, M. D., Vickery, H. B. 1948. *J. Biol. Chem.* 172:579
14. Pucher, G. W., Vickery, H. B., Leavenworth, C. S. 1934. *Ind. Eng. Chem. Anal. Ed.* 6:190
15. Hargreaves, C. A. II, Abrahams, M. D., Vickery, H. B. 1951. *Anal. Chem.* 23:467
16. Pucher, G. W., Sherman, C. C., Vickery, H. B. 1936. *J. Biol. Chem.* 113:235
17. Pucher, G. W., Vickery, H. B., Wakeman, A. J. 1934. *Ind. Eng. Chem. Anal. Ed.* 6:288
18. Vickery, H. B., Osborne, T. B. 1928. *Physiol. Rev.* 8:393
19. Vickery, H. B., Schmidt, C. L. A. 1931. *Chem. Rev.* 9:169
20. Vickery, H. B. 1950. *Yale J. Biol. Med.* 22:387
21. Vickery, H. B. 1947. *J. Biol. Chem.* 169:237
21a. Vickery, H. B. 1952. *Chem. Eng. News* 30:4522
22. Vickery, H. B. 1963. *J. Org. Chem.* 28:291
23. Pucher, G. W., Leavenworth, C. S., Vickery, H. B. 1948. *Anal. Chem.* 20:850

24. Pucher, G. W., Vickery, H. B. 1949. *J. Biol. Chem.* 178:557
25. Vickery, H. B., Leavenworth, C. S., Bliss, C. I. 1949. *Plant Physiol.* 24:335
26. Vickery, H. B. 1952. *Plant Physiol.* 27:231. 1957. 32:220
27. Klein, A. O. 1964. *Plant Physiol.* 39:290
28. Vickery, H. B. 1953. *J. Biol. Chem.* 205:369
29. Vickery, H. B. 1954. *Plant Physiol.* 29:520
30. Vickery, H. B. 1956. *Plant Physiol.* 31:455
31. Vickery, H. B. 1959. *Plant Physiol.* 34:418
32. Palmer, J. K. 1955. *Conn. Agr. Exp. Sta. Bull. 589*
33. Vickery, H. B., Hargreaves, C. A. II 1952. *J. Biol. Chem.* 197:121
34. Vickery, H. B. 1955. *J. Biol. Chem.* 214:323
35. Vickery, H. B. 1952. *J. Biol. Chem.* 196:409
36. Vickery, H. B., Palmer, J. K. 1956. *J. Biol. Chem.* 218:225
37. Vickery, H. B., Palmer, J. K. 1956. *J. Biol. Chem.* 221:79
38. Vickery, H. B., Palmer, J. K. 1957. *J. Biol. Chem.* 225:629
39. Vickery, H. B. 1959. *J. Biol. Chem.* 234:1363
40. Vickery, H. B., Palmer, J. K. 1954. *J. Biol. Chem.* 207:275
41. Vickery, H. B. 1957. *J. Biol. Chem.* 227:943
42. Vickery, H. B., Palmer, J. K. 1957. *J. Biol. Chem.* 227:69
43. Vickery, H. B., Levy, C. C. 1958. *J. Biol. Chem.* 233:1304
44. Vickery, H. B., Zelitch, I. 1960. *J. Biol. Chem.* 235:1871
45. Wilson, D. G. 1963. *Can. J. Biochem. Physi*ol. 41:1571
46. Vickery, H. B., Wilson, D. G. 1958. *J. Biol. Chem.* 233:14
47. Vickery, H. B., Hanson, K. R. 1961. *J. Biol. Chem.* 236:2370
48. Vickery, H. B. 1963. *J. Biol. Chem.* 238:2453
49. Vickery, H. B., Meiss, A. N. 1953. *Conn. Agr. Exp. Sta. Bull. 569*
50. Vickery, H. B. 1963. *J. Biol. Chem.* 238:3700
51. Nolan, L. S. 1949. *Anal. Chem.* 21: 1116

Ann. Rev. Plant Physiol. 1972. 23:29–50

PLANT CELL PROTOPLASTS—ISOLATION AND DEVELOPMENT

7521

Edward C. Cocking

Department of Botany, University of Nottingham
University Park, Nottingham, England

CONTENTS

INTRODUCTION

Before discussing the actual isolation of protoplasts from various plant cells it is important to consider first the general usage of the term "protoplast." Unfortunately, as pointed out by Frey Wyssling (31), the meaning of this general cytological term has been narrowed by considering the protoplast as the result of the removal of the cell wall, whereas in fact the protoplast represents the totality of the living cell constituents quite independent of whether a wall is present or not. It has even been suggested (9), for instance, that in general botanical usage the term "protoplast" describes that part of the cell which lies within the cell wall and which in certain cells can be plasmolyzed away from it. Hanstein in 1880 (43) was probably the first to use the term protoplast (protos—first, plastos—formed), and his drawings of protoplasts within cells of *Equisetum, Alyssum,* and *Paeonia* together with his drawings of *Vaucheria* protoplasts, and the release of protoplast fragments from these cells, served generally to detract from the use of the term protoplast to cover a living uninucleate primitive protoplasmic unit. This tended to suggest a meaning for the term protoplast in relation to that part of the cell which lies within the cell wall. Physiologically, as Mills & Krebs (71) have emphasized, the protoplast certainly cannot be considered simply as a plant cell from which the cell wall has been removed by digestion. Küster

29

(63) attempted to clarify the situation by calling naked protoplasts "gymnoplasts" in contrast to "dermoplasts," the normal plant cells with their cell walls. As a result, Frey Wyssling (31) was prompted to advocate the use of the term "gymnoplasts" instead of "protoplasts." It does seem, however, very important not to add yet another term which is not fully understood by workers studying bacterial and fungal protoplasts. Even distinguished plant physiologists have been known to confuse protoplasts and chloroplasts, at least in their terminology! Klercker in 1892 (58) described "Eine Methode zur Isolierung Lebender Protoplasten" in which he referred to them as isolated protoplasts (die Isolierten-Protoplasten). Clearly, therefore, isolated protoplasts are obtained from cells from which the wall has been removed by enzymatic or physical means, and generally most workers refer to such as isolated protoplasts or naked protoplasts. The basic concept is that these isolated protoplasts are naked cells, and as we shall see later, being cells they can under suitable conditions be induced to grow and divide.

ISOLATION OF PLANT CELL PROTOPLASTS

Mechanical methods for the isolation of protoplasts.—In the plant kingdom the Angiospermae have been extensively studied in relation to the isolation of protoplasts from plasmolyzed cells using mechanical means. When such plant cells become plasmolyzed the plasmalemma retracts from the cell wall, and as a result of plasmolysis the protoplast takes on a variety of different appearances depending on the shape of the cell being plasmolyzed (112). Plasmolysis of the largely isodiametric cells of leaf mesophyll, callus, and suspension culture cells with hypertonic solutions most frequently results in a major shrinkage in the overall volume of the protoplast which rounds off within the cell wall. With more elongated cells plasmolysis may result in the protoplast dividing up, without rupture of the plasmalemma, into a number of subprotoplasts, one of which contains the nucleus (18). The plasmodesmata which connected the protoplasts of contiguous cells to form the tissue symplasm are apparently self-sealing, as the protoplast retracts from the cell wall during plasmolysis. When cells of *Elodea* leaf are kept plasmolyzed in solutions of calcium chloride or calcium nitrate (0.25–0.3 *M*) for up to 50 days, the cytoplasm still shows typical protoplasmic streaming (139). If plasmolyzed cells such as these are cut, then sometimes it is possible to cut through the cell wall without damaging the structure within, and as a result mechanically isolated protoplasts can be obtained. This was essentially the method first used by Klercker in 1892 (58) for the mechanical isolation of protoplasts from plasmolyzed cells of the water warrior (*Stratiotes aloides*) and which later formed the basis for the pioneering microdissection studies of Seifriz (109). These studies were later extended by Chambers & Höfler (11), who were able to isolate a few protoplasts by using thin slices of epidermis of onion bulb scale immersed in 1.0 *M* sucrose until the protoplasts had shrunk away from their enclosing walls, and then cutting sheets of the epidermis with a sharp knife. They showed that the protoplasm of the isolated protoplast could be dispersed while the integrity of the central vacuole was maintained. Isolated protoplasts have been obtained from beet tissue (131) by first plasmolyzing small cubes of beet in slightly hypertonic sucrose solution for 2 hours and then gradually increasing the molarity of the external medium to about 1 M. A thin slice is then cut from the cube which is then replaced in about 0.5 M sucrose. The undamaged protoplasts swell and are pushed out of the ends of the cut cells by osmotic forces. The tissues which can be successfully employed using this mechanical method of isolation are necessarily those in which marked plasmolysis and good separation of the protoplast from the cell wall takes place. It has also been successfully applied to the epidermis of radishes (122) and to the mesocarp of cucumber (125). However, the method has always yielded only a relatively small number of isolated protoplasts, mainly from storage tissues. A considerable amount of new and important information, nevertheless, has been gleaned from a detailed study of individual isolated protoplasts. This includes most of our detailed knowledge of the expansion capacity of naked plant protoplasts (122), the first basic study of possible effects of fungal toxins on the plasmalemma (10, 125), the first basic study of possible effects of auxins on the plasmalemma (60), and detailed studies of the osmotic relations of plant cells (64, 130). For a general review of this earlier work see Cocking (19). The few protoplasts obtained, however, has always been a major drawback of this method. Only when extensive plasmolysis is possible is there any chance of releasing protoplasts using this procedure, and as a result it is not possible to obtain protoplasts from more meristematic plant cells. As we shall see later, one of the advantages of the enzymatic method is that it is applicable in many instances to both mature and meristematic plant cells. The one possible advantage of the mechanical method for the isolation of protoplasts is that harmful effects of added enzymes on the structure and metabolic activity of isolated protoplasts are eliminated (82, 92), although of course there is always the chance of enzymes released from broken cells influencing the metabolism even of mechanically isolated protoplasts.

Mechanical isolation of algal protoplasts is readily achieved when the algae are filamentous. Indeed the isolation of protoplasts from sparingly branched filaments of *Vaucheria* using mechanical methods of isolation was one of the first descriptions of isolated protoplasts (58). During plasmolysis, and associated with the isolation, numerous subprotoplasts (18) are released. These frequently contain several nuclei since *Vaucheria* is coenocytic. Studies on the mechanical isolation of fungal and bacterial protoplasts have been much more limited, notably in the case of bacteria because of the difficulty inherent in their small size. It should be noted, however, that at the intermediate stage of the mosses protoplasts have been isolated and their properties extensively investigated (5).

Enzymatic methods for the isolation of protoplasts.—Until 1960, studies on higher plant protoplasts were largely hampered by the fact that only a few protoplasts could be isolated by mechanical methods, and as we have seen, studies which were carried out were limited mainly to investigations of osmotic behavior. No detailed studies were carried out on cell wall regeneration by these mechanically isolated protplasts, nor was any really valid indication obtained of their viability and ability to grow and divide, with the exception of the pioneering studies of Binding on moss protoplasts previously mentioned (5).

Ever since Giaja in 1919 (33) isolated protoplasts from yeast cells by digesting the cell wall by means of gastric juice obtained from the snail *Helix pomatia,* other workers have attempted to isolate protoplasts enzymatically from bacteria (67, 111), algae (32), fungi (128, 129), mosses, and higher plants in preference to mechanical methods of isolation. Typically growing cell walls of higher plants contain, in varying proportions, α cellulose, hemicellulose, pectin, and protein, but there are major changes in the relative proportions of these as differentiation of the cell proceeds; particularly noteworthy is the presence of lignin. For a general discussion of cell walls see Rogers & Perkins (98).

Ruesink (99) has listed the three major advantages of an enzymatic over a cutting method. These are: (*a*) a larger number of protoplasts may be obtained; (*b*) less osmotic shrinking of the cytoplasm is required; and (*c*) no cells are broken as a result of cutting. Perhaps the chief advantage really lies in the ability to use enzymatic methods to degrade the cell wall from the more meristematic type of cell in which plasmolysis does not readily occur. Of course the larger number of protoplasts which can be isolated using enzymatic methods is also a very major advantage. Another key advantage of enzymatic methods is that the use of cell wall degrading enzymes provides the opportunity of isolating multinucleate protoplasts which have arisen as a result of spontaneous fusion during the actual treatment of the tissue with wall-degrading enzymes (90). It has been shown that in this process of spontaneous fusion, which requires no inducing agent for fusion, the protoplasts of contiguous cells fuse together as a result of the expansion of the plasmodesmatal connections which linked the protoplasts together in the tissue (137). Such spontaneously fused multinucleate protoplasts are, as we shall see, particularly useful in elucidating nuclear interactions in growing and dividing multinucleate cells which may serve as "starter cultures" for polyploid studies. It should be noted that such spontaneously fused protoplasts cannot be isolated if the cells of a tissue are first separated by the sequential enzyme isolation method (see later).

The enzymes used for cell wall degradation are often impure, and harmful effects on the protoplasts may result, although there is good evidence that when protoplasts are in the plasmolyzed state these effects are minimal. Tribe (125) showed that the killing of cells normally associated with cell separation is greatly retarded when the cells are first plasmolyzed. Hall & Wood (41) have found that prior plasmolysis decreases markedly the leakage of electrolytes that would otherwise occur. There is no doubt, however, that plasmolysis has profound effects physiologically on plant tissues, altering protein and starch synthesis and solute uptake as well as respiratory activity generally (38). Many of the effects appear to be reversible, but it is clear that plasmolysis, cell wall removal, and the separation of protoplasts from the symplasm of the plant tissue are rather traumatic. Cell and protopast separation, however, may be stimulatory to cell wall regeneration and also to division. There are also effects of osmotic pressure on mitosis as well. González-Bernáldez et al (35) observed that the rate of the cell cycle in *Allium* roots, expressed as the percentage of cells passing through any given point in the cycle in one hour, was roughly a linear function of the osmotic pressure. At an osmotic pressure of 12 atm the rate of the cell cycle was reduced to 80% of what it was at zero osmotic pressure.

It should be noted that exogenous addition of cell wall degrading enzymes may not always be necessary for the enzymatic isolation of higher plant protoplasts. In the fruits of several plants (e.g. *Solanum nigrum* and *Lycopersicon esculentum*) the cell walls are hydrolyzed during the ripening process so that free protoplasts and protoplasmic units are left (24). Küster (62) preferred this physiological (enzymatic) method for isolating protoplasts to the mechanical one because the latter might easily damage the protoplasts; but as pointed out by Vreugdenhil (130), it might be argued equally that the enzymes used to degrade the cell walls might change the properties of the protoplasts.

It is interesting to recall that the initial objective of the studies of Cocking (15) which resulted in the first exogenous enzymatic isolation of higher plant protoplasts was the prior separation of cells and their culture. Extensive studies were carried out using EDTA and other chelating agents to remove Ca^{++} selectively from the middle lamella of root tips, but while separation readily took place, the chelating agents were found to have a deleterious effect on the metabolism of the separated cells, and this effect could not be reversed (16). Earlier Chayen (12) had noted that following a 1-hour digestion with 10% pectinase, it was possible to separate root meristem cells of *Vicia* provided he teased them apart with a needle. This work, however, was carried out under nonplasmolyzing conditions, and the general impression obtained from this paper was that the cells obtained were not viable. Subsequently Zaitlin & Coltrin (140) reported the use of pectinase from several sources to separate single cells of tobacco, but the method would not work for leaves of other species. Later Jyung, Wittwer & Bukovac (52, 53) obtained single cells from bean and tobacco leaves using pectinase with sucrose to aid in osmotic stabilization. This latter modification may have been of particular significance, especially in the light of Tribe's earlier work (125) which showed that plasmolysis had a stabilizing effect, making the protoplasts less susceptible to the effects of possible deleterious enzymes. This work was later extended by Takebe, Otsuki & Aoki (121), who developed a procedure using a fungal pectinase (macerozyme) to release mesophyll cells rapidly from tobacco leaves. The method necessitated the removal of the lower epidermis to facilitate penetration of the enzyme solution into the mesophyll of the leaf [a procedure which has since been found tedious and unnecessary (49)]. Cell separation was considerably enhanced by the presence of potassium dextran sulphate (121). In the sequential enzyme isolation procedure, protoplasts were subsequently isolated from the separated cells by treatment with a *Trichoderma viride* cellulase solution. More recently this procedure has been extended to the leaves of many species of dicotyledons and to some monocotyledons. As noted by Otsuki & Takebe (81), the results reported do not necessarily represent the best obtainable results for individual species. Both the rate of maceration and the yield of intact cells vary considerably depending on the age and the physiological condition of the plants. Each worker is well advised to investigate systematically the situation in the particular species and plant tissue he is employing, particularly with respect to the physiological condition of the plant.

In many instances, as we shall discuss later, it may be preferable not to use this sequential method of isolation but rather to use the cellulase enzyme preparation alone or mixed with pectinase (88).

These studies which involve the separation of the cells of various tissues highlight the major problem often encountered in the isolation of higher plant protoplasts, namely the adequate penetration of the wall-degrading enzymes. It is obvious that prior separation of the cells of a tissue greatly facilitates the ease with which cell wall degradation takes place, because the very nature of the degradation of an insoluble substrate such as cellulose involves marked adsorption of the enzymes to the substrate. As discussed by Mandels & Reese (65), the enzymatic degradation of cellulose is far from being understood; the most difficult question involves the nature of the initial enzymatic attack on the crystalline portions of the cellulosic fiber. The macromolecular crystalline structure of cellulose greatly limits accessibility of the glucosidic bonds in cellulose to the hydrolytic Cx enzymes. As a result, amorphous cellulose is more readily degraded than crystalline cellulose. The results obtained by Preston and his co-workers in studies on algae, elm, and pinewood seem consistent with the idea that hemicellulose chains (having β-1, 3 and β-1, 4 linkages) are intimately adsorbed onto the cellulose microfibrils (79). It would seem likely, therefore, that the degree of crystallinity of the cellulose in plant cell walls and the presence of hemicellulose may greatly modify the extent of wall degradation by exogenously supplied cellulase enzymes. Any lignification of the wall will also restrict accessibility of the cellulase complex. It should be noted that the stage of differentiation of the plant cell will also influence the situation, since it is clear that in more meristematic cells endogenous cellulase synthesis plays a major role in the plasticization of the wall (14). Cell wall degrading enzyme mixtures of varying enzymatic complexity, therefore, will be required for adequate wall degradation and the subsequent isolation of protoplasts if it is desired to isolate them from a wide range of higher plant cells. Albersheim (personal communication) has suggested that protein enzyme inhibitors present in cell walls may be important; these possibly could be removed by plasmolysis with salt solutions, thereby enabling cell wall degradation to proceed.

However, it should not be assumed that the action of the complex of enzymes present in Macerozyme, as used by Takebe, Otsuki & Aoki (121), only results in cell separation. The cell walls frequently appear thinner than initially, as judged by electron microscopic observations, and often the cells are osmotically labile, bursting as the osmotic pressure is reduced. With tissue systems such as that of the root meristematic region, separation of cells frequently is very slight. The use of more concentrated pectinase or Macerozyme seldom helps because harmful effects of contaminating substances, including enzymes, cause death of the cells since toxic materials are then present at greatly increased levels. Purification of the Macerozyme by a simple Biogel treatment (28) can often help greatly in this respect. It may also be advantageous to use much more highly purified polygalacturonic acid degrading enzymes (56). Sometimes crude cellulase enzymes will release protoplasts directly from the cells of certain tissues without the necessity of carrying out a sequential treatment of first separating cells and then degrading the cell walls of the separate cells. Indeed this was the procedure adopted by Cocking (15), who used concentrated solutions of crude *Myrothecium verrucaria* cellulase on tomato root tips. The procedure was later extended by Ruesink & Thimann (102), again using a concentrated *Myrothecium verrucaria* cellulase, to the isolation of protoplasts from leaf, coleoptile, and callus tissue. Interestingly, in this work the addition of polyuronidases did not increase the yield. Glusulase, a commercially available snail enzyme preparation, has been used (83) for the isolation of protoplasts from a wide variety of tissues. The enzyme mixture, while sometimes effective, is not recommended for obtaining protoplasts for cultural studies. In 1969, cellulase Onozuka P1500, a crude cellulase preparation from *Trichoderma viride* which had then become commercially available in Japan, was used for the isolation of protoplasts from cell suspensions of *Haplopappus gracilis* (27). The commercial availability of such cellulases greatly stimulated work on the isolation of protoplasts, since before 1968 no satisfactory cellulase preparation was commercially available. More recently the use of high-potency commercial cellulases has been extended to the direct isolation of protoplasts from a wide range of plant cells in liquid culture by Schenk & Hildebrandt (107). These workers obtained evidence for the presence of toxic substances, some of which may have been peroxidases, in these commercial preparations. They advocated a simple purification procedure involving initial precipitation of high molecular material, followed by removal of low molecular weight solutes by elution through Sephadex G-25, followed by treatment with activated charcoal and evaporation to dryness in vacuo. These workers found that viable protoplasts could be obtained from all of the 19 species of callus routinely grown in their laboratory but that the quality and stability of the protoplasts varied greatly between species. Such slightly purified cellulases were also used to isolate protoplasts from suspension culture cells of peanut and red kidney bean.

Schenk and Hildebrandt also noted that the cytoplasmic streaming in isolated protoplasts was decreased at low pH levels and at high osmolarity and moreover that toxic substances in the unpurified commercial cellulases induced almost immediate cessation of cyclosis.

It is difficult to generalize about the differing susceptibility of protoplasts to the action of toxic substances present in crude enzyme preparations. These toxic substances may be enzymes such as various ribonucleases, proteolytic enzymes, or lipases, as well as peroxidases, and nonenzymatic material present may also be toxic. The toxicity may reside in direct action on the plasmalemma, causing its degradation or an alteration in its permeability sufficiently to make the protoplasts osmotically unstable and thereby bring about their bursting. It may affect cyclosis, perhaps by alterations in the viscosity of the cytoplasm as a result of penetration of enzymes and other high and low molecular weight material into the cytoplasm probably as a result of micropinocytosis (21, 135) or direct penetration through the plasmalemma (8). It would seem, however, that peroxidase is not toxic to certain protoplasts (100); and while cyclosis is often assumed to be a good indication of protoplast viability, the lack of cyclosis in protoplasts is not an infallible guide to their nonviable state. Quite often cyclosis can be resumed. It is evident, at least initially, that in the turgid cell of higher plants plasmolysis results in a marked infolding of the plasmalemma, as a result of which vesicles are formed in the cytoplasm. These vesicles are basically different in origin from the much smaller vesicles formed as a result of micropinocytosis (endocytosis) in protoplasts (21, 135). Therefore, it may be advantageous to plasmolyze initially the protoplasts of cells, from which one is attempting to isolate protoplasts, in the plasmolyticum only (i.e. without any cell wall degrading enzymes being added) and only later treat the plasmolyzed tissue or cell suspension with the plasmolyticum plus cell wall degrading enzymes (137). In this way extensive uptake of crude enzyme solutions into vesicles in the cytoplasm of the protoplasts perhaps largely can be avoided, since in the plasmolyzed condition uptake is considerably reduced and is restricted to very small vesicles formed in the cytoplasm as a result of endocytosis. *Myrothecium verrucaria* cellulase, which was the cellulase first used for the exogenous enzymatic degradation of higher plant cell walls leading to liberation of isolated protoplasts (15), is the cellulase obtained from culture filtrates of the mold *Myrothecium verrucaria*. The enzyme complex is secreted into the medium as a result of the growth of the fungus; and as noted by Whitaker (132), the crude enzyme precipitated from solution contains virtually all the protein originally present in the culture medium. It is clear that even the more recent improved procedures for the preparation and characterization of *Myrothecium* cellulase involve only a few-fold increase in the specific activity of the enzyme, and as reported by Whitaker, Hanson & Datta (133), the requirements for this enzyme to be used for the preparation of protoplasts is often no more than freedom from gross dialyzable impurities. Enzymes meeting this requirement can be obtained simply by precipitation with saturated ammonium sulphate and elution through Sephadex G-25. It is interesting that this is basically the procedure which has been adopted for the partial purification of both cellulases and pectinases by other workers (28, 54, 55, 107). More extensive purification of other cellulase preparations, particularly that of the cellulase preparation from the fungus *Trichoderma viride,* has been carried out by Keller et al (57). These workers found that there was a fair degree of correlation between protoplast-forming ability and cell wall composition. Cellulase on its own formed protoplasts from species with cell walls containing the least amounts of arabinose, galactose, and xylose present as hemicelluloses. Considerations such as these may explain why in certain instances cellulase alone may yield protoplasts while with other tissues of the same species or the same tissue of different species it will do so only if mixed with pectinase of Macerozyme. For a detailed consideration of this aspect see Ruesink (101).

The commercially available cellulases, Cellulase Onozuka (All Japan Biochemicals Co. Ltd., Nishinomiya, Japan) or Meicelase (Meiji Seika Kaisha Ltd., 2-Chome, Kyobashi Chuo-ku, Tokyo, Japan), are both derived from the fungus *Trichoderma viride.* This mold produces a highly active cellulolytic enzyme capable of decomposing native cellulose such as cotton, and the fungus is famous (or infamous) for its attack on the mushroom (*Lentinus edodes*) which is cultivated on logwood in Japan. The mushrooms are killed off by the activity of the *Trichoderma viride* fungus growing on the logwood. The usual preparation of cellulase Onozuka has an activity of 1500 units/mg which is based on the rate of degradation of filter paper pieces (124). Both cellulase preparations contain cellulase C_1 (attacking native and crystalline cellulose), cellulase Cx (attacking amorphous cellulose), cellobiase, xylanase, glucanase, pectinase, lipase, phospholipase, nucleases, and several other enzymes, as well as potent mycolytic enzyme activity consisting of β-1,3-glucanase and chitinase. Macerozyme (All Japan Biochemicals Co. Ltd.) is derived from a *Rhizopus* sp. It has cell-separating activity and degrades potato pieces into unicells within 4 hr (124). It has little cellulolytic enzyme activity and no mycolytic enzyme activity. Its activity and its impurities are probably largely comparable to other pectinases. Both Macerozyme and other pectinases contain many enzymes including nucleases (25). Pectinase has been partially purified, probably resulting in the removal of various phenolics and ribonucleases, by elution through Sephadex G-25, and Macerozyme by elution through Biogel (28). Relatively simple purification procedures such as these often greatly facilitate the ease with which viable protoplasts can be released, either with the purified cellulase preparations alone, or by mixtures of cellulase and Macerozyme (pectinases) or by Macerozyme (pectinases) followed by cellulase when using the sequential method of protoplast isolation. Perhaps the best indication of whether protoplasts are viable, after enzymatic isolation, is whether they will regenerate a wall and divide, or divide mitotically without wall regeneration (27, 97). Kao, Keller & Miller (55) briefly reported that soybean suspension cultures can release protoplasts if treated with partially purified cellulase and Sigma pectinase dissolved in their B5 medium enriched with 0.1 *M* sucrose and 0.3 *M* sorbitol. These soybean protoplasts regenerate a new cell wall and then divide in a conditioned medium (55) or a completely synthetic medium (54). Nagata & Takebe (72) have described cell wall regeneration and cell division in isolated tobacco mesophyll protoplasts. These workers used 0.5% Macerozyme to separate the leaf cells and then 2% cellulase Onozuka to release protoplasts from the cells, using 0.7 M mannitol as the plasmolyticum. No purification of the enzymes was attempted. Power & Cocking (88) pioneered the use of a mixture of cellulase and Macerozyme (or pectinase) for the ready isolation of leaf mesophyll protoplasts, and here again no purification of the enzymes was attempted. Chupeau & Morel (13) obtained protoplasts from carrot callus using a mixture of crude enzymes, 5% cellulase Onozuka, 5% cellulase No. 5 (Industrie Biologique Française), 1% Sigma pectinase, and 0.1% Helicase (suc. digestif de *Helix pomatia*). In certain tissues, particularly those of fruits where there is a high pectin content, protoplasts can be released by the action of pectinase alone. The pectinase apparently destroys the structural integrity of the cell walls, and as a result the partially digested cell walls become disorganized, effecting release of the protoplasts. Gregory & Cocking (39) were able to isolate protoplasts in very large quantity from the locule tissue of mature green tomato fruit using 20% (w/v) Rohm and Haas Co. Pectinol R.10 in 25% (w/v) sucrose, and Raj & Herr (95) readily isolated protoplasts from the interplacental regions of *Solanum* berries by digesting the walls with 12% Sigma pectinase in plasmolyzing sucrose solution. *Myrothecium verrucaria* cellulase by itself will also release protoplasts from the locule tissue of tomato fruit (18).

Thus, while the enzymatic isolation of higher plant protoplasts has now become an acceptable laboratory procedure, each tissue must be investigated systematically in relation to the optimum conditions for protoplast release. Moreover, since the usefulness of protoplasts lies in their further manipulation and culture, it is most important to expose the protoplast for the minimum period to the action of crude cell wall degrading enzymes, and to use plasmolytica both ionic or nonionic, or mixtures, which experimentally have been determined most favorable for the subsequent wall regeneration and division of the isolated protoplasts. Such naked cells are variable in their survival, depending on the nature of the plasmolyticum (30, 108), and workers generally have paid insufficient attention to this aspect of their isolation of protoplasts. When working with protoplasts isolated from cultured cells, this aspect may be critically important (27). The other factor which is often not sufficiently considered is the actual osmotic pressure of the plasmolyticum employed during the isolation. This can have a very significant effect on the general stability of protoplasts. For detailed practical considerations see Evans & Cocking (28).

GROWTH AND DEVELOPMENT OF ISOLATED PROTOPLASTS

It is important before beginning any discussion of the growth and development of isolated protoplasts to try to assess why plant physiologists interested in plant cell culture should be so particularly interested in the culture of these naked cells. As we have seen, these naked cells can be isolated either from already cultured tissues or cells, callus or suspension culture cells, or directly from the various tissues of plant organs including leaves, petals, roots, and coleoptiles. If we look back at the early studies of Haberlandt (61), we see the first attempts to culture single cells isolated directly from the plant, mainly from leaves. Haberlandt noted that only plant parts which had cells loosely organized in tissues so that they were easy to isolate by mechanical means could be used for his experiments; the leaves subtending the bracts of *Lamium* proved to be most suitable. From this early beginning, plant physiologists have long appreciated the possible advantage of isolating somatic cells directly from the tissues of plants and of culturing these under carefully controlled conditions. As plant tissue culture developed, single plant cells were plated (4), but these cells were from tissues already cultured and were not isolated directly from the plant. Moreover, as recently discussed by Nagata & Takebe (73), the frequency of colony formation by these cultured single cells tends to be very low, and there is a wide range of karyological aberrations frequently encountered in such cultured cells. It is interesting to survey the methods adopted for the isolation of somatic cells directly from the tissues of plants. Both Kohlenbach (59) and Joshi & Ball (50, 51) isolated cells mechanically from the mesophyll of leaves and showed that the differentiated nonmeristematic mesophyll cells could be induced to divide and grow when isolated as single cells and incubated in suitable culture media. As we have already discussed, the work of Chayen (12) paved the way for the enzymatic isolation of cells by degradation of the middle lamella with suitable pectinase enzyme preparations. This led to the isolation of single cells from bean and tobacco leaves using pectinase with sucrose to aid as an osmotic stabilizer (52, 53). Even when these methods were further elaborated by others (121), it was clear that a major proportion of the cells, approximately 50%, were damaged during isolation. It was possible to culture these isolated cells, and they grew and divided in a culture medium containing 0.7 M mannitol (126). It was clear, however, that the walls of these cells were considerably weakened during the enzymatic isolation. More recently Nagata & Takebe (73) have obtained a 40% plating efficiency with cells from tobacco leaf palisade. Enzymatically isolated protoplasts from the palisade layer gave over 60% plating efficiency (73), and since, as we shall see later, these naked cells can be more readily modified than cells and can be cultured to regenerate plants in certain instances (73, 78), they are ideal starting material for the cloning of somatic plant cells. Somewhat similar results have recently been obtained by Grambow et al (37) using carrot protoplasts isolated from carrot suspension cultures. These results with naked carrot cells undergoing embryogenesis are particularly exciting in view of the already well-established work on carrot cell embryogenesis by Steward et al (115).

Cell wall regeneration.—Cell wall regeneration by plant cell protoplasts was first described by Townsend in 1897 (123). Townsend showed that in cells of *Elodea canadensis* and *Gaillardia lanceolata* that had been plasmolyzed, the plasmolyzed protoplast, still within the original cell wall, regenerated a new wall when maintained in 20% sucrose. Townsend was particularly interested in whether the presence of the nucleus was essential for cell wall regeneration, and since in these cells plasmolysis often resulted in the protoplast separating into two halves, one of which was nucleate and the other enucleate, these plasmolyzed cells were also of use in this respect as well. Thus he was able to show that only the nucleate subprotoplast within the leaf cell regenerated a new cell wall. He was also able to show that subprotoplasts isolated by the plasmolysis (mechanical) method of Klercker (58) from protonema of mosses, prothallia of ferns, hairs of stems, and leaves of higher plants also regenerated a new cell wall provided they were nucleate. No special culture media were necessary; sucrose and the necessary illumination provided adequate conditions. Binding (5) isolated protoplasts from various mosses, again using the plasmolysis (mechanical) method of Klercker. He found that wall formation was only carried out by nucleate protoplasts. A significant observation made by Binding was that some of the moss protoplasts which had regenerated a wall germinated to protonema. This perhaps would indicate that higher plant protoplasts might first regenerate some kind of cell wall and then divide, and under suitable conditions even regenerate into whole plants. With the introduction of enzymatic methods for the isolation of protoplasts, it was, as has been discussed fully earlier, possible to isolate protoplasts from a wide range of plant tissues and cultured cells. The early studies of Cocking (17) using protoplasts isolated from tomato roots made it seem likely that particularly those protoplasts isolated from the more meristematic regions of the root "first develop some form of primary cell wall and then, parallel to their later development, they acquire some aspects of the behavior of cultured freely suspended isolated cells."

In thin-sectioning studies it is difficult to be certain that the cell wall is entirely absent from spherical units which are sensitive to osmotic shock. Streiblova (116) applied freeze etching methods to the study of the yeast cell during protoplast formation, and she was able to show that at least in some cases the entire wall substance was not removed from the surface of the protoplasts. It would seem likely, however, in the case of higher plant protoplasts that little, if any, of the original cell wall remains, provided initially the cell has been markedly plasmolyzed and that suitable enzyme treatment has been given. If marked plasmolysis has not taken place—and this will depend not only on whether the cell is more meristematic, and therefore less vacuolated, but also on the osmolarity and nature of the plasmolyticum (108)—and if enzymatic degradation is incomplete, there is sometimes the possibility that a fine network of cellulose fibrils remains around the protoplast. These faint remnants of the wall are not visible optically either in bright field, phase contrast, or Normaski interference, nor do they fluoresce adequately with calcofluor (Roland, personal communication). They are not readily detected in thin section in the electron microscope after the usual fixation with glutaraldehyde and osmium tetroxide, staining during dehydration with uranyl acetate, and post staining with lead citrate. Their detection in thin section necessitates the use of the periodate, silver proteinate stain (93); however, they are readily detected by surface replica studies (85) and by freeze etching (136).

From the earlier studies in which cell wall regeneration was observed within the plasmolyzed cell, while the regenerating protoplast was still confined by the original cell wall, it was evident that special physical conditions for cell wall regeneration would be unnecessary. Special physical conditions, in which there is a certain and often critical degree of solidity including viscosity of the external medium surrounding protoplasts, have been found necessary for the successful regeneration of yeast protoplasts (74–76, 117). Recently, the de novo synthesis of wall structures by *Candida utilis* protoplasts incubated in a liquid medium (80) has been described so that a semisolid medium is not an absolute requirement for wall regeneration in the yeasts generally. The exact mechanism of the effects of such polysaccharides on cell wall regeneration is not yet fully understood. No critical studies have so far been carried out on higher plant protoplast regeneration in agar or gelatin. As we shall see later, many higher plant protoplasts will regenerate cell walls readily solely in liquid nutrient media, and therefore the effects of polysaccharides on the range and rate of wall regeneration in higher plant protoplasts might well repay careful investigating, particularly since it is known that contiguous higher plant protoplasts readily regenerate (84). Under such conditions it would seem likely that cell wall regeneration is being stabilized as a result of middle lamella formation between the cells (23). One of the features which the regeneration of fungal protoplasts and higher plant protoplasts share in common is that regeneration is a relatively slow process taking several days, and as a result it is possible to follow stages of biosynthesis of the new wall.

Given suitable conditions, bacterial protoplasts can be induced to grow and multiply without their cell wall being reformed (110). Such so-called L forms of bacteria therefore superficially resemble the *Haplopappus* protoplasts isolated from suspension cultures of this higher plant (27); but as we shall see later, isolated protoplasts of higher plants more normally first regenerate some form of wall and then divide.

Just as fungal and bacterial protoplasts do not require the presence of what could be described as extra special culture media during the actual wall regeneration process, a simple White's medium is all that is required for cell wall regeneration by most isolated higher plant protoplasts; and if they are photosynthetic, illumination often facilitates regeneration. The culture medium usually contains added sucrose, mannitol, or sorbitol or a mixture of all three to provide an external plasmolyzing solution. Added growth substances are usually not required for wall regeneration (48, 85) and in fact at certain levels of plasmolyticum, auxins may cause steady progressive swelling of protoplasts or even bursting (17, 42).

Tomato fruit protoplasts isolated from the green locule tissue comprised the first isolated protoplast system in which cell wall regeneration was established using both light and electron microscopic methods (85), and the pattern of cell wall regeneration in this system has been elucidated fairly extensively. Wall regeneration is very slow when protoplasts are cultured in sucrose alone, but when sucrose is combined with a modified White's medium, early stages of wall regeneration can be detected after a few hours. Progressively during the next few days a multilamellar system is formed at the surface of the plasmalemma. Experiments were carried out in liquid medium, and preliminary studies have indicated no marked effects from keeping protoplasts in high concentrations of gelatin during this regeneration process. After 1 week in culture, fibrils could be detected readily by the freeze-etch method between the plasmalemma and this lamellar pile, and subsequently masses of randomly oriented fibrils of cellulose appeared in this region. After 2 weeks protoplasts which had now fully regenerated could be transferred gradually and subsequently cultured in a culture medium containing only a few percent sucrose. Occasionally, with protoplast systems undergoing regeneration osmotic blowout of weaker areas of the wall may result if the regenerated wall is not sufficiently rigid to resist the osmotic stresses associated with any vacuole system present (2, 85). This has also been observed in the regenerating protoplasts from yeast when only the glucan stage of regeneration has been reached (128).

Division.—Somewhat different culture media (or conditioned media) are usually required for the subsequent division of regenerated protoplasts. The cell density as well as other factors are probably of key importance (27, 54, 55, 72). Normally most workers are convinced that they are studying the growth and development of strictly naked cells. When, as we have seen, the objective is to obtain high plating efficiency for cloning, the presence of cells may not be a serious complication. However, when claims are made that dividing cells, callus, or embryoids, or even whole plants, have been regenerated from isolated protoplasts, it is essential that really adequate methods have been employed to establish that in the isolated protoplast undergoing the supposed regeneration and division the original cell wall has in fact been removed, and that regeneration therefore has not taken place from cells. It has also been established that when isolated protoplasts are regenerating a wall and are in contact with each other, a middle lamella is formed between them resulting in cell aggregate formation (21, 23, 84). This sometimes occurs with isolated leaf protoplasts at high density (10^5/ml) in liquid culture, and these small cell aggregates could serve for regeneration of callus and even whole plants. Such cell aggregates do not form so readily when isolated protoplasts are cultured in agar. In the elegant study of Nagata & Takebe (73), who used protoplasts isolated from diploid tobacco leaves cultured in agar, and who observed the regeneration, growth, and division of single naked cells, it is clear that in this instance regeneration, division, and callus, and whole plant regeneration was taking place from single isolated protoplasts. In earlier work isolated protoplasts were cultured at high density, often in liquid media, and the regeneration and division of individual protoplasts were not fully recorded (68, 119). Grambow et al (37), when studying isolated carrot protoplasts and their regeneration into whole plants, state that embryos are formed from single protoplasts, but the technique employed did not permit continuous observation of individual protoplasts.

Several other isolated higher plant protoplast systems have been investigated in relation to their growth and division. Those workers principally interested in division have only briefly discussed cell wall regeneration (37, 54, 55). Nagata & Takebe (72) have attempted to follow cell wall regeneration using the semispecific Calcofluor, which if attached to any cellulose in the cell wall will fluoresce in the ultraviolet. This is a useful method, but burst protoplasts which have not regenerated a wall do take up Calcofluor (often near the plasmalemma) and fluoresce; and as we have already seen, early stages of wall regeneration probably do not involve cellulose synthesis. Chupeau & Morel (13) have also observed cell wall regeneration in protoplasts isolated from carrot callus, but others have not (45). The general pattern emerging from these studies seems to be that with suitable protoplasts cell wall regeneration, while it may not necessarily produce a completely typical wall (48), yet under the correct nutritional and other conditions it is a prelude to division. At this stage one begins to recapitulate the sequence of events involved in the growth and division of single isolated cultured cells generally (17, 115). It is of course possible for the division of the cytoplasmic portion of the protoplast (cytokinesis) to be variously correlated in time with mitosis, and in some plant cells no cytokinesis follows mitosis. This is sometimes the case in the fertilized egg of higher plants where even before cell division begins the development of the endosperm has taken place. This commences with the division of the primary endosperm nucleus, and quite frequently several of these mitotic divisions can occur before cytokinesis takes place, so that a multinucleate coenocytic state results as an intermediate stage. The removal of cell wall from cultured cells giving rise to protoplasts may result in a similar inbalance between mitosis and cytokinesis, as suggested from the studies of Eriksson & Jonasson (27) on the division of isolated *Haplopappus* protoplasts. These workers obtained some evidence that nuclear division (mitosis) was taking place while the isolated protoplasts were remaining as naked cells and before any cell wall regeneration had taken place. The possibility does exist therefore that protoplasts may undergo mitosis without cytokinesis (23). It would also seem likely that once wall regeneration has been initiated, mitosis and cytokinesis temporal relationships may become more closely similar until the situation is similar to that found normally in cultured cells, in root meristems, and in other somatic tissues of higher plants. It has been observed recently that isolated *Petunia* leaf protoplasts cultured in the medium described by Nagata & Takebe (72) undergo nuclear division giving rise to binucleate protoplasts, and then later, following wall regeneration, they undergo both mitosis and cytokinesis (Frearson, personal communication).

Binucleate cells in culture are sometimes observed, and this could have resulted from an inhibition of cytokinesis. Experimentally binucleate, trinucleate, and tetranucleate cells can be induced in root tip cells by means of caffeine (34), but this experimental induction of the multinucleate condition has not as yet been extended to cultured cells. Gonzalez-Fernandez et al (36) have observed that in binucleate cells the nuclei, formed in a mitosis which is not followed by later cytokinesis, divide synchronously.

Experimentally it is now possible to try to follow comparable nuclear events in cultured multinucleate cells by utilizing protoplasts which have been fused together either spontaneously (29, 91) or by sodium salt treatment (86, 87, 90). A prerequisite for spontaneous fusion is the maintenance of contact of adjacent protoplasts via plasmodesmata during the enzyme treatment in which the cell walls of the cultured cells are being degraded by cellulase action (137). In such intraspecies fusion bodies, the number of nuclei can be from 2 to 20 or more, and it is observed that after culture for a few days there is pronounced systrophy in most of the multinucleate protoplasts (91). The nuclei mass together in the center of the cell and are surrounded by the chloroplasts. In such protoplasts, containing two or more nuclei, some nuclear division is synchronous and concomitant with the initiation of wall regeneration. A wide spectrum of different nuclear behaviors was evident in these intraspecies fusion bodies during the mitotic division, with the possibility that nuclear fusion could occur during these divisions (23, 70).

Development and possible genetic modification of isolated protoplasts and plants.—It is interesting to recall that Philip White (134), in his address to the seminar on Plant Cell, Tissue and Organ Culture in Delhi in 1967, emphasized that the challenges for the future were still technical ones. One major requirement would be an easy and reliable method of reducing a plant to a single cell without destroying its viability: "ideally we should isolate cells directly from the tissues of the plants for there are scores of problems which demand the availability of single cell suspensions without a long previous history of in vitro cultivation." White emphasized the need for improved cloning methods generally, and it is clear that the ability to grow single cell clones of higher plants in a sufficient number (4) would greatly enlarge the scope of genetic investigations on somatic plant cells, just as the plating technique of Puck (94) has enlarged the scope of genetic investigations using cultured animal cells. This same theme has also been discussed by Steward et al (114), since clonal cultivation of cells offers a way of preserving a genome threatened by extinction (for example, the American elm).

The other major requirement for the future would be for a much better experimentally based plant cell karyology. White emphasized that somatic animal cells could be hybridized, but at that time nothing was known about the corresponding process in plants. There was a need to be able to define karyotypes with much better precision and to be able to recognize hybrids if they occurred. White was aware of the slow but steady progress being made in this general direction with isolated protoplasts, for he also highlighted the importance of more effective methods of studying the pathology of cells, such as the penetration of a virus through the cell membrane, for example, and its establishment in the protoplast (20).

Workers were generally quick to appreciate the key role that single isolated protoplasts could play in the improved cloning of plant cells once techniques for their enzymatic isolation had been perfected. They were also quick to recognize the importance of the fusion of isolated protoplasts (89, 90) for the ultimate somatic hybridization of plants (22, 26, 47, 77, 106), and the uptake of macromolecules by protoplasts (21) for studies on virus infection and transformation (25, 46, 118, 120). The isolation of tobacco leaf protoplasts directly from leaves and the plating of these on a fully defined medium at 60% efficiency (73), as well as the differentiation of these colonies giving shoots and roots (73, 78) and the eventual regeneration into whole plants both diploid (73) and haploid (78), will now further extend these new and exciting vistas. Clearly there is much work required, since each species and even variety of the same species has its own special problems, not the least of which are the special nutritional requirements for the adequate growth and development of the naked cell as it regenerates its cell wall and reestablishes itself as a growing and dividing cell, quite apart from the special nutritional requirements for any regeneration of plants which may take place. Such difficulties should not be minimized; yet it is clear that the foundations have been laid, for these isolated protoplasts are no longer merely interesting cytological curiosities but naked cells which if isolated under suitable conditions and cultured satisfactorily can be induced to grow and regenerate a wall, and divide, and acquire some aspects of the behavior of freely suspended isolated cells in culture. As we have seen, isolated protoplasts differ from cells insofar as aggregates of such naked cells can be formed, so that experimentally it is possible to obtain chimaeral aggregates from isolated protoplasts of different species. As they regenerate a wall between contiguous protoplasts a tissue is formed. Such studies could help greatly in our fuller understanding of stock scion interrelationships.

Major advances have been made in the development of techniques for the somatic hybridization of plants. The absence of the cell wall from isolated protoplasts makes these somatic naked cells ideally suited for fusion studies. Power, Cummins & Cocking (90) have shown that it is possible to fuse meristematic protoplasts from roots with the aid of an inducing agent. Sodium salts, preferably sodium nitrate, act as inducing agents. Sodium nitrate does not affect the viability of the protoplasts, and the overall effect is somewhat similar to the induced fusion of animal cells by Sendai virus (40). When isolated protoplasts of the same species are induced to fuse, homokaryons are formed, and when protoplasts of different varieties or different species are induced to fuse, heterokaryons result. Both intra (86, 90, 137) and inter (87, 90, 138) species induced fusion can now be achieved. The rate of fusion is largely dependent on the degree of vacuolation of the naked cells being induced to fuse (89, 137). Normally mature isolated leaf protoplasts and most callus and suspension culture isolated protoplasts have one large vacuole with only a thin peripheral layer of cytoplasm and numerous transvacuolar strands. Fusion of adpressed plasmalemmae readily occurs, and at this stage the protoplasts may only optically appear to be adhering to each other (137). Electron microscopic studies, however, show clearly that this is the prelude, in the presence of the inducing agent, to the ultimate overall fusion of the protoplasts. This may take many hours to be completed if large vacuoles are present in the protoplasts, since these large vacuoles impede the mixing together of the cytoplasm (137). As we have seen, induced fusion results in the formation of homokaryons or heterokaryons; and as is the case with animal cells which have been induced to fuse, using Sendai virus, nuclear fusion does not occur at this stage. In animal cells nuclear fusion results during the subsequent mitotic divisions (40). Homokaryons can also be formed readily by the spontaneous fusion of protoplasts [for a detailed analysis of the clear distinction between spontaneous and induced fusion see Withers & Cocking (137)].These homokaryons which have arisen by spontaneous fusion of the protoplasts of contiguous cells readily regenerate a wall; and the nuclei enter mitosis synchronously and appear to fuse (70, 91). It would therefore seem likely that hybrid cells may result, provided isolated protoplasts are selected from different species which have requirements for growth and development that are not too dissimilar. Cell wall regeneration is not likely to be a complicating factor, since there is a common pattern of wall synthesis in protoplasts of different species (3, 21, 23, 48, 93). Such cells will either be hybrid cells or cytoplasmic hybrid cells. Power & Cocking (89) have extensively outlined the implications of these developments in the somatic hybridization of cells for the improved breeding of plants and in agriculture generally. What is needed now is what White appreciated with characteristic prescience (134), namely the development of an improved plant genetics and the development of methods to detect the formation of hybrid cells and their behavior in culture. Will chromosome elimination occur in such plant cell somatic hybrids as it does in certain somatic animal hybrid cells (40)? And what will be the nature of the cytoplasmic interaction effect in hybrids formed somatically, between the chloroplasts of maize and those of wheat, for instance? Cytoplasmic heterokaryons can be identified optically (46, 87, 90) and electron microscopically using nuclear and chloroplast characteristics as markers (Davey & Short, personal communication), and the behavior of nuclei can be studied optically (86). The ability to regenerate plants from haploid protoplasts of tobacco (78) may facilitate the detection of diploid cells after fusion and any subsequent division, and of course also facilitate the identification of the hybrid plant. Chromosome markers need to be employed for the detection of hybrid cells and any plants that ultimately may be regenerated. Culture media need to be developed for the selection of nutritionally dependent hybrids as has been done for hybrid animal cells in culture (40). Initially somatic hybridization needs to be carried out with varieties of the same species or closely related species which can be crossed sexually and which are able to be cultured satisfactorily as cells. Tobacco clearly commends itself since its genetics and cell culture are so well documented (68, 69, 73, 78, 104, 127). Also carrot (6, 7, 37, 96, 113, 115) and sugar cane may be very suitable (1, 44, 66), and perhaps *Haplopappus* (6, 7, 26, 27) and *Crepis* (103, 105).

In the future one will undoubtedly see the growth and development of isolated protoplasts from an ever-increasing range of species, including the cereals. One may also see the development of a new cell genetics, including the use of protoplasts to take up informational molecules, perhaps thereby undergoing transformations of various types (46). These developments probably will not be restricted to higher plant protoplasts but will involve fungal and algal protoplasts as well. They will become increasingly orientated towards the general modification of plant cells, and perhaps ultimately whole plants. Both nuclear and cytoplasmic modifications will be involved, including perhaps a modification of cells to enable them to fix atmospheric nitrogen. This could be through hybrid cell formation between leguminous and nonleguminous isolated protoplasts or by the fusion of nitrogen fixing bacterial protoplasts with isolated protoplasts of plants such as the cereals which do not fix nitrogen.

ACKNOWLEDGMENTS

The author is indebted to many colleagues from Canada, France, Germany, Japan, Sweden, the United Kingdom, and the USA, who have supplied information in advance of publication so that this review could be as up to date as possible. He is also grateful to his colleagues in the Agricultural Research Council Group, University of Nottingham, for helpful suggestions.

LITERATURE CITED

1. Barba, R., Nickell, L. G. 1969. *Planta* 89:299–302
2. Bawa, B. S., Torrey, J. G. 1972. *Bot. Gaz.* 132. In press
3. Benbadis, A. 1971. *C. R. Acad. Sci. Ser. D* 273:797–800
4. Bergmann, L. 1960. *J. Gen. Physiol.* 43:841–51
5. Binding, H. 1966. *Z. Pflanzenphysiol.* 55:305–21
6. Blakely, L. M., Steward, F. C. 1964. *Am. J. Bot.* 51(7): 780–91
7. Ibid. 51(8): 809–20
8. Brachet, J. 1955. *Biochim. Biophys. Acta* 16:611–13
9. Brenner, S. et al 1958. *Nature* 181:1713–15
10. Brown, W. 1965. *Ann. Rev. Phytopathol.* 3:1–18
11. Chambers, R., Höfler, K. 1931. *Protoplasma* 12:338–55
12. Chayen, J. 1952. *Nature* 170:1070
13. Chupeau, Y., Morel, G. 1970. *C. R. Acad. Sci Ser. D* 270:2659–62
14. Cleland, R. 1971. *Ann. Rev. Plant Physiol.* 22:197–222
15. Cocking, E. C. 1960. *Nature* 187: 927
16. Cocking, E. C. 1960. *Biochem. J.* 76:51P
17. Cocking, E. C. 1961. *Nature* 191: 780–82
18. Cocking, E. C. 1963. *Biochem. J.* 88:31–32P
19. Cocking, E. C. 1965. *Viewpoints Biol.* 4: 170–203
20. Cocking, E. C. 1966. *Planta* 68: 206–14
21. Cocking, E. C. 1970. *Int. Rev. Cytol.* 28: 89–124
22. Cocking, E. C. 1970. *Colloq. Int. Cent. Nat. Rech. Sci.* 193:303–17

23. Cocking, E. C. 1972. In *Dynamic Aspects of Plant Ultrastructure,* ed. A. W. Robards. In press
24. Cocking, E. C., Gregroy, D. W. 1963. *J. Exp. Bot.* 14:504–11
25. Cocking, E. C., Pojnar, E. 1969. *J. Gen. Virol.* 4: 305–12
26. Eriksson, T. 1970. *Colloq. Int. Cent. Nat. Rech. Sci.* 193:297–302
27. Eriksson, T., Jonasson, K. 1969. *Planta* 89:85–89
28. Evans, P. K., Cocking, E. C. 1972. In *Plant Tissue and Cell Culture,* ed. H. Street. Oxford: Blackwell. In press
29. Fodil, Y., Esnault, R., Trapy, G. 1971. *C. R. Acad. Sci. Ser D* 273:727–29
30. Ibid. 272:948–51
31. Frey-Wyssling, A. 1967. *Nature* 216:516
32. Gabriel, M. 1970. *Protoplasma* 70:135–38
33. Giaja, J. 1919. *C. R. Soc. Biol. Fil. Paris* 82:719–20
34. Giménez-Martin, G., López-Sáez, J. F., Moreno, P., González-Fernández, A. 1968. *Chromosoma* 25:282–96
35. González-Bernáldez, F., López-Sáez, J. F., Garcia-Ferrero, G. 1968. *Protoplasma* 65:255–62
36. González-Fernandez, A., López-Sáez, J. F., Gimenez-Martin, G. 1964. *Phyton* 21:157–65
37. Grambow, H. J., Kao, K. N., Miller, R. A., Gamborg, O. L. 1972. *Planta.* In press
38. Greenway, H. 1970. *Plant Physiol.* 46:254–58
39. Gregory, D. W., Cocking, E. C. 1965. *J. Cell Biol.* 24:143–46
40. Harris, H. 1970. *Cell Fusion (The Dunham Lectures).* Oxford: Clarendon
41. Hall, J. A., Wood, R. K. S. 1970. *Nature* 227:1266–67
42. Hall, M. D., Cocking, E. C. 1971. *Biochem. J.* 124:33P
43. Hanstein, J. 1880. *Bot. Abhandlungen* 4:1–56
44. Heinz, D. J., Mee, G. W. P. 1971. *Am. J. Bot.* 58(3):257–62
45. Hellmann, S., Reinert, J. 1971. *Protoplasma* 72:479–84
46. Hess, D. 1970. *Ber. Deut. Bot. Ges.* 83:279–300
47. Hildebrandt, A. C., Schenk, R. U. 1970. *Colloq. Int. Cent. Nat. Rech. Sci.* 193: 319–31
48. Horine, R. K., Ruesink, A. W. 1972. *Plant Physiol.* In press
49. Jensen, R. G., Francki, R. I. B., Zaitlin, M. 1971. *Plant Physiol.* 48:9–13
50. Joshi, P. C., Ball, E. 1968. *Develop. Biol.* 17:308–25
51. Joshi, P. C., Ball, E. 1968. *Z. Pflanzenphysiol.* 59:109–23
52. Jyung, W. H., Wittwer, S. H., Bukovac, M. J. 1965. *Plant Physiol.* 40:410–14
53. Jyung, W. H., Wittwer, S. H., Bukovac, M. J. 1965. *Nature* 205:921–22
54. Kao, K. N., Gamborg, O. L., Miller, R. A., Keller, W. A. 1971. *Nature* 232:124
55. Kao, K. N., Keller, W. A., Miller, R. A. 1970. *Exp. Cell Res.* 62: 338–40
56. Karr, A. L., Albersheim, P. 1970. *Plant Physiol.* 46:69–80
57. Keller, W. A., Harvey, B., Gamborg, O. L., Miller, R. A., Eveleigh, D. E. 1970. *Nature* 226: 280–82
58. Klercker, J. 1892. *Ofvers Vetensk Akad. Forh. Stockholm* 49: 463–75
59. Kohlenbach, H. W. 1966. *Z. Pflanzenphysiol.* 55:142–57
60. Koningsberger, V. J. 1947. *Meded. Kon. Vlaam. Acad. Wetensch. Belg.* 9:5–28
61. Krikorian, A. D., Berquam, D. L. 1969. *Bot. Rev.* 35:59–88
62. Küster, E. 1927. *Protoplasma* 3: 223–33
63. Küster, E. 1935. *Die Pflanzenzelle,* 568. Jena: Fischer
64. Levitt, J., Scarth, G. W., Gibbs, R. D. 1936. *Protoplasma* 26: 237–48
65. Mandels, M., Reese, E. T. 1965. *Ann. Rev. Phytopathol.* 3:85–102
66. Maretzki, A. 1970. *Ann. Rep. Exp. Stat. Hawaii. Sugar Plant. Assoc.* 64–65
67. Martin, H. H. 1963. *J. Theor. Biol.* 5:1–34
68. Melchers, G. 1971. *Mitt. Max-Planck Inst. Ges., München* 2: 72–92
69. Melchers, G., Labib, G. 1970. *Ber. Deut. Bot. Ges.* 83:129–50
70. Miller, R. A., Gamborg, O. L., Keller, W. A., Kao, K. N. 1971. *Genetics.* In press
71. Mills, A. K., Krebs, H. 1968. *As-*

pects of Yeast Metabolism, 12. Oxford: Blackwell
72. Nagata, T., Takebe, I. 1970. *Planta* 92:301–8
73. Ibid 1971. 99:12–20
74. Nečas, O. 1961. *Nature* 192:580–81
75. Nečas, O., Svoboda, A. 1967. *Symposium über Hefeprotoplasten,* ed. R. Müller, 67–71. Berlin: Akad-Verlag
76. Nečas, O., Svoboda, A. 1969. *Proc. 2nd Symp. Yeast,* ed. A. Kocková-Kratochvilová, 213–18. Slovak Acad. Sci., Bratislava
77. Nickell, L. G., Torrey, J. G. 1969. *Science* 166:1068–70
78. Nitsch, J. P., Ohyama, K. 1971. *C. R. Acad. Sci. Ser. D* 273: 801–3
79. Norkrans, B. 1963. *Ann. Rev. Phytopathol.* 1:325–50
80. Novaes-Ledieu, M., Garcia-Mendoza, C. 1970. *J. Gen. Microbiol.* 61:335–45
81. Otsuki, Y., Takebe, I. 1969. *Plant Cell Physiol.* 10:917–21
82. Pilet, P. E., Prat, R., Roland, J. C. 1972. *Plant Cell Physiol.* In press
83. Pinto da Silva, P. G. 1969. *Naturwissenschaften* 56:41
84. Pojnar, E., Cocking, E. C. 1968. *Nature* 218:289
85. Pojnar, E., Willison, J. H. M., Cocking, E. C. 1967. *Protoplasma* 64:460–80
86. Potrykus, I. 1971. *Naturwissenschaften* 58(6):328
87. Potrykus, I. 1971. *Nature (New Biology)* 231:57
88. Power, J. B., Cocking, E. C. 1970. *J. Exp. Bot.* 21:64–70
89. Power, J. B., Cocking, E. C. 1971. *Sci. Progr. Oxford* 59: 181–98
90. Power, J. B., Cummins, S. E., Cocking, E. C. 1970. *Nature* 225:1016–18
91. Power, J. B., Frearson, E. M., Cocking, E. C. 1971. *Biochem. J.* 123:29–30P
92. Prat, R., Roland, J. C. 1970. *C. R. Acad. Sci. Ser. D* 271:1862-65
93. Ibid 1971. 273:165–68
94. Puck, T. T. 1957. *Cellular Biology, Nucleic Acids and Viruses,* ed. T. M. Rivers. New York Acad. Sci.
95. Raj, B., Herr, J. M. 1970. *Protoplasma* 69:291–300

96. Reinert, J. 1966. *Planta* 68:375–78
97. Reinert, J., Hellmann, S. 1971. *Naturwissenschaften* 58:419
98. Rogers, H. J., Perkins, H. R. 1968. *Cell Walls and Membranes,* ed. C. Long. London: Spon
99. Ruesink, A. W. 1971. *Methods Enzymol.* 23A:197–209
100. Ruesink, A. W. 1971. *Plant Physiol.* 47:192–95
101. Ruesink, A. W. 1972. *Am. J. Bot.* In press
102. Ruesink, A. W., Thimann, K. V. 1966. *Science* 154:280–81
103. Sacristán, M. D. 1971. *Chromosoma* 33:273–83
104. Sacristán, M. D., Melchers, G. 1969. *Mol. Gen. Genet.* 105: 317–33
105. Sacristán, M. D., Wendt-Gallitelli, M. F. 1971. *Mol. Gen. Genet.* 110:355–60
106. Schenk, R. U., Hildebrandt, A. C. 1968. *Am. J. Bot.* 55:731
107. Schenk, R. U., Hildebrandt, A. C. 1969. *Crop Sci.* 9:629–31
108. Schenk, R. U., Hildebrandt, A. C. 1969. *Øyton* 26(2):155–66
109. Seifriz, W. 1928. *Protoplasma* 3: 191–96
110. Smith, D. G. 1969. *Sci. Progr. Oxford* 57:169–92
111. Spizizen, J. 1962. *Methods Enzymol.* 5:122–34
112. Stadelmann, E. J. 1966. *Methods Cell Physiol.* 2:143–216
113. Steward, F. C. 1970. *Proc. Roy. Soc. B* 175:1–30
114. Steward, F. C., Ammirato, P. V., Mapes, M. O. 1970. *Ann. Bot.* 34:761–87
115. Steward, F. C., Mapes, M. O., Ammirato, P. V. 1969. *Plant Physiol. Treatise* 5B:329–76
116. Streiblova, E. 1968. *J. Bacteriol.* 95:700–7
117. Svoboda, A. 1966. *Exp. Cell Res.* 44:640–42
118. Takebe, I. 1970. *Colloq. Int. Cent. Nat. Rech. Sci.* 193:503–11
119. Takebe, I., Labib, G., Melchers, G. 1971. *Naturwissenschaften* 58:318–20
120. Takebe, I., Otsuki, Y. 1969. *Proc. Nat. Acad. Sci.* 64:843–48
121. Takebe, I., Otsuki, Y., Aoki, S. 1968. *Plant Cell Physiol.* 9: 115–24
122. Törnävä, S. R. 1939. *Protoplasma* 32:329–41

123. Townsend, C. O. 1897. *Jahrb. Wiss. Bot.* 30:484–510
124. Toyama, N., Fujii, N., Ogawa, K. 1970. *Cellulase, cell separating enzyme and mycolytic enzyme.* (Appl. Microbiol. Lab., Dep. Agr. Chem., Univ. Miyazaki, Japan)
125. Tribe, H. T. 1955. *Ann. Bot.* 19: 351–68
126. Usui, H., Takebe, I. 1969. *Development Growth and Differentiation* 11:143–47
127. Vasil, V., Hildebrandt, A. C. 1967. *Planta* 75:139–51
128. Villanueva, J. R. 1966. *The Fungi,* ed. G. C. Ainsworth, A. S. Sussman, 2:3–62. New York, London: Academic
129. Villanueva, J. R., Acha, I. G. 1971. *Methods Enzymol.* 4: 665–718
130. Vreugdenhil, D. 1957. *Acta Bot. Neerl.* 6:472–542
131. Whatley, F. R. 1956. *Modern Methods of Plant Analysis,* ed. K. Paech, M. V. Tracey, 1:452–67. Göttingen: Springer
132. Whitaker, D. R. 1959. *Marine Boring and Fouling Organisms,* ed. D. L. Ray. Friday Harber Symp., Univ. Washington, Seattle
133. Whitaker, D. R., Hanson, K. R., Datta, P. K. 1963. *Can. J. Biochem. Physiol.* 41:671–96
134. White, P. R. 1967. *Seminar on Plant Cell, Tissue and Organ Culture,* ed. B. M. Johri. Univ. Dehli: U.G.C. Cent. Advan. Study Bot. Plant Morphol., Embryol.
135. Willison, J. H. M., Grout, B. W. W., Cocking, E. C. 1971. *J. Bioenerg.* 2. In press
136. Willison, J. H. M., Cocking, E. C. 1972. *Planta.* In press
137. Withers, L. A. W., Cocking, E. C. 1972. *J. Cell Sci.* In press
138. Withers, L. A. W., Power, J. B., Cocking, E. C. 1971. *Biochem. J.* 124:47P
139. Yotsuynagi, Y. 1953. *Cytologia* 18:146–56
140. Zaitlin, M., Coltrin, D. 1964. *Plant Physiol.* 39:91–95

Ann. Rev. Plant Physiol. 1972. 23:51–72

ION TRANSPORT IN THE CELLS OF HIGHER PLANT TISSUES

7522

W. P. ANDERSON

Department of Botany, University of Liverpool
Liverpool, England

CONTENTS

The rate of advance in elucidating the processes of ion transport in higher plant cells has been depressingly slow, but the cause of this relatively poor progress is established. Critical and unequivocal experiments are difficult to devise and perform because of the nature of the cells and tissues in higher plants.

However, the parallel comparison often made between ion transport work in the algal coenocytes on the one hand and higher plant cells on the other is to an extent misleading, if the implication is taken that the only difference is in practical experimental difficulty. It is not only possible to isolate and work on a single coenocytic cell, it is also meaningful to discuss the observations in terms of the behavior of that cell; in many species the mature coenocytes may be considered as individual entities. In higher plant tissues it is not only experimentally difficult to isolate and work on a single cell, but it is also of dubious physiological validity to discuss the behavior of that cell without reference to the other cells of the tissue. The cells comprising higher plant tissues are integrated in function so that the tissue overall has distinct physiological behavior. Thus two factors are involved: the practical experimental problems are greater in higher plant work, and adequate interpretation is necessarily more complex.

The sophistication of ion transport studies in certain algal coenocytes is now very great. Any review here of this area would be redundant because of the recent article by the leading authority, Dr. MacRobbie (88), but insofar as these algal coenocytes are the plant cells in which ion transport is best understood, it is worthwhile to sketch briefly the historical development. The hope is that the progression of systematic investigation which has been used will provide guidelines for workers with higher plants.

First the physicochemical situation was described by assuming that a cell and its environment could be treated as a simple thermodynamic system of three phases. The basic thermodynamics involved have been set out in detail in an earlier review (25) and in many specialist papers (e.g. 39, 117, 118, 127) and need no further elaboration here. Measurements of the partial ion fluxes across each membrane, the ionic concentrations in each phase, and the electric potential differences at each membrane, when inserted into certain predictor relationships, principally the Ussing-Teorell flux ratio equation (127, 130), result in decisions of whether an ion is actively or passively distributed across plasmalemma or tonoplast. On the basis of these criteria active transport is as defined by Ussing (130): active transport is the process by which an ion is moved against an electrochemical potential gradient. Recent redefinition in terms of irreversible thermodynamic formalism (67) requires tests which are less convenient. The discussion given by MacRobbie (88) on the merits of these definitions of active transport is worth consulting. This thermodynamic descriptive stage is now essentially complete for several algal species under appropriate environmental conditions (20, 54, 85, 86, 112, 122, 125), and the transport of the major ions at plasmalemma and tonoplast has been characterized.

In higher plants only relatively few attempts have been made to set about this systematic characterization of the fluxes as active or passive influxes or effluxes. An appraisal of this work will be given later, but it is clear that the experimental difficulties are great, particularly in obtaining information about the cytoplasm. Therefore, it may be worthwhile to consider whether it is possible to somehow circumvent the thermodynamic characterization stage as used in the algal work. There is no implication in this that an alternative, experimentally more accessible, phenomenological description method can be devised; rather it is speculation about whether the ultimate goal of ion transport investigations—elucidation of the mechanisms of ion penetration of biological membranes and of the coupling and control of that penetration to the other life activities of the cell—can be pursued without prior thermodynamic characterization of the membrane situation.

The answer to this speculation seems to be that it both can and cannot. To deal with the "cannot" first, the reason is this: whatever molecular, physicochemical mechanism is eventually deduced as the transport mechanism for K^+, say, it will require for its application knowledge of the state of K^+ on either side of the membrane. This is for the future when useful descriptors such as "K^+carrier" are resolved in physicochemical terms. The reason for the "can" answer to the speculation is that useful progress can be made, using descriptor terms like the one just given, by judicious flux measurements under varying conditions. Difficulties abound without prior thermodynamic characterization, but a retrogression to the trivial arguments of the past, whether a high Q_{10} for a flux means that flux is active and others of like genre, can be avoided by careful interpretation of experiments together with an awareness of the physicochemical fundamentals of the situation, however poorly they are experimentally known. Nevertheless, one need only notice how present progress in the algal studies (87, 90, 113, 114, 123) depends heavily on the established flux characterization in order to be convinced of the need for work on similar characterization in higher plant cells to proceed with urgency.

As has been indicated previously, the chief experimental difficulty in work on higher plant cells is resolving the cytoplasmic and vacuolar phases. In order to measure the electrochemical state of an ion in these phases actual mechanical resolution is required; a microelectrode must be inserted in the cytoplasm and then again in the vacuole. The difficulty is that an average higher plant cell of perhaps $30\mu m$ diameter will have only an outer cytoplasm layer, perhaps $1\mu m$ thick. Presently available mechanical-optical systems cannot locate a microelectrode tip with such precision, and the best that can be done to date is to insert the electrode into the vacuole (35, 84, 95, 97, 103, 120). Since microelectrode membrane penetration inevitably is preceded by distortion of the membrane as it yields before the advancing electrode tip, there may be grounds for believing that penetrating the plasmalemma alone is nearly impossible. The displacement of the plasmalemma may exceed the normal thickness of the cytoplasm layer before penetration occurs. Both the distorted plasmalemma and the tonoplast with which it is now contingent may be ruptured simultaneously by the electrode. Whether or not it will ever be technically possible to achieve such mechanical resolution is difficult to assess; the dimensions of the cytoplasm are often close to the optical resolution limit of the light microscope, and, as has been said, the system is highly deformable. However, a properly designed mechanical apparatus with a controlled high-speed electrode thrust, possibly by refinement of the work already done on electrode movement by the piezoelectric effect, eventually may enable measurements to be made on the cytoplasm and vacuole of at least certain selected cell types.

Resolving the plasmalemma and tonoplast fluxes is a routine matter resting on the fact that the half-times of penetration of the two membranes are so dissimilar. The analysis for tissues of uniform cell type at flux equilibrium is well established (22, 29, 98). Briefly, the time course of an ion tracer efflux from a tissue segment of suitable size has three distinct components, each with a characteristic rate constant for exchange. From largest rate constant to smallest they refer respectively to extracellular space, cytoplasm, and vacuole, and by the usual analysis the individual membrane fluxes and the ionic content of the various compartments can be determined. The subdivision of extracellular space into Donnan free space and water free space (16) seems to have fallen into disuse. The physicochemical interpretation of this subdivision has been given (26). However, in many experiments no attempt is made to establish the compartmentation in the tissue. In this instance measurements are made of what is generically known as "ion uptake." The assay procedure is by monitoring, either chemically or by radioactive counting, representative tissue segments before and after exposure for a given length of time to a solution of known composition. In such experiments the tissue is usually examined during a transient in flux status so that compartmental analysis of the results is more difficult and is not normally attempted, but see also (24).

Two other assay techniques also in common use in ion uptake experiments are microautoradiography (15, 74, 82) and electron probe measurements (73–75, 111). Both are capable of cellular, perhaps subcellular, resolution of ion distributions but there are disadvantages. Neither is easily made quantitative with respect to the specific activity or concentration of the loading solution, although comparisons within a given tissue preparation are of acceptable accuracy with both techniques. The difficulties of calibration of these techniques have been dealt with to some extent (74), but it is true to say that as yet neither of these assays yields uptake values which can be properly expressed in units such as moles cm^{-2} s^{-1}, the units one should consider as normal for defining an ion flux across a membrane.

The point to be taken from this introductory preamble is that the pressing need is for quantitative elucidation of the situation at plasmalemma and tonoplast, with proper respect being given to the overall condition of the tissue. In this connection it is worth remarking again that the cells of higher plant tissues are integrated in function so that the tissue itself has distinct function. In ion transport studies on roots, for example, one must first characterize the ion transport behavior of the root before one attempts to understand the functioning of the individual cells. Without a fairly comprehensive view of the whole situation in one's mind, much effort can be wasted.

Ion Transport in Various Plant Tissues

The work will be reviewed under five headings of tissue types on which most of the experimental evidence has been obtained. It is hoped that this presentation will not mask the essential unity of the subject. It is merely convenient, and there is much cross-referring in the discussion of various topics. Indeed the choice of section in which to include a particular topic is often quite arbitrary. The selection of published material reflects nothing more than the author's interests and his rather arrogant desire to put across a particular point of view.

Storage tissue.—The cells of such tissue are principally parenchyma, and those most commonly used are underground storage organs such as beet, carrot, potato, etc. The advantages are convenience of supply and experimentation and homogeneity of cell type. The preparation is usually discs, 10–20 mm in diameter and 1 mm thick, cut from the tissue. Freshly cut discs are not very active in ion uptake. Washing the discs for several days in aerated, frequently changed water results in the development of an ion uptake capability by the discs.

The enhancement of ion uptake is only one of the changes induced by "aging" the discs. Respiration rates increase several fold, considered by Laties in a very useful early review (69) to be "the consequence of the development of vigorous phosphorylation ... dependent upon the upsurge in activity of the tricarboxylic cycle." A more recent report (8) disputes this interpretation of the increase of respiration with aging and ascribes it to a primary increase in protein synthesis. There is now a body of evidence for an aging-induced increase in protein synthesis, enzyme induction, and nucleic acid synthesis in storage tissue discs (7, 8, 17, 83), and there seems no doubt that these are the primary events which cause the observed enhancement in ion uptake. A very useful generalization which helps in understanding the aging phenomenon is given by Bryant & Ap Rees (17): "in general terms the aging phenomenon may be viewed as a reversal of the changes that accompany dormancy."

Similar aging effects are found in excised tissue slices from a variety of plant parts. In the present context, however, the reports of the effect of aging on the ion transport properties of isolated steles of roots are most pertinent. This topic will be discussed in detail later, and it is sufficient here to say that a recent report shows that aging in steles has very little in common with aging in storage tissue. Hall, Sexton & Baker (42) find that the levels of ATPase, glycerophosphatase, malic dehydrogenase, glucose-6-P dehydrogenase and 6-phospho-gluconate dehydrogenase fall during aging in steles. Total protein falls and only the levels of peroxidase and alcohol dehydrogenase increase. The authors rightly reject any comparison with the aging effects in storage tissue, although both aged steles and aged storage tissue discs show enhanced respiration rates and ion uptake rates. They consider the effect in steles to be comparable to the climacteric in respiration found in fruits at the onset of ripening (14). Thus aging in storage tissue is in a sense a break in dormancy; aging in steles is the onset of necrosis.

Ion transport studies in storage tissue have been conducted in the main by two types of experiment. In the first type, the tissue discs, immediately after aging in distilled water or in a dilute Ca^{++} solution, are placed in the experimental solution. The relationship between the concentration of this solution and the initial rate of uptake of the ion (which in most of the reported experiments is also a good measure of the initial ion influx) follows the pattern commonly called "the dual isotherm of ion uptake." A most comprehensive review of this topic is available from Epstein (30), whose work laid down the fundamentals of the process. In beet discs K^+ and Cl^- uptake exhibit this pattern (94). In fresh discs System 1 is barely discernible although System 2 is present; aging the discs for 72 hr in 0.5 mM $CaSO_4$ results in the establishment of a clearly defined System 1 and in modifications to System 2 as compared with fresh discs.

In experiments of this type one is dealing with tissue in which there is a large net influx of the ion. Note that this influx cannot be ascribed straightforwardly to either plasmalemma or tonoplast (24). It is obvious that storage tissue discs cannot continue indefinitely accumulating an ion. The initial net influx (uptake rate) will decrease with time of exposure to the loading solution, even if the tissue does not deteriorate metabolically. This is simply because the rate of net accumulation decreases as the tissue approaches equilibrium with the loading solution. In fact, the only real steady state possible in such tissue is equilibrium, and in a strict sense all the work on dual isotherms should be seen as establishing descriptors of the overall tissue behavior of ion uptake either for an extrapolated initial uptake rate (94) or on the implicit assumption that over a reasonable period of time the tissue is in a pseudo-steady state where the "initial" uptake rate does not vary appreciably (30–33).

The second type of experiment is conducted on tissue which is at or near flux equilibrium. Aged discs are exposed for suitable times to solutions of identical compositions to those of the experimental solutions, so that there is approximately zero net flux of the ion to be studied at the time the experiment starts. The only attempt to study tissue discs during the transient from the initial state to the flux equilibrium state has been by Osmond & Laties (94). They believe their analysis of the fluxes during the transient give evidence that System 1 is located at the plasmalemma of the beet parenchyma cells, and that System 2 is at the tonoplast. It is difficult to assess this evidence because no formal flux analysis is given; it seems unlikely that complicated, time-dependent processes can be analyzed without recourse to algebraic methods.

In situations where flux equilibrium is established, an exact compartmental analysis has been given for a system of three compartments in series, the usual model of the vacuole-cytoplasm-environment sequence of a single parenchyma cell (22, 29, 95, 98). Pitman (98) first investigated beet tissue by this technique and made estimates of the cytoplasmic and vacuolar ionic contents, as well as investigating the fluxes under various conditions. He showed that the establishment of salt equilibrium is brought about not by a decrease in influx but by an increase in efflux (see also 132). This therefore is justification for reading more into the results of initial net uptake experiments of the "dual isotherm" type than simply a description of initial influx. It is not unlikely that influx continues to be described by the dual isotherm kinetics, and that the decrease in net uptake as the tissue approaches equilibrium is caused by an increase in efflux. Indeed the results of Pitman (98) and of Cram (22) on beet and carrot discs respectively confirm in steady-state experiments the duality in the relationship of ion influx to solution concentration. On this point of cross-confirmation of dual isotherm observations and zero net flux compartmental analysis, it should be noted that the one explicit attempt at this confirmation (78) is based on very uncertain ground. The tissues used by Lüttge & Bauer (78) were not in a condition of zero net flux, and furthermore, as pointed out by Cram (22), their results imply that the tonoplast flux is rate-limiting at low external concentrations and the plasmalemma flux is rate-limiting at high external concentrations. The observations of Pitman (98) and of Cram (22) on their respective tissues are just the reverse, in support of the proposed location of System 1 at the plasmalemma and System 2 at the tonoplast, as first suggested by Torii & Laties (128) and later forcefully argued by Laties (70). The whole question of the location of Systems 1 and 2 will be discussed later.

There is some information of the nature of the ion fluxes in storage tissue. In beet tissue a fine series of investigations has been carried out by Poole (106–109). His early work made no attempt to resolve events at plasmalemma and tonoplast but showed that there is an overall active uptake of KCl in aged discs, possibly by the action of a neutral KCl pump. Addition of bicarbonate to the external solution promoted an increase in passive K^+ influx, in exchange for an active H^+ extrusion. The influx/efflux ratio for K^+ does not behave in the manner expected for diffusion, so the conclusion is that the K^+ efflux is carrier mediated or is a 1:1 exchange diffusion of K^+ across the membrane system (106). Recently Poole has used a modified compartmental analysis to show that cation selectivity occurs at the plasmalemma; in aged discs Na^+ strongly inhibits K^+ influx and slightly increases K^+ efflux at this membrane (108). In fresh discs there is no transport system for Na^+ which develops with aging. Both System 1 and 2 characteristics appear in Na^+ transport, and cation selectivity apparently is not affected by the presence or absence of Ca^{++} (109). Cram (23) has studied the effects of ouabain on the fluxes of K^+ and Na^+ in carrot discs and has found no effect on the influx of either but a pronounced inhibition of the Na^+ efflux. He suggests that this is prima facie evidence against the existence of a K^+-Na^+ linked exchange pump of the type known to operate in algae (88).

Investigations of Cl^- fluxes have been carried out in two tissues: in carrot discs by Cram (22) and in potato discs by Laties, MacDonald & Dainty (72) and again by Macklon & MacDonald (84). In carrot Cram has resolved the separate membrane fluxes, and he tentatively concludes that Cl^- influxes at both plasmalemma and tonoplast are active and that most of the Cl^- efflux is coupled to Cl^- influx. Indeed most of the changes in $^{36}Cl^-$ fluxes at external concentrations above 8 mM is due to a 1:1 exchange of Cl^- at the plasmalemma. In fresh potato discs it seems that Cl^- influx is passive (72), a decision reached after assuming the plasmalemma electric potential. In aged discs or in fresh discs in the presence of Ca^{++} where the Cl^- uptake is substantially different, it was again concluded that Cl^- influx was essentially passive and that the difference was primarily due to an increased K^+ permeability. An experimental value for the membrane potential has been obtained (84) which agrees well with the assumed value except when Ca^{++} is present. This divergence leads Macklon & MacDonald to suppose that there is an active Cl^- influx in the presence of Ca^{++}. Jennings (62) contends that the electric potential discrepancy is due to the fact that the inserted electrodes measured the plasmalemma potential and not the overall potential from vacuole to cell exterior as Macklon & MacDonald supposed. This latter explanation seems no more likely than that given by the workers who made the measurements, and it may be concluded that there is an active Cl^- influx in aged potato discs in the presence of Ca^{++}. Attempts to work in the absence of Ca^{++} often lead to membrane anomalies.

The studies of Van Steveninck (58, 132–135) on K^+ and Cl^- fluxes in red beet tissue have made an important contribution, since in many ways they are the most comprehensive investigations undertaken on the tissue. He has considered in some detail the role of organic anions in the tissue in relation to ion uptake from an external salt solution, a consideration without which no interpretation can be complete. The enhancement of KCl uptake with aging is due to a marked increase in the Cl^- influx, the K^+ influx remaining substantially unaltered (132). The presence of tris buffer in the external solution enhances the K^+ influx by promoting a K^+-H^+ exchange with concomitant increases in vacuolar malate and citrate levels (134). The Cl^- influx remains negligibly small until the requisite transport mechanism develops during the process of aging.

There is evidence of the coupling of exergonic metabolic reactions to transport in storage tissue. The early observation that ion accumulation is accompanied by an increase in tissue respiration rate (76) has been pursued in more detail recently. In carrot discs there is conflicting evidence. Atkinson and co-workers (11, 12) think that ion transport is not driven by ATP, while a more recent report (79) shows that respiration is stimulated specifically by ion transport at the tonoplast, and that this transport is driven by high energy intermediates of the cytochrome chain or by ATP. In beet, Polya & Atkinson (105) found that the cessation of ion transport under anaerobic conditions is not accompanied by a decrease in ATP level in the tissue. They argue that ion transport is directly coupled to electron transfer along the cytochrome chain. The arguments on this possibility have not advanced since the early classic reviews (76, 115, 116), and the reader is referred again to these. The difficulties associated with direct linkage of electron transfer at the mitochondrial inner membrane and ion transport at plasmalemma and tonoplast are discussed by Polya & Atkinson (105). It may be said that at present no suitable explanation of an apparently direct biochemical linkage at such a relatively large physical separation has been proposed. A similar problem exists in the coupling of ion transport and photosynthetic electron transfer in algal coenocytes (89, 113).

To conclude this section, it seems that a consistent pattern is emerging, especially for Cl^- fluxes. In fresh discs Cl^- fluxes are small and passive. During aging when general protein synthesis is in progress, Cl^- carriers are produced and function by deriving energy directly from cytochrome electron transfer. Cation uptake is more complex, perhaps because of the involvement of organic acid production and K^+-H^+ exchange. The K^+-Na^+ exchange system is unlikely to operate in carrot; in beet there is evidence that the K^+ and Na^+ carriers are separate and that Na^+ inhibition of K^+ influx is by competition for electrically balancing transported anions (108). The K^+ carriers in beet are present in fresh discs, but the Na^+ carriers only develop during aging. Both cation and anion influxes show dual isotherm kinetics with variation in external concentration.

Roots.—Roots and salt glands are the prime examples of plant tissues which normally maintain a steady state of ion transport other than flux equilibrium. Both these tissue types are capable of accumulating salt from their environments—the bathing solution or soil solution in the case of roots and the neighboring parenchyma and mesophyll cells of the leaf in the case of salt glands—and exuding these accumulated ions at some other point, usually as a hypertonic solution. This steady-state net transport of ions is achieved by the integrated physiological functioning of all the cells of the tissue. Elucidating the mechanisms involved requires consideration not only of the molecular transport system of any one cell, but also the anatomical arrangement of the cells in the tissue. The geometrical interrelationships of the cells is central to the overall function of the tissue.

The steady gathering of data in description of the gross ion uptake and translocation performed by roots has not produced much that is controversial. Root systems which have been excised from the plant provide an easily manageable experimental system for the study of xylem sap composition under various external conditions. A body of information is available (e.g. 3, 4, 6, 13, 40, 41, 43, 55, 61, 68, 91, 92, 93, 96, 99). The data generally are sound and there is widespread agreement on the chief characteristics of the explanatory model. The root maintains net fluxes of ions from the external solution to the xylem sap, and the resulting accumulation of ions causes an osmotic pressure gradient which drives water into the xylem. In an excised root system the osmotically driven water is free to flow along the xylem vessels and to exude from the cut end. The composition of the xylem sap depends upon the nutritional history of the plant, upon the species, and upon the composition of the external medium.

It is difficult to imagine a more convenient system for a phenomenological description of a biological process; the external medium and the exuded fluid are both easily accessible to assay. But convenience in a system is not synonymous with simplicity, and there are several examples of overstretched interpretation of the results of exudation studies. The first (55) of these involves a discrepancy in the observed and calculated water fluxes and is not relevant in detail to the present review except insofar as it points a general caution. Although it has been claimed (37) that this discrepancy is due to the presence of an active water flux, a more likely explanation has been given by Anderson, Aikman & Meiri (3). There is no discrepancy if the calculation of the osmotic water flux in an excised root of *Zea mays* is made with allowance for the hydraulic conductivity (osmotic permeability) to vary along the root length. In other words, if the root tip and the young tissue behind it have a different permeability from the tissues many centimeters from the tip and several days older, then an apparent anomaly in water exudation behavior is removed. An excised root system, even if it be as apparently uniform as the terminal 10 cm segment of a 4-day-old maize primary root, does not have uniform properties along its length.

What may well be a similar example of the inadequacy of considering a root system as a homogeneous cylinder has been recently reported. Minchin & Baker (93) believe they have evidence that the K$^+$ flux to the exudation stream in excised *Ricinus* root systems is comprised of two components, a water-dependent and a water-independent flux. The basis of their argument is that there is an extrapolated nonzero K$^+$ flux when the water flux is zero. This extrapolation has several features in common with the situation which led to the suggestion that there is a nonzero water flux when the osmotic driving force is zero in *Zea mays* roots (55) and may well be similarly explained. It is regrettable that so much arithmetical ingenuity has gone into extrapolations and curve fittings when the basic fundamentals of the interpretation are inadequate. It is most unlikely that one can meaningfully consider the whole root system of a 6-week-old *Ricinus* plant to have properties which are more uniform along its length than those of a 4-day-old maize plant. However, Minchin & Baker argue the ramifications of their case with a certain insight of root function, and their publications are worth study (92, 93).

Although it is true that the great majority of collected data on exuding root systems has produced little controversy, it is also true that interpreting these data to propose mechanisms for the exudation phenomenon has produced a great deal of controversy. The most favored point of view at present is a combination of the proposals of Crafts & Broyer (21) and of Arisz (9). The viewpoint is well known and can be stated succinctly: ions from the root environment are transported across the plasmalemmata of the epidermal and cortical cells. A radial flux of these ions is maintained through the symplasm, perhaps by diffusion (129), into the stele. The endodermis effectively prevents back-diffusion of the accumulated ions from the stele. In the stele the ions leak passively into the xylem vessels because the cells in the stele are 'leaky' due to low oxygen levels. The osmotic pressure gradient for water exudation is set up across the endodermis. It is probably correct to say that most exudation studies are interpreted as being broadly in agreement with this mechanism. That is not surprising for most exudation experiments are not designed to test any aspect of the mechanism. Of the few which have been so designed, most give rather equivocal results but tend to suggest that the osmotic barrier lies within the endodermis (5, 10, 56).

Let us examine this mechanism and first consider the stele. Laties & Budd (71) demonstrated that only limited Cl$^-$ uptake occurred in fresh, isolated steles. After aging for 24 hr, Cl uptake was much enhanced, at first sight reminiscent of the situation in storage tissue. Recently Hall, Sexton & Baker (42) in a comprehensive study have shown that Laties & Budd's finding is essentially correct, although they think that bacteria were responsible for much of the enhancement found in the original work. But biochemical assays suggest that aging in steles, with a decrease in total protein and an increase only in peroxidase and alcohol dehydrogenase, is quite different from aging in storage tissue. This point has been made in more detail in the previous section. Surprisingly, in view of their fine experimental work, Hall, Sexton & Baker (42) conclude that their results "clearly support a more passive role for stelar tissue." In the author's opinion, quite the reverse conclusion must be drawn from the work.

The reported values of respiration rate in fresh isolated steles is quite high, 192 μl g^{-1} h^{-1} (42), which may be compared with 68 μl g^{-1} h^{-1} (8) in fresh carrot discs, for example. This rate for fresh isolated steles may be limited for steles in situ by low oxygen levels as has been supposed (21), but it is certain that the respiratory capacity is present to supply energy for the exudation ion fluxes, even if all the work in transporting the ions from the root environment to the xylem were performed by the cells of the stele. Furthermore, it seems unlikely that fresh isolated steles, being nondormant tissue, would have the operational enzyme pathways to enable them to respire at the observed rate if in situ they were in an essentially anaerobic environment. An earlier report of respiration rates in stele and cortex (140, 141) gives much higher values for fresh steles, and also finds that steles respire more actively than the cortex, the inverse of the situation reported by Hall et al (42). Whichever set of measurements is closer to the in situ values, the salient fact is that both agree that fresh isolated steles have considerable respiration rates.

Second, the low rates of Cl⁻ uptake in fresh isolated steles (42, 71) does not necessarily imply that the cells are "leaky." It could equally well imply that the cells are very impermeable to Cl⁻, or that the membrane transport mechanisms are designed to produce ion effluxes from the cells. In any event, there is conflicting evidence on the uptake rates of fresh steles; Yu & Kramer (140) find considerable rates of phosphate uptake by fresh, isolated steles and by steles in situ. It is difficult to see how these divergent results have been obtained, and no comment will be given on which ought to be preferred. The point to be emphasized is that low uptake rates in themselves do not necessarily imply that the cells are "leaky."

In summary of the condition of the stele as far as it is known, it seems to the author to be clearly demonstrated that a fresh stele is not a "leaky," anaerobically suffering tissue as is required by the Crafts & Broyer proposal (21). Consideration of the changes induced in steles by aging seems to be irrelevant; it is certain that these are the effects of advancing necrosis and have no relevance to the function of the stele in the intact root. Indeed, if the reason given by Hall et al (42) for the enhancement of respiration rate in aged steles, where they liken it to the climacteric shown in aged fruits and thought to be "associated with, or a result of, increased membrane permeability (14)," then the enhancement of Cl⁻ uptake with aging must also be seen as the result of an increase in "leakiness" with aging rather than the reverse.

If one concludes that the stele is not in the state supposed so long ago by Crafts & Broyer (21), what role in exudation does one then ascribe to it?

Various proposals have been made in the past (5, 57, 126). At best these are all correlations and at worst speculations, a none too hopeful state of affairs. The experimental difficulties of investigating the stele are considerable; in a typical cereal root, for example, the stele occupies approximately 10% of the total root volume. Attempts to study the stele in an intact root are usually frustrated by the very geometry of the system, and the behavior of isolated steles may not be a good indication of the functioning in some respects of the stele in situ. Much is made by advocates of a passive, anaerobic stele that an isolated stele cannot produce a root pressure exudation. In his experience of preparing isolated steles, the author finds that the stele is caused to be extended by between 5–10% of its original length during its removal from the cortex. This extension may severely dislocate the vascular system. In this connection the preparation which most closely mirrors the functioning of an isolated, undislocated stele may be a refinement of that used by Anderson & Reilly (6), in which part of the cortex was removed by "shaving" it with a razor blade. This preparation, which admittedly has several layers of cortical cells still present, exudes well and in an interesting manner. At low external concentrations it produces a smaller salt flux than does a "normal" excised root, but at high concentrations it produces a similar or possibly greater exudation salt flux than control. The simplest explanation of these observations is that the final site of active transport of ions into the xylem stream is in the stele, and that the cortex acts as a primary accumulatory system. At low external ion concentrations this primary accumulation is necessary, but at high external concentrations (a situation not normally experienced by mesophyte plants) the free space diffusional supply of ions to the stele is not rate-limiting on the exudation fluxes.

Discussion of the cortex will be taken up in a subsequent paragraph, but first it seems appropriate to outline the author's present credo of the function of the stele. It is that the xylem parenchyma cells actively transport ions into the xylem vessels from the point at which any particular vessel matures until some later point, separated from the first by a distance which will vary with species, growth rate, nutritional status, etc. The reported correlation between salt exudation rates from root segments and the presence of membranous xylem vessels (5), a report which has received direct and indirect confirmation (18, 27), is now seen in this light. The correlation was not exact, being only good to within 1 cm of root length, and whereas it was previously supposed (5) that the vessels were most active in ion transport just before they reached maturity, now it is supposed that the parenchyma cells are active in transporting ions into the vessels just after the vessels have matured. It may be remarked in passing that the report by Smith (124) which seeks to throw doubt on the original correlation of Anderson & House (5) is based on an erroneous interpretation of his assay technique. The rate of appearance of ^{86}Rb in the exuded fluid is not simply related to the Rb transport of the root segments; under his experimental conditions there will be ^{86}Rb exchange for tissue K^+ and his assumptions of $^{86}Rb:Rb$ specific activity are invalid.

In a root from an actively growing plant, almost all the ions transported into the xylem by the parenchyma will be provided by the cortical symplasm from the environment, but in a plant not actively growing, the main supply of ions to the xylem may be ions delivered through the parenchyma in a recirculation from the phloem. Such a mechanism would provide a rather direct and simple means of control of ion uptake by the root system; those ions in the leaves and growing points which are in excess of the growth requirements are simply recirculated round the plant in the xylem-phloem system. This mechanism, together with variations in hormonal levels, could then control the rate of ion uptake of the root system and effect an integration of the salt balance of the entire plant. The evidence available is that such a feedback control, if it exists, is not capable of large compensatory action. The salt status of mesophytes grown in high saline conditions, or indeed of halophytes grown in their natural environments, is high, and progressive increase in the soil salinity eventually leads to lethally high levels of salt in the plants.

Consider now the role of the root cortex. The most commonly held view is that suggested by Arisz (9), in which the cortex is composed of cells of uniform type, the cytoplasm of each being in physical contact with its nearest neighbors through the plasmodesmata. The arrangement and frequency of these are such as to imply that the cortex as a whole is capable of radial cell to cell transport of solutes. A quantitative treatment of symplasmic movement has been given (129); the calculations show that ion fluxes of the magnitude required for root pressure exudation can be accommodated, even when very conservative values are taken for the physical dimensions of the plasmodesmata pores. To the author's knowledge there are no reports of experimental investigations of symplasmic transport as such in the root cortex. However, there is a fair amount of rather indirect evidence to suggest that the cortical symplasm does function as a primary accumulatory system for exudation in the manner supposed by Arisz (6, 13, 40, 61, 68, 120). Its function is particularly significant in low salt environments; it should be noted in this connection that there is increasing evidence against considering the vacuoles of the cortical cells as "diversionary sumps" for the symplasm. The half-times for cytoplasm-vacuole exchange in the cortex are not so large as to allow the vacuolar contents to be ignored in even first approximations of symplasmic function (2, 43).

Let us now leave the exuding root system and examine the many reports on the ion uptake of root segments. The experimental techniques of this work are by and large identical with those used on storage tissue discs. They are essentially investigations of ion uptake into the cortex, which in cereal roots, for example, occupies approximately 90% of the root tissue volume. Indeed it is in root segments that most of the observations on dual isotherm kinetics have been made (31–33, 80, 81, 128). The distinctions drawn between initial uptake rates and steady-state rates in the section on Storage Tissues should be borne in mind in what follows. All investigators agree that root segments show the dual isotherm pattern of initial uptake rates versus loading concentration. The details are well known and have been the subject of extensive reviews (30, 70). Furthermore, most investigators interpret the coincidence of the observed uptake rates with the rates predicted by Michaelis-Menten enzyme kinetics to imply that there are in the cell membranes two types of enzyme carrier for each ion. There is, however, a sharp difference of opinion about which cell membranes, plasmalemma or tonoplast, contain these enzyme systems. The most direct information perhaps is to be found in the work of Torii & Laties (128), who demonstrated that root tips containing cells without large central vacuoles show only System 1 uptake. Arguing that the absence of the large central vacuole is the reason why System 2 is absent, Torii & Laties consider that System 1 is located at the plasmalemma and System 2 at the tonoplast. Root segments at a distance from the tip, having mature central vacuoles, show "normal" System 1 and 2 characteristics. System 2 uptake is not rate controlled by System 1 uptake because in the view of Laties and his co-workers (70, 78, 80, 81, 94) the plasmalemma permeability is sufficient that large passive ion fluxes cross the plasmalemma at high external concentrations. As has been noted in the previous section, there is evidence in favor of this possibility (22, 98). An adjunct of this interpretation is the report (80, 81) that long distance xylem transport of ions shows only System 1 kinetics. This observation, together with siting System 2 at the tonoplast, leads Lüttge & Laties (80, 81) to uphold the "accumulating cortex-leaky stele" model of root pressure exudation, with the cortex vacuoles as "diversionary sumps" for ions.

The interpretation of Lüttge & Laties (80) seems suspect. Xylem transport in a maize seedling with a leaf blade still attached is primarily controlled by the transpiration rate. As this varies, the rate of appearance of tracer, which will be exchanging with the cells surrounding the xylem during the transport, will vary. It seems to the author that the agreement with System 1 kinetics and the nonexistence of System 2 is purely fortuitous.[1] The experimental system is too complex to be meaningfully treated in the simplistic manner they suggest. Furthermore, their model is strained in explaining how an excised root can accumulate salt and produce a hypertonic xylem exudation when bathed in solutions where System 1 is well saturated and where the large passive plasmalemma fluxes are supposedly operating. A model of "accumulating cortex-leaky stele," with predominantly passive fluxes operating for both cations and anions at the plasmalemmata of the cortical cells, cannot produce an exudate of 30 mM KCl from a bathing solution of 10 mM KCl; yet such values are regularly found in root pressure exudation studies.

The alternative location of Systems 1 and 2 at the plasmalemma has been forcefully argued by Epstein and his co-workers (30, 137, 138). A parallel arrangement of the carriers removes the necessity for large passive fluxes to avoid System 1 rate control at high concentration, but unfortunately it cannot be the whole story. Are we to assume that the tonoplast fluxes are passive and

---

[1] For a detailed criticism, see Weigl, J. 1969. *Planta* 84:311–23.

never rate-limiting and that loading the cytoplasm is the only active process involved? This would be surprising in the light of the knowledge about algae. Cram's note (24) shows how difficult it is to observe only the plasmalemma influx in even very short term loading. Pitman (100) has conducted a very interesting computer simulation of Cl^- uptake by low salt barley roots in which he argues that neither locating System 1 at the plasmalemma and System 2 at the tonoplast, nor locating Systems 1 and 2 at the plasmalemma can be accommodated without including rather unrealistic values for plasmalemma permeability in the first case and for tonoplast permeability in the second. It should be added that this paper points to the future when extensive use will be made of computer simulations to replace the rather loose verbalizing so commonly used in model building at the present time.

Steady-state flux measurements have been made on root segments (95, 104, 120), but the interpretations of all of these are based on a fallacy resulting from adoption of the analysis of similar measurements on storage tissue (1). In a recent paper Pitman (102) realized the problem and produced an experimental design to cope with the situation. The difficulty has been that the tracer efflux from a root segment is not the efflux from the cortical cells but is comprised of two components. One component is the cortical cell efflux and the other is the flux exuded from the xylem at the cut ends of the segments. This point has also been made by Weigl (136), who believes that the xylem flux may be resolved into two distinct phases. It is difficult to quantify the correction necessary, but all reported cortical cell efflux values are overestimates. Pitman (102) sets the xylem exudate contribution to the total Cl^- efflux from barley root segments at 75%. However, at face value the reports show that Na^+ is actively extruded in bean (120), barley (104), and pea (34), and that K^+ is actively accumulated in barley (104) and is close to equilibrium in bean, except near the root tip where it is actively accumulated (121). Jeschke (65) has shown that the Na^+ efflux is coupled to K^+ influx in barley.

There is good evidence that ion uptake by root tissue is connected with organic acid levels in the cells. In root exudation experiments (4, 19) it is clear that organic acids may be exuded in the xylem sap under conditions where the inorganic anion supplied in the bathing solution is only slowly transported. In studies on root segments it is known that organic acid levels rise in parallel with the cation-anion imbalance in uptake from the bathing medium (44–46). These organic acids must be ionized to maintain overall charge neutrality in the tissue, and zero net electric current flow across the membranes is achieved by H^+ extrusion. Pitman (101) showed that this H^+ extrusion is against the electrochemical gradient in barley roots. The overall phenomenon has been known since the earliest days of liquid culture of plants (53). There is no doubt that the ion relations of an intact, naturally growing plant invariably involve organic acid levels and rates of synthesis and breakdown, together with H^+ movements. In the completed picture of ion uptake it is certain that this mechanism will occupy a central role.

Salt glands.—A detailed review of the functioning of salt glands is not intended here since that topic has been covered in a recent review (77). The subject rather is used as a vehicle for discussion of some of the most interesting electrical measurements which have been made with higher plant tissue. Hill has published a very fine series of papers (49–52, 119) dealing with electrical and other measurements on the salt glands of the leaves of the common sea lavendar, *Limonium vulgare.* His work involves the use of the electrical techniques employed effectively on frog skin by numerous animal physiologists and used originally by Ussing (131). In animal physiology the measurement of short circuit currents and the use of voltage clamp experiments have made major contributions to understanding ion transport. In plant physiology these techniques have also been employed with various members of the Characeae (e.g. 36, 139).

The salt glands on the leaves of salt marsh plants exude concentrated salt solutions, the exuded ions being accumulated by the gland cells from the surrounding cells of the leaf tissue. The similarity in basic function of salt glands and roots has already been noted; a net flux of salt across both tissues can be maintained. In both tissues the structure and function are intimately linked and in both complete understanding of salt secretion can be attained only by a parallel approach of physiological and anatomical investigations.

In *Limonium* the major cations and anions—K^+, Na^+, and Cl^-—are actively transported into the secretion fluid (49). The Cl^- pump is thought to be electrogenic but to have affinity for both K^+ and Na^+. Thus in external solutions of choline chloride a large short circuit current is measured, while in normal solutions with both K^+ and Na^+ present, the short circuit current is negligible (50). The existence of electrogenic ion pumps in higher plant cells, a divergence from the situation normally reported in algae, is being demonstrated increasingly and will be taken up in the next section. Hill's short circuit demonstration is perhaps the most unequivocal. Short circuit current measurements have also been attempted in roots (38), but it seems that the time and space constants in roots are too long to allow convenient and meaningful experimentation and interpretation. This is most unfortunate in view of the power of the method, and the situation should be reexamined on short segments. Recently the technique has been extended in *Limonium* in an interesting direction (51, 52, 119). The plant is euryhaline and can tolerate a large range of salinities, but transfer of leaf discs from one salinity to another results in rather slow transients as the secreted ion fluxes adjust from the original to a new steady-state value. By comparing these observed transient responses with those predicted on a compartmental analysis of the leaf, the conclusion is reached that the transport systems of the glands are induced by the levels of Na^+ and/or Cl^- in the external solution. This induction is blocked by inhibitors of RNA or protein synthesis (119). The conditions for induction are apparently complex, but from the inhibitor studies the period of protein synthesis appears to last for about 1 hr from the onset of the salt load. Were this situation to be repeated in salt-starved root tissue, on which much of the dual isotherm work has been done, the interpretation of the observations would be more complicated.

Coleoptiles.—The most comprehensive electrophysiological investigations of higher plant cells have been carried out in coleoptile cells, with several notable exceptions in roots, and it is this aspect of the ion relations of coleoptile tissue which will be examined here. Higinbotham and his colleagues, in particular, have produced over the years a fairly complete characterization of the situation. The difficulties of inserting an electrode into the cytoplasm of a higher plant cell (see the introductory preamble) should be borne in mind. For this reason it is apparent that assumptions of how the measured electric potential value is resolved into plasmalemma and tonoplast potentials must be made throughout the work. Expositions of the basis of these assumptions have been given (35, 97).

Perhaps the chief, general conclusion is that the plasmalemma is the site of the main electropotential barrier in the cell and of ion selection mechanisms. On the basis of the Ussing-Teorell flux ratio equation, it is found that Na^+ is pumped out at the plasmalemma in *Avena* coleoptile (97), in *Pisum* roots (34), in *Pisum* epicotyl (47), in barley roots (104) and in bean roots (120). It should be noted at this point that although the efflux estimates in most of the work in root segments are wrong, it is not clear that flux ratio tests of active ion fluxes are necessarily invalid. In all the tissues mentioned above, the condition of the K^+ fluxes is uncertain, usually being close to passive equilibrium. In the *Avena* coleoptile Cl^- is pumped inward at the plasmalemma (97), and K^+, Na^+, and Cl^- are pumped inwards at the tonoplast. Recently a cautionary note (103) has been published, in which there is evidence that the measured electric potential across a higher plant cell varies with the length of time from excision of the experimental tissue. However, it seems unlikely that the electric potential variations are large enough to upset the previously reported conclusions which are based on the flux ratio equation. As MacRobbie has pointed out (88), decisions on active transport by this criterion are usually based on the logarithm of the flux ratio having the "wrong" sign.

There has been a major development in plant cell electrophysiology in recent years concerning the existence of so-called electrogenic ion pumps. For a time the consensus of opinion was that the membrane potentials measured across the membranes of plant cells arose solely because of the different rates at which the various ions tended to diffuse across the membrane (25). Such potentials are known generically as diffusion potentials; a good approximate predictor of the magnitude of such potentials is the Goldman equation (25, 39). However, it is now clear from a variety of sources (28, 48, 97, 103, 121) that the observed membrane potential is composed of two components, the major one being a diffusion potential and the second one being due to the operation of electrogenic ion pumps. These pumps produce a membrane potential as a direct result of the pumping action. The simplest mechanism for an electrogenic pump is one where an ion is bound to a neutral carrier at the exterior membrane surface, say, and the electrically charged carrier-ion complex diffuses across the membrane to release the ion at the interior surface; the carrier, now electrically neutral again, diffuses back to the exterior surface to bind another ion. If the transport process is active, there will be a metabolic energy input at some stage or stages of this cycle. The important point for the present discussion, however, is that there is net transfer of electric charge across the membranes as a primary result of the pump action in one complete cycle. This electric current flow across a resistance (the membrane) results, as in Ohm's law, in the production of an electric potential difference. The total membrane potential is simply the algebraic sum of the diffusion potential and the electrogenic pump potential. Note that in evaluating the diffusion potential the condition of zero net current flow in the absence of an electrogenic component, $\Sigma_i z_i \phi = 0$, where z_i is the valency of ion i, and ϕ is the flux of ion i, must be replaced with $\Sigma z \phi + I = 0$ where I is the electric current due to the operation of the electrogenic pump. Evidence indicating an electrogenic pump has been clearly set forth elsewhere (48). Briefly the two most important factors are (a) a marked and immediate change in the membrane potential upon poisoning, and (b) a progressively worsening fit of the observed membrane potential to the Goldman equation prediction as the external ion concentration increases.

Leaf tissue.—Ion uptake studies of leaf discs from terrestrial plants are confounded by the resistance of the surface cuticles, which offer by far the largest permeability barriers to ion movement into the disc. However, by using thin slices of leaf tissue (59, 60, 110) this problem can be overcome, and genuine cellular ion absorption is then thought to be observed. It seems fair to speculate on the relative proportion of undamaged cells in such a preparation and on the extent to which the results are meaningful, but be that as it may, the dual isotherm pattern of ion uptake has been reported in several tissues. In the light of the exhaustive work relating ion uptake in algae to electron transfer in the photosynthetic pathways (86, 88, 113, 114), it would be interesting to have further systematic investigation of a like occurrence in higher plant leaves.

The leaves of aquatic plants are not cuticilized and serve as far superior experimental material for ion uptake investigations. In many such plants the root system is vestigial and the leaves are the natural ion absorption regions. The most comprehensive studies available are the results of the fine work by Jeschke (63–66). He finds that K^+ influx into *Elodea densa* leaves shows the dual isotherm relationship to external concentration and that the influx increase is some 40-fold in the light. He interprets the light enhancement of K^+ influx to be due to an increase in intracellular ATP concentration primarily, but additionally to be caused by a hyperpolarization of the tonoplast potential and a possible increase in K^+ permeability. The uptake of Cl^- in *Elodea* is also enhanced in light, but the enhancement is saturated at low intensities in a CO_2-free atmosphere. The presence of CO_2 inhibits Cl^- uptake but removes the saturation of light enhancement, and both these effects of CO_2 are explained (66) by assuming competition for ATP between CO_2 assimilation and Cl^- uptake. In *Hydrodictyon* (113) there is an interesting difference. Raven demonstrated competition for ATP between CO_2 assimilation and cation uptake, but found no such competition for the energy supply to the Cl^- uptake. Indeed it is thought that the Cl^- influx in many algae is directly coupled to electron flow either in the mitochondria or in the chloroplasts. It may well be that there are several basic differences in ion transport in algal coencytes in the one hand and in the cells of higher plant tissues on the other.

Acknowledgments

I wish to record my appreciation for the help I have had from discussions with David Jennings and Julian Collins.

LITERATURE CITED

1. Anderson, W. P. 1969. *Proc. Int. Bot. Congr., 11th*, Seattle, Wash.
2. Anderson, W. P. 1971. In press
3. Anderson, W. P., Aikman, D. P., Meiri, A. 1970. *Proc. Roy. Soc. London* B174:445–58
4. Anderson, W. P., Collins, J. C. 1969. *J. Exp. Bot.* 20:72–80
5. Anderson, W. P., House, C. R. 1967. *J. Exp. Bot.* 18:544–55
6. Anderson, W. P., Reilly, E. J. 1968. *J. Exp. Bot.* 19:19–30
7. Ap Rees, T., Bryant, J. A. 1971. *Phytochemistry* 10:1183–90
8. Ap Rees, T., Royston, B. J. 1971. *Phytochemistry* 10:1199–1206
9. Arisz, W. H. 1956. *Protoplasma* 46:1–62
10. Arisz, W. H., Helder, R. J., Van Nie, R. 1951. *J. Exp. Bot.* 2:257–97
11. Atkinson, M. R., Eckerman, G., Grant, M., Robertson, R. N. 1966. *Proc. Nat. Acad. Sci. USA* 55:560–64
12. Atkinson, M. R., Polya, G. M. 1968. *Aust. J. Biol. Sci.* 21:409
13. Baker, D. A., Weatherley, P. E. 1969. *J. Exp. Bot.* 20:485–96
14. Baur, J. R., Workman, M. 1964. *Plant Physiol.* 39:540–43
15. Boyd, G. A. 1955. *Autoradiography in Biology and Medicine.* New York:Academic
16. Briggs, G. E., Robertson, R. N. 1957. *Ann. Rev. Plant Physiol.* 8:11–32
17. Bryant, J. A., Ap Rees, T. 1971. *Phytochemistry* 10:1191–97
18. Burley, J. W. A., Nwoke, F. I. O., Leister, G. L., Popham, R. A. 1970. *Am. J. Bot.* 57:504–11
19. Collins, J. C., Reilly, E. J. 1969. *Planta* 83:218–22
20. Coster, H. G. L., Hope, A. B. 1968. *Aust. J. Biol. Sci.* 21:243–54
21. Crafts, A. S., Broyer, T. C. 1938. *Am. J. Bot.* 25:529–35
22. Cram, W. J. 1968. *Biochim. Biophys. Acta* 163:339–53
23. Cram, W. J. 1968. *J. Exp. Bot.* 19:611–16
24. Cram, W. J. 1969. *Plant Physiol.* 44:1013–15
25. Dainty, J. 1962. *Ann. Rev. Plant Physiol.* 13:379–402
26. Dainty, J., Hope, A. B. 1961. *Aust. J. Biol. Sci.* 14:541–51
27. Davis, R. F. 1968. *Electropotentials and ion transport across excised corn roots.* PhD thesis. Washington State Univ., Pullman
28. Davis, R. F., Higinbotham, N. 1969. *Plant Physiol.* 44:1383–92
29. Dodd, W. A., Pitman, M. G., West, K. R. 1966. *Aust. J. Biol. Sci.* 19:341–54
30. Epstein, E. 1966. *Nature (London)* 212:1324–27
31. Epstein, E., Rains, D. W., Elzam, O. E. 1963. *Proc. Nat. Acad. Sci. USA* 49:684–92
32. Epstein, E., Rains, D. W., Schmid, W. E. 1962. *Science* 136:1051–52
33. Epstein, E., Schmid, W. E., Rains, D. W. 1963. *Plant Cell Physiol.* 4:79–84
34. Etherton, B. 1967. *Plant Physiol.* 42:685–90
35. Etherton, B., Higinbotham, N. 1960. *Science* 131:409–10
36. Findlay, G. P., Hope, A. B. 1964. *Aust. J. Biol. Sci.* 17:62–71
37. Ginsburg, H., Ginzburg, B. Z. 1971. *J. Membrane Biol.* 4:29–41
38. Ginzburg, B. Z., Hogg, J. 1967. *J. Theor. Biol.* 14:316–22
39. Goldman, D. E. 1943. *J. Gen. Physiol.* 27:37–60
40. Greenshpan, H., Kessler, B. 1970. *J. Exp. Bot.* 21:360–70
41. Greenway, H. 1967. *Physiol. Plant.* 20:903–10
42. Hall, J. L., Sexton, R., Baker, D. A. 1971. *Planta* 96:54–61
43. Hay, R. K. M. 1970. *Root pressure exudation in Alium cepa.* PhD thesis. Univ. East Anglia, Norwich, England
44. Hiatt, A. J. 1967. *Z. Pflanzenphysiol.* 56:233–45
45. Hiatt, A. J. 1967. *Plant Physiol.* 42:294–98
46. Ibid 1968. 43:893–901
47. Higinbotham, N., Etherton, B., Foster, R. J. 1967. *Plant Physiol.* 42:37–46
48. Higinbotham, N., Graves, J. S., Davis, R. F. 1970. *J. Membrane Biol.* 3:210–22
49. Hill, A. E. 1967. *Biochim. Biophys. Acta* 135:454–60
50. Ibid, 461–65

51. Hill, A. E. 1970. *Biochim. Biophys. Acta* 196:66–72
52. Ibid, 73–79
53. Hoagland, D. R., Broyer, T. C. 1940. *Am. J. Bot.* 27:173–85
54. Hope, A. B., Simpson, A., Walker, N. A. 1966. *Aust. J. Biol. Sci.* 19:355–62
55. House, C. R., Findlay, N. 1966. *J. Exp. Bot.* 17:344–54
56. Ibid, 627–40
57. Hylmö, B. 1953. *Physiol. Plant.* 6:333–405
58. Jackman, M. E., Van Steveninck, R. F. M. 1967. *Aust. J. Biol. Sci.* 20:1063–67
59. Jacoby, B., Dagan, J. 1967. *Protoplasma* 64:325–29
60. Jacoby, B., Plessner, O. E. 1970. *Ann. Bot. (London)* 34:177–82
61. Jarvis, P., House, C. R. 1970. *J. Exp. Bot.* 21:83–90
62. Jennings, D. H. 1969. *J. Exp. Bot.* 19:13–18
63. Jeschke, W. D. 1967. *Planta* 73:161–74
64. Ibid 1970. 91:111–28
65. Ibid. 94:240–45
66. Jeschke, W. D., Simonis, W. 1969. *Planta* 88:157–71
67. Kedem, O. 1961. *Criteria of Active Transport. Proc. Symp. Transp. Metab.*, 87. New York: Academic
68. Klepper, B., Greenway, H. 1968. *Planta* 80:142–46
69. Laties, G. G. 1963. *Control Mechanisms in Respiration and Fermentation*, ed. B. Wright, 129. New York: Ronald
70. Laties, G. G. 1969. *Ann. Rev. Plant Physiol.* 20:89–116
71. Laties, G. G., Budd, K. 1964. *Proc. Nat. Acad. Sci. USA* 52:462–69
72. Laties, G. G., MacDonald, I. R., Dainty, J. 1964. *Plant Physiol.* 39:254–62
73. Läuchli, A. 1967. *Planta* 75:185–206
74. Läuchli, A., Lüttge, U. 1968. *Planta* 83:80–98
75. Läuchli, A., Spurr, A. R., Wittkopp, R. W. 1970. *Planta* 95:341–50
76. Lundegårdh, H. 1955. *Ann. Rev. Plant Physiol.* 6:1–18
77. Lüttge, U. 1971. *Ann. Rev. Plant Physiol.* 22:23–44
78. Lüttge, U., Bauer, K. 1968. *Planta* 80:52–64
79. Lüttge, U., Cram, W. J., Laties,

G. G. 1971. *Z. Pflanzenphysiol.* 64:418–26
80. Lüttge, U., Laties, G. G. 1966. *Plant Physiol.* 41:1531–39
81. Lüttge, U., Laties, G. G. 1967. *Plant Physiol.* 42:181–85
82. Lüttge, U., Weigl, J. 1965. *Planta* 64:28–36
83. MacDonald, I. R., Knight, A. H., De Kock, P. C. 1961. *Physiol. Plant.* 14:7–19
84. Macklon, A. E. S., MacDonald, I. R. 1967. *J. Exp. Bot.* 17:703–17
85. MacRobbie, E. A. C. 1962. *J. Gen. Physiol.* 45:861–78
86. Ibid 1964. 47:859–77
87. MacRobbie, E. A. C. 1969. *J. Exp. Bot.* 20:236–56
88. MacRobbie, E. A. C. 1970. *Quart. Rev. Biophys.* 3:251–93
89. MacRobbie, E. A. C. 1971. *Ann. Rev. Plant Physiol.* 22:75–96
90. MacRobbie, E. A. C. 1971. *J. Exp. Bot.* In press
91. Meiri, A., Anderson, W. P. 1970. *J. Exp. Bot.* 21:908–14
92. Minchin, F. R., Baker, D. A. 1969. *Planta* 89:212–23
93. Ibid 1970. 94:16–26
94. Osmond, C. B., Laties, G. G. 1968. *Plant Physiol.* 43:747–55
95. Pallaghy, C. K., Scott, B. I. H. 1969. *Aust. J. Biol. Sci.* 22:585–600
96. Pettersson, S. 1960. *Physiol. Plant.* 13:133–47
97. Pierce, W. S., Higinbotham, N. 1970. *Plant Physiol.* 46:666–73
98. Pitman, M. G. 1963. *Aust. J. Biol. Sci.* 16:647–68
99. Ibid 1966. 19:257–69
100. Pitman, M. G. 1969. *Plant Physiol.* 44:1417–27
101. Ibid 1970. 45:787–90
102. Pitman, M. G. 1971. *Aust. J. Biol. Sci.* 24:407–21
103. Pittman, M. G., Mertz, S. M. Jr., Graves, J. S., Pierce, W. S., Higinbotham, N. 1970. *Plant Physiol.* 47:76–80
104. Pitman, M. G., Saddler, H. D. W. 1967. *Proc. Nat. Acad. Sci. USA* 57:44–49
105. Polya, G. M., Atkinson, M. R. 1969. *Aust. J. Biol. Sci.* 22:573–84
106. Poole, R. J. 1966. *J. Gen. Physiol.* 49:551–63
107. Poole, R. J. 1969. *Plant Physiol.* 44:485–90
108. Ibid 1971. 47:731–34
109. Ibid, 735–39

110. Rains, D. W. 1968. *Plant Physiol.* 43:394–400
111. Rasmussen, H. P., Shull, V. E., Dryer, H. T. 1968. *Develop. Appl. Spectrosc.* 6:29–42
112. Raven, J. A. 1967. *J. Gen. Physiol.* 50:1607–25
113. Raven, J. A. 1969. *New Phytol.* 68:45–62
114. Ibid, 1089–1113
115. Robertson, R. N. 1960. *Biol. Rev.* 35:231–64
116. Robertson, R. N. 1968. *Protons, Electrons, Phosphorylation and Active Transport.* Cambridge Univ. Press. 97 pp.
117. Sandblom, J. P. 1967. *Biophys. J.* 7:243–65
118. Sandblom, J. P., Eisenman, G. 1967. *Biophys. J.* 7:217–42
119. Schacher-Hill, B., Hill, A. E. 1970. *Biochim. Biophys. Acta* 211:313–17
120. Scott, B. I. H., Gulline, H., Pallaghy, C. K. 1968. *Aust. J. Biol. Sci.* 21:185–200
121. Shone, M. G. T. 1969. *J. Exp. Bot.* 20:698–716
122. Smith, F. A. 1965. *Links between solute uptake and metabolism in Characean cells.* PhD thesis. Univ. Cambridge, England
123. Smith, F. A. 1970. *New Phytol.* 69:903–17
124. Smith, R. C. 1970. *Plant Physiol.* 45:571–75
125. Spanswick, R. M., Williams, E. J. 1964. *J. Exp. Bot.* 15:193–200
126. Steward, F. C., Sutcliffe, J. F. 1959. In *Plant Physiology,* ed. F. C. Steward, 2:253–465. New York: Academic
127. Teorell, T. 1949. *Arch. Sci. Physiol.* 3:205–19
128. Torii, K., Laties, G. G. 1966. *Plant Physiol.* 41:863–70
129. Tyree, M. T. 1970. *J. Theor. Biol.* 26:181–214
130. Ussing, H. H. 1949. *Acta Physiol. Scand.* 19:43–56
131. Ussing, H. H., Zerhan, K. 1951. *Acta Physiol. Scand.* 23:110–19
132. Van Steveninck, R. F. M. 1964. *Physiol. Plant.* 17:757–70
133. Van Steveninck, R. F. M. 1965. *Aust. J. Biol. Sci.* 19:271–81
134. Ibid, 283–90
135. Van Steveninck, R. F. M. 1965. *Physiol. Plant.* 18:54–69
136. Weigl, J. 1971. *Z. Pflanzenphysiol.* 64:77–79
137. Welch, R. M., Epstein, E. 1968. *Proc. Nat. Acad. Sci. USA* 61:447–53
138. Welch, R. M., Epstein, E. 1969. *Plant Physiol.* 44:1301–4
139. Williams, E. J., Bradley, J. 1968. *Biochim. Biophys. Acta* 150:626–39
140. Yu, G. H., Kramer, P. J. 1967. *Plant Physiol.* 42:985–90
141. Ibid 1969. 44:1095–1100

Ann. Rev. Plant Physiol. 1972. 23:73–86

FORMS OF CHLOROPHYLL IN VIVO 7523

JEANETTE S. BROWN

Carnegie Institution of Washington [1]
Stanford, California

CONTENTS

Chlorophyll *a* is the essential pigment for the conversion of radiant energy to chemical energy by photosynthesis in green plants. Although chlorophyll can be extracted and purified, its function in photosynthesis ceases upon removal from its normal environment where it is attached to lipoproteins in chloroplast membranes. Along with this loss of activity on extraction is a striking change in the absorption spectrum. The absorption maximum of chlorophyll *a* in most plants is near 678 nm, and the band has a complex structure obviously composed of several overlapping components, whereas the maximum of chlorophyll in solution is blue-shifted by about 15 nm and the band is nearly symmetrical. The component bands seen in plant material have been called "the biological forms of chlorophyll." Individual components have been referred to as C*a*670 or C*a*680 depending upon their supposed peak positions, but not one of them has been separated completely from the others. The precise cause of the variability in absorption is not yet known, but knowledge of this cause will undoubtedly contribute toward an understanding of how chlorophyll functions in photosynthesis. This review will discuss the relevant literature that has appeared since the reviews by Smith & French (83) and Brown (7) in 1963. A great deal of information has been brought together already in *The Chlorophylls*, edited by Vernon & Seely (93) in 1966.

[1] C.I.W.—D.P.B. Publication No. 486.

SPECTROSCOPY

Absorption and curve analysis.—An active interest in the state of chlorophyll in vivo has long been maintained by Krasnovskii and his associates. Much of the Russian work before 1968 is summarized in *Spectroscopy of Chlorophyll and Related Compounds* by G. P. Gurinovich et al (50). Gulyaev & Litvin (48) measured absorption spectra of a variety of plants and of model systems containing different concentrations of chlorophyll dissolved in nonpolar solvents and deposited in films. All the plants had maxima or shoulders in their spectra indicating discrete component forms at 648–650, 664–666, 670–672, 678–680, and 692–694 nm. Cooling to −196°C did not change the positions of the maxima, but did sharpen the spectra by narrowing individual component bands. By increasing the chlorophyll concentration in model systems they were able to produce longer wavelength absorption maxima that corresponded to maxima observed in vivo. Their results suggest that the maxima between 665 and 693 nm are due to discrete aggregates of chlorophyll *a*.

These data, as well as earlier work by Krasnovskii, Brody, and others, suggest the working hypothesis that the maxima in the main red absorption band are dimers and small polymers of chlorophyll *a* whereas the maxima in the longwave region (> 690 nm) may be due to crystalline formations. The activity and spectral properties of the forms may be determined by the chemical bonds between molecules in the aggregates and by interaction with other kinds of molecules in the lipoprotein matrix. Other work pertaining to this hypothesis will be discussed under *Chlorophyll in vitro*.

In a more recent paper Gulyaev & Litvin (49) measured absorption spectra at 20°C of cells from 14 different higher plants and algae and performed curve analyses of the first and second derivatives of these spectra. The first derivative spectra showed maxima near 664–665, 671–673, 679–680, 692–693, 703–709, 712–718, and 720–734 nm in most of the plants. The second derivative spectra defined these same bands except that those near 679–680 mm were shifted to 681–683 nm. The halfwidth of the 683 nm band in *Chlorella* was 12–13 nm at 20°C and 7.5–9.0 nm at −196°C; the shorter wavelength bands were wider and the longer wavelength ones somewhat narrower. These papers also presented data on the absorption of chlorophyll and the carotenoids in the blue spectral region and of bacteriovirdin and bacteriochlorophyll.

Cramer & Butler (28) measured the absorption spectra at −196°C of spinach chloroplasts and fractions 1 and 2, enriched in photosystems 1 and 2 respectively, prepared from them with digitonin. The first derivative of each spectrum was obtained by electronic differentiation. The derivative spectrum of the chloroplasts clearly showed two components of chlorophyll *b* and six of chlorophyll *a*. Fraction 2 had a higher proportion of the shorter wavelength components and fraction 1 of the longer wavelength forms.

Butler & Hopkins (23) analyzed higher derivatives of complex absorption spectra with the aid of a small on-line computer. They tested the ability of the first through fourth derivative analyses to detect different mixtures of Gaussian and Lorentzian input bands spaced at various intervals from each other. Narrow, closely spaced bands are best detected in the fourth derivative spectra although a small artificial shift in their peak positions may result. They found an appreciable improvement in the signal-to-noise ratio when the four differentiating intervals were nearly but not exactly equal. Small broad bands may be smoothed out and lost in the fourth derivative, indicating the necessity for a combination of approaches when analyzing complex experimental spectra.

Fourth derivative curve analyses of the spectra of spinach chloroplasts recorded at 23° and −196°C were presented. At 23°C only the 670 and 682 nm bands of chlorophyll *a* were visible, but at −196°C, maxima in the fourth derivative occurred at 661, 669, 677, 684, 691, and 698 nm.

French and his colleagues at the Carnegie Institution have been investigating both the spectroscopy and curve analysis of chlorophyll absorption in vivo. For an experimental curve to be worth analyzing, it must be as free as possible from artificial errors due to such factors as light scattering, the sieve effect, and deviation from Beer's law (20). Brown (17) has recorded spectra of very small particles prepared from a variety of algae with a spectrophotometer constructed to minimize error due to light scattering. Chloroplasts from a number of kinds of leaves and algae have also been fractionated by a mechanical procedure (61) and their spectra recorded (16). The absorption spectra of these fractions were similar to corresponding spectra of fractions obtained by detergent treatment (28).

For several years French has been applying a computer program to analyze the complex spectra. At first he attempted to fit the main red band with two components having peaks near 670 and 680 nm. When only two components were used, their peak positions varied from one spectrum to another, and the 670 nm band was very broad. Some spectra, notably from *Euglena,* also required a 695 nm component. Discussions of these first attempts at analysis with an adapted RESOL computer program may be found in (37, 40, 42).

Cederstrand et al (24) analyzed the red absorption band of several algae and spinach chloroplasts measured at room temperature. They were able to obtain reasonably good fits with two Gaussian components at ca 668 and 683 nm, each having a half bandwidth of ca 18 nm. However, the peak positions had to be waived by as much as 5 nm and the bandwidths by 8 nm in order to fit the spectra of different algae. Also their results showed that the simple hypothesis of Ca668 being the chlorophyll *a* in photosystem 2 and Ca683 the form in system 1 is not tenable.

All of these results showed that if, in fact, there are only two major components of chlorophyll *a,* they each must vary considerably in peak position and halfwidth from one plant to another. Another possibility is that there are more than two major forms. Evidence from difference spectra for a real component near 665 nm comes from the work of Briantais (6) and of Thomas & Bretschneider (89). The component near 685 nm can be seen most clearly in low temperature spectra of fractions enriched in P700 measured by Ogawa & Vernon (69).

These observations and others from unusual spectra of various algae led French to try to fit spectra with a minimum number of components greater than two (38, 39, 41, 95). For this effort 65 spectra of chloroplast homogenates or fractions from leaves and algae that had been measured at $-196°C$ were studied. Fifteen spectra were selected that fell into groups having characteristic shapes, and these were analyzed repeatedly using different program specifications.

Results from this study point clearly to the existence of four major chlorophyll forms in place of the two originally hypothesized. These four occur within 0.5 nm of 661.6, 669.6, 677.1 and 683.7 nm in all of the spectra. The different components have half bandwidths between 7.6 and 12 nm, and the width of each component varies by 1 to 2 nm in different spectra of the same type. Generally the halfwidths of $Ca662$ and $Ca670$ are greater than those of $Ca677$ and $Ca684$, and the latter two are greater in fraction 1 than in fraction 2. An explanation for this variation in halfwidth may be important for an understanding of the chemical nature and binding of the chlorophyll forms, or the differences may be produced only by variations in the size or shape of the chloroplast fragments measured. The possibility that the same chlorophyll form may have a different halfwidth depending upon its environment would complicate analysis of difference spectra.

Forms of chlorophyll that absorb at wavelengths greater than 685 nm can be seen clearly in spectra of some algae and will be discussed in more detail under a separate heading. Here, it is noted that the recent analyses by French et al (39) placed $Ca695$ at 691.5 nm with a half width of 13–18 nm. The peak wavelength of $Ca705$ varied from 700 to 706 nm, and its halfwidth from 14 to 25 nm.

Thus we now have corroborative evidence from three laboratories for more than two and probably four forms of chlorophyll a absorbing between 660 and 685 nm. A comparison of the components in spectra from different plants discounts the possibility that any two of the components are from a single dimeric form with two peaks due to exciton splitting (60) because no pair maintains a constant ratio of peak heights from one spectrum to another.

The absorption maximum of chlorophyll b in vivo near 650 nm has been known for many years (36), and a second absorption band near 640 nm was noted in the marine alga *Ulva*. The curve analyses by Butler & Hopkins (23) and French et al (39) also show the presence of the 640 nm band in other plants. Thomas has recently studied the effect of photobleaching and differential acetone extraction on the absorption of chlorophyll b in *Ulva* (87, 88). His results suggest that the 640 nm band may not be due to either chlorophyll a or b.

Fluorescence.—Only a few papers relating to the fluorescence of the forms of chlorophyll will be mentioned here because this subject will be covered in the review by Goedheer in this volume. One attempt to identify the form of chlorophyll *a* responsible for each emission band by comparing absorption and fluorescence spectra measured at −196°C of material having different proportions of the forms of chlorophyll *a* gave some evidence that an increased amount of Ca684 correlated with relatively greater emission between 711 and 735 nm (16). However, absorption spectra of some algae such as *Scenedesmus D₃*, *Euglena*, *Phaeodactylum*, *Plectonema*, and *Anabaena* showed definite long wavelength absorbing forms of chlorophyll between 695 and 710 nm that correlated very well with the emission between 710 and 740 nm. The peak position of the long wavelength emission was extremely variable from one plant to another, and the band was broad, suggesting that it may be composed of two bands (94). Perhaps the emission between 710 and 720 nm comes from Ca684 and the still longer wavelength emission from Ca692 or Ca704.

Cho & Govindjee (26) measured the absorption, fluorescence, and fluorescence excitation spectra of *Chlorella* and *Anacystis* at 4°K as well as at higher temperatures. No more peaks or shoulders were seen at 4°K than at 77°K (−196°C). They concluded that Ca670 fluoresces at 681 nm at low temperature, but under physiological conditions transfers nearly all of its energy to Ca678 which in turn fluoresces at 687 nm.

Chlorophyll Formation

Thus far the discussion has been about the direct observation of absorption spectra which may be composed of bands that have been called "biological forms of chlorophyll." However, there are other kinds of information that demonstrate the heterogeneity of chlorophyll in the plant. When chlorophyll *a* is first formed from protochlorophyllide in vivo, various changes in absorption can be followed for several hours. The "ide" suffix means the absence of phytol in the molecule. There are two, perhaps three spectral forms of protochlorophyll (ide); these can be transformed directly by light into chlorophyll (ide) forms absorbing at approximately 672, 678, and 682 nm. Ca678 is a transient that immediately changes to Ca682, that in turn transforms during 20 to 30 min in darkness to Ca672 (34, 43, 52). The newly formed Ca672 may be the same as Ca670 discussed above, but the transient at 678 nm is probably not the same as the stable form, Ca677, seen in green material.

From their results with freezing or grinding etiolated leaves or those exposed only briefly to light, Butler & Briggs (22) suggested that Ca682 is converted to Ca672 by a disaggregation process. On the other hand, Kahn et al (52) found that when these mechanical treatments were applied to the prolamellar body instead of leaves, this shift did not occur. They suggested that the release of a substance or enzyme in the leaf cells causes the absorption changes. Sironval et al (81) found that the first-formed Ca683 is chlorophyllide *a*. In darkness this is phytolated to chlorophyll *a* absorbing at 672 nm. They suggested that the chlorophyll form absorbing near 683 nm in green leaves might be chlorophyllide *a*. However, Schneider (74) showed that the amount of chlorophyllide in green leaves is far too low to support that hypothesis.

Shultz (79, 80) measured the absorption and circular dichroism (CD) spectra of protochlorophyllide and newly formed chlorophyll in finely divided homogenates of etiolated leaves before and after various periods of illumination. The CD spectra showed that protochlorophyllide *a* exists as an aggregate of at least two chromophores. The first formed chlorophyllide *a* absorbing near 680 nm is also an aggregate. This is rapidly converted to monomers as shown by the CD spectra, and more slowly the absorption peak shifts to 670 nm. If the homogenate containing protochlorophyllide *a* is suspended in 2 M sucrose during the phototransformation, the chlorophyllide *a* remains aggregated, and the shift in absorption does not occur. After the leaves have been illuminated for about 2 hours, new aggregates containing a repeating structure of at least two chlorophyll molecules begin to form, the absorption spectrum becomes complex and undergoes a red shift. The data are also consistent with the hypothesis that the red shift is caused by a change in the environment of the chlorophyll as well as an aggregation.

De Greef et al (29) were able to extend the greening process by irradiation with far red light. They observed by low temperature spectroscopy of intact bean leaves that the chlorophyll *a* was differentiated into three forms with maxima near 670, 677, and 683 nm during the first few hours, and that these forms accumulated during the greening process. Small amounts of longer wavelength forms near 690 and 698 nm appeared at about the same time as photosynthetic activity.

Butler (21) has discussed some of the earlier work of Shlyk & Nikolayeva (76, 77), who showed by differential extraction with solvents of different polarity and differential labeling with $^{14}CO_2$ that both chlorophylls *a* and *b* may exist in more than one state in vivo. This work has been confirmed and extended by using ^3H_2O to a study of the carotenoids by Deroche (30–32).

A pool of newly formed chlorophyll exists even in fully green leaves, and this pool can be differentially extracted from the leaves by nonpolar solvents. New molecules of chlorophyll *a* are in a more lipophilic condition, preferentially attacked by chlorophyllase, and more readily destroyed by heat, intense light, or prolonged darkening (78). Following any of these differential extractions or destructive procedures, a decrease in long wavelength absorption of the residue occurs, suggesting that the new chlorophyll molecules absorb near 680 nm. Although it seems plausible, there is some danger in this extrapolation because the shift in absorption may be due to a change in state of the material because of protein denaturation. This, in turn, might cause a change in light scattering or other optical properties of the membraneous material unrelated to true chlorophyll absorption. However, other evidence that Ca684 or one of the longer wavelength forms may be the newly formed chlorophyll comes from the finding that either a mechanical or detergent method of fractionating the two photosystems produces small membrane fragments that are enriched in the pool of newly formed chlorophyll as well as in longer wavelength absorbing chlorophyll forms and photosystem 1 activity. Shlyk et al have developed the hypothesis that chlorophyll formation occurs throughout the life of the leaf in special centers (35, 75). Since these centers are in fraction 1, it follows (70) that they are located in the intergrana or stroma lamellae of the chloroplast.

Some of the evidence from the "greening" and *Euglena* (18) experiments suggests that the longer wavelength forms of chlorophyll are increased in proportion as the chlorophyll concentration in the lamellae increases. On the other hand, *Chlorella prototophecoides* forms chlorophyll in direct proportion to the nitrogen-carbon ratio in its growth medium, and a comparison between spectra of nearly colorless, partially and fully green cells revealed no difference in the proportion of the usual chlorophyll forms (11).

DESTRUCTION AND TRANSFORMATION

Most destructive treatments of chloroplasts initially cause a preferential decrease in absorption on the long wavelength side of the chlorophyll band, and in some cases a corresponding increase at shorter wavelengths can be observed. Nathanson & Brody (65) incubated chloroplast fragments with *Ricinus* leaf extract and observed a decrease in absorption near 682 and 705 nm and an increase near 674 nm. They suggested that a proteinaceous component of the extract caused a disaggregation of the two longer wavelength forms of chlorophyll to a special monomeric form that fluoresces at 698 nm. Michel-Wolwertz (19, 62) incubated chloroplast fragments from *Euglena* and *Chlorella* with several enzymes including a protease and a lipase. The protease caused a decrease in absorption at 695 nm in *Euglena* and at 680 nm in *Chlorella* and an increase near 670 nm. The lipase preferentially destroyed Ca670. These data support the hypothesis that the distinctive absorption of the chlorophyll forms is related to different lipoprotein environments. Brown (15, 38) has shown that a relatively low concentration of Triton X-100 converts Ca684, 692, and 700 into Ca662 and 670. This was found by curve analysis of the spectra which also indicated that no new forms of chlorophyll were made with the triton. The photochemical activity of the chloroplast fragments at 680 nm remained high immediately after the absorption shift, but gradually declined with further detergent action.

LONG WAVELENGTH FORMS

There has been a special interest in the forms of chlorophyll *a* absorbing at longer wavelengths (> 685 nm). Since absorbed light energy can be transferred from one chlorophyll molecule to another having absorption at longer wavelengths, a special long wavelength chlorophyll form may act as a trap or sink where photoconversion of radiant to chemical energy occurs. Clayton (27) discussed the concept of light-harvesting chlorophylls that absorb and transfer light energy to reaction-center chlorophylls. The reaction center for photosystem 1 is probably P700 which, although it may be a type of chlorophyll, is not one of the long wavelength forms considered here because it occurs in amounts too small to detect in absorption spectra. Earlier studies on the various long wavelength chlorophylls were reviewed by Smith & French (83) and by Butler (21).

Brown found that when *Euglena* was aged in darkness, some of its chlorophyll *a* was converted to pheophorbide *a* absorbing at 710 nm (8). Kunieda & Takamiya (59) observed a similar conversion of chlorophyll *a* to a 710 nm absorbing form of pheophorbide *a* in aqueous extracts of young *Ginkgo* leaves. In solution phaeophytin has a red absorption maximum at only slightly longer wavelengths than chlorophyll (82), and in damaged cells of *Ochromonas danica* phaeophytin *a* absorbs at 671 nm (10). Katz (53) found that some mixtures of chlorophyll *a*, phaeophytin *a*, and water had absorption maxima at 712 and 720 nm. The occasional, apparently enzymatic conversion of chlorophyll to a long wavelength absorbing form of phaeophytin is an interesting anomaly, but it may have little or no significance in photosynthesis.

A form of chlorophyll absorbing near 700 nm has been seen in spectra of two diatoms, *Phaeodactylum tricornutum* (9) and *Detonula* sp. (51) even at 20°C. In both cases the relative height of the long wavelength band increased inversely with the light intensity during growth.

The measurement of the absorption of chlorophyllide *a* in vivo was possible because a mutant of *Chlorella*, "SCA," accumulates this pigment (13). The peak position was at 690 nm, showing that the phytol tail is not necessary for the arrangement of chlorophyll molecules that absorb at long wavelengths.

The blue-green algae *Plectonema* and to a lesser extent *Anabaena* have long wavelength absorption bands near 710 nm, observable in spectra recorded at −196°C (12). Further study of *Plectonema* indicated that the extra light absorbed at long wavelengths did not enhance photosystem 1 activity (14). So far none of these long wavelength absorbing forms of chlorophyll have been observed to undergo reversible absorbance changes or any other activity which might indicate some specific function in photosynthesis other than as additional light-harvesting pigments. Hypotheses concerning the possible reasons for the long wavelength absorption will be discussed in another section. The notion that the long wavelength absorption band of phytochrome might be detectable in some plant material and mistaken for chlorophyll cannot be entirely ruled out.

CHLOROPHYLL-PROTEIN COMPLEXES

The literature about chlorophyll-protein complexes up to 1965 has been reviewed by Goedheer (46). He discussed the complex obtained from *Chenopodium album* (96) that will undergo a light-induced reversible absorption change. The chlorophyll maximum at 668 nm disappears, and a new band appears at 743 nm. Terpstra (86) investigated this complex (CP 668) further. She concluded that CP 668 in its extracted form is an artifact, and that the 740 nm band arises from the interaction of photooxidized chlorophyll with protein.

Murata et al (64) isolated a different soluble chlorophyll-protein from the inflorescence of *Brassica oleraceae* (cauliflower) that has red absorption peaks near 674 and 700 nm due to chlorophyll *a* and fluorescence maxima at 683, 706 and 744 nm. This protein could be separated from the pigment and recombined with purified chlorophyll to make a complex very similar to the original except for the lack of the 700 nm band that the authors suggest may be a decomposition product of chlorophyll *a*. A similar chlorophyll protein complex was also isolated from the leaves of wild mustard, *Brassica nigra* (63). Curve analysis of the red absorption band of this complex by French revealed four components with peaks (and halfwidths) at 662 (12), 670 (11), 676 (10), and 684 (9) nm. Thus this soluble chlorophyll-protein complex with a molecular weight of ca 93,000 appears to be composed of nearly the same four absorbing components as chloroplast membrane fragments from a variety of plants.

Murata & Murata (63) have also prepared a soluble chlorophyll complex from *Lepidium virginicum*. It has a red absorption maximum at 660 nm and a fluorescence maximum at 672 and shoulder at 730 nm. Curve analysis suggested that the absorption band is composed of two components at 659 (12) and 668 (11) nm.

Thornber (90) has prepared a chlorophyll *a*-protein complex from the blue-green alga *Phormidium luridum* that has a red absorption peak at 677 nm. Low temperature absorption spectra of this complex having about one P700 per 100 chlorophylls showed peaks at 676 and 710 nm (33). The 710 nm component did not respond to oxidizing or reducing agents by a change in magnitude. It probably is the same as Ca710 in *Plectonema* mentioned above (14).

Ogawa & Vernon (69) have prepared very small membrane fragments enriched in the reaction center P700 by using Triton X-100 on *Anabaena, Scenedesmus,* and spinach. Some variation in the spectra is apparent, but all are complex with peaks or shoulders near 670, 678, 683, 695, and 705 nm.

Chlorophyll in Vitro

The foregoing has been concerned with chlorophyll in plant material or still attached to lipoproteins. A number of people have attempted to simulate and thereby explain the complex spectra observed in vivo by studying pure chlorophyll *a* in different chemical or physical states using different solvents and thin films. Some of the work in the Soviet Union has already been mentioned. The visible absorption and fluorescence of chlorophyll and its aggregates in solution was reviewed by Goedheer (45).

Katz et al (54) reviewed the infrared and NMR spectroscopy of chlorophyll. They favor the hypothesis that the infrared bands that appear when chlorophyll is dissolved in nonpolar solvents are due to intermolecular aggregation involving coordination of ketone and aldehyde carbonyl oxygen atoms of one molecule with the central magnesium atom of another (4, 5). In certain solvents chlorophyll can form oligomers with aggregation numbers as large as 20. Aggregation of chlorophyll in solution has also been studied by Tomita (92) and Quinlan (71). Komissarov et al (57) found that when pure chlorophyll was adsorbed to proteins, the absorption maximum did not shift appreciably, but when the pigment was concentrated 30 times a shift in the maximum from 665 to 672 nm occurred.

The interaction of chlorophyll and water in various solvents has also been investigated by the Argonne group (3, 55). Their results suggest that water can be coordinated to the Mg atom of one chlorophyll molecule and hydrogen bonded to the ketone oxygen and carbomethoxy carbonyl functions of another. Thus chlorophyll-water aggregates may be formed that are very different in structure from chlorophyll-chlorophyll dimers or oligomers. Since the interaction of chlorophyll and water can produce long wavelength absorption bands somewhat similar to those produced by chlorophyll aggregation, care must be taken when interpreting the experiments of those authors who may not have had sufficiently dry chlorophyll preparations. The aggregating effect of water on chlorophyll in lipoprotein complexes was also studied by Giller et al (44). Complexes of chlorophyll and pheophytin with water have been prepared that absorb at longer wavelengths than either pigment alone in monomeric form (66).

Sauer (73) compared the optical rotary dispersion of chlorophyll in solution and in chloroplast subunits. The spectra suggested that the pigment-pigment interaction of dimers is a sufficient explanation for the large Cotton effects observed in the chloroplast fragments, although interaction with the lipoproteins of the lamellar matrix could not be ruled out.

Ke (56) has reviewed the literature concerning chlorophyll monolayers and crystalline chlorophyll. The red absorption maximum may be considerably shifted towards longer wavelengths in either of these physical states.

All of this work on chlorophyll in vitro, including the most recently available papers from the Argonne group (2, 53, 67), argue for the concept that the light-harvesting or bulk chlorophyll *a* in vivo exists as aggregates of dimers or oligomers. The electron acceptor and donor properties of chlorophyll dimers and different sized oligomers may account for differences in activity with wavelength of absorption. Evidence from ESR spectroscopy suggests that charge separation may occur in a reaction center composed of two chlorophyll molecules separated by a water molecule. The orientation of particular chlorophyll molecules in vivo probably also determines the wavelength position of their absorption (84). New techniques for studying the absorption of chlorophyll in lipid membranes offer further promise for simulating chlorophyll in vivo (25, 91).

State of Chlorophyll in Vivo

The evidence in favor of interpreting the absorption bands of chlorophyll in vivo in terms of aggregation between chlorophyll molecules or with water is almost overwhelming. An alternate hypothesis has been that chlorophyll may be loosely attached to different proteins or bound to one lipoprotein in different ways. Various orientations of different chlorophyll-lipoprotein complexes in the chloroplast lamellae would be the reason for the several absorption bands. Most of the evidence for this hypothesis has been reviewed (83). More recently Steffen & Calvin (85) found that a long wavelength absorption band of bacteriochlorophyll in vivo was due to a specific interaction between monomeric bacteriochlorophyll and certain functional groups in the protein with which it is associated. Gregory et al (47) observed that *Scenedesmus* mutant 8, that lacks a functional photosystem 1 and a long wavelength absorption band near 700 nm, also lacks a specific chlorophyll-protein complex.

Since it has been possible to fractionate chloroplasts into grana stack fragments enriched in chlorophyll *b* and the short wavelength forms of chlorophyll *a* and stroma lamellae fragments enriched in the long wavelength forms, several workers have analyzed the insoluble proteins in the two types of lamellae by gel electrophoresis (1, 68, 72). Specific proteins associated with the chlorophyll *a* or *b* enriched fractions have been found. Although all these results show clearly that different proteins are associated with different parts of the photosynthetic apparatus, they do not prove any relationship between a specific protein and chlorophyll. However, on the basis of X-ray diffraction studies of the lamellar proteins, Kreutz (58) has constructed an elaborate but tentative model of a thylakoid membrane with dimers or polymers of $Ca673$, $Ca683$, and $Ca705$ oriented in or on a protein matrix.

In summary, the available data about the forms of chlorophyll are insufficient to prove a definite model, but they do suggest that the several absorption bands may be from different dimers and oligomers of chlorophyll and chlorophyll adducts with bifunctional ligands such as water. The specific orientation or geometry imposed on the polymers by the ligand in the lipoprotein matrix is probably also important in determining their absorption and function.

LITERATURE CITED

1. Bailey, J. L., Kreutz, W. 1969. In *Progress in Photosynthesis Research*, ed. H. Metzner, 1:149–58. Munich: Goldmann-Verlag
2. Ballschmiter, K., Katz, J. J. 1968. *Nature* 220:1231–33
3. Ballschmiter, K., Katz, J. J. 1969. *J. Am. Chem. Soc.* 91:2661–77
4. Ballschmiter, K., Truesdell, K., Katz, J. J. 1969. *Biochim. Biophys. Acta* 184:604–13
5. Boucher, L. J., Strain, H. H., Katz, J. J. 1966. *J. Am. Chem. Soc.* 88:1341–46
6. Briantais, J. M. 1969. *Physiol. Veg.* 7:135–80
7. Brown, J. S. 1963. *Photochem. Photobiol.* 2:159–73
8. Brown, J. S. 1963. *Biochim. Biophys. Acta* 75:299–305
9. Ibid 1967. 143:391–98
10. Ibid 1968. 153:901–2
11. Brown, J. 1968. *Carnegie Inst. Washington Yearb.* 67:528–34
12. Ibid 1969. 68:566–70
13. Ibid, 570–72
14. Ibid 1970. 69:678–82
15. Ibid 1971. 70:499–504
16. Brown, J. S. 1969. *Biophys. J.* 9:1542–52
17. Brown, J. S. 1971. In *Photosynthesis*, ed. A. San Pietro, *Methods in Enzymology* 23:477–87. New York: Academic
18. Brown, J. S., French, C. S. 1961. *Biophys. J.* 1:539–50
19. Brown, J. S., Michel-Wolwertz, M.-R. 1968. *Biochim. Biophys. Acta* 153:288–90
20. Butler, W. L. 1964. *Ann. Rev. Plant Physiol.* 15:451–70
21. Butler, W. L. See Ref. 93. 343–79
22. Butler, W. L., Briggs, W. R. 1966. *Biochim. Biophys. Acta* 112:45–53
23. Butler, W. L., Hopkins, D. W. 1970. *Photochem. Photobiol.* 12:439–50, 451–56
24. Cederstrand, C. N., Rabinowitch, E., Govindjee 1966. *Biochim. Biophys. Acta* 126:1–12
25. Cherry, R. J., Hsu, K., Chapman, D. 1971. *Biochem. Biophys. Res. Commun.* 43:351–58
26. Cho, F., Govindjee 1970. *Biochim. Biophys. Acta* 216:139–50, 151–61
27. Clayton, R. See Ref. 93, 609–41
28. Cramer, W. A., Butler, W. L. 1968. *Biochim. Biophys. Acta* 153:889–91
29. De Greef, J., Butler, W. L., Roth, T. F. 1971. *Plant Physiol.* 47:457–64
30. Deroche, M. E. 1968. *Bull. Soc. Fr. Physiol. Veg.* 14:459–72
31. Deroche, M. E. 1969. *Physiol. Veg.* 7:335–89
32. Deroche-Laborie, M. E., Costes, C., Ferron, F. 1964. *Ann. Physiol. Veg.* 6:187–209
33. Dietrich, W. E., Thornber, J. P. 1971. *Biochim. Biophys. Acta* 245:482–93
34. Dujardin, E., Sironval, C. 1970. *Photosynthetica* 4:129–38
35. Fradkin, L. I., Kalinina, L. M., Shlyk, A. A. 1971. *Abstr. 2nd Int. Congr. Photosyn. Res.* (Stresa) p. 28
36. French, C. S. 1958. In *The Photochemical Apparatus. Brookhaven Symp. Biol.* 11:65–73
37. French, C. S. 1969. *Carnegie Inst. Washington Yearb.* 68:578–87
38. French, C. S., Brown, J., Lawrence, M. 1971. *Carnegie Inst. Washington Yearb.* 70:487–95
39. French, C. S., Brown, J., Lawrence, M. 1972. *Plant Physiol.* In press
40. French, C. S., Brown, J., Prager, L., Lawrence, M. 1968. *Carnegie Inst. Washington Yearb.* 67:536–46
41. French, C. S., Brown, J., Wiessner, W., Lawrence, M. 1970. *Carnegie Inst. Washington Yearb.* 69:662–70
42. French, C. S., Michel-Wolwertz, M. R., Michel, J.-M., Brown, J., Prager, L. 1969. *Porphyrins and Related Compounds. Biochem. Soc. Symp.*, ed. T. W. Goodwin, 28:147–62. London:Academic
43. Gassman, M., Granick, S., Mauzerall, D. 1968. *Biochem. Biophys. Res. Commun.* 32:295–300
44. Giller, Y. E., Krasichkova, G. V., Sapozhnikov, D. I. 1968. *Dokl. Akad. Nauk SSSR* 182:1230–33
45. Goedheer, J. See Ref. 93, 147–84
46. Ibid, 399–411
47. Gregory, R. P. F., Raps, S., Bertsch, W. 1971. *Biochim. Biophys. Acta* 234:330–34
48. Gulyaev, B. A., Litvin, F. F. 1967. *Biofizika* 12:845–54
49. Ibid 1970. 15:670–80

50. Gurinovich, G. P., Sevchenko, A.
N., Solov'ev, K. N. 1968. *Spectroscopy of Chlorophyll and Related Compounds.* Minsk:Izdatel'stvo Nauka i Tekhnika. 520
pp. Transl. No. AECtr-7199
available from Nat. Tech. Inform. Serv., US Dep. Commerce,
Springfield, Va. 22151
51. Jupin, H., Giraud, G. 1971. *Biochim. Biophys. Acta* 226:98–102
52. Kahn, A., Boardman, N. K.,
Thorne, S. W. 1970. *J. Mol. Biol.*
48:85–101
53. Katz, J. J. 1972. In *Inorganic Biochemistry,* ed. G. Eichorn. Amsterdam:Elsevier. In press
54. Katz, J. J., Dougherty, R. C.,
Boucher, L. J. See Ref. 93, 185–251
55. Katz, J. J., Ballschmiter, K., Garcia-Moran, M., Strain, H. H.,
Uphaus, R. A. 1968. *Proc. Nat.
Acad. Sci. USA* 60:100–7
56. Ke, B. See Ref. 93, 253–79
57. Komissarov, G. G., Nekrasov, L. I.,
Kobozev, N. I. 1964. *Biofizika* 9:
625–27
58. Kreutz, W. See Ref. 1, 91–105
59. Kunieda, R., Takamiya, A. 1965.
Plant Cell Physiol. 6:431–39
60. McRae, E. G., Kasha, M. 1958. *J.
Chem. Phys.* 28:721–22
61. Michel, J.-M., Michel-Wolwertz,
M.-R. 1967. *Carnegie Inst.
Washington Yearb.* 67:508–20
62. Michel-Wolwertz, M.-R. 1968. *Carnegie Inst. Washington Yearb.*
67:505–8
63. Murata, T., Murata, N. 1971. *Carnegie Inst. Washington Yearb.* In
press
64. Murata, T., Toda, F., Uchino, K.,
Yakushiji, E. 1971. *Biochim. Biophys. Acta* 245:208–15
65. Nathanson, B., Brody, M. 1970.
Photochem. Photobiol. 12:469–79
66. Norris, J. R., Uphaus, R. A., Cotton, T. M., Katz, J. J. 1970. *Biochim. Biophys. Acta* 223:446–49
67. Norris, J. R., Uphaus, R. A.,
Crespi, H. L., Katz, J. J. 1971.
Proc. Nat. Acad. Sci. USA 68:
625–28
68. Ogawa, T., Obata, F., Shibata, K.
1966. *Biochim. Biophys. Acta*
112:223–34
69. Ogawa, T., Vernon, L. P. 1970.
Biochim. Biophys. Acta 197:
332–34
70. Park, R. B., Sane, P. V. 1971. *Ann.

Rev. Plant Physiol.* 22:395–430
71. Quinlan, K. P. 1968. *Arch. Biochem. Biophys.* 127:31–36
72. Remy, R., 1971. *Fed. Eur. Biochem. Soc. Lett.* 13:313–17
73. Sauer, K. 1965. *Proc. Nat. Acad.
Sci. USA* 53:716–22
74. Schneider, H. 1968. *Phytochemistry*
7:885–86
75. Shlyk, A. A. 1971. *Abstr. 2nd Int.
Congr. Photosyn. Res.* (Stresa)
p. 29
76. Shlyk, A. A., Nikolayeva, G. N.
1963. *Biofizika* 8:261–73
77. Shlyk, A. A., Nikolayeva, G. N.
1963. In *La Photosynthese,* 119:
301–4. Paris:Centre Nat. Rech.
Sci.
78. Shlyk, A. A., Prudnikova, I. V.,
Fradkin, L. I., Nikolayeva, G.
N., Savchenko, G. E. See Ref. 1,
2:572–91
79. Shultz, A. J. 1970. *The development and organization of photosynthetic pigment systems.* PhD
thesis. Univ. California, Berkeley. 142 pp. UCRL-20202
80. Shultz, A. J., Sauer, K., 1971. *Biochim. Biophys. Acta.* In press
81. Sironval, C., Michel-Wolwertz, M.-R., Madsen, A. 1965. *Biochim.
Biophys. Acta* 94:344–54
82. Smith, J. H. C., Benitez, A. 1955.
In *Modern Methods of Plant
Analysis,* ed. K. Paech, M. V.
Tracey, 4:142–96. Berlin:
Springer-Verlag
83. Smith, J. H. C., French, C. S. 1963.
Ann. Rev. Plant Physiol. 14:
181–224
84. Sperling, W., Ke, B. 1966. *Photochem. Photobiol.* 5:865–76
85. Steffen, H., Calvin, M. 1970. *Biochem. Biophys. Res. Commun.*
41:282–86
86. Terpstra, W. 1966. *Biochim. Biophys. Acta* 120:317–25
87. Thomas, J. B. 1971. *Fed. Eur. Biochem. Soc. Lett.* 14:61–64
88. Thomas, J. B. 1971. *Proc. Eur. Biophys. Congr. 1st, Baden* 4:37–41
89. Thomas, J. B., Bretschneider, F.
1970. *Biochim. Biophys. Acta*
205:390–400
90. Thornber, J. P. 1969. *Biochim. Biophys. Acta* 172:230–41
91. Ting, H. P., Huemoeller, W. A.,
Lalitha, S., Diana, A. L., Tien,
H. T. 1968. *Biochim. Biophys.
Acta* 163:439–50
92. Tomita, G. 1968. *Biophysik* 4:296–301

93. Vernon, L. P., Seely, G. R., Eds. 1966. *The Chlorophylls*. New York:Academic. 679 pp.
94. Vredenberg, W. J., Slooten, L. 1967. *Biochim. Biophys. Acta* 143:583–94
95. Wiessner, W., French, C. S. 1970. *Planta* 94:78–90
96. Yakushiji, E., Uchino, K., Sugimura, Y., Shiratori, I., Takamiya, F. 1963. *Biochim. Biophys. Acta* 75:293–98

Ann. Rev. Plant Physiol. 1972. 23:87–112

FLUORESCENCE IN RELATION TO PHOTOSYNTHESIS

7524

J. C. GOEDHEER

Biophysical Research Group, Physics Institute, State University Utrecht, The Netherlands

CONTENTS

INTRODUCTION

A large number of papers in the field of photosynthesis have been dedicated to the measurement of fluorescence phenomena. Under natural conditions this fluorescence is emitted mainly by chlorophyll *a* or bacteriochlorophyll, but other chloroplast components such as phycobilins, protochlorophyll, or pyridine nucleotides also can be studied by using fluorescence methods.

One of the main advantages of studying fluorescence is that it enables the in vivo measurement of a large number of phenomena with a high precision and high time resolution. Fluorescence emission and action spectra, fluorescence intensity or yield, mean lifetime, and polarization of fluorescence can all be measured as a function of various parameters such as temperature, pretreatment, time of illumination, or addition of large variety of chemical reagents. This can be done with intact cells as well as with chloroplast preparations treated in various ways. The measured phenomena then give information about transfer of energy within the photosynthetic apparatus, distribution of chloroplast pigments over photochemical centers, kinetics in the electron carrier chain, and various other problems.

The chloroplast is of a complex structure and has locally a very high pigment content, which results in all kinds of interactions between adjacent molecules, influencing the fluorescence properties. These interactions may also be time dependent, due to changes in the redox state of chloroplast components or to conformational changes. For interpretation of the results obtained in vivo, adequate measurement of fluorescence properties in vitro is required.

FLUORESCENCE EMISSION SPECTRA OF CHLOROPHYLLS

In vitro.—The shape of the fluorescence spectra of chlorophylls in various solvents at low concentration does not depend much on the nature of the solvent (64, 80, 164), though the shape of the corresponding absorption spectra, especially in the region of the minor red bands, differs markedly. As fluorescence is generally assumed to be caused by the transition from the first excited state to the ground state and is in first approximation mirror symmetrical to the absorption spectrum ascribed to this transition, it indicates that the red part of the absorption spectrum should be due to more transitions. This is confirmed by consideration of the spectra at liquid nitrogen temperature ($-196°C$). In glassy solvents the fluorescence bands are sharpened but no new bands occur, while in the absorption spectrum, especially in ethanol, methanol, and pyridine, a new band around 645 nm occurs (28, 177). In acetone a chlorophyll *a* with maximum at 698 nm was reported (34).

Cooling of a chlorophyll *a* solution in a mixture of 80% ethanol and 20% water results in the appearance of a long wave shoulder in the absorption spectrum, which is ascribed to pigment aggregation. Concurrently a sharp decline in fluorescence intensity without change in spectral shape occurs (8), indicating absence of fluorescence of these aggregates.

Similar chlorophyll aggregates probably occur in "colloidal" suspensions of chlorophyll in water or a water/glycerol mixture. The fluorescence of these suspensions is extremely low, at 20°C as well as at −196°C.

With chlorophyll *a* in solid solutions at −269°C the fluorescence spectrum was found to be similar to that at room temperature. Apart from some reabsorption effects, this was true at low as well as at high pigment concentration (90). More or less ordered chlorophyll or chlorophyllide aggregates, according to the long wavelength shift of the absorption maximum, may occur in crystals (102), in condensed monolayers (110), in films (129), in isoctane (4), and possibly in chloroplasts treated with 50% methanol (1). The reports about fluorescence in these aggregates range from no fluorescence at all (102, 105) to an intensity comparable to that of chlorophyll monomers (129). The absorption maximum of the red band may be shifted from about 660 nm in dilute solution to about 740 nm in ordered aggregates, and corresponding fluorescence bands are reported (129).

The differences in capacity to fluorescence for various preparations might be due to influence of water (152), presence of some residual plant lipids, or to reabsorption effects in the concentrated films. The last effect may also explain the influence of cooling on fluorescence, but not on absorption spectrum of artificial chlorophyll-lipoprotein complexes (71). In view of the possible presence of chlorophyll aggregates in plants, the precise conditions under which chlorophyll aggregates are able to emit fluorescence need further investigation.

A single fluorescence maximum about 780 nm is measured with bacteriochlorophyll in organic solvents. Absorption maxima, observed at 800 and 850 to 890 nm with solid films of bacteriochlorophyll, are ascribed to aggregated forms. Only one fluorescence maximum at 875 nm (shifting towards 920 nm due to cooling to −196°C) was measured (118).

The fluorescence spectrum of *Chlorobium* chlorophyll in dilute solution is similar to that of chlorphyll *a*. The absorption spectrum, however, may show differences, especially in the minor red bands (80). In solid films of *Chlorobium* chlorophyll an absorption maximum at 740 nm was measured, while two fluorescence maxima, at about 740 and 770 nm (at 20°C as well as at −196°C), were ascribed to emission by pigment aggregates (119). Colloidal suspensions, prepared by dilution with water of *Chlorobium* chlorophyll dissolved in acetone, were found to have a very low fluorescence yield.

In green plants and algae—In vivo chlorophyll fluorescence is emitted only by chlorophyll *a*. Light absorbed by chlorophyll *b* is efficiently transferred to chlorophyll *a*, and no chlorophyll *b* fluorescence, either at 20°C or at −196°C, can be measured (55, 72). At room temperature the fluorescence spectra are usually of a shape similar to those in vitro, with a main peak at

685 nm and a lower one, due to vibrational levels, around 740 nm. With intact leaves high pigment absorption may cause strong reabsorption of fluorescence. As a result a decrease and a shift towards longer wavelength of the main maximum occurs (191).

With some green algae like *Ochromonas danica, Euglena gracilis* (30) *Stichochrisis* sp. (73), and *Vischeria stellata* the main maximum at room temperature may occur at wavelengths as long as 708 nm, while the usual 685 nm band is seen as a shoulder.

Cooling chlorophyll-containing cells and chloroplasts to −196°C results in the appearance of two additional bands, one at around 695 nm (called here F 695) and one at 717–740 nm (called here F 725), their exact location depending upon the species investigated. In green algae F 725 is generally located around 720 nm, while in higher plants it is around 735 nm (78).

Results of the last two decades suggest that in vivo chlorophyll *a* is divided over two pigment systems, one of which (photosystem 2) is responsible for water decomposition and oxygen production, the other (photosystem 1) for production of high energy compounds needed for the reduction of CO_2. Many measurements of fluorescence spectra deal with the designation of fluorescence bands to specific chlorophyll *a* forms participating in one of the two pigment systems.

Treatment of spinach chloroplasts with digitonin (15, 202) yields fractions with different chlorophyll "forms" and different photochemical activity. A light fraction (144,000 *g*) was assumed to contain mainly photosystem 1 pigments. It had a much lower fluorescence yield at room temperature than the heavier one (10,000 *g*), though the fluorescence spectra were similar (18). At −196°C, F 725 was relatively lower as compared to F 685 in the "heavy," and relatively higher in the "light" fraction than in intact chloroplasts. F 695 was reduced to a shoulder by the separation procedure. The supernatant of the "light" fraction showed fluorescence with a high yield and a maximum at 680 nm, which was probably caused by some chlorophyll solubilized by digitonin.

Further measurements confirmed the above results (39, 116, 197), whereas other detergents like Triton X-100 (112) or sodium desoxycholate (DOC) (23) could be used also. With the latter a good separation of partial fractions—as judged on the basis of −196°C fluorescence spectra—was obtained. Also the F 695 band was little affected by the procedure used. These data are reproduced in Figure 1.

Separation into fractions with similar fluorescence characteristics was also tried with sonication and pressure gradient application. After 15 min sonication differences were found between fractions precipitated at 13,000 and at 30,000 *g*, the heaviest one having the highest relative amount of F 725 (27). Other experiments did not give differences in fluorescence spectra after sonication (17). Further investigation showed that only after a short period of sonication in a medium of high salt concentration could particles with system 1 activity be separated (101). Probably two different types of particles with

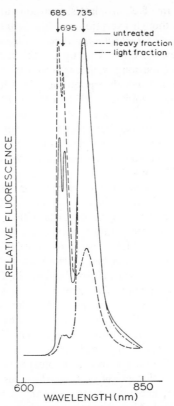

685   735

↓695 ↓

—— untreated
---- heavy fraction
—·— light fraction

RELATIVE FLUORESCENCE

600   850
WAVELENGTH (nm)

FIGURE 1. Fluorescence spectra, measured at −196°C, of untreated spinach chloroplasts (———), and the "heavy" fraction (— — —) and "light" fraction (. — .) obtained by DOC treatment (after Bril, van der Horst, Poort & Thomas 23).

this activity occur, one firmly connected to the electron transport chain and the other relatively free of it. Similar conclusions from experiments based upon pressure gradient separation (French press) were drawn (170). Whether the fluorescence properties of the system 1 particles obtained with these methods differ has not as yet been reported. Particles with a fluorescence spectrum similar to that of the light fraction (photosystem 1) isolated with detergents can be obtained from untreated leaf homogenates by differential and sucrose gradient centrifugation (181). With the French press, fragmentation of chloroplasts from a large number of plant species, except blue-green algae, yielded particles with system 2 or 1 absorption and fluorescence properties, though the resemblance of the fluorescence spectra to those obtained with detergents was less marked than that of the absorption spectra (32).

Fluorescence spectra of chloroplast suspensions are generally similar to those of intact cells, although exceptions to this include chloroplast preparations from *Euglena gracilis, Ochromonas danica, Pheodactylum tricornutum,* and some other species. Here breaking up the cell may result in disappearance of at least part of the long wavelength fluorescence, correlated with a disappearance of the long wave absorption shoulder (30).

The nature of the low temperature F 695 and F 725 bands and the function of the chlorophyll forms by which they are emitted is still a matter of discussion. Provided no artifacts such as multiple scattering occur at low temperature, increase in fluorescence as a result of cooling may be caused by an increase in (*a*) absorption, (*b*) fluorescence yield, or (*c*) energy transfer to the emitting chlorophyll "form."

The changes in the absorption spectrum during cooling are too small to account for the marked increase in F 725 and F 695. Comparison of the fluorescence spectrum of spinach chloroplasts with that of isolated chlorophylls shows that at room temperature part of the fluorescence around 725 nm (about 20% with intact chloroplasts and about 50% with isolated photosystem 1 particles) should be ascribed as room temperature F 725 emission. Cooling to −196°C results in a gradual increase in F 725, which is much more marked with spinach chloroplasts and *Chlorella* than with *Euglena* and other species where a long wave absorption shoulder is observed (78). This suggests that the increase in F 725 due to cooling is caused by an increase in energy transfer to a long wave chlorophyll form. This form is usually present in low concentration but has a high intrinsic fluorescence yield. It is probably not identical to P700, the energy trap of photosystem 1. At room temperature F 725 in spinach chloroplasts or *Chlorella* does not follow induction phenomena (126, 166), and its presence appears to be independent of the state of oxidation of P700. Also, Bishop's *Scenedesmus* mutant No. 8, which lacks P700 and photosystem 1 activity, contains F 725 in an amount comparable to that of the wild strain (115). Possibly the chlorophyll form emitting F 725 acts as an energy sink, protecting the photochemical systems for photooxidation under unfavorable conditions.

Emission spectra measured at various temperatures between 20 and −196°C show that F 725 increases gradually during cooling, while F 695 is not visible above −120°C and increases rapidly below this temperature (89, 93). This increase seems due to an increase in fluorescence yield rather than in energy transfer. At very low temperatures (−196 to −269°C), F 695 and F 725 in *Chlorella* increase much less rapidly than F 685 (43).

The correlation of the appearance of F 695 with a phase transition in ice crystals (40) does not seem a valid explanation for its occurrence only below −120°C. This phase transition is irreversible upon cooling, while F 695 can be observed many times upon repeatedly cooling and warming, provided warming occurs in the dark and does not proceed above 0°C. A relation with the state of water in or near the protein nevertheless may be possible (89).

With respect to the relation of F 695 to photosynthetic properties, it was suggested that this band represents emission from energy traps of photosystem 2 (115), a suggestion which fits in with results obtained by analyzing fluorescence changes at $-196°C$ (53). Also, it was found that F 695 could be enhanced by addition of a protein factor occurring in a *Ricinus* extract (25, 29).

From consideration of the Stokes shift bandwidth and bandshape of the F 685 room temperature band in various algae and isolated chloroplast fragments, it may be suggested that this band is emitted by a shortwave form of photosystem 2 chlorophyll *a* (203). Little difference was found between the fluorescence spectra of fully deuterated *Chlorella* and cells grown in hydrogen (70).

The fluorescence spectrum of a chlorophyll-protein complex derived from *Chenopodium album* was found to have a maximum at 676 before and 747 nm after phototransformation (182). The value of the Stokes shift for the far red form is low, indicating a weak coupling between pigment and protein.

In blue-green and red algae.—Apart from the presence of bands ascribed to phycocyanin and phycoerythrin, the room temperature fluorescence spectrum of blue-green and red algae is similar to that of green algae. At $-196°C$ the shape of the fluorescence spectrum differs greatly with the various algae (83), and with the light intensity during growth (69), preillumination (140), and wavelength of excitation (11). F 685 and F 695 are primarily excited by phycobilin absorption (121), while F 725 is primarily excited by absorption in chlorophyll *a* and β carotene (84). The emission band of phycocyanin, especially in the red algae, is often doubled at low temperature (78). As the location of the fluorescence maxima in vitro is not markedly affected by cooling (65), this probably indicates the presence of more than one phycocyanin form. At $-196°C$ the spectra show more detail (37).

Attempts to separate the photosynthetic organelles of blue-green algae into fractions containing primarily photosystem 1 or 2 have met with less success than with chloroplasts of higher plants. Treatment of *Synechococcus* with DOC (84), of *Anabaena* with Triton X-100 (153), and of *Anacystis* with digitonin (174) showed that generally more F 725 (and also a somewhat higher long wavelength absorption) occurred in the fraction precipitated at high speed, while the fraction precipitated at a lower speed contained relatively more of the shorter wavebands. Fragmentation with the French press method so far did not yield fractions with different absorption or fluorescence characteristics (32).

Fluorescence difference spectra at room temperature with *Porphyridium* (120) and *Anacystis* (161) illuminated at high and at low light intensity showed a shoulder at about 695 nm, but such a shoulder was not observed in difference spectra made by illumination with blue and red light (57). In view of its temperature dependency, it is unlikely that this shoulder represents room temperature emission of F 695. It is possible that it is caused by overlapping phycocyanin fluorescence.

Although with some red algae a large fraction of fluorescence beyond 700 nm is due to room temperature F 725 emission, their main maximum is around 685 nm (121). No species are known which have, like *Euglena* (a green algae) or *Pheodactylum* (a diatom), a room temperature maximum beyond 700 nm.

In brown algae and diatoms.—Brown algae and diatoms contain chlorophyll *c* as an accessory pigment. Like chlorophyll *b* in green algae, this pigment does not emit fluorescence in vivo.

As shown in Figure 2, at room temperature the fluorescence maximum is usually located at 681 nm, a shorter wavelength than with most other photosynthetic organisms. At −196°C only one maximum at around 697 nm is found with several species of this group, while shoulders occur at 715 and 750 nm (85). The shoulder at 750 nm is most probably caused by the vibrational band of F 695. The asymmetry on the shortwave side suggests the presence of a shoulder around 685 nm. As in green algae, and in contrast to blue-green and red algae, the −196°C fluorescence spectrum is little influenced by wavelength of excitation. In aging cells, especially with *Fucus serrata* and some diatoms, a separate F 725 can be measured.

With the diatom *Pheodactylum tricornutum* the room temperature fluorescence maximum can be at about 710 nm (31). As in *Euglena,* preparation of chloroplast suspensions usually results in loss of an appreciable fraction of room temperature 710 nm fluorescence and of absorption around 695 nm.

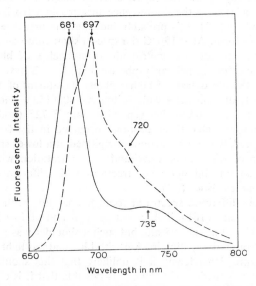

FIGURE 2. Fluorescence spectra of chloroplasts of the brown alga *Laminaria digitata,* measured at 20° (——) and −196°C (— — —) (after Goedheer 85).

In organisms containing protochlorophyll.—At room temperature the major fraction of protochlorophyll(ide) in etiolated leaves is usually transformed into chlorophyll(ide) with such a high efficiency [about one molecule transformed for each one or two quanta absorbed (175)] that no reliable fluorescence spectrum of active in vivo protochlorophyll can be measured. Below −90°C phototransformation does not occur (176), and the fluorescence spectrum shows a maximum at 656 nm at −196°C (82, 130, 173), corresponding with an absorption maximum at 650 nm. The Stokes shift of this pigment thus is only 5–7 nm, while the half-width value of the fluorescence band is about 9 nm. Of the newly formed chlorophyll(ide) these values are 10–12 and 19 nm respectively. As in organic solvents the values of the Stokes shift half-width of protochlorophyll and chlorophyll *a* are about equal; the in vivo differences suggest that structural changes follow phototransformation. In "holochromes" the 656 nm band is located at 643 nm (54a).

With intact etiolated leaves or cotyledons a fluorescence band at about 630 nm is observed besides the one at 656 nm. This band does not change upon illumination. The ratio 630/656 emission at −196°C depends upon the age of the leaves or cotyledons (89, 183). The shape of the 630 nm band differs appreciably from that of active protochlorophyll and inactive protochlorophyll formed as a result of grinding the leaves, repeatedly thawing, or heating to 45°C (172).

Algae in general do not synthesize protochlorophyll in the dark. With dark-grown *Euglena,* however, protochlorophyll(ide) with fluorescence maxima at 635 and 655 nm was observed at −196°C (26), while the 656 nm band was transformed into one at 673 nm by illumination at room temperature.

In photosynthetic bacteria.—Although the absorption spectrum of red bacteria consists of a complex system of bands between 790 and 890 nm, the fluorescence spectrum shows only a single band at about 900 nm, whereas in contrast to organisms containing chlorophyll *a,* no new bands occur after cooling to −196°C (24a). The existing band is sharpened and shifted about 20 nm towards longer wavelength, while the absorption band, assumed to be responsible for this emission, is shifted a like amount (195). This implies that the Stokes shift is little altered.

The absence of fluorescence emitted by the lower wavelength bands between 790 and 890 nm is ascribed to an efficient energy transfer (196). In *Chromatium* this energy transfer can be abolished by detergent action (22), while restoration occurs after removal of detergent by dialysis. Also additional bands may be produced by treatment with the detergent Triton X-100 (111).

With chromatophores of *Rhodopseudomonas spheroides* a fluorescence band at 870 nm, ascribed to emission by the 850 nm bacteriochlorophyll form, is measurable if the main fluorescence band is decreased by addition of oxidants (81). With intact bacteria of this species a shoulder around 870 nm is found with low intensities of exciting light (208). The results suggest that in this bacterium energy transfer to the longest wavelength form does not proceed with nearly 100% efficiency.

Some fluorescence, about 1% of total emission, was measured at 803 nm and excited by light absorbed in the main maximum (208). From this it was concluded that "uphill" energy transfer is measurable in these bacteria.

With *Rhodopseudomonas viridis,* which contains the bacteriochlorophyll *b,* the main absorption band is at 1014 nm and the fluorescence maximum is located at about 1044 nm (156, 157).

Photosynthetic green bacteria emit fluorescence at about 750 nm, which is ascribed to the major *Chlorobium* chlorophyll pigment, and fluorescence at about 800 nm, which is ascribed to a minor bacteriochlorophyll pigment. Cooling to −196°C results in sharpening of the bands and a change in the mutual intensity ratio, while both bands are shifted towards longer wavelength (119, 158). Only the bacteriochlorophyll band is measured in a protein complex derived from green bacteria (155).

FLUORESCENCE ACTION SPECTRA OF CHLOROPHYLLS

In vitro.—In dilute solution the fluorescence action spectra of chlorophyll and bacteriochlorophyll are found to be similar to the absorption spectra in the same solvents. This should indicate that all light quanta, absorbed by different vectors in a single molecule, are transferred without loss to the one responsible for fluorescence emission. In the region beyond the red absorption maximum, however, the quantum efficiency in organic solvents shows a steep drop (63).

In higher plants and green algae.—A recent review of action spectra in general was given by Fork & Amesz (62). Due to a considerable overlap of red absorption and fluorescence bands, fluorescence action spectra of the 685 nm emission have to be made without the red band. The complete action spectrum can be measured only for emission beyond 700 nm. The percentage F 725 emission at room temperature, however, usually is not negligible. Consequently, additional bands or shoulders which are not characteristic of F 685 fluorescence may occur.

In the action spectra of spinach chloroplasts and *Chlorella* cells at 20°C the ratio of chlorophyll *b*/chlorophyll *a* is appreciably higher than in the absorption spectrum, in the Soret as well as in the red band (79). This indicates that light absorbed by chlorophyll *b* is primarily transferred to that chlorophyll *a* fraction which is responsible for F 685 emission. This holds also for the action spectrum of F 685 + F 695 at −196°C, while the ratio is approaching that in the absorption spectrum with the F 725 action spectrum (145).

At the temperature of liquid helium the fluorescence action spectra are but little different from the ones measured at −196°C (41). Also the action spectra of *Chlorella* grown in a medium containing deuterium closely resemble those of cells grown in a medium containing hydrogen (70).

The fluorescence action spectra indicate that light absorbed by carotenoids is also transferred to chlorophyll, at 20° as well as at −196°C. Experiments with greening leaves, with chloroplasts after petroleum ether extraction, and with algae deficient in chlorophyll *b* suggest that light absorbed by carotenes is transferred with a high efficiency to chlorophyll, both of photosystems 1 and 2, while transfer from xanthophylls is much less efficient (84). As both types of pigment are nonfluorescing in vitro, this indicates that carotenes are closer to chlorophyll in the photosynthetic unit than are xanthophylls.

Fluorescence action spectra of fractions from spinach chloroplasts obtained by digitonin treatment show at −196°C a somewhat higher ratio of chlorophyll *b*/chlorophyll *a* in the "heavy" than in the "light" fraction (18, 21). In the blue region a difference between the 10,000 *g* and 80,000 *g* fraction obtained with Triton X-100 was recorded (112).

In some cultures of *Euglena* and *Vischeria,* where long wavelength emission is predominant at 20°C, the fluorescence action spectra resemble the absorption spectra. This indicates that an energy transfer occurs to the chlorophyll form responsible for this emission (79).

Fluorescence action spectra with barley chloroplasts deficient in chlorophyll *b* showed an unusually high activity of light quanta absorbed around 540 nm (16).

In blue-green and red algae.—High efficiency of phycobilins is shown in the action spectra obtained with these algae. Often no separate chlorophyll bands can be seen (55, 79). In room temperature fluorescence action spectra of chromatophores from blue-green algae from which phycobilins have been removed by washing, only bands at 670 and 435 nm are observed, which indicates that no energy transfer from carotenoids to chlorophyll occurs (79). At −196°C the action spectra of F 685 + F 695 with intact cells resemble the 20°C spectrum, while in the action spectrum of F 725 the chlorophyll bands are much more pronounced than at room temperature, and bands ascribed to carotenoid absorption are observed. The latter are evident also in the action spectra of phycobilin-free chromatophores of the blue-green and chloroplast fragments of the red algae, measured at −196°C (74, 79).

In brown algae and diatoms.—Excitation of chlorophyll fluorescence by fucoxanthin, the main carotenoid of this group, was one of the earliest indications of energy transfer within the photosynthetic system (199). The fluorescence action spectrum at −196°C is but little different from the one measured at room temperature, though slightly better resolved maxima are seen. These maxima indicate an efficient energy transfer from fucoxanthin and chlorophyll *c* to fluorescing chlorophyll *a*. Both pigments most probably belong primarily to photosystem 2 (85). The absence of a marked F 725 with most species of this group in the −196°C emission spectrum prevents the determination of its action spectrum.

With the diatom *Pheodactylum tricornutum,* however, a high room and low temperature fluorescence band around 710 nm is observed. Fluorescence action spectra indicate here that energy is transferred from shortwave chlorophyll *a,* chlorophyll *c,* and fucoxanthin to long wave chlorophyll *a,* but not with 100% efficiency.

In organisms containing protochlorophyll.—In the fluorescence action spectrum of etiolated leaves measured immediately after phototransformation, only chlorophyll(ide) bands are observed, while carotenoid bands, though dominating the absorption spectrum, are absent (35, 37, 77, 131). This is also true at −196°C, where no transformation occurs by the light used for excitation of fluorescence (67).

Fluorescence action spectra of protochlorophyll(ide) holochrome preparations measured in the blue and ultraviolet region show only bands ascribed to protochlorophyll(ide) or chlorophyll(ide) (12). This indicates that little or no energy transfer occurs from the protein moiety of the holochrome— with absorption bands in the ultraviolet—to protochlorophyll(ide) or chlorophyll(ide).

From fluorescence action spectra in the red region and fluorescence emission spectra of protochlorophyll(ide) holochromes illuminated at 0°C, it was concluded that energy transfer occurs between several protochlorophyll(ide) molecules on each holochrome or, after partial photoconversion, from protochlorophyll(ide) to chlorophyll(ide) (106, 184).

In photosynthetic bacteria.—Most far red bands, present in the absorption spectrum of red photosynthetic bacteria, are also present in the fluorescence action spectrum. With some species the action spectra do not coincide with the absorption spectra, indicating that the efficiency of energy transfer to the longest wavelength bacteriochlorophyll form emitting fluorescence is less than 100% (3, 154).

With *Rhodopseudomonas spheroides* strain R 26 the weak 800 nm absorption band was not present in the fluorescence action spectrum. Addition of sodium dithionite resulted in a slight change in the shape of the action spectrum, which may indicate a dependence of energy transfer on redox conditions in the chromatophore (207).

Fluorescence action spectra in the region of carotenoid absorption indicate that usually 20–40% of light absorbed by carotenoids is transferred to bacteriochlorophyll (55). The percentage differs among the species used: with *Rhodopseudomonas palustris* it is about 60% and with *Rhodopseudomonas spheroides* about 90% (76, 150).

The action spectrum of fluorescence quenching due to strong illumination of bacterial chromatophores shows the activity of carotenoids (88).

FLUORESCENCE YIELD OF CHLOROPHYLLS

In vitro.—In various solvents, except in the extreme long wavelength region, the quantum yield of fluorescence of chlorophyll *a* was found to be about 0.24 (63). With chlorophyll *b,* values in diethylether of 0.10 and in methanol of 0.06 were found, though somewhat higher values are also reported (122). Little dependence on the viscosity of the solvent was observed (66), but at high pigment concentration fluorescence quenching occurs [50% in a $10^{-2}M$ solution in ether (200)]. This effect is also observed in chlorophyll monolayers (190).

Addition of a number of reagents including oxygen, quinones, and vitamin K quenches the fluorescence of chlorophyll and its analogs both in solution and in monolayers (127, 190).

Most organic solvents used in chlorophyll fluorescence measurements contain traces of water. The fluorescence yield of chlorophyll solutions in nonpolar solvents in the complete absence of water is very low (61, 128). This may be due to the formation of nonfluorescing aggregates.

In higher plants and green algae.—Most of the papers dedicated to fluorescence of photosynthetic organisms are concerned with changes in fluorescence yield related to intermediate steps in the photosynthetic process. Due to the presence of various accessory pigments which transfer absorbed energy more or less efficiently to the fluorescing chlorophyll *a* fraction, the absolute fluorescence yield in whole cells and chloroplasts depends upon wavelength of excitation. Also pretreatment, influencing the state of oxidation of various quenchers, is important. With *Chlorella* a value of 0.04 is reported for the absolute fluorescence yield when excitation occurs with strong light at 600 nm (147). This value is appreciably lower than the one of chlorophyll *a* in organic solvents, which may be caused either by the presence of a chlorophyll *a* fraction with low fluorescence yield, or by a low intrinsic fluorescence yield of all chlorophyll molecules.

From measurements of chlorophyll fluorescence and photochemical activity of spinach chloroplast fractions obtained by digitonin treatment, it was suggested that the fluorescence yield at room temperature of photosystem 2 chlorophyll *a* is somewhat more than twice that of photosystem 1 chlorophyll *a* (197). Appreciably higher values are derived from other measurements (18, 20, 21). With a chlorophyll-protein fraction prepared from spinach chloroplasts, a fluorescence yield not exceeding 10% of that of chlorophyll *a* in acetone was observed (107).

The fluorescence yield of the long wave chlorophyll *a* form (or forms) responsible for F 725 emission is most probably high at room as well as at liquid nitrogen temperature, but energy transfer to this pigment form is assumed to be low at room temperature. With a spinach fraction (obtained upon treatment with Triton X-100) containing mainly photosystem 1, about 30 to 50% of room temperature fluorescence should be ascribed to this long wave pigment. In the absorption spectrum only a weak shoulder is observed (20, 21). In *Euglena* and the cells with high F 725 emission at room temperature a higher percentage of long wavelength chlorophyll *a* absorption was found (33).

The relation between changes in fluorescence yield and in condition of cells and chloroplast were studied in a number of ways. Addition of far red intermittent light to shorter wavelength actinic light resulted with *Chlorella* and green leaves in a decrease in fluorescence yield (36, 92, 136). As light absorbed by photosystem 1 resulted in oxidation of a cytochrome, and that cytochrome could be reduced by absorption of photosystem 2 light, Duysens & Sweers assumed that the reaction centers of photosystem 1 oxidized a quenching unit Q, while light absorbed by system 2 results in reduction of Q to QH (57). A side reaction, which changed QH into a different quencher Q′, should be activated by photosystem 1 light, possibly as a result of formation of high energy compounds by cyclic photophosphorylation (178).

The complex fluorescence induction phenomena, occurring after onset of continuous illumination, were studied simultaneously with oxygen evolution (14, 48, 51, 104). A fast phase (changes in fluorescence yield occurring between 0 and 0.5 sec) and a slow phase (changes occurring longer than 0.5 sec after onset illumination) could be observed. During the fast phase fluorescence intensity and oxygen production were parallel, while during the slow phase they were antiparallel. A model was derived which deviated somewhat from the one of Duysens & Sweers (50, 104).

Study of the fast phase showed that a dip in the fluorescence time curve, observable after 0.2 sec in anaerobic conditions (109), could also be observed under normal oxygen pressure, while a peak occurred after 0.4 sec (138, 139). Fast changes in the fluorescence yield of *Chlorella* were also observed after a sudden decrease of strong illumination (108). Slow changes may be caused by a wide variety of events. Uncouplers of photophosphorylation, probably in relation to conformational changes of the chloroplast (151, 162), water content of the cells (52, 60), and state of cell development can affect the fluorescence yield and cause its time dependent changes.

A few recent studies were made with intact cells of other algae and higher plants. The induction curves of various leaves, starting 2 msec after onset of illumination, showed no dip as mentioned with *Chlorella* (167). Also *Scenedesmus* under anaerobic and aerobic conditions did not show this phenomenon (72). Simultaneous time courses of fluorescence and absorption changes were measured with *Chlamydomonas* (48). No variable fluorescence, ascribed to photosystem 2, was found with bean leaves grown under flash illumination (54).

Measurement of fluorescence induction curves with isolated spinach chloroplasts shows the influence of addition of electron acceptors, photosynthesis inhibitors, and various physical treatments. Addition of electron acceptors decreases fluorescence, inhibitors of oxygen evolution shorten the induction period without appreciably affecting the final fluorescence level, while a 5 min heating at 45°C abolishes all induction phenomena. From the results of the experiments some authors calculated pool sizes for the various intermediates in the electron transport chain (7, 10, 24a, 49, 95, 132–134, 143, 144, 146). Addition of DCMU [3,(3,4-dichlorophenyl)-1,1-dimethylurea] caused immediate loss of Hill reaction capacity and decrease in fluorescence yield, but ascorbate photooxidation was not affected (160).

A study of the relation between fluorescence yield and chlorophyll afterglow (44, 47, 117, 124, 125) or redox potential (46) provides additional information, while conclusions about the electron flow in the photosynthetic chain were drawn from the effect of temperature between 40 and −196°C (185), the influence of pH (163), Mn (5, 113, 141), Cl⁻ (96, 97), Mg⁺⁺ (100) deficiency (5, 113, 141), and from measurements on chloroplasts with impaired photosystem 2 (98).

Simultaneous measurements of fluorescence changes and changes in electric potential in membranes of *Nitella* cells show a close correlation between these phenomena (193). The fluorescence of *Chlorella* can be enhanced 4–9% if the excitation light beam is parallel to an external magnetic field. This may be caused by orientation of the pigment molecules in this field (68).

In blue-green and red algae.—The absolute quantum yield of fluorescence of the blue-green alga *Anacystis* was found to be higher than that of *Chlorella* (147). Considering the low temperature fluorescence spectra of both species, this may be due to a different ratio of photosystem 2/photosystem 1 chlorophyll *a* in both species.

The absorption bands of the accessory pigments in blue-green and red algae are better separated from those of chlorophyll *a* than the ones of green algae. This may imply that the excitation of photosystems 1 and 2 can be better separated. The fluorescence intensity changes with the red algae *Porphyra* and *Porphyridium* were explained along the same lines as those with *Chlorella* (2, 57). The reactivation of the quencher Q′ by accumulation of oxidized products formed by photosystem 1 activity was studied in the blue-green algae *Schizotrix calcicola* (58). With the red alga *Porphyra*, light-induced changes in fluorescence yield and pH (194) as well as changes in fluorescence yield caused by alternate illumination with light absorbed by phycoerythrin and chlorophyll *a* were studied (142).

In brown algae and diatoms.—Although fluorescence and absorption spectra of brown algae and diatoms may be quite different from those of other groups, also with respect to chlorophyll *a* (85), few recent papers on fluorescence yield changes of algae or chloroplasts of these organisms were found. With the diatom *Pheodactilum tricornutum* the room temperature emission beyond 700 nm follows induction phenomena similar to the emission at 680 nm, which suggests that here this emission is also influenced by light absorbed by photosystem 2 pigments (31). Although after French press treatment this long wave emission disappears, and the room temperature fluorescence spectra of the fractions obtained by differential centrifugation were found to be similar, the fluorescence yield of the "'light" fraction was about ⅕ that of the "heavy" fraction, similar to the yield ratios of fractions obtained from species of green plants upon French press treatment or detergent action.

In organisms containing protochlorophyll.—The fluorescence intensity at 656 nm of etiolated leaves decreases gradually upon warming from −196°C to −90°C. Phototransformation in this temperature region is inhibited. Extrapolation to higher temperatures leads to a very low fluorescence yield of "active" protochlorophyll (which is able to transform in the light) at room temperature (89). The intensity of the fluorescence at 630 nm with intact leaves, or the one at 635 nm with leaves which had been heated to 45°C or repeatedly thawed, is much less temperature dependent.

In photosynthetic bacteria.—The absolute quantum yield of fluorescence of the red photosynthetic bacteria *Chromatium* and *Rhodopseudomonas spheroides* was found to be between 0.01 and 0.05, depending upon the conditions used (198). With the bacteriochlorophyll-protein complex derived from the green bacterium *Chlorobium,* yields of 0.19 at 20°C and 0.29 at −196°C were found to result from excitation at 366 nm (179).

Fluorescence yield changes show that if the energy flow between the light harvesting pigments and the reaction centers is impaired, the fluorescence yield of the light harvesting pigments increases (45, 196). The fluorescence yield is also affected by the presence of quenchers, natural as well as artificial. High intensity illumination of bacterial chromatophores, or of intact bacteria cooled to about −40°C, results in a strong decrease in fluorescence. The original intensity is to a large extent restored during the dark (88, 135), but at temperatures below −60°C the effect, brought about mainly by carotenoid absorption, is irreversible.

Fluorescence quenching is also observed by addition of some oxidants, while addition of reductants may restore the fluorescence yield and prevent the light induced fluorescence decrease. This indicates that the fluorescence yield is also affected by changes in redox state of the environment of the fluorescing molecules.

In photosynthetic green bacteria, the yield of 814 nm fluorescence, ascribed to bacteriochlorophyll, increases if the energy trap molecules absorbing at 840 nm are photobleached. The fluorescence yield of the major pigment *Chlorobium* chlorophyll is unaffected (158).

MEAN LIFETIME OF FLUORESCENCE OF CHLOROPHYLLS

In vitro.—The mean lifetime of fluorescence is defined as the time in which fluorescence is reduced to about 0.4 its initial value, provided an exponential decay occurs. With chlorophyll *a* in organic solvents the mean fluorescence lifetime was found to depend upon the pigment concentration. In dilute solution ($10^{-3}M$) the measured mean fluorescence lifetime of chlorophyll *a* dissolved in ethanol was 6.3 nsec, while at higher concentrations ($8.10^{-3} M$) a shorter lifetime of 2.5 nsec was found (38). As the mean life-time is proportional to the fluorescence yield of the emitting molecules, the decrease in fluorescence yield with increasing concentration (concentra-tion quenching) is caused by a decrease in lifetime of all molecules, and not by an increase in the number of nonfluorescing aggregates with similar ab-sorption spectrum. In ethyl ether for dilute solutions lifetime values of 2.1 nsec (29) and 4.9 nsec (171) were reported for chlorophyll *a*, and 3.9 nsec for chlorophyll *b* (29). With protochlorophyll in acetone, extracted from eti-olated leaves, a value of 6.2 nsec was found (168). At $-196°C$ the mean lifetime of fluorescence of chlorophyll *a* dissolved in ethanol in dilute solu-tion was measured to be about equal to that at room temperature (38), which suggests a weak temperature dependency in this temperature range. The mean fluorescence lifetime of bacteriochlorophyll in acetone solution was measured to be 4.7 nsec (169).

In higher plants and green algae.—The mean lifetime of *Chlorella* fluorescence at room temperature was found to depend upon intensity but not on wavelength of exciting light (137). At low intensities (300–500 ergs/cm² sec) it was 0.3 to 0.5 nsec, while at saturating light intensities (about 30.000 ergs/cm² sec) the mean lifetime was between 1.1 and 1.7 nsec. The value with cells treated with DCMU, a photosystem 2 inhibitor, was 1.8 to 2.2 nsec and independent of intensity of exciting light. Most likely earlier measured lifetime values of about 1.7 nsec, obtained with the flash technique, refer to relatively high intensities (29, 186). Redetermination with improved experi-mental conditions and lower light intensities leads to values of about 0.7 nsec for spinach chloroplasts and *Chlorella* cells (168).

The in vivo lifetimes measured with DCMU treated cells or at saturating light intensity are about one-third the length of those measured with chloro-phyll in organic solvents. This implies that, provided the red absorption bands in vivo and in vitro are similar, the yield of the fluorescing fraction of in vivo chlorophyll *a*, present mainly in photosystem 2, is about 0.08 at room temperature.

If more than one fluorescing chlorophyll *a* form with a different fluores-cence yield exists, the fluorescence decay curve may be expected to show de-viations from the exponential shape. Within the accuracy of the measure-ments such deviations were not detected (149). This result indicates that at room temperature either only one fluorescing chlorophyll *a* form is present, or more forms are present with a mean lifetime differing not more than 0.2 to 0.3 nsec from each other. Also the possibility that part of the measured fluorescence should be considered an afterglow with short lifetime (in the order of microseconds) was found to be unlikely (148).

With bean leaves at $-196°C$ a lifetime of 3.1 nsec for fluorescence emit-ted beyond 720 nm (F 725) was measured, while the value at 20°C was 0.7

nsec (38, 168). This indicates that the long wave chlorophyll a form, assumed to emit F 725, has a long fluorescence lifetime and hence probably a high fluorescence yield at $-196°C$. No fluorescence lifetime measurements were reported from *Euglena* or *Pheodactylum*, with high long wave emission at $20°C$, nor for photosystem 1 particles, where room temperature F 725 emission is an appreciable fraction of total emission. Such measurements would decide whether the strong increase of F 725 upon cooling is due to increase in energy transfer or increase in quantum yield.

The fluorescence lifetime of greening bean leaves decreases markedly during greening. This occurs, like the decrease in quantum yield, after a time-lag phase following first illumination in which no new chlorophyll is formed (131, 189). It indicates that this decrease in fluorescence yield is not caused by formation of nonfluorescing chlorophyll a fractions, but results from energy transfer or is due to internal reorganization in the plastid.

In blue-green and red algae.—The fluorescence lifetime of chlorophyll a in *Anacystis* and *Porphyridium* was found to be approximately equal to that of *Chlorella* (171). The lifetime of the highly fluorescing phycobilins was reported to be 7.1 nsec for phycoerythrin and 1.8 nsec for phycocyanin (29). Recent experiments gave values of 3.5 and 1.8 nsec respectively (47).

In organisms containing protochlorophyll.—The lifetime of the 656 nm form of protochlorophyll in etiolated leaves was measured to be 5.0 nsec at $-196°C$, equal to that of the 630 nm form. The lifetime of the 630 nm form at $20°C$ was equal to that at $196°C$, while that of the 656 nm form was about half this value (168). This decrease in mean lifetime of fluorescence, occurring mainly between -160 and $-100°C$, corresponds to a decrease in fluorescence yield in the same temperature region. The temperature dependence of the fluorescence yield suggests an even stronger decrease in mean lifetime, also in view of overlap of fluorescence spectra of the 656 and 630 nm forms.

In photosynthetic bacteria.—The mean lifetime of fluorescence of bacteriochlorophyll in *Chromatium* cells was found to be 0.8 to 1.0 nsec, about one-fifth of the value found in organic solvents (169). The lifetime depended on the condition of the cells; in the initial phase of growth the values were greater than for cells in the logarithmic growth phase. Lifetime measurements with the bacterium *Ectothiorhodospira shapozhnikovii* indicated the presence of two types of emission: a background fluorescence with a lifetime of 1.35 nsec independent of light intensity, and a fluorescence correlated with the state of oxidation of reaction centers, with lifetime values ranging from about $4 \cdot 10^{-2}$ nsec at very low to 0.5 nsec at saturating light intensities (19).

FLUORESCENCE POLARIZATION OF CHLOROPHYLLS

In vitro.—The degree of fluorescence polarization p, resulting from excitation with linearly polarized light, depends upon the wavelength of excitation, viscosity of the solvent, pigment concentration, and fluorescence lifetime. In viscous solvents (to prevent the pigment molecules from rotating during their mean fluorescence lifetime) and with low pigment concentration (to prevent depolarization due to energy transfer to differently oriented neighbor molecules), the theoretical maximal value, occuring when the vector for absorption is an ideal linear oscillator and identical to, or oriented parallel to the vector emitting fluorescence), is $p = 0.50$. In practice the measured maximal values with chlorophylls in solution do not exceed $p = 0.42$.

If the vector for light absorption is perpendicular to that responsible for fluorescence emission, a value of $p = 0.27$ should be expected in view of the measured maximal values. Overlap in absorption due to differently oriented absorption vectors in a single molecule accounts for various maxima and minima in the fluorescence polarization spectra. These spectra show that only in the region of the main red band can high values approaching $p = 0.42$ be found (9, 75, 94). The structures of the polarization spectra measured at $-196°C$ in glassy solvents or with chlorophylls attached to detergent micelles do not differ in shape from those measured at $20°C$, though the absorption spectra may show marked differences (87).

At a fixed viscosity value of the solvent and of the size of the pigmented particle, the polarization values are dependent on the lifetime of fluorescence, hence from the relation between p and the viscosity the lifetime values of in vitro chlorophyll can be calculated. These values do not differ much from the ones measured directly (75).

The average number of times that energy transfer between pigment molecules occurs can be calculated from the relation between fluorescence polarization and pigment concentration in random oriented pigment solutions (114). With mixed monolayers of chlorophyll and chloroplast glycolipids the fluorescence polarization decreases strongly when the films are compressed, while the fluorescence yield also drops (188). This indicates that the pigments are not orientated in these layers. Measurements of fluorescence polarization in solid solutions at $-267°C$ indicate that energy migration between pigment molecules is not very temperature dependent (90).

The values of polarization from protochlorophyll are lower and vary less with wavelength of excitation than the ones of chlorophyll a, at least if the polarization spectrum of the second fluorescence band is measured (80, 188). The fluorescence polarization spectrum of the first fluorescence band shows more structure (99).

The fluorescence polarization spectrum of bacteriochlorophyll shows a negative value of $p = 0.23$ in the orange band (580 nm in ricinus oil and 625 nm in solid ethanol at $-196°C$), which is close to the value for perpendicularly oriented oscillators (59, 75). In the far red band a value of $p = 0.40$ is found. With *Chlorobium* chlorophyll the fluorescence polarization spectrum is of shape similar to that of chlorophyll a (80).

In higher plants and green algae.—With fully intact green cells the fluorescence polarization is, when measured at 20°C, generally low. Values of $p = 0.03$ resulting from 630 nm and $p = 0.06$ from 650 nm exciting light were recorded (6, 180). According to the in vitro measurements, polarization of chlorophyll a fluorescence should be about half the value obtained with exciting light in the red maximum. Also, depolarization is expected to occur by reflections at cell walls and other cell constituents, while absorption at the mentioned wavelengths is partly due to chlorophyll b. Therefore, it might be expected that the polarization values of chloroplast suspensions, which show much less scattering than intact cells, when measured with about 670 nm incident light would be appreciably higher. However, with small chloroplast fragments even lower polarization values are measured. Very low values were also found, at 20°C as well as at −196°C, if spinach chloroplasts were fractionated with DOC treatment and the fluorescence polarization of the fraction assumed to contain system 2 was measured. The low values of fluorescence polarization with in vivo chlorophyll is probably caused by energy transfer between differently oriented pigment molecules. Depolarization due to molecular rotation is unlikely, in view of the low temperature measurements (201). The results so far obtained suggest that the fluorescence polarization of F 685 or F 685 + F 695 does not exceed $p = 0.03$, indicating an efficient energy transfer between a large number of random or nearly random oriented pigment molecules.

The spectra of polarized emission show that higher polarization values are found beyond 700 nm (123). These higher values were found to correlate with the "constant fraction" of fluorescence (the fraction which does not change during induction phenomena and which might be ascribed to F 725), while the value of the "changing fraction" (probably corresponding with F 685) was appreciably lower (21, 123).

Measurements with the photosystem 1 fraction of DOC-treated spinach chloroplasts, *Euglena,* and *Vischeria* cells showed that a polarization of F 725 emission, at 20°C as well as at −196°C, is brought about by absorption on the long wave side of the red chlorophyll a band.

These measurements were all done with random-oriented pigmented particles either of complete cells or chloroplast fractions. Provided the size of the particles is such that rotational depolarization is unlikely, depolarization occurs if energy is transferred to differently oriented neighbor molecules. The extent of depolarization then depends on the average number of transfers, and the relatively high value of F 725 fluorescence polarization may be brought about by the presence of only one or two F 725 emitting molecules on each particle, or a larger number of molecules oriented parallel to each other. From measurements of bifluorescence (polarized emission from oriented particles) with intact cells of *Euglena* and *Mougeotia* which were observed in the fluorescence microscope it was found that emission was higher when the light vector was vibrating parallel than when it was vibrating perpendicular to the plane of the chloroplast (159). With these algae room temperature emission is mainly beyond 700 nm, and it was concluded that a high degree of orientation of the pigment responsible for emission (F 725) occurs. In view of the polarization spectra of photosystem 1 containing chloroplast fractions, it might be suggested that in general the molecules of the chlorophyll *a* fraction emitting F 725 are oriented.

With holochrome preparations of etiolated leaves the polarization values, measured 10 min after grinding the leaves, were about $p = 0.40$ in the red band (87). In the Soret region the polarization spectrum was similar to that of chlorophyll *a* in ricinus oil. The results suggest that either only one or more perfectly parallel oriented chlorophyll(ide) molecules are present on each holochrome at the time of measurement. The polarization values drop gradually during greening to a value of $p = 0.02$, found with chloroplasts of green leaves at 20°C.

In blue-green and red algae.—The fluorescence polarization spectrum measured with *Porphyridium* resembles polarization in the spectrum of chlorophyll *a* in vitro (91).

In brown algae and diatoms.—Due probably to a low percentage of F 725, the polarization values measured with chloroplasts of brown algae are low also at −196°C. With the diatom *Pheodactylum*, however, the fluorescence action spectra at −196°C in polarized light show a polarization of about $p = 0.18$ in the marked 707 nm band and 688 nm shoulder, but less in the main band at 670 nm.

In photosynthetic bacteria.—The values of fluorescence polarization of bacteriochlorophyll in chromatophores of *Rhodopseudomonas spheroides* are, when excitation occurs in the far red bands, found to be about 0.08, while a value of $p = 0.16$ is measured with 590 nm excitation (59). These values are for bacteriochlorophyll in viscous solvents 0.40 and −0.23 respectively. In view of the size of the chromatophore, depolarization from the in vitro values to the ones measured with chromatophores is most likely caused by energy transfer between differently oriented molecules. If a complete random orientation of pigments on each chromatophore occurs, the average number of transfers necessary to explain the in vivo values at 590 nm can be calculated to be of the order of 2–8, depending on the species and wavelength used. If the pigment molecules are oriented parallel, the low in vivo polarization occurs only after a much higher number of transfers. Thus from the polarization values indication about size of photosynthetic unit and pigment orientation is obtained.

The disproportionally large negative value of polarization around 590 nm is difficult to explain. Such a disproportionality was also found with chromatophores of other bacteria (59). Negative polarization is also found when light is absorbed by carotenoids (500–540 nm). The latter suggests that a certain orientation of the carotenoids occurs or that the effect is caused by dichroism of shape.

FLUORESCENCE NOT EMITTED BY CHLOROPHYLLOUS PIGMENTS

In the photosystems of green plants the chlorophylls are accompanied by carotenoids. These pigments are considered to be nonfluorescent. Reports that a weak fluorescence occurs with a maximum around 330 nm, resulting from absorption in a sharp band around 285 nm with some plant carotenoids (204), could not be confirmed (187). A weak fluorescence around 520 nm, from its behavior during photobleaching of the main pigment, was concluded to be due to a trace of impurities (187).

Since effective energy transfer may occur between the nonfluorescing carotenoids and chlorophyll *a,* a close spatial relationship between these pigments should occur. This need not be the case with the highly fluorescent phycobilins, which show fluorescence both in vitro and in vivo (66). They diffuse easily from the photosynthetic system after damage of the cells, and they have such a high in vitro quantum yield of fluorescence that energy transfer by inductive resonance can occur over a relatively wide distance. Consideration of fluorescence polarization as a function of pH (86) and study of the fluorescence polarization spectra (47) gave information about the number of different subunits in the intact biliprotein.

With chloroplasts or bacterial chromatophores fluorescence emitted in the blue region of the spectrum can be due to a number of compounds. The fluorescence changes at about 440 nm in *Rhodospirillum rubrum* were assumed to be due to changes in the redox state of NAD (nicotinamide adenine dinucleotide) (56). Apart from this cofactor, a fluorescent protein factor with about the same optical properties was isolated from spinach chloroplasts (205). It showed optical characteristics of unconjugated pteridines (206) and phosphodoxin (13). The subcellular localization of pteridines in a strain of *Rhodopseudomonas* was also studied with the fluorescence technique (165).

LITERATURE CITED

1. Aghion, J., Bourret, R. L. 1969. *Physiol. Plant Veg.* 7:297–303
2. Amesz, J., Nooteboom, W., Spaargaren, D. H. 1969. *Progress in Photosynthesis Research,* ed. H. Metzner, 2:1064–72. Tübingen. 1127 pp.
3. Amesz, J., Vredenberg, W. J. 1966. *Biochim. Biophys. Acta* 126:254–61
4. Anderson, A. F. H., Calvin, M. 1964. *Arch. Biochem. Biophys.* 107:251–59
5. Anderson, J. M., Thorne, S. W. 1968. *Biochim. Biophys. Acta* 162:122–34
6. Arnold, W., Meck, E. S. 1956. *Arch. Biochem. Biophys.* 60:82–90
7. Arnon, D. L., Tsujimoto, H. Y., McSwain, B.D. 1965. *Proc. Nat. Acad. Sci. USA* 54:927-34
8. Balny, C., Brody, S. S., Hui Bon Hoa, G. 1969. *Photochem. Photobiol.* 9:445–54
9. Bär, F., Laang, H., Schnabel, E., Kuhn, H. 1961. *Elektrochemie* 65:346–54
10. Bennoun, P. 1970. *Biochim. Biophys. Acta* 216:357–63
11. Bergeron, J. A., Olson, J.M. 1967. *Biochim. Biophys. Acta* 131:401–4
12. Björn, L. O. 1969. *Physiol. Plant.* 22:1–17
13. Black, C. C., San Pietro, A., Limbach, D., Norris, G. 1963. *Proc. Nat. Acad. Sci. USA* 50:37–43

14. Bonaventura, C., Myers, J. 1969. *Biochim. Biophys. Acta* 189: 366–83
15. Boardman, N. K., Anderson, J. M. 1964. *Nature* 203:166–67
16. Boardman, N. K., Thorne, S. W. 1968. *Biochim. Biophys. Acta* 153:448–58
17. Ibid 1969. 189:294–97
18. Boardman, N. K., Thorne, S. W., Anderson, J. M. 1966. *Proc. Nat. Acad. Sci. USA* 56:586–93
19. Borisov, A. Y., Godik, V. I. 1970. *Biochim. Biophys. Acta* 223: 441–43
20. Briantais, J. M. 1967. *Photochem. Photobiol.* 6:155–62
21. Briantais, J. M. 1969. *Physiol. Veg.* 7:135–80
22. Bril, C. 1963. *Biochim. Biophys. Acta* 66:50–60
23. Bril, C., van der Horst, B. J., Poort, S. R., Thomas, J. B. 1969. *Biochim. Biophys. Acta* 172:345–48
24. Brody, M. 1971. *Biophys. J.* 11: 189–204
24a. Brody, M., Linschitz, H. 1961. *Science* 133:728–29
25. Brody, M., Nathanson, B., Cohen, W. S. 1969. *Biochim. Biophys. Acta* 172:340–42
26. Brody, S. S. 1969. *Photosynthetica* 3:279–84
27. Brody, S. S., Brody, M., Levine, J. M. 1965. *Biochim. Biophys. Acta* 94:310–12
28. Brody, S. S., Broyde, S. B. 1968. *Biophys. J.* 8:1511–33
29. Brody, S. S., Rabinowitch, E. 1957. *Science* 125:555–56
30. Brown, J. S. 1966. *Biochim. Biophys. Acta* 120:303–7
31. Ibid 1967. 143:391–98
32. Brown, J. S. 1969. *Biophys. J.* 9: 1542–52
33. Brown, J. S., French, C. S. 1961. *Biophys. J.* 7:537–49
34. Broyde, S. B., Brody, S. S. 1966. *Biophys. J.* 6:353–65
35. Butler, W. L. 1961. *Biochem. Biophys. Res. Commun.* 2:419–22
36. Butler, W. L. 1962. *Biochim. Biophys. Acta* 64:309–17
37. Ibid 1965. 102:1–8
38. Butler, W. L., Norris, K. H. 1963. *Biochim. Biophys. Acta* 66:72–77
39. Cederstrand, C. N., Govindjee 1966. *Biochim. Biophys. Acta* 120:177–80
40. Cho, F., Govindjee 1970. *Biochim. Biophys. Acta* 205:371–78
41. Ibid 1970. 216:139–51
42. Ibid, 151–61
43. Cho, F., Spencer, J., Govindjee 1966. *Biochim. Biophys. Acta* 126:174–75
44. Clayton, R. K. 1969. *Biophys. J.* 9:60–76
45. Clayton, R. K. 1966. *Photochem. Photobiol.* 5:679–88
46. Cramer, W. A., Butler, W. L. 1969. *Biochim. Biophys. Acta* 172:503–10
47. Dale, R. E., Teale, F. W. J. 1970. *Photochem. Photobiol.* 12:99–117
48. DeKouchkovsky, Y. See Ref. 2, 959–70
49. DeKouchkovsky, Y., Joliot, P. 1967. *Photochem. Photobiol.* 6: 567–87
50. Delosme, R. 1967. *Biochim. Biophys. Acta* 143:108–28
51. Delosme, R., Joliot, P., Lavorel, J. 1959. *C. R. H. Acad. Sci.* 249: 1409–11
52. Döhler, G., Ried, A. 1963. *Arch. Mikrobiol.* 46:190–216
53. Donze, M., Duysens, L. N. M. See Ref. 2, 991–95
54. Dujardin, E., DeKouchkovsky, Y., Sironval, C. 1970. *Photosynthetica* 4:223–27
54a. Dujardin, E., Sironval, C. 1970. *Photosynthetica* 4:129–38
55. Duysens, L. N. M. 1951. PhD thesis. Univ. Utrecht. 94 pp.
56. Duysens, L. N. M., Sweep, G. 1957. *Biochim. Biophys. Acta* 26:13–16
57. Duysens, L. N. M., Sweers, H. E. 1963. *Studies in Microalgae and Photosynthetic Bacteria*, 353–72. Univ. Tokyo Press. 636 pp.
58. Duysens, L. N. M., Talens, A. See Ref. 2, 1073–82
59. Ebrey, T. G., Clayton, R. K. 1969. *Photochem. Photobiol.* 10:109–17
60. Egle, K., Döhler, G. 1962. *Beitr. Biol. Pflanz.* 38:99–136
61. Evstigneev, V. B., Gavrilova, V. A., Krasnovskii, A. A. 1949. *Dokl. Akad. Nauk SSSR* 66: 1133–38
62. Fork, D. C., Amesz, J. 1969. *Ann. Rev. Plant. Physiol.* 20:305–28
63. Forster, L. S., Livingston, R. 1952. *J. Chem. Phys.* 20:1315–16
64. French, C. S., Smith, J. H. C.,

Virgin, H. I., Airth, R. L. 1956. *Plant Physiol.* 31:369–74

65. Frackowiak, D., Grabowski, J. 1971. *Photosynthetica* 5:146–52
66. Frackowiak, D., Marszalek, T. 1960. *Bull. Acad. Pol. Sci.* 8:713–18
67. Fradkin, L. I., Shlyk, A. A., Kalinina, L. M., Faludi-Daniel, A. 1969. *Photosynthetica* 3:326–37
68. Geacintov, N. E., Van Nostrand, F., Pope, M., Tinkel, J. B. 1971. *Biochim. Biophys. Acta* 226:486–91
69. Ghosh, A. K., Govindjee 1966. *Biophys. J.* 6:611–19
70. Ghosh, A. K., Govindjee, Crespi, H. L., Katz, J. J. 1966. *Biochim. Biophys. Acta* 120:19–22
71. Giller, Yu. Y., Krasichkova, G. V., Sapozhnikov, D. I. 1970. *Biofizika* 15:38–46
72. Gingras, G., Lavorel, J. 1965. *Physiol. Veg.* 3:109–30
73. Giraud, G. 1964. *Proc Int. Seaweed Symp., 4th, Biarritz,* 327–30. Oxford: Pergamon
74. Goedheer, J. C. 1969. *Biochim. Biophys. Acta* 172:252–65
75. Goedheer, J. C. 1957. PhD thesis. Univ. Utrecht. 90 pp.
76. Goedheer, J. C. 1959. *Biochim. Biophys. Acta* 53:1–8
77. Ibid 1961. 51:494–504
78. Ibid 1964. 88:304–17
79. Ibid 1965. 102:75–89
80. Goedheer, J. C. 1966. *The Chlorophylls,* ed. L. P. Vernon, G. R. Seely, 147–84. New York, London: Academic. 679 pp.
81. Goedheer, J. C. 1966. *Currents in Photosynthesis,* ed. J. C. Goedheer, J. B. Thomas, 177–88. Rotterdam: Donker. 487 pp.
82. Goedheer, J. C. 1967. *Le Chloroplasts,* ed. C. Sironval, 77–85. Paris: Masson
83. Goedheer, J. C. 1968. *Biochim. Biophys. Acta* 153:903–6
84. Goedheer, J. C. See Ref. 2, 811–17
85. Goedheer, J. C. 1970. *Photosynthetica* 4:97–106
86. Goedheer, J. C., Birnie, F. 1965. *Biochim. Biophys. Acta* 94:579–81
87. Goedheer, J. C., Gulyaev, B. 1971. *Abstr. 1st Eur. Biophys. Congr. Vienna*
88. Goedheer, J. C., van der Tuin, A. K. 1967. *Biochim. Biophys. Acta* 143:399–407

89. Goedheer, J. C., Verhulsdonk, C. A. H. 1970. *Biochem. Biophys. Res. Commun.* 39:260–66
90. Gorshkov, V. K. 1969. *Biofizika* 14:28–32
91. Govindjee 1966. See Ref. 81, 93–102
92. Govindjee, Ichimura, S., Cederstrand, C., Rabinowitch, E. 1960. *Arch. Biochem. Biophys.* 89:322–23
93. Govindjee, Yang, L. 1966. *J. Gen. Physiol.* 49:763–80
94. Gouterman, M., Stryer, L. 1962. *J. Chem. Phys.* 37:2260–69
95. Heath, R. L. 1970. *Biophys. J.* 10:1173–88
96. Heath, R. L., Hind, G. 1969. *Biochim. Biophys. Acta* 172:290–99
97. Ibid 1970. 189:222–53
98. Homann, P. H. See Ref. 2, 932–37
99. Houssier, C., Sauer, K. 1969. *Biochim. Biophys. Acta* 172:492–502
100. Homann, P. 1969. *Plant. Physiol.* 44:931–36
101. Jacobi, G. 1969. *Z. Pflanzenphysiol.* 61:203–17
102. Jacobs, E. E., Holt, A. S., Rabinowitch, E. 1954. *Arch. Biochem. Biophys.* 53:228
103. Jacobs, E. E., Vatter, A. E., Holt, A. S. 1954. *Arch. Biochem. Biophys.* 53:228
104. Joliot, P. 1965. *Biochim. Biophys. Acta* 102:135–48
105. Journeaux, R., Hochapfel, A., Viovy, R. 1969. *J. Chim. Phys.* 66:1474
106. Kahn, A., Boardman, N. K., Thorne, S. W. 1969. *J. Mol. Biol.* 48:85–101
107. Kahn, J. S., Bannister, T. T. 1965. *Photochem. Photobiol.* 4:27–32
108. Kamrin, M. 1966. *Biochim. Biophys. Acta* 153:262–68
109. Kautsky, H., Apfel, W., Amann, H. 1960. *Biochem. Z.* 332:272–92
110. Ke, B. See Ref. 80, 253–78
111. Ke, B., Chanay, T. H. 1971. *Biochim. Biophys. Acta* 226:341–54
112. Ke, B., Vernon, L. P. 1967. *Biochem. J.* 6:2221–26
113. Kessler, E. 1970. *Planta* 92:222–34
114. Knox, R. S. 1968. *Physica* 39:361–86
115. Kok, B. 1963. *Photosynthetic*

Mechanisms of Green Plants, 45–55. Nat. Acad. Sci.-Nat. Res. Counc. Publ. 1145, Washington, D.C.
116. Kok, B., Rurainski, H. J. 1966. Biochim. Biophys. Acta 126: 587–90
117. Kraan, G. P. B., Amesz, J., Veldhuis, B. R., Steemers, R. G. 1970. Biochim. Biophys. Acta 223:129–45
118. Krasnovsky, A. A., Erokhin, Yu. E., Hung, Y.-Ch. 1962. Dokl. Acad. Nauk SSSR 143:456–60
119. Krasnovsky, A. A., Erokhin, Yu. E., Gulyaev, B. A. 1963. Dokl. Akad. Nauk SSSR 152:1231–34
120. Krey, A., Govindjee 1964. Proc. Nat. Acad. Sci USA 52:1568–72
121. Krey, A., Govindjee 1966. Biochim. Biophys. Acta 120:1–18
122. Latimer, P., Bannister, T. T., Rabinowitch, E. 1956. Science 124:585–86
123. Lavorel, J. 1964. Biochim. Biophys. Acta 88:20–36
124. Ibid 1968. 153:727–30
125. Lavorel, J. See Ref. 2, 905–12
126. Lavorel, J. 1962. Biochim. Biophys. Acta 60:510–23
127. Livingston, R., Thompson, L., Ramarao, M. V. 1952. J. Am. Chem. Soc. 74:1073
128. Livingston, R., Watson, W. F., McArdle, J. 1949. J. Am. Chem. Soc. 71:1542
129. Litvin, F. F., Gulyaev, B. A. 1964. Dokl. Akad. Nauk SSSR 158: 460–63
130. Litvin, F. F., Krasnovsky, A. A. 1957. Dokl. Akad. Nauk SSSR 117:105–9
131. Losev, A. P., Gurinovitch, G. P. 1969. Biofizika 14:110–18
132. Malkin, S. 1966. Biochim. Biophys. Acta 126:433–42
133. Ibid 1971. 234:415–27
134. Malkin, S., Kok, B. 1966. Biochim. Biophys. Acta 126:413–32
135. Mayne, B. C. 1965. Biochim. Biophys. Acta 189:59–66
136. Mohanty, P., Munday, J. C., Govindjee 1970. Biochim. Biophys. Acta 223:198–200
137. Müller, A., Lumry, R., Walker, M. S. 1969. Photochem. Photobiol. 9:113–26
138. Munday, J. C., Govindjee 1969. Biophys. J. 9:1–21
139. Ibid, 22–35
140. Murata, N. 1969. Biochim. Biophys. Acta 172:242–51
141. Ibid 1969. 189:171–81
142. Ibid 1970. 203:379–89
143. Ibid 1971. 226:422–32
144. Murata, N., Nishimura, M., Takamiya, A., 1966. Biochim. Biophys. Acta 120:23–33
145. Ibid 1966. 126:234–43
146. Murata, N., Sugahara, K. 1969. Biochim. Biophys. Acta 189: 182–92
147. Murty, N. R., Cederstrand, C., Rabinowitch, E. 1965. Photochem. Photobiol. 4:917–1
148. Nicholson, W. J. See Ref. 2, 943–46
149. Nicholson, W. J., Fortoul, J. I. 1967. Biochim. Biophys. Acta 143:577–82
150. Nishimura, M., Takamiya, A. 1966. Biochim. Biophys. Acta 120:34–44
151. Nobel, P. S. 1968. Biochim. Biophys. Acta 153:170–82
152. Norris, J. R., Uphaus, R. A., Cotton, T. M., Katz, J. J. 1970. Biochim. Biophys. Acta 223: 446–50
153. Ogawa, T., Vernon, L. P., Mollenhauer, H. H. 1969. Biochim. Biophys. Acta 172:216–29
154. Olson, J. M., Clayton, R. K. 1966. Photochem. Photobiol. 5:655–60
155. Olson, J. M., Filmer, D., Radloff, R., Romano, C. A., Sybesma, C. 1963. Bact. Photosyn. Symp., Yellow Springs, Ohio, 423–31
156. Olson, J. M., Nadler, K. D. 1965. Photochem. Photobiol. 4:783–97
157. Olson, J. M., Stanton, E. K. See Ref. 80, 381–97
158. Olson, J. M., Sybesma, C. 1963. Bact. Photosyn. Symp., 413–22
159. Olson, R. A., Jennings, W. H., Butler, W. L. 1964. Biochim. Biophys. Acta 88:331–37
160. Okayama, S. 1967. Plant Cell Physiol. 8:47–59
161. Papageorgiou, G., Govindjee 1967. Biophys. J. 7:375–89
162. Papageorgiou, G., Govindjee. See Ref. 2, 905–12
163. Papageorgiou, G., Govindjee 1971. Biochim. Biophys. Acta 234:428–32
164. Rabinowitch, E. 1951. Photosynthesis and Related Processes,

740–833. New York: Interscience. 601 pp.
165. Reed, D. W., Mayne, B. C. 1971.
Biochim. Biophys. Acta 226:
477–80
166. Rosenberg, J. L., Bigat, T. See
Ref. 115, 122–30
167. Rosenberg, J. L., Bigat, T., Dejae-
gere, S. 1964. *Biochim. Biophys. Acta* 79:9–19
168. Rubin, A. B., Minchenkova, L. E.,
Krasnovsky, A. A., Tumerman,
L. A. 1962. *Biofizika* 7:571–77
169. Rubin, A. B., Osnitskaja, L. K.
1963. *Mikrobiologia* 32:200–3
170. Sane, P. V., Goodchild, D. J.,
Park, R. B. 1970. *Biochim. Biophys. Acta* 216:162–79
171. Singhal, G. S., Rabinowitch, E.
1969. *Biophys. J.* 9:586–91
172. Sironval, C., Brouers, M. 1970.
Photosynthetica 4:38–47
173. Sironval, C., Brouers, M., Michel,
J. M., Kuyper, Y. 1968. *Photosynthetica* 2:268–87
174. Shimony, C., Spencer, J., Govindjee 1967. *Photosynthetica* 1:
113–25
175. Smith, J. H. C. 1958. *Brookhaven
Symp. Biol.* 11:296–302
176. Smith, J. H. C., Benitez, A. 1954.
Plant Physiol. 29:135–43
177. Stensby, P. S., Rosenberg, J. L.
1961. *J. Phys. Chem.* 65:906
178. Sybesma, C., Duysens, L. N. M.
See Ref. 81, 85–92
179. Sybesma, C., Olson, J. M. 1963.
Proc. Nat. Acad. Sci. USA 49:
248–53
180. Teale, F. W. J. 1960. *Biochim.
Biophys. Acta* 42:69–75
181. Terpstra, W. 1970. *Biochim. Biophys. Acta* 216:179–91
182. Terpstra, W., Goedheer, J. C.
1966. *Biochim. Biophys. Acta*
120:326–31
183. Thorne, S. W. 1971. *Biochim. Biophys. Acta* 226:113–27
184. Ibid, 128–34
185. Thorne, S. W., Boardman, N. K.
1971. *Biochim. Biophys. Acta*
234:113–23
186. Tomita, G., Rabinowitch, E. 1962.
Biophys. J. 2:483–97

187. Tric, C., Lejeune, V. 1970. *Photochem. Photobiol.* 12:331–43
188. Trosper, T., Park, R. B., Sauer, K.
1968. *Photochem. Photobiol.* 7:
451–69
189. Tumerman, L. A., Rubin, A. B.
1952. *Dokl. Akad. Nauk SSSR*
145:202
190. Tweet, A. G., Bellamy, W. D.,
Gaines, G. L. 1964. *J. Chem.
Phys.* 41:2068–77
191. Virgin, H. I. 1956. *Physiol. Plant.*
9:679–81
193. Vredenberg, W. J. 1970. *Biochim.
Biophys. Acta* 223:230–34
194. Vredenberg, W. J. See Ref. 2,
923–32
195. Vredenberg, W. J., Amesz, J.
1967. *Biochim. Biophys. Acta*
126:244–253; 254–261
196. Vredenberg, W. J., Duysens, L. N.
M. 1963. *Nature* 197:355–57
197. Vredenberg, W. J., Slooten, L.
1967. *Biochim. Biophys. Acta*
143:583–91
198. Wang, R. T., Clayton, R. K. 1971.
Photochem. Photobiol. 13:215–
24
199. Wassink, E. C., Kersten, J. A. H.
1946. *Enzymologia* 12:3–32
200. Watson, W. F., Livingston, R.
1950. *J. Chem. Phys.* 18:1174–
78
201. Weber, G. 1960. *Comparative Biochemical Photoreactive Systems,*
ed. M. B. Allen, 395–411. New
York: Academic
202. Wessels, J. H. C. 1962. *Biochim.
Biophys. Acta* 65:561–69
203. Williams, W. P., Murty, N. R.,
Rabinowitch, E. 1969. *Photochem. Photobiol.* 9:455–69
204. Wolf, F. T., Stevens, M. V. 1967.
Photochem. Photobiol. 6:597–
99
205. Wu, M., Myers, J. 1969. *Arch.
Biochem. Biophys.* 132:430–35
206. Ibid 1970. 140:391–98
207. Zankel, K. L. 1969. *Photochem.
Photobiol.* 10:259–66
208. Zankel, K. L., Clayton, R. K.
1969. *Photochem. Photobiol.* 9:
7–15

Ann. Rev. Plant Physiol. 1972. 23:113–32

CHEMISTRY OF THE PLANT CELL WALL 7525

D. H. Northcote

Department of Biochemistry
University of Cambridge, England

CONTENTS

INTRODUCTION

This review will consider the cell wall as a growing, constantly changing, composite material consisting of a dispersed phase of microfibrils within a complex continuous matrix (33). It is thus possible by studying their roles in the composite to relate the chemical and physical properties of the individual constituents to those of the whole wall. During the growth of the cell the polymers of the wall interact and change, and the resulting alteration in the properties of the wall can be correlated with a variation in its function. A change in properties of the wall can also occur in response to a variation in the environment of the growing cell and can be brought about by interactions between cells, by gaseous or aqueous changes in the surrounding medium, or by an increase or decrease in the stresses and strains applied to the cell wall.

113

Although the properties of the wall become modified, the basic composite structure consisting of microfibrils dispersed in a complex matrix is retained, but there are alterations in the orientation of the microfibrils, changes in the chemical composition of the matrix, or variations in the interaction of the fibrillar and matrix components (71, 72).

The interrelationships of the individual components of the wall with each other will be discussed and their contributions to the properties of the total composite material will be indicated. An explanation will be given for the presence of the very diverse types of branched and linear polysaccharides that make up the plant cell wall (73), especially as the conformations and detailed fine structures of these polymers are now being investigated, and this has made it feasible to assess their contribution to the properties of the composite.

The cell wall is a dynamic structure whose composition and properties constantly respond to the growth, the stage of differentiation, and the environment of the cell, and thus its formation is a continuous part of the growth processes of the cell. However, for simplicity there is some advantage in comparing and contrasting it with a static man-made composite material such as glass-fiber-reinforced-plastic. In this way certain components of the wall can be regarded as keying materials, others as fillers, and some as wetting or as swelling agents, and it is evident that both the natural and the man-made composite may have skins. The differences between the two composite materials, one static and the other continually being formed and changed, are also significant for an understanding of the chemical, physical, and biochemical properties of the wall.

The characteristics of a glass fiber composite depend upon such properties as the length and area of cross section of the fibers, the fraction of the total area of a cross section of the material occupied by the fibers, and upon the elastic constants (Young's modulus, shear modulus, Poisson's ratio) of the fibers and of the matrix (40, 49, 54). It is also important to consider the geometry of the fibers in the matrix and, in addition, the interaction of the fibers and the matrix at their interfaces. An alteration of the matrix-fiber interaction brings about an alteration in the important stress transmission characteristics between the fibers and the matrix, and this has a very important effect on the properties of the composite. Thus, compounds which form bridges between the polymers of the fibers and the matrix, although present in relatively small amounts, can have a major influence upon the properties of the material.

In the cell wall, polymers compatible with those of the microfibrils and of the matrix and situated at the interface can provide points of entanglement, and these keying agents, by altering the adhesion, consequently change the stress transfer between the matrix and the microfibrils. One of the important consequences of this type of study is that the function of water as a cell wall constituent is made apparent, and thus the role of water in the cell wall will be discussed as a distinct major cell wall component.

THE MICROFIBRILLAR COMPONENT

Microfibrils are a remarkably constant feature of the cell walls of all green plants, and in nearly all species these microfibrils are made up of cellulose. In some exceptional instances certain algae have microfibrils which may be composed of xylans [$\beta(1\rightarrow3)$ linked xylose residues] or mannans [$\beta(1\rightarrow4)$ linked mannose residues] (14, 31, 32). Cellulose does not occur in yeast and some other fungi.

The structure of cellulose is characterized by long chains of $\beta(1\rightarrow4)$ linked glucose residues consisting of 8,000–12,000 units (i.e. a degree of polymerization, D.P., of 8,000–12,000) (39, 64). Glucose polymers which are made up of $\beta(1\rightarrow4)$ links exist as extended chains with a twofold screw axis, and these are arranged in an ordered manner within the microfibril (58, 67). The position of the chains of glucose relative to one another are so ordered that in some parts of the microfibril, crystalline regions occur which give a distinct X-ray diagram. From this can be constructed a unit cell (53, 67) that shows the position of the chains relative to one another within the crystalline region. The conformation and crystalline lattice of cellulose in the microfibril is shown in Figure 1. It cannot be predicted from the X-ray diagram that the glucose chains are arranged in an antiparallel manner, but comparisons with the structure of chitin make it seem likely that the orientation of the cellulose

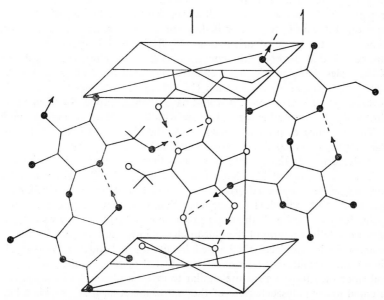

FIGURE 1. Diagram to illustrate a possible arrangement of the glucose chains in a unit cell of native cellulose. The chains are drawn antiparallel, and inter- and intramolecular hydrogen bonds are depicted by arrows (53).

chains is similar to that of the acetyl glucosamine chains of chitin and that they are antiparallel (53, 58, 67). Within the cellobiose units of the individual chains there are intramolecular hydrogen bonds between 0–3′ and 0–5 atoms which are 2.7 Å apart (58). These observations suggest that the cellobiose units must have a bent conformation along the chain. There are intermolecular hydrogen bonds between the antiparallel chains at 0–6 and 0–1′ (58). It has been difficult to reconcile the antiparallel arrangement of the chains with possible synthetic mechanisms for the interweaving microfibrillar network of the wall. In order to overcome this difficulty folded structures have been suggested for the arrangement of the chains in the microfibril (56, 65), but the calculated modulus of elasticity of one of these folded structures is not compatible with the experimental observations of the mechanical properties of mature tracheids, and therefore it cannot be correct (61, 62).

The structure, size, and orientation of the microfibrils is obviously of some considerable significance for the properties of the cell wall. The diameter of the microfibrils varies with the plant, and usually they are larger in the algae than in higher plants. The exact measurement is in dispute at present. Normally the diameter is said to be about 10 nm–30 nm, but recently results obtained from material investigated by the negative staining technique have been interpreted to suggest that microfibrils of 3.5 nm diameter occur (44, 45, 68). However, Preston has shown clearly that measurements of fibrillar diameters can give results which are at least two or three times too small when they are calculated from measurements made on a negatively stained preparation of a portion of the wall where there is overlapping of the microfibrils, one on top of another, or if they are made on a section not cut pependicularly to the plane of orientation (25, 81). The importance of the measurement is that the crystallite contained in the microfibril (i.e. the diameter of the ordered region of glucan chains) measured by X-ray scattering indicates a diameter of about 4–12 nm (70, 81), and if the microfibril consisted of this crystalline array only, it would have different properties from a microfibril where the crystallite formed a core with less ordered chains surrounding it. The evidence at present indicates that the microfibril is about 10 nm in diameter with a crystallite core of about 4 nm diameter (81).

Polymeric material containing sugars other than glucose is always found with cellulose when it is isolated from the wall (71, 72). These nonglucan polysaccharides may be adsorbed onto the surface of the microfibril, or they may be part of individual heteroglucan chains, or they may exist as separate chains intermingled with the outer glucan chains (71). In any case, because of the crystalline nature of the glucan chains in the microfibrils (which must be strongly intermolecularly bonded by hydrogen bonds in the central regions) the nonglucan chains must be situated at or near the surface (82), and since they have a structure similar to or even identical with some of the oriented linear polysaccharides of the matrix material, they provide a keying material for the entanglement of the microfibrils with the matrix.

During the initial stages of cell growth the matrix of the wall is not rigid and the microfibrils may be grouped in bands within the matrix. They are oriented at a fairly large angle with respect to the long axis of the cell, and they form a network that is interwoven fairly loosely although there are some regions of parallel arrangement. The secondary wall is formed by microfibrils that are more closely packed, lie more parallel to one another, and are oriented with a smaller angle to the long axis of the cell. Hence the fractional volume of fibrils is increased in the composite and the elastic modulus of the wall is increased.

The secondary thickening is also made up of distinct layers in which the orientation of the microfibrils is different in adjacent layers so that, for instance, an inner (S_3) and outer layer (S_1) may carry closely packed microfibrils oriented in a more transverse direction than those in a middle layer (S_2). Thus the secondary thickening achieves strength in more than one direction with a maximum packing of the microfibrils within the composite. This method is now used for the manufacture of articles made from glass fiber and other composites. Sheets of the composite are bonded together as laminates, and in each sheet the fibers are aligned uniaxially but in different directions in the various layers so that strength is maintained in more than one direction (26).

MATRIX PHASE

The polysaccharides (hemicellulose and pectic substances).—The polysaccharides of the matrix are made up of linear oriented polymers which are present at all stages of the development of the wall and also of highly branched polysaccharides that are deposited at particular stages of the growth.

The structure of these polysaccharides varies from plant to plant, and the general pattern of their structure is different in relation to their changing functions. However, they can be considered as groups or families of polysaccharides for which a basic structure can be described which with minor variations will designate any member of the group (2). A particular polysaccharide fraction prepared from a plant tissue does not contain identical molecules even though the preparation does not necessarily contain two or more different polysaccharides (8, 23, 30, 87–89); the preparation is polydisperse but it need not necessarily contain diverse molecules. This concept can be more easily discussed by reference to the carbohydrate moiety of glycoproteins such as those found in blood plasma since in these compounds the limits of the polysaccharide from its reducing end to its nonreducing ends are clearly defined. The reducing end is combined as a glycosidic bond from N-acetyl glucosamine to aspartamide, and the nonreducing ends are terminated by sialic acid or fucose. A preparation of these glycoproteins contains a range of molecules which vary in the extent of completion of the carbohydrate prosthetic groups, and the preparation is polydisperse with respect to the carbohydrate moiety of the molecules. This phenomenon is sometimes called microheterogeneity (94). Thus a biological polymer which is not directly coded for by the nucleic acid of the cell may be polydisperse as a result of the incompleteness of its formation within the tissue, and the variation is a result of the biochemical rather than the genetic process. A distinction can thus be made between a primary and a secondary gene product. It is clear that polysaccharides formed as a continuous process within the cytoplasm and exported outside the cell into the wall can also be polydisperse for similar reasons to those indicated for the glycoproteins, and further variations can occur in positions of branching. A fair amount of randomness in the sequence of two or more different monomers in a chain is possible in most polysaccharides. The criteria of Gibbons (38) can be used to make distinctions so that a preparation is said to be pure or homogeneous although it may be polydisperse if a property that is measured for the total number of molecules varies as a unimodal function, but it is said to be heterogeneous if it varies as a bi- or polymodal function.

Variations in polysaccharide preparations made from cell walls therefore occur because of: (a) the essential polydisperse nature of the particular polysaccharide in the cell wall; (b) changes that may occur in the polysaccharide as it is synthesized at various stages of cell wall development so that, for instance, the preparation is extracted from young primary and mature secondary walls in the cells of the tissue; (c) differences in the polysaccharides deposited in lateral and end walls of the same cell; and (d) the heterogeneous nature of most plant tissues which contain cells of different types and which have cell walls of various chemical composition.

Water.—Water is an extremely important constituent of the cell wall, and the water content of the wall is one of its most variable features. The amount of water within the wall matrix can be controlled to some extent by the deposition of polysaccharide filler material which forms close intermolecular associations and gel-like structures, or by a nonwettable filler such as lignin.

The function of the water as a wall constituent can be considered from four points of view. 1. Structural component. As part of a gel structure the water can play a direct role in the form and in the supporting nature of the wall, and in this way it becomes a structural component of the matrix material. The conformation of polysaccharides such as pectins which form these gels can be altered reversibly so that the gel changes to a viscous solution (83), and this type of change can bring about marked differences in the texture of the wall because of the difference in the physical properties of the matrix. 2. Wetting agent. The strength of various polysaccharide associations depends upon hydrogen bonding. Cellulose microfibrillar structures are built up from bundles of glucan chains which are held together laterally by intermolecular hydrogen bonds, and hydrogen bonds also form connections between the surface of the microfibrils and the nonglucan chains that are in the matrix. Thus either the penetration of water into the microfibrillar structure or the wetting of the surface can greatly reduce both the cohesion of the cellulose microfibrils and the adhesion at the surface between the matrix and the microfibril. Alterations of either of these can have a very great effect upon the physical properties of the wall. 3. Lattice component. A conformation of some polysaccharides, in particular an important matrix material like xylan, is known to be stabilized by an association of water molecules within the crystalline lattice of the complex (69, 97). The shape of the molecules of the matrix will have a major influence upon the properties of the matrix. 4. Besides these structural properties, the water of the wall has the more obvious effect upon the permeability of the wall, and it also allows the presence of metallic ions such as Ca^{++} which may form salts with acidic cell wall constituents. The inclusion of soluble material into the wall structure will also allow any possible biosynthetic and enzymic reactions to occur within the wall structure itself, and these metabolic reactions on any of the wall constituents will considerably alter its properties (92).

Many of the different types and structures of the polysaccharides that form the matrix of the cell wall alter the hydrophilic or hydrophobic nature of the matrix and contribute to the properties of the wall by their relationship to water molecules. At late stages in development the space occupied by the water in the wall becomes progressively filled by lignin. This preserves the tensile strength of the microfibrils and makes a rigid matrix phase.

MATRIX POLYSACCHARIDES WHICH ARE LAID DOWN
THROUGHOUT THE GROWTH OF THE WALL
(PRIMARY AND SECONDARY WALL)

Xylans.—The xylans are laid down throughout the growth of the wall and form the bulk of the hemicellulose fraction (alkaline soluble polysaccharides) of the angiosperms.

The general structure of the xylans in higher plants is that of a main chain of D-xylopyranose residues joined by $\beta(1\rightarrow4)$ links (1). The xylans are large molecules with a degree of polymerization of about 150–200. Attached to the backbone are short terminal side chains. In the angiosperms these are 4-0-methyl-D-glucuronic acid units which are joined to the xylose main chain by $\alpha(1\rightarrow2)$ bonds, and they are probably distributed randomly along the chain. They occur in the ratio of about one uronic acid radical for every ten xylose residues (99) and are present for the most part as esters and not as free acids (101). In vivo about half of the xylose groups in the polysaccharide chain are acetylated. The bulk of the acetylation occurs at C-3 although there is some at C-2, and certain xylose residues are acetylated at both C-2 and C-3 (18, 19). When the acetylated xylans are isolated they are found to be soluble in water, especially in comparison with the deacetylated polymers obtained from the wall by alkaline extraction. The acetyl groups are numerous enough therefore to prevent alignment of the molecular chains, and molecular aggregation cannot take place. The presence of the acetyl groups must thus influence the association of these chains with each other and with other polysaccharide complexes within the wall structure.

The grasses have large amounts of xylans in the hemicellulose fractions prepared from their cell walls. The xylans have the characteristic main chain of $\beta(1\rightarrow4)$ linked D-xylopyranose units, but the side units are L-arabinofura-nosyl groups attached by $\alpha(1\rightarrow3)$ links in addition to the 4-0-methyl-D-glu-curonic acid residues which are $\alpha(1\rightarrow2)$ linked (1).

Arabino-(4-0-methylglucurono) xylans are also present in the cell walls of gymnosperms but in much smaller amounts than the xylans of the angiosperms (100). They have the same basic structure as that of the xylans of monocotyledons: the xylose:arabinose ratio in the polysaccharide is approximately 7:1 to 12:1, and the xylose:uronic acid ratio is approximately 5:1 to 6:1. These polysaccharides are not considered to be acetylated in vivo.

The conformation of xylan preparations has been investigated by X-ray analysis, and the presence of hydrogen bonding has been studied by polarized infrared investigations (69, 97). It has been shown that the molecules exist as extended chains with a threefold screw axis, but unlike cellulose the chains are not stabilized by intermolecular hydrogen bonds. Nevertheless, chain groupings do occur and have been shown to be stabilized by the inclusion of water molecules in the crystalline structure, and there is a continuous range of hydrated xylans. The structure of the xylan hydrate can be represented by a crystalline lattice in which a site within the lattice is occupied by a column of water molecules that stabilizes the structure (Figure 2) (69). This hydrophilic site within the lattice can also accommodate the 4-0-methyl-D-glucuronic acid and arabinofuranose side chains that can hold the water molecules in this position (85).

In the matrix of the wall the xylan molecules are oriented in the direction of the microfibrils as a paracrystalline array; that is, they are oriented parallel to the cellulose chains but between the microfibrils. Xylan molecules are also closely applied to the surface of the microfibrils, and these can act as keying substances between the matrix and the fibrils. According to some theories of

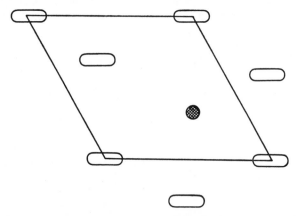

FIGURE 2. Diagram of a base-plane projection to illustrate a possible arrangement of the xylose chains in a unit cell of xylan hydrate. The xylan chains are represented by the flattened circles and the water column by the circle (69).

polymer adsorption, at surfaces only a fraction of the total number of chain monomers of a flexible macromolecule become attached at the sorption sites of the surface, the remainder of the molecules form bridges and loose ends which enter into the surrounding medium (55). Xylan molecules have been shown experimentally to be adsorbed to cellulosic fibers from a solution, and this adsorption is a result of van der Waals forces or of hydrogen bonds (24, 55, 102). Within the wall structure the adsorption would be expected to be controlled by the presence of water molecules at the surface of the microfibrils and upon the degree of hydration of the xylan molecules which is dependent upon the amount of uronic acid and arabinose side chains.

Gluco- and galactoglucomannans.—The gluco- and galactoglucomannans usually form the bulk of the hemicellulose fraction of gymnosperms. They consist of chains of randomly arranged D-glucose and D-mannose radicals joined by $\beta(1\rightarrow4)$ links (4, 100); this main chain may be branched once or twice. In gymnosperms the glucomannans can be divided into the two groups according to their solubility. The delignified tissue is extracted with aqueous potassium hydroxide solution in which the more soluble galactoglucomannans go into solution and can be fractionated and purified by precipitation with barium hydroxide. Left behind in the residue after the alkaline extraction are α-cellulose and glucomannans; the latter are extracted with solutions of borate mixed with sodium hydroxide solution and subsequently fractionated with barium hydroxide.

Galactoglucomannans are triheteropolymers and once isolated tend to be soluble in water. The D-galactose residues are attached to the main mannose-glucose chain as terminal branches linked $\alpha(1\rightarrow6)$ (4, 100). The ratios galactose:glucose:mannose in the polymer are approximately 1:1:3. In situ the polysaccharide molecules are large, usually with a degree of polymerization of about 100, and are probably acetylated at C-2 or C-3 of the mannose residues in the chain.

When the glucomannans are prepared from gymnosperms they always contain some D-galactose (approximately 1-2%) which may be joined on to the main chain as $\alpha(1\rightarrow6)$ linked terminal units. The mannose:glucose ratio in the polysaccharide is about 3:1, and the molecules tend to be larger than those of the galacto-glucomannans, having a degree of polymerization greater than 200. In situ the chain is probably acetylated at C-2 or C-3 of the D-mannose units.

Glucomannans occur in very small amounts in the hemicellulose fraction of the cell walls of angiosperms, usually accounting for 3–5% of the total cell-wall material. No galactose is present, and the mannose:glucose ratio in the material is approximately 2:1, while the degree of polymerization of the chains is usually found to be about 70 (99).

Glucomannan chains have a very similar conformation to that of cellulose. They exist as extended chains with a twofold screw axis, but because of the change in orientation of the hydroxyl group at C_2 of mannose to an axial position, an unfavorable interaction between C_6 and O_2' atoms of contiguous residues is prevented, and this loosens the chain and weakens the packing and organization compared with cellulose (85, 98). However, the chains are organized in a paracrystalline array between the cellulose microfibrils and are strongly adsorbed at the microfibrillar surface (24, 93).

The galactose groups that are joined to the main glucomannan chain as terminal $\alpha(1\to6)$ linkages are flexible groups (84, 85) which can provide non-covalent connecting bridges with water and with other matrix polysaccharides.

Matrix Polysaccharides Which Are Laid Down During the Initial Phases of Cell Wall Growth (the Cell Plate and Primary Wall)

Larch arabinogalactans.—Arabinogalactans occur as a general hemicellulose constituent of conifers such as larches (103, 104), where they comprise a large proportion of the hemicellulose. They also occur as a constituent of the pectic substances of all cell walls.

Larch arabinogalactans are water-soluble, highly branched polysaccharides. Two closely related neutral polymers (arabinogalactans A and B) are usually found in each extract. They can be separated by electrophoresis in borate buffer or by ultracentrifugation and have been found to have similar chemical structures (17, 20) (Figure 3). The arabinose:galactose ratio in the polysaccharides is about 1:6, and they are composed of interior chains of $\beta(1\to3)$ linked D-galactopyranose units to which are attached side chains by $\beta(1\to6)$ links. The side chains are $\beta(1\to6)$ linked D-galactose oligosaccharides which may carry L-arabinofuranose units linked $\beta(1\to3)$ or $\beta(1\to6)$ to the galactose. Arabinose groups are also found attached in a similar manner to the galactose residues of the main chain (10). The arabinose sometimes may be present as a disaccharide composed of L-arabinopyranose linked $\beta(1\to3)$ to L-arabinofuranose (Figure 3) (10).

The pectic substances.—The pectin polysaccharides are a complex mixture which can be extracted from the wall with water or with aqueous solutions of chelating agents such as ethylene-diaminetetra-acetate or sodium hexametaphosphate (95). Various neutral (arabinogalactans) and acidic (galacturonorhamnans) polysaccharides can be obtained from the mixture.

Galacturonorhamnans are the major constituents of the pectic substances, occurring together with the neutral arabinogalactans. The acidic polysaccharides have a basic chemical structure which consists of a chain of $(1\to4)$ α-D-galacturonopyranosyl units in which L-rhamnopyranosyl units occur linked as shown in Figure 4. A variable number of the carboxyl groups are esterified as methyl esters. Attached to the main chains are side chains that contain L-fucose, D-xylose, and D-galactose. D-glucuronic acid is also present in the polysaccharide and is found attached by $\beta(1\to4)$ links to fucose and by $\beta(1\to6)$ links to galactose where it probably terminates some of the side

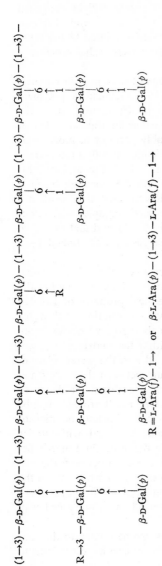

FIGURE 3. Formula to illustrate the general features of the structure of the arabinogalactans of larch. In this and in Fig. 4 (p) following the abbreviation for a monosaccharide denotes the pyranose form and (f) the furanose form.

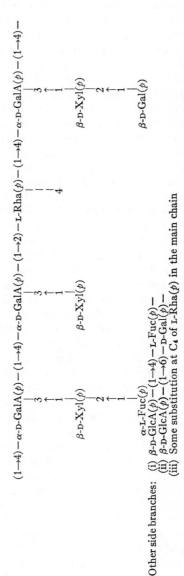

FIGURE 4. Formula to illustrate the general features of galacturonorhamnans that occur in pectic substances.

chains (3). Some of the xylose units are attached by $\beta(1{\to}3)$ links to the galacturonic acid residues of the main chain (7, 9, 11, 12).

In the more mature cell walls, blocks of large molecular weight composed of L-arabinose and D-galactose are attached to the acidic galacturonorhamnan main chain (16). This neutral material is metabolically related to the separate neutral polymers containing L-arabinose and D-galactose that are always found in the pectic fraction (74, 92, 96).

While the arabinogalactan of pectin can be separated as neutral material during electrophoresis of any pectin fraction (16), the neutral polymers might also contain separate arabinans and galactans. From some plant tissues, degraded material has been used to prepare either an arabinan (47) or a methylated galactan (48), but these were probably portions of more complex compounds. Work on these degraded materials indicates that the L-arabinose forms a highly branched polysaccharide in which the arabinofuranose units are linked $\alpha(1{\to}3)$ and $\alpha(1{\to}5)$, whereas the galactose forms a main chain of $\beta(1{\to}4)$ linked galactose units. Recent analytical work has indicated that arabinogalactans exist as part of the pectin complex with structures comparable to that of larch arabinogalactan (6, 13) and that branched arabinogalactans can occur in which the galactose residues are linked $1{\to}3$ and $1{\to}6$ in addition to the $1{\to}4$ linkages (5).

The pectin polysaccharides of the wall are thus a complex mixture which can be metabolically altered by transglycosylation reactions even within the wall matrix (16, 92, 95). A striking physicochemical property of isolated pectins is their ability to form reversible gels and viscous solutions with water (83). These properties are very much related to their role within the wall, and they can be regarded as filler substances within the matrix which will influence the water distribution within the wall and also the water relationship between the polysaccharides of the matrix and between the matrix and the microfibrils. The packing of pectin into the wall, therefore, will alter both the texture and mechanical properties of the wall. One result probably will be that the hydrogen bonds between the matrix polysaccharides and the microfibrils will be weakened so that the wall will become less rigid, while under stress the microfibrils will slip over one another in the more fluid matrix because of the considerably decreased mechanical interaction between the two phases. This is very necessary during the early stages of wall growth when the microfibrils in the expanding cell wall change their orientation to accommodate the increase in area, and it is during the period of primary cell wall growth that pectin is deposited in the wall.

The chemical structure of the pectin molecules can be related to the conformation of the polymers, and the shapes of the molecules contribute to their physical properties. The principal part of the polysaccharide complex is the polygalacturonic acid backbone that exists as an extended ribbon-like chain with a threefold screw axis (5, 79, 83, 85, 86). Within the chain rhamnose radicals occur, and at these points the chain is bent or kinked. In the formation of gels, the straight parts of the chains probably are associated in bundles with the rest of the chains forming an interconnecting network so that water molecules are held in a structural framework (86).

The galactose radicals are present in chains of $\beta(1\rightarrow4)$ linked units that are flexible polymers able to form networks (85). The galactan carries a branched arabinose shell, and the arabinofuranose residues of this part of the structure form a hydrophilic network so that these molecules of the complex hold the water within the gel structures formed by the polygalacturonan.

The properties of the pectin can be further modified by methylation of the ionic carboxyl groups of the galacturonic acid residues and by acetylation of the hydroxyl groups. The ionic nature of the polygalacturonic acid molecules allows the molecules to form salts, and the physical state of the pectin will be different in the presence of monovalent from that in the presence of divalent cations.

MATERIAL LAID DOWN DURING SECONDARY THICKENING ONLY: LIGNIN

It is now possible to define lignin in terms of the aromatic p-hydroxycinnamyl alcohols from which it is produced. Lignin may be described as the insoluble constituent of the cell wall that is aromatic, of high molecular weight, and derived by the enzymic dehydrogenation and subsequent polymerization of coumaryl, coniferyl, and sinapyl alcohols (36). The relative proportions of these alcohols taking part in the synthesis of the lignin vary in different plants; thus the principal characteristic of the lignin is its methoxyl content. If any preparation of lignin is oxidized with nitrobenzene, it will give a mixture of p-hydroxybenzaldehyde, vanillin, and syringaldehyde; if it is degraded by a nonoxidative procedure such as ethanolysis, a mixture of watersoluble phenylpropanoid ketones (Hibbert's ketones) is produced (46).

Any formulae for lignin should accommodate all the 100 or more units which go to make up the molecule and should show the correct but very variable types of linkage between them. All that is necessary in practice, however, is to display a part of the molecule to show the types of unit which occur in the particular variety of lignin and the types of linkage which could possibly arise between them by the known mechanism of its biological synthesis. The precise linkages can be demonstrated only by a detailed chemical study (36). Freudenberg (36) has suggested that the structure shown in Figure 5 represents a fragment of spruce lignin in which the units are interlinked in a fashion corresponding to the biochemical growth of the molecule. The complete molecule extends as a three-dimensional covalently linked network.

The lignin penetrates the wall from the outside (primary wall) inwards at a very early stage of secondary thickening (28, 29, 43, 105). It is a hydrophobic filler material that replaces the water in the cell wall and finally encrusts the microfibrils and the matrix polysaccharides so that the wall may become thicker. Since water is displaced, strong hydrogen bonds can occur between the polysaccharides both at the microfibrillar-matrix interface and between the components of the matrix. In addition, covalent bonds may be formed between the carbohydrates and the lignin (22, 66), and the system

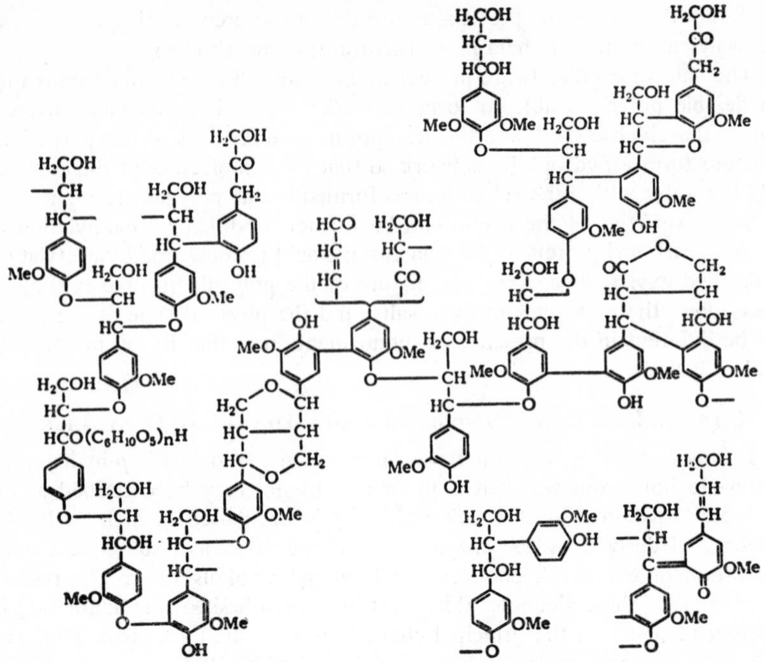

FIGURE 5. Formula suggested by Freudenberg (36) for the structure of a portion of spruce lignin. It shows the types of linkages which could arise by the known methods of biosynthesis.

therefore becomes one in which the linear polysaccharide polymers are enclosed in a cross-linked polymer cage (80). At this stage the wall becomes very similar to a synthetic glass fiber composite with great tensile strength because of the presence of the microfibrils, and a rigid structure because of the lignified matrix.

There is a very good contact between the microfibrils and the matrix in the lignified wall; this ensures a good stress transfer between the components and also ensures that the layers and components of the wall will not slip with respect to one another.

GLYCOPROTEIN OF THE WALL

It is difficult to ascertain whether protein found in an isolated cell wall preparation is a constituent of the wall or whether it is present because of an unspecific adsorption from the cytoplasm. However, part of the protein of the wall carries hydroxyproline, and the distribution of this amino acid between the cytoplasm and cell wall preparations of broken cells makes it clear that a protein containing hydroxyproline is present as a cell wall constituent since only traces of the amino acid are found in the cytoplasm (27, 52). This is made even more certain when the wall is fractionated into pectin, hemicellulose, and α-cellulose by extraction of the wall with water and alkali, since the bulk of the hydroxyproline remains with the α-cellulose residue (42).

A crude cellulase preparation containing proteolytic activity has been used to render glycopeptides soluble from cell wall material, and this work indicates that the polysaccharide portion of the molecule contains arabinose and galactose (50, 51). However, it is probable that these enzyme preparations degrade both the protein and the carbohydrate parts of the original glycoprotein. It is possible to degrade protein by hydrazinolysis, leaving glycosidic bonds intact (15, 106), and this method has been used to isolate glycopeptides from sycamore callus cells (42). The glycopeptide prepared by hydrazinolysis was shown to consist of a series of hydroxyproline residues, each of which carried an oligosaccharide containing galactose and arabinose with an average degree of polymerization of approximately eight carbohydrate residues (42). These short oligosaccharides cannot be involved in covalent cross-links between the major large polysaccharides of the cell wall, and since over 80% of the hydroxyproline of the wall can be extracted with hydrazine, it is unlikely that any protein containing hydroxyproline is involved in cross links of this type.

Since the glycoprotein is so closely bound to the α-cellulose fraction of the wall, its function might be that of bonding the microfibrils together by means of hydrogen bonds between the oligosaccharides of the glycoprotein and the outer polysaccharide chains of the microfibrils, or it may take part in the interaction of the matrix and the microfibrils. Its localization with the α-cellulose during the chemical fractionation of the wall would suggest that it is part of the microfibrillar component of the wall rather than of the matrix. At what stage of the formation and development of the wall it is laid down is not known at present.

THE SKINS AND PROTECTIVE COVERINGS OF THE WALL

The outer walls of the epidermal cells of leaves and many other aerial organs of plants are covered with a protective film of wax and cutin. This not only is deposited over individual cells but is spread to form a protective covering over the whole surface area of the leaf or other organ. This cuticle thus binds the cells together, preventing the plant from excessive water loss and at the same time largely preventing entry of water into the tissues and also serving as an aid against mechanical injury (63).

Generally the covering is made up of an outer layer of wax which overlays a skin of cutin, and the lipid material of the cutin penetrates and is intermingled with the outer polysaccharides of the wall. The waxes are a complex mixture containing long-chain alkanes, alcohols, ketones, and fatty acids. The alcohols and fatty acids occur in the form of long-chain esters and are also found uncombined. Some hydroxy fatty acids occur, and these are present as polymeric ester-linked compounds known as estolides (63).

Cutin is a complex mixture of polymeric fatty acids mainly interconnected by ester, peroxide, and ether linkages probably as a three-dimensional network. Hydroxy fatty acids usually predominate in the mixture so that the polymers are mainly estolides (63).

It has been suggested (37) from optical analysis of the outer layers of a cutinized wall that a distinct layer of wax occurs at the outside over a layer of cutin. The molecules of the cutin carry methyl, hydroxyl, and carboxylic acid groups, and it is thought that these are oriented within the cutin complex so that the hydrophobic methyl groups are directed towards the wax while the hydrophilic, hydroxyl, and carboxylic groups are directed towards the polysaccharide layers of the wall. The cutin penetrates the outer layers of the wall and firmly bonds the hydrophobic material to the polysaccharide wall.

Suberin, which occurs in barks and cork deposits, has a very similar composition to that of cutin (63). The barks form the outer tissues of stems and surround the wood. The cell walls of cork cells which are produced by the meristem of the periderm contain suberin, lignin, fats, and tannins.

THE CELL WALL AS A COMPOSITE STRUCTURE

The rigid cell wall of a lignified tracheid is complicated both in its chemical composition and in its ultrastructure, but because it resembles a glass fiber composite it can be analyzed mechanically in terms of the elastic constants of its two major constituents, microfibrils and lignin. This has been studied in detail by Mark and his colleagues (59–61). They have determined the levels of stress reached in the various components of the unaltered wall of the tracheids in the late springwood of *Juniperus virginiana* L. as it undergoes tensile loading to fracture. For the stress analysis it is assumed that (*a*) the cell wall is elastic, that is, it obeys Hooke's law; (*b*) a layer of the cell wall is mechanically homogeneous; and (*c*) a layer of the wall is orthotropic, that is, its elastic and strength properties can be resolved along two perpendicular axes. The calculations need values for the elastic constants of the cell wall constituents which may be theoretically calculated from the molecular structure of the constituents or may be measured on the purified materials. In order to apply the theories of stress analysis which have been developed to study composite materials it is necessary to measure the cross-sectional area of the wall perpendicular to the direction of the applied tension, the proportion of this area assignable to the various layers of the wall, the proportion of these layers assignable to the microfibrils and matrix component, and the orientation of the microfibrils with respect to the tracheid axes within each wall layer. With these data it is possible to calculate the levels of stress reached in each component in each cell wall layer, and if these values are compared with the calculated theoretical strength of the components, that is the crystalline cellulose microfibrils and the lignin (matrix), then the site of the first fracture in the tracheid wall under a given load can be predicted. This is thought to be a shear fracture in the outer (S_1) layer of the secondary thickening. A microscopic examination of the experimentally fractured tissue indicates that this is the site of the initial fracture of the cells (60). The method therefore does demonstrate in a very precise way the analogy between the rigid tracheid wall and a glass fiber composite.

However, in the growing cell wall the application of the analogy is much more difficult. At the early stages of growth the constituents of the matrix cannot be equated with a single substance, the wall components can slip and move relative to one another under an applied stress, and the stress transfer between the microfibril and the matrix can vary with changes in the adhesion between the components. At the stage of development of the primary wall when the matrix is much less rigid than in the secondary lignified wall and is in all probability a fluid plastic structure, the load is transmitted to the microfibrils by the viscous drag of the plastic deformation of the matrix. This can alter very much with the composition and physical state of the matrix materials, especially with that of the pectin complex and with the amount of water present.

SITES OF SYNTHESIS OF THE CELL WALL COMPONENTS

During growth and development the main feature of the wall that changes is the composition of the matrix. The microfibrillar reinforcement is maintained as a constant chemical and morphological component of the wall although the geometry of its deposition alters. Thus by controlling the synthesis and packing of the various matrix polysaccharides and the amount and orientation of the microfibrils the cell is able to bring about changes which range from the plastic flexible growing wall of the cambial tissue to the rigid thick wall of the xylem tracheid. This is achieved by two related but different mechanisms for the incorporation of the components into the wall during its growth.

The pectins and hemicellulose polysaccharides of the matrix are initially synthesized so that the polymerization of the monosaccharides takes place within the membrane complex of the Golgi body and its associated vesicles (41, 75, 76). The membrane-bounded material within the vesicles is moved through the cytoplasm and across the plasmalemma by reverse pinocytosis and the polysaccharides are packed into the wall. This process involves incorporation of the membrane of the vesicles into the plasmalemma (73, 78, 90).

The cellulose microfibrils, on the other hand, are probably spun out into the wall by an enzymic complex which is at the plasmalemma surface and which may be mobile (73).

Both types of polymer are therefore synthesized in association with membranes, and in addition the membranes are part of the same transport and membrane-synthesizing system. This is made up by the physical flow of material from the endoplasmic reticulum → vesicles → Golgi body → vesicles → plasmalemma (76–78). Thus the enzymic complexes for the synthesis of both the matrix polysaccharides and the cellulose are held within or on the same membrane system of the cell, but the amount of polymerization that occurs before the membrane is incorporated as the plasmalemma is different for the two phases. In higher plants most of the synthesis of the hemicelluloses and pectins occurs within the cytoplasm, but very little synthesis of the cellulose occurs until the plasmalemma stage. In some algal species it is possible for a particular type of cellulose that is found in extracellular scales (57) to be synthesized to some extent within the Golgi body—vesicle part of the system before it forms part of the plasmalemma (21).

At the final stage of wall formation lignification occurs. This takes place by a polymerization process which is in a large part nonenzymic and involves the joining together of free radicals. The free radicals are formed by an enzymic dehydrogenation of monomers such as coniferyl alcohol (35, 36), and the very short half-life of these radicals makes it clear that their polymerization occurs within the matrix of the wall at the site of lignification. The monomers are probably transported into the matrix of the secondary thickened wall from other cells in the actively growing and differentiating regions of the tissue (34, 36, 91). The polymerization to the three-dimensional cage-like lignin polymer can be compared to the curing of the resin in a glass fiber composite and is the final process in the formation of the cell wall as a living growing structural material.

LITERATURE CITED

1. Aspinall, G. O. 1959. *Advan. Carbohyd. Chem.* 14:429–68
2. Aspinall, G. O. 1964. Chimie et biochimie de la lignine de la cellulose et des hemicelluloses. *Acta Symp. Int. Grenoble*, 421–33
3. Aspinall, G. O. 1967. *Pure Appl. Chem.* 14:43–55
4. Aspinall, G. O., Begbie, R., McKay, J. E. 1962. *J. Chem. Soc.* 214–19
5. Aspinall, G. O., Cottrell, I. W. 1970. *Can. J. Chem.* 48:1283–89
6. Ibid 1971. 49:1019–22
7. Aspinall, G. O., Cottrell, I. W., Egan, S. V., Morrison, I. M., Whyte, J. N. C. 1967. *J. Chem. Soc. C* 1071–80
8. Aspinall, G. O., Cottrell, I. W., Molloy, J. A., Uddin, M. 1970. *Can. J. Chem.* 48:1290–95
9. Aspinall, G. O., Craig, J. W. T., Whyte, J. L. 1968. *Carbohyd. Res.* 7:442–52
10. Aspinall, G. O., Fairweather, R. M., Wood, T. M. 1968. *J. Chem. Soc. C* 2174–79
11. Aspinall, G. O., Gestetner, B., Molloy, J. A., Uddin, M. 1968. *J. Chem. Soc. C* 2554–59
12. Aspinall, G. O., Hunt, K., Morrison, I. M. 1967. *J. Chem. Soc. C* 1080–86
13. Aspinall, G. O., Molloy, J. A., Craig, J. W. T. 1969. *Can. J. Chem.* 47:1063–70
14. Atkins, E. D. T., Parker, K. D., Preston, R. D. 1969. *Proc. Roy. Soc. London, Ser. B* 173:209–21
15. Aston, W. P., Donald, A. S. R., Morgan, W. T. J. 1968. *Biochem. Biophys. Res. Commun.* 30:1–6
16. Barrett, A. J., Northcote, D. H. 1965. *Biochem. J.* 94:617–27
17. Bouveng, H. O. 1959. *Acta Chem. Scand.* 13:1869–76
18. Ibid 1961. 15:87–95
19. Ibid, 96–100
20. Bouveng, H. O., Lindberg, B. 1958. *Acta Chem. Scand.* 12:1977–84
21. Brown, R. M. 1969. *J. Cell Biol.* 41:109–23
22. Brownell, H. H. 1971. *Tappi* 54:66–71
23. Buchala, A. J., Wilkie, K. C. B. 1970. *Naturwissenschaften* 57:496
24. Clayton, D. W., Phelps, G. R. 1965. *J. Polym. Sci.* 11:197–220
25. Colvin, J. R. 1963. *J. Cell Biol.* 17:105–9
26. Cottrell, A. H., Kelly, A. 1966. *Endeavour* 25:27–32
27. Dougall, D. K., Shimbayashi, K.

1960. *Plant Physiol.* 35:396–404
28. Esau, K., Cheadle, V. I., Gill, R. H. 1966. *Am. J. Bot.* 53:756–64
29. Ibid, 765–71
30. Fraser, C. G., Wilkie, K. C. B. 1971. *Phytochemistry* 10:1539–42
31. Frei, E., Preston, R. D. 1961. *Nature* 183:1152–55
32. Frei, E., Preston, R. D. 1964. *Proc. Roy. Soc. London Ser. B* 160:293–313; 314–27
33. Freudenberg, K. 1932. *J. Chem. Educ.* 9:1171–80
34. Freudenberg, K. 1959. *Nature* 183:1152–55
35. Freudenberg, K. 1965. *Science* 148:595–600
36. Freudenberg, K. 1968. Constitution and biosynthesis of lignin. *Molecular Biology, Biochemistry and Biophysics*, ed. A. Kleinzeller et al, 2:45–122. Berlin: Springer
37. Frey-Wyssling, A. 1948. *Submicroscopic Morphology of Protoplasm and its Derivatives*. Amsterdam, London: Elsevier
38. Gibbons, R. A. 1963. *Nature* 200:665–66
39. Goring, D. A. I., Timell, T. E. 1962. *Tappi* 45:454–60
40. Holliday, L., Ed. 1966. *Composite Materials*, 1–27. Amsterdam, New York: Elsevier
41. Harris, P. J., Northcote, D. H. 1971. *Biochim. Biophys. Acta* 237:56–64
42. Heath, M. F., Northcote, D. H. 1971. *Biochem. J.* 125:953–61
43. Hepler, P. K., Fosket, D. E., Newcomb, E. H. 1970. *Am. J. Bot.* 57:85–96
44. Heyn, A. N. J. 1966. *J. Cell Biol.* 29:181–97
45. Heyn, A. N. J. 1968. *J. Ultrastruct. Res.* 26:52–68
46. Hibbert, H. 1942. *Ann. Rev. Biochem.* 11:183–202
47. Hirst, E. L., Jones, J. K. N. 1947. *J. Chem. Soc.* 1221–25
48. Hirst, E. L., Jones, J. K. N., Walder, W. O. 1947. *J. Chem. Soc.* 1225–29
49. Kelly, A. 1966. *Strong Solids.* Oxford: Clarendon
50. Lamport, D. T. A. 1969. *Am. Chem. Soc. 158th Meet., N.Y.,* Abstr. 47

51. Lamport, D. T. A. 1970. *Ann. Rev. Plant Physiol.* 21:235–70
52. Lamport, D. T. A., Northcote, D. H. 1960. *Nature* 188:665–66
53. Liang, C. Y., Marchessault, R. H. 1959. *J. Polym. Sci.* 37:385–95
54. Loewenstein, K. L. See Ref. 40, 129–218
55. Luce, J. E., Robertson, A. A. 1961. *J. Polym. Sci.* 51:317–35
56. Manley, R. St. J. 1964. *Nature* 204:1155–57
57. Manton, I. 1967. *J. Cell Sci.* 2:411–18
58. Marchessault, R. H., Sarko, A. 1967. *Advan. Carbohyd. Chem.* 22:421–82
59. Mark, R. 1965. *Cellular Ultrastructure of Woody Plants*, ed. W. A. Côté, 493–533. Syracuse Univ. Press
60. Mark, R. E. 1967. *Cell Wall Mechanics of Tracheids*. New Haven, London: Yale Univ. Press
61. Mark, R. E., Kaloni, P. N., Tang, R. C., Gillis, P. P. 1969. *Text. Res. J.* 39:203–12
62. Mark, R. E., Kaloni, P. N., Tang, R. C., Gillis, P. P. 1969. *Science* 164:72–73
63. Martin, J. T., Juniper, B. E. 1970. *The Cuticles of Plants.* U.K.: Arnold
64. Marx-Figini, M., Schulz, G. V. 1963. *Makromol. Chem.* 62:49–65
65. Marx-Figini, M., Schulz, G. V. 1966. *Biochim. Biophys. Acta* 112:81–101
66. Merewether, J. W. T. 1959. *Holzforschung* 11:65–80
67. Meyer, K. H., Misch, L. 1937. *Helv. Chim. Acta* 20:232–44
68. Mühlethaler, K. 1967. *Ann. Rev. Plant Physiol.* 18:1–24
69. Nieduszynski, I., Marchessault, R. H. 1971. *Nature* 232:46–47
70. Nieduszynski, I., Preston, R. D. 1970. *Nature* 225:273–74
71. Northcote, D. H. 1953. *Biol. Rev. Cambridge Phil. Soc.* 33:53–102
72. Northcote, D. H. 1963. *Int. Rev. Cytol.* 14:223–65
73. Northcote, D. H. 1969. *Essays Biochem.* 5:89–137
74. Northcote, D. H. 1969. *Symp. Soc. Gen. Microbiol.* 19:333–49
75. Northcote, D. H. 1969. *Proc. Roy. Soc. Ser. B* 173:21–30

76. Northcote, D. H. 1970. *Endeavour* 30:26–33
77. Northcote, D. H. 1971. *Symp. Soc. Exp. Biol.* 25:51–69
78. Northcote, D. H. 1972. *Cell Biology in Medicine,* ed. E. E. Bittar. New York: Wiley
79. Palmer, K. J., Hartzog, M. B. 1945. *J. Am. Chem. Soc.* 67: 2122–27
80. Pew, J. C., Weyna, P. 1962. *Tappi* 45:247–56
81. Preston, R. D. 1971. *J. Microsc.* 93:7–13
82. Preston, R. D., Cronshaw, J. 1958. *Nature* 181:248–50
83. Rees, D. A. 1969. *Advan. Carbohyd. Chem.* 24:267–332
84. Rees, D. A., Scott, W. E. 1969. *Chem. Commun.* 1037–38
85. Rees, D. A., Scott, W. E. 1971. *J. Chem. Soc. B* 469–79
86. Rees, D. A., Wight, A. W. 1972. *J. Chem. Soc.* In press
87. Reid, J. S. G., Wilkie, K. C. B. 1969. *Phytochemistry* 8:2045–51
88. Ibid, 2053–58
89. Ibid, 2059–65
90. Roberts, K., Northcote, D. H. 1970. *J. Cell Sci.* 6:299–321
91. Rubery, P. H., Northcote, D. H. 1968. *Nature* 219:1230–34
92. Rubery, P. H., Northcote, D. H.
1970. *Biochim. Biophys. Acta* 222:95–108
93. Rydholm, S. A. 1965. *Pulping Processes.* New York: Interscience
94. Spiro, R. G. 1970. *Ann. Rev. Biochem.* 39:599–638
95. Stoddart, R. W., Barrett, A. J., Northcote, D. H. 1967. *Biochem. J.* 102:194–204
96. Stoddart, R. W., Northcote, D. H. 1967. *Biochem. J.* 105:45–59
97. Sundararajan, P. R., Rao, V. S. R. 1969. *Biopolymers* 8:305–12
98. Ibid 1970. 9:1239–47
99. Timell, T. E. 1964. *Advan. Carbohyd. Chem.* 19:247–302
100. Ibid 1965. 20:409–83
101. Wang, P. Y., Bolker, H. I., Purves, C. B. 1967. *Tappi* 50: 123–24
102. Watson, A. J., Stewart, C. M., Dadswell, H. E. 1956. *Tappi* 39: 318–21
103. White, E. V. 1942. *J. Am. Chem. Soc.* 64:1507–11
104. Ibid, 2838–42
105. Wooding, F. B. P., Northcote, D. H. 1964. *J. Cell Biol.* 23:327–37
106. Yosizawa, Z., Sato, T., Schmid, K. 1966. *Biochim. Biophys. Acta* 121:417–20

Ann. Rev. Plant Physiol. 1972. 23:133–56

LIGHT AND ENZYMES[1] 7526

MILTON ZUCKER

*Department of Agricultural Chemistry, Washington State University
Pullman, Washington*

CONTENTS

This review deals with the photoregulation of enzyme synthesis in plants. Photosynthesis will not be treated directly, although the photocontrol of chloroplast development and the light activation of photosynthetic enzymes are to be discussed. The chapter is presented in the spirit of a workshop on photoregulation of enzymes rather than as a storage bin of information. Therefore, background literature has not been cited in many cases. Ample refer-

[1] Scientific Paper No. 3785, College of Agriculture, Washington State University, Pullman, Washington 99163.

ence to previous studies usually has been provided in the more recent works discussed below.

There are good reasons, theoretical and agricultural, for examining the photocontrol of enzymes. First, photoregulation exists and is an important control mechanism in plant growth (135). Secondly, the ability to trigger a whole series of specific biochemical events by the gentle procedure of shining visible light into the cell provides a powerful biological tool for investigation of biochemical regulation in living organisms. Agriculturally speaking, the development of laser technology offers a new approach to the regulation of crop growth. Heretofore, consideration of effects of light on agricultural productivity has been limited almost exclusively to the relationship between photosynthetic rates and crop yield (see 127). However, light may also exert profound effects on crop yield and quality by regulating other biochemical processes including enzyme synthesis. That photomorphogenic control can be affected experimentally on a field scale has already been demonstrated (146). To take full advantage of this new technology in terms of agricultural application and of breeding crops with improved responses to both natural and artificial illumination, we must become better acquainted with the mechanisms whereby the photoregulation of enzymes in plants is affected. It is to such a goal that this chapter speaks.

The pervading importance of light-regulated development in plants (photormorphogenesis) must in part be related to the fact that plant tissues possess a photosynthetic capacity and can therefore respond biochemically to light in an instantaneous and very dramatic way. This fact, that profound alterations in the metabolism of the plant cell can be occassioned simply by turning on the light, is an oft-repeated truism. Nevertheless, photosynthetic capacity is a basic characteristic of plant cells and establishes a fundamental need for the biochemical regulation of responses to light. Not only must the plant be able to maintain a smooth, coordinated development in the face of dramatic diurnal fluctuations in its metabolism, but the emergence into light of etiolated seedlings from the soil or of etiolated leaflets from the bud requires coordination in development to allow for an orderly transformation from a nonphotosynthetic to a photosynthetic existence. Furthermore, the ability of plants to perceive the changing seasons by responding to changes in daylength is an obvious evolutionary advantage. At one time photomorphogenesis could be considered only in terms of morphological reactions. However, it is becoming more feasible to examine photoregulation of development in terms of the photocontrol of enzyme synthesis and activation (135).

EVOLUTIONARY CONSIDERATIONS OF PHOTOREGULATION

Aside from chlorophyll, the two well-characterized photomorphogenic receptors in plants are the closely related protochlorophyllide holochrome and another tetrapyrrole, phytochrome (see 122 for a summary of recent structural work). The cytochromes, which are porphyrins as well, have been implicated in the blue light inhibition of respiration in *Prototheca* (51). Flavins also may be involved in blue light effects (125, 172). Although carotenoids and other model compounds (see below) have been suggested as photoreceptors, tetrapyrrole derivatives dominate the picture and attest to an evolutionary relationship between photosynthesis and photoregulation of morphogenesis.

Some interesting patterns emerge from a consideration of the photomorphogenic roles of phytochrome and protochlorophyllide holochrome among the phyla of plants (171). In angiosperms, chloroplast development involves both phytochrome and protochlorophyllide holochrome as photoreceptors. Gymnosperms, on the other hand, are reported to develop chloroplasts in complete darkness even though they contain both phytochrome and protochlorophyll and show a variety of photomorphogenic responses. Green algae, like gymnosperms, develop chloroplasts in darkness with the exception of a few light-sensitive mutants such as those of *Chlorella* (9). Although phytochrome pigments have been isolated from green algae (78, 184), only very limited phytochrome responses have been observed, i.e. rotation of chloroplasts (78). *Euglena* apparently contains no phytochrome (171), but chloroplast development in this organism does involve photoregulation by a protochlorophyllide holochrome photoreceptor (90).

A very interesting report by Scheibe (168) that a phytochrome-like pigment has been isolated from the blue-green alga *Tholupothryx* adds significantly to our knowledge of phytochrome distribution. The pigment from the blue-green alga shows a reversible green-red shift instead of the familiar red-far red shift associated with phytochrome from higher plants. This new phytochrome-like pigment appears to control the phenomenon of chromatic adaptation in blue-green algae (59). Under certain conditions some taxa of blue-greens can be made to synthesize either phycocyanin or phycoerythrin in the dark, depending on the color of light (green or red) last seen by the organism. The photoreversible effects of green and red light on *Nostoc* morphology (see 168) may be another aspect of photoregulation by this pigment. Thus the phytochrome-like pigment in blue-green algae appears to govern a much broader category of responses than that found in green algae. In this respect, the blue-green algae are closer to higher plants. The proposal that higher plant chloroplasts were derived from blue-green algae is supported by an imposing list of biochemical relationships (31). Quite possibly, further comparison of photoregulation in this group of algae and in higher plants may add other correlations.

MODEL SYSTEMS OF PHOTOREGULATION

Most model systems are composed of enzymes that can be photosensitized by light-absorbing chromophores. Where the chromophores occur naturally the systems have some potential for participation in photoregulated events in vivo. For example, pyridoxal phosphate, which has an absorption in the near uv, can bind to the epsilon amino group of lysine in proteins and can serve as a chromophore in the photosensitization of specific histidine residues in spinach leaf aldolase and other enzymes (35). The photoactivation of a purified bacterial urocanase by near uv has recently been reported (95). Apparently no chromophore absorption could be detected in the purified enzyme, but difference spectra between active and inactive enzyme were not presented. Perhaps photosensitivity may involve light absorption by amino acid residues activated by configurational strains imposed on them by the conformation of the enzyme (41). Participation of an FMN-containing amino acid oxidase in the blue light stimulation of respiration is another example of a natural model system (125).

A recently described system in frog photoreceptors (17, 131) provides a model much more analogous to that proposed for the phytochrome control of membrane permeability in plants (81). In the frog photoreceptor system, adenyl cyclase serves as a link between photon capture by rhodopsin and changes in membrane permeability. The light-induced *cis-trans* isomerization and bleaching of rhodopsin produce conformational changes in the molecule which inactivate the closely associated adenyl cyclase. Since the latter enzyme is involved in ion-induced changes in membrane potential, the system provides a model for photoregulation of membrane permeability. As such, this system holds great interest for proposed models of phytochrome regulation of permeability.

A model of photoregulation governing synthesis rather than activity of an enzyme has also been described. The furocoumarin psoralens are activated and become DNA intercalating agents when irradiated with long wave uv light, and as such they induce synthesis of PAL in irradiated pea pods (65).

PHOTOREGULATION OF CHLOROPLAST DEVELOPMENT

Protochlorophyllide Holochrome Transformation

In etioplasts, protochlorophyllide holochrome is known to be a membrane component of the crystalline prolamellar bodies (104, 114). Irradiation of the protochlorophyllide holochrome initiates a complex chain of events which lead to profound changes in ultrastructure (82, 83, 115, 147). Studies of etioplast development thus can provide a detailed model of biochemical events associated with photocontrol of development. Since photoreduction of protochlorophyllide holochrome is almost instantaneous, only millisecond flashes of light are needed to trigger these changes (105). Subsequent to the initial rapid shift in absorption spectrum signaling the photoconversion of protochlorophyllide to chlorophyllide holochrome, a further series of less rapid shifts in the chlorophyllide holochrome absorption occur (see 114). These slower shifts in absorption can be correlated with a series of ultrastuructural changes in the membranes of the prolamellar body (82, 83). Changes are observed within 2 minutes of the initial period of illumination in barley etioplasts but continue for some time thereafter. The time period is thus quite analogous to that for rapid phytochrome responses (55, 100, 167).

The in vitro photoconversion of protochlorophyllide holochrome to chlorophyllide holochrome is accompanied by a series of changes in absorption spectra (173) analogous to those observed in vivo. Recently changes have been detected in vitro in circular dichroism (CD) at 590 nm of the initial conversion product (56). These changes in optical rotation of chromophore groups occur simultaneously with the slow shifts in absorption of the chlorophyllide. The concomitant changes in spectra and circular dichroism have suggested that conformational changes of the holochrome occur as a result of photoreduction. Analogous changes in optical rotation have been reported for phytochrome (92, 122). The chlorophyllide holochrome initially produced on photoconversion of the protochlorophyllide appears to be in an unstable conformational state. The series of simultaneous changes in spectrum and CD observed represent a "conformational relaxation" of the holochrome to a more stable state (56). Such a series of light-induced conformational changes in a membrane component have been invoked as the initial reactions in the light-induced dispersion of the crystalline structure of the prolamellar body (82).

Although this sequence of events provides a biochemical picture of light-induced changes in ultrastructure, much in the way of detail remains to be uncovered. The photoactive protochlorophyllide protein unit itself may be a complex structure containing at least four chromophores per holochrome molecule (106) in some plants but not in others (84). Further complications involve the functioning of the holochrome protein as a recycling carrier of protochlorophyllide in its photoreduction (60, 183).

Binding of protochlorophyllide holochrome into the membrane structure is another crucial aspect of the photocontrol of plastid development. Plastid mutants in barley such as those studied by von Wettstein and his group (196) provide some evidence for the importance of other membrane proteins. Analysis of one barley mutant, infrared-5, has shown that it requires ten times the energy needed for phoroconversion of protochlorophyllide holochrome in normal dark-grown plants. Yet the protochlorophyllide holochrome isolated from the mutant behaves quite normally in vitro (56). These results, as well as the presence of abnormal membrane structures in the mutant plastids, suggest that the mutation has affected the binding of the holochrome to the membranes of the prolamellar body rather than altered the pigment-protein complex itself. The availability of regulatory and structural mutants in barley should aid materially in the elucidation of the mechanism of photoregulation involving protochlorophyllide holochrome as a photoreceptor.

Photoregulation of Lag Phase Enzymes

Subsequent to the initial photoconversion of protochlorphyllide holochrome there is a lag before a rapid or linear synthesis of chlorophyll and further development of chloroplast structure and photosynthetic competence occur. Analyses of these events suggest that at least two types of photocontrol are involved (19). One governs the synthesis of specific proteins during the lag phase and involves low intensity illumination of brief duration. The other requires high intensity irradiation of long duration for continued synthesis of chloroplast proteins (not necessarily synthesized in chloroplasts) after the lag ends.

Enzymes of chlorophyll synthesis.—Recent studies of the lag in *Euglena* have shed light on events which occur during this period (90, 171). Brief illumination of dark-grown *Euglena* 12 hr prior to placing the culture in continuous light completely eliminates the lag in chlorophyll synthesis (potentiation). During the 12 hr period of darkness following brief ilumination, synthesis occurs of proteins which are required for subsequent chlorophyll formation (174) and development of thylakoid membranes (13, 114). Cycloheximide and chloramphenicol can inhibit synthesis of some of these proteins during the period of potentiation (13, 114, 119). Since cycloheximide inhibits synthesis of proteins required for chloroplast development, potentiation appears to involve the participation of cytoplasmic as well as chloroplast ribosomes. The apparent synthesis of chloroplast enzymes on cytoplasmic ribosomes is particularly evident in the case of NADP-glyceraldehyde-3-phosphate dehydrogenase (171). Streptomycin bleaching in *Euglena* inhibits synthesis of many chloroplast enzymes involved in photosynthetic fixation of carbon dioxide (171). However, it does not inhibit synthesis of the NADP-linked triose phosphate dehydrogenase which is also a chloroplast enzyme. Therefore Schiff (171) concludes that this enzyme must be synthesized on cytoplasmic ribosomes and is subsequently incorporated into the chloroplast.

The action spectrum for photocontrol of potentiation in *Euglena* is that of protochlorophyllide (90). However, the probable involvement of cytoplasmic ribosomes and the observation of light-induced changes in *Euglena* mutant containing no detectable chlorophyll (37, 171) raise the possibility that another photoreceptor exists.

In higher plants, photocontrol of enzyme synthesis during the lag phase appears to involve phytochrome rather than protochlorophyllide holochrome as the photoreceptor (116). The lag in rapid chlorophyll synthesis, as well as the inhibition of greening by cycloheximide or chloramphenicol, can be overcome by feeding δ-aminolevulinic acid (ALA) to etiolated seedlings before they are exposed to light (139). Therefore the lag in rapid chlorophyll production may involve the phytochrome-controlled synthesis of enzymes necessary for ALA formation (60, 139). Synthesis (cycloheximide sensitive) of succinyl CoA synthetase is induced by irradiating etiolated bean leaves briefly with red or blue light (180), suggesting phytochrome as a possible photoreceptor in this system. Since this enzyme may be involved in ALA formation, its synthesis is of particular interest in photoregulation of chloroplast development. Synthesis of δ-aminolevulinic acid dehydrase, on the other hand, required a long period of illumination (181) and therefore did not appear to be part of the lag phase.

Cycloheximide added to greening barley seedlings at the end of the lag phase stops linear chlorophyll synthesis within a few hours, suggesting that the phytochrome-induced enzymes of ALA formation have a rapid turnover (139). Recent simple but elegant experiments on synthesis of chlorophyll in etiolated barley seedlings exposed to light flashes suggest that the half-life of the phytochrome-induced enzymes is less than 45 min (183). Studies of the cycloheximide inhibition of greening in a light-requiring mutant of *Chlorella* also suggest that an enzyme with a very short half-life (about 30 min) limits ALA synthesis in darkness (9). Since the mutant is blocked at a step which is phytochrome controlled in higher plants, it would be of considerable interest to determine the action spectrum and possible reversibility of photoconversion of the algal receptor involved in the light requirement. Thus far phytochrome has been found to affect only chloroplast rotation in green algae (78).

Inhibitors of transcription have no effect in some higher plant tissues on synthesis of the lag phase enzymes with a short half-life (139). Therefore mRNA for such enzymes must be more stable than the enzymes themselves. Very long-lived but inactive mRNA of plant origin has been isolated from wheat seeds (26). Isolation of specific enzymes such as an ALA synthetase whose turnover is photoregulated will greatly aid further in understanding photocontrol of the lag phase in chloroplast development. It is possible that, as in some other light-sensitive induction systems (205, 206), photoregulation of the rate of inactivation rather than synthesis determines the level of the enzyme in the tissue.

Other lag phase enzymes.—Enzymes of chlorophyll formation are not the only chloroplast enzymes whose synthesis may be controlled by phytochrome during the lag phase. The synthesis of Calvin cycle enzymes such as ribulose diphosphate carboxylase (RUDP-carboxylase) and NADP-linked glyceraldehyde-3-phosphate dehydrogenase is phytochrome controlled and will occur in darkness after a brief period of illumination (36, 53, 54, 62, 118). Brief exposure to red light also stimulates synthesis of enzymes of the C_4-dicarboxylic acid pathway of photosynthesis in detached corn leaves (63). Since far red reversibility was not examined, phytochrome participation may be questioned, especially in view of the fact that brief exposure of chloroplasts to red light will produce dark effects such as recrystallization of prolamellar bodies which do not involve phytochrome (14). However, classical phytochrome effects have been demonstrated in the photoregulated, coordinant induction of two C_4 pathway enzymes in corn leaves, chloroplast adenylate kinase, and inorganic pyrophosphatase (24, 175).

Perhaps the most intensively studied of the phytochrome-induced chloroplast enzymes is RUDP carboxylase. Its synthesis has been unequivocally demonstrated by immunological techniques (120) as well as by incorporation studies with ^{14}C amino acids (120, 148, 188). However, attempts to localize its site of synthesis have produced a complex picture. In an organism like *Chlamydomonas*, inhibitor studies suggested that synthesis of RUDP carboxylase involves both chloroplast and cytoplasmic ribosomes as well as nuclear and chloroplast DNA (5, 177). Since RUDP carboxylase is such a large enzyme composed of two subunits that are not formed synchronously (112), the complexity of its synthesis should not be too surprising (5). The phytochrome controlled step(s) in synthesis of the enzyme remains to be determined.

Photoinduction of RUDP carboxylase in etiolated barley is biphasic; a decrease in the level of enzyme activity occurs after the initial induction, followed thereafter by a steady increase (107). The loss of extractable enzyme may represent an incorporation of RUDP carboxylase into an insoluble chloroplast structure (107) or an initial overshoot in the rate of turnover. The level of the carboxylase diminishes in mature leaves if they are transferred from high to low light intensity during growth (18). This sun-shade effect is also noted for enzymes of the C_4 dicarboxylic acid pathway in several plants (76). As in the case of biphasic induction, the photoregulation of enzyme levels produced by changes in light intensity could involve incorporation of enzymes into insoluble membrane structures or could represent adjustments in rates of enzyme synthesis and inactivation.

Light activation of enzymes.—Another type of photocontrol of chloroplast enzymes involves light activation of enzymes rather than regulation of enzyme synthesis. Factors have been isolated from chloroplasts which when added back produce stimulation of photosynthesis at wavelengths not normally effective (61). One light-activating factor (LAF) was shown to photosensitize RUDP carboxylase in vitro (4, 197). The maximum photoactivation was originally attained at 325 nm, but recent reports (120) suggest that this photoactivation may involve photosynthetic reactions in the visible range. A number of other chloroplast enzymes can be light activated (see 91, 150). These include enzymes of the C_4 dicarboxylic acid pathway such as phosphoenol pyruvate (PEP) synthetase and PEP carboxylase (176), pyruvate dikinase (75), and NADP linked malic dehydrogenase in chloroplasts (103), as well as other Calvin cycle enzymes such as ribulose-5-phosphate kinase (123, 124) and NADP linked triosphosphate dehydrogenase (200). All of these enzymes, including RUDP carboxylase, are sulfhydryl enzymes and can be activated in vitro by light in the presence of chloroplasts (91). Some are activated in darkness by reagents that reduce disulfides. The mechanism of activation appears to involve reduction of the "photosensitive" enzymes by a disulfide reductase (77) in the presence of photosynthetically generated NADH or NADPH (91). The reductase is O_2 and arsenite sensitive. Probably the inhibition of light activation by O_2 is the basis of the Warburg effect in photosynthesis (124). Light activation of a chloroplast peroxidase has also been reported (99), but the mechanism of this activation is not known.

PHYTOCHROME MEDIATED ENZYME SYNTHESIS

The effects of transcription inhibitors on anthocyanin synthesis coupled with a variety of biological observations led Mohr (134, 135) to propose a theory of phytochrome regulation involving gene activation. Subsequently, the discovery that phytochrome controlled the synthesis of phenylalanine ammonia-lyase (PAL) and hence anthocyanin pigmentation in mustard seedlings (40) provided more direct evidence for the theory of gene activation. Since then Mohr and his colleagues have shown that phytochrome also controls the synthesis of α-amylase (38, 137), lipoxidase (109, 142), glycolic acid oxidase, and glyoxylate reductase (189) in mustard seedlings. Synthesis of lipoxidase is inhibited rather than stimulated by phytochrome and thus requires that the gene activation hypothesis be extended to include gene repression. The phytochrome-induced disappearance of PAL from jerusalem artichoke (39) suggests another possible explanation for the far red effect on lipoxidase that does not involve repression. That is, phytochrome may induce the synthesis of a lipoxidase inactivating system which maintains lipoxidase activity at a low level in the light. Several researchers have attempted to gain further evidence for the gene activation theory by demonstrating that phytochrome regulates the synthesis of nucleic acids (144) and of RNA polymerases (21).

These studies of phytochrome-regulated changes in enzyme activity (see section on chloroplast development for other examples), coupled with the use of inhibitors of transcription and translation, suggest strongly that synthesis of specific enzymes can be regulated by phytochrome. Such regulation could well involve gene activation.

Another theory of phytochrome action, promulgated by Hendricks and Borthwick (81), is based on rapid responses of plant cells to phytochrome photoconversion (55, 100, 167). These responses usually occur within a few minutes of exposure to irradiation but may be as fast as 10 seconds (140). They are probably too fast to allow for processes of translation and transcription to intervene. Most rapid responses involve an ion flux (55, 100, 140, 167), a fact which has led to the theory that phytochrome is a membrane component whose conformational change resulting from light absorption alters the permeability of the membrane (81, 178). That conformational changes occur when phytochrome is converted by light absorption from one form to another has been demonstrated by immunochemical (92) and spectroscopic techniques (51, 122) involving circular dichroism and optical rotary dispersion measurements of purified phytochrome. Conversely, absorption of light at 280 nm (probably by the protein moiety) is sufficient to transform phytochrome from one form to the other (149).

The similarity between this proposed mechanism of phytochrome action and that of photochlorophyllide holochrome (see above) is striking. In both systems conformational changes produced by light absorption of membrane bound components initiate a chain of events. The frog photoreceptor system involving rhodopsin-linked adenyl cyclase (17, 131) is another example of a photosensitive membrane system. The membrane theory of phytochrome action is very attractive because it involves conformational changes in the photoreceptor which are analogous to those known to occur in other systems of photoregulation. Photoinduced conformational changes in phytochrome can also play a role in the theory of direct gene activation. Thus, a common mechanism but different sites of action of phytochrome are proposed. Mohr points out (135) that phytochrome very likely has more than one mode of action. Certainly there appear to be a number of different active forms of the pigment (11, 20, 52).

INDUCTION OF PHENYLALANINE AMMONIA-LYASE (PAL)

Phenolic biosynthesis in plants is influenced by a variety of environmental factors, particularly conditions of illumination. The discovery of the enzyme phenylalanine ammonia-lyase (PAL) in barley by Koukol & Conn (121) has opened the door to an understanding of the molecular basis of the effects of light and other factors on synthesis of phenolic compounds derived from phenylalanine. PAL catalyzes the direct deamination of L-phenylalanine to trans-cinnamic acid. The carbon skeleton of trans-cinnamic acid can be incorporated into a wide variety of phenolic compounds. PAL synthesis is induced in potato tubers by wounding. The stimulation of induced PAL synthesis by light which was first observed in the potato system (202) prompted reinvestigations of other systems involving photoregulation of phenolic synthesis (40). As a result, many reports of light-induced synthesis of PAL have appeared in the literature.

In most of the studies of PAL formation, evidence for synthesis of the enzyme has been obtained by use of inhibitors of protein synthesis such as cycloheximide (3, 32, 43, 64, 96, 133, 155, 169, 170, 190, 203, 204), puromycin (133, 156, 190), ethionine (16, 170, 202), and fluorophenylalanine (93, 190), or, in a few cases, inhibitors of transcription such as actinomycin D (96, 133, 190). Attempts to demonstrate PAL synthesis directly by labeling the protein with radioactive amino acids (204-206) (successful) or with heavy water (3) (unsuccessful) have been reported in only a few systems. Many of the PAL induction systems discussed here involve inactivation as well as synthesis of the enzyme. A temperature dependent activation of the enzyme in gherkin seedlings has also been described by Engelsma (46). Consequently it becomes increasingly important to obtain direct measurement of PAL synthesis using incorporation studies or immunological techniques.

Factors Controlling PAL Synthesis

Phytochrome controls the synthesis of PAL in many (3, 6, 11, 12, 39, 40, 49, 156, 169, 179, 195) but not all seedlings. Some seedlings show no light dependent PAL synthesis (3, 130, 191), or they may possess a blue light HER system (43, 141, 169). PAL synthesis is not necessarily under phytochrome control in all tissues of the seedling, the roots often being insensitive (3, 12). Tuber slices of jerusalem artichoke are the only nonseedling material thus far to show a phytochrome-regulated synthesis of PAL (39). In contrast to the etiolated seedling, PAL synthesis in green leaf tissue is often controlled by photosynthetic reactions (3, 32, 33, 43, 204). All of the PAL inductions dependent on photosynthesis also require exogenously supplied sugars. However, the photosynthetic requirement can be distinguished sharply from effects of carbohydrates in several of the systems (3, 204). Ultraviolet light (68) and gamma radiation (143, 154) have also been used to induce PAL synthesis.

Hormonal influences on PAL induction have been demonstrated in a few tissues. The effective hormones include gibberellin (28, 153), abscisin (191), ethylene (50, 96, 98, 155), and kinetin (158). Excision of the tissue is another requirement for PAL induction in some light-sensitive systems (32, 45, 141, 158, 202, 204). Wounding alone without excision can serve to induce PAL synthesis, but in these cases wounding followed by infection with fungal pathogens stimulates synthesis greatly (16, 57, 132).

Relation between PAL induction and phenolic biosynthesis.—The variety of factors that influence PAL induction usually stimulate production of phenolic compounds. In fact, in many induction systems, synthesis of PAL can be correlated with subsequent appearance in the tissue of phenolic compounds whose carbon skeleton is derived in whole or in part from phenylalanine. These components include lignin (28, 57, 86, 87, 158, 159, 199), flavonoids (3, 32, 33, 70, 128, 154, 169), and hydroxycinnamoyl conjugates (3, 43, 44, 47, 113, 132). The correlation between PAL synthesis and formation of isoflavonoid phytoalexins in legumes provides an interesting functional relationship. Phytoalexins are compounds of plant origin, usually phenolic in nature, that are toxic to fungi and are synthesized by plants in response to fungal invasion (34). In peas (64), beans (85), and soybean seedlings (16) synthesis of phenolic phytoalexins is dependent on PAL induction in the tissue. On the basis of his studies with intercalating agents, Hadwiger (67) has proposed that PAL induction in legumes is controlled at the transcriptional level. These studies of PAL induction have formed the basis of an ingenious biochemical model of gene for gene disease resistance in plants (66). Induction of PAL synthesis and subsequent lignification of the wounded surfaces has also been proposed as a defense mechanism in potato tubers (57, 207).

The correlation between PAL induction and phenolic biosynthesis cannot be established in all situations. Low temperature induction of PAL synthesis in gherkin seedlings does not produce a corresponding increase in phenolic production (48). At low temperature, it is possible that subsequent steps in the biosynthesis of phenolic compounds may limit their formation. Just the opposite situation is noted in *Eucalyptus* tissue where very high levels of phenolic compounds occur but only very low levels of PAL activity have been observed (88). Of course, the difficulty of extracting enzymes, even phenolic enzymes, from tissues rich in phenolic compounds is well known. Incorporation studies with radioactive precursors of phenolic compounds have been used to suggest dual pathways of phenolic biosynthesis that may not involve PAL (74). However, the fact that phenolic compounds are subject to extensive turnover (7, 8, 58) makes these labeling studies difficult to interpret. The rate of turnover of some phenolic compounds can be influenced by conditions of illumination that affect PAL synthesis (8). Thus enzymes of phenolic metabolism in general may be induced by light.

The PAL operon.—A new aspect of PAL induction which in part accounts for the close relation between PAL synthesis and phenolic production has come to light recently. The syntheses of several enzymes catalyzing subsequent steps in the pathway of phenolic biosynthesis are induced coordinantly with PAL. Cinnamic acid-4-hydroxylase (160) catalyzes the hydroxylation of *trans*-cinnamic acid at the 4-position yielding the first phenolic acid of the pathway, *p*-coumaric acid. This second enzyme of phenolic biosynthesis is induced simultaneously with PAL in buckwheat seedlings (1–3) and in parsley tissue culture (70) by light and in peas (97) by ethylene. The hydroxylase is microsomal while PAL is a soluble enzyme (3, 160). Nevertheless, their coordinant induction and a common sensitivity of synthesis to cycloheximide suggests that both are formed on cytoplasmic ribosomes, the hydroxylase being incorporated into the microsomal fraction subsequently. A third enzyme of phenolic biosynthesis, *p*-coumarate: CoA ligase (cinnamic acid:CoA synthetase), has been isolated recently from several plants (69, 192, 193). Hahlbrock & Grisebach (69) have demonstrated that the apparent synthesis of this enzyme in parsley, like that of PAL and the hydroxylase, is induced by light. These observations suggest that a light-induced phenolic operon consisting of structural genes for at least three enzymes exists in plant tissues. A subsequent enzyme in the pathway of flavonoid biosynthesis, the chalcone-flavanone isomerase (72), is not induced by light in buckwheat (3) under conditions that induce synthesis of PAL and the hydroxylase. Another light-sensitive enzyme involved in glycosylation of flavonoids, UDP-apiose synthetase, has also been isolated from parsley (71). It catalyzes one of the final steps in flavonoid biosynthesis. Its synthesis is induced under the same conditions of illumination required for PAL induction. However, the lag period of synthesis is much longer for the synthetase than for PAL. Consequently the two induction systems are well separated temporally (71).

Isolation and characterization of PAL.—Since the original purification from barley (121), PAL has been isolated and purified from sweet potato roots (133), oak (22), tobacco (145), *Xanthium* leaves (204), corn (130 and buckwheat seedlings (3), and potato tubers (79). PAL has also been purified almost to homogeneity from a yeast (89) and purified to a lesser extent from other fungi (15, 25, 182, 198) where the enzyme appears to serve a catabolic rather than biosynthetic role as in higher plants. The molecular weight of PAL isolated from plant tissues is about 300,000 (79, 130). Although reported once (79), the presence of an active form of PAL with a very large molecular weight on the order of 600,000 has not been observed in subsequent preparations from potatoes or from other plants.

Some, but not all, of the ammonia-lyases from higher plants show a sulfhydryl requirement for maximal activity (43, 121, 145). Unlike bacterial histidine ammonia-lyase, none of the PAL enzymes can be reversibly depolymerized by sulfhydryl reagents (117). However, kinetic studies of PAL isolated from potatoes (80), corn (130), and *Xanthium* leaves (unpublished observations) indicate that the enzyme contains an allosteric binding site for the substrate in addition to the active site. Dehydroalanine has been implicated as part of the active site in both the potato (73) and the yeast (89) enzyme.

Although depolymerization of PAL has not been achieved, several isozymes have been isolated by DEAE column chromatography of preparations from sweet potatoes (133) and from oak leaves (22). The oak leaf isozymes show quite different sensitivities to benzoic acid and cinnamic acid which are inhibitors of PAL in vitro (22). These studies, along with a report (94) that PAL preparations from radish show differential heat stabilities, are the only indications that pathways of phenolic biosynthesis leading to different end products may involve different PAL isozymes. However, many attempts to obtain isozymes of PAL have been unsuccessful. Even when synthesis of the enzyme is induced in buckwheat by two mutually exclusive processes such as illumination of whole seedlings and excision of hypocotyls in darkness, only one form of PAL can be detected by electrophoresis (3).

The Induction Process

Although PAL induction may involve any one of several different photoreceptors or none at all, most induction systems do have features in common. The induction process involves at least two of three overlapping aspects (47–49, 156, 195, 203, 204): (*a*) synthesis of PAL with or without a lag; (*b*) repression of PAL synthesis; and (*c*) inactivation of PAL. PAL synthesis and its eventual repression appear to occur in all of the induction processes described above, but inactivation of PAL is not a universal characteristic. In those light-sensitive systems where inactivation does not occur (3, 12, 39, 49), PAL induction must involve light-induced changes in the rate of PAL synthesis. Most evidence for induction of PAL synthesis by light rests on inhibition of changes in PAL activity by inhibitors of translation and transcription. Inducers of PAL synthesis have not been identified in higher plants, although stimulatory substances have been found in cotyledons (43). In fungi, PAL synthesis can be induced readily by its substrate, L-phenylalanine (15, 25, 182). Higher plant systems show a repression of PAL synthesis when phenylalanine is added to the tissue (43, 190, 202). The repression can be explained in part by the rapid accumulation of cinnamic acid and other phenolic repressors in the presence of excess substrate (44, 45).

The lag of several hours in PAL synthesis which usually occurs after induction (43, 136, 156), the sensitivity of some induction systems to actinomycin D, and the superinduction obtained with intercalating agents suggest that synthesis of mRNA is required for induction. The suggested requirement for mRNA synthesis and the possible existence of an operon in phenolic biosynthesis are consistent with the gene activation theory of phytochrome action (134, 135). However, so little is yet known of the initial events occurring during PAL induction in higher plants that little concrete support has been provided for any theory of photoregulation.

Repression of PAL synthesis.—The abrupt cessation of increase in PAL activity that occurs sometime after induction begins (3, 12, 43, 71, 179, 191, 202) first suggested that repression of PAL synthesis occurred. Since cinnamic acid (202) and subsequent phenolic products (44, 45) inhibited PAL synthesis, they were proposed as likely natural repressors that accumulated in the tissue as a result of PAL induction. The loss of light sensitivity after an initial burst of PAL synthesis in gherkin seedlings was ascribed by Engelsma (43-45) to the accumulation of these phenolic repressors in the tissue. Excision of the gherkin hypocotyls allowed the repressors to diffuse from the excised tissue and thus reestablish a light sensitivity (45). When the temperature of induction is lowered from 32° to 12° the rate of PAL synthesis decreases in gherkin seedlings and the length of time to reach maximum activity increases (44). The rates of accumulation of phenolic compounds is correspondingly lowered under these conditions. This inverse relationship established between length of the period of PAL synthesis and rate of phenolic accumulation in the tissues is that predicted if phenolic products at high enough concentrations can serve as repressors of PAL synthesis.

Inactivation of PAL.—The rapid decay of PAL activity which occurs as the final stage of PAL induction in some tissues (3, 6, 12, 43, 71, 154, 191, 202), or which can be induced in other tissues by changing conditions of illumination, usually transfer from light to dark (3, 39, 195, 204), suggests that inactivation is a part of the PAL induction process.

Inactivation of PAL can only be detected in some systems after the initial induction has produced a relatively high level of PAL in the tissue (195, 203). The appearance of an inactivating system during the course of PAL synthesis is thought to involve the induction of synthesis of a protein inactivator because inactivation of PAL can be prevented by cycloheximide (42, 43, 96, 204). Sequential induction of an enzyme and a protein inactivator is not limited to inducible PAL synthesis in plants. Inactivation of nitrate reductase in barley (186) and of a neutral phosphatase in *Euglena* (126) is inhibited by cycloheximide. In addition, synthesis of protein inactivators of invertase follows synthesis of the enzyme in potatoes (151) and in corn endosperm (102).

The mechanism of inactivation of PAL is not known. Engelsma (45) has obtained evidence that phenolic acids are required for PAL inactivation, in addition to a protein component. The rate of accumulation of phenolic acids in gherkin seedlings receiving various exposures to low temperature and light can be correlated directly with the rate of PAL inactivation (44). Engelsma (46) has also obtained evidence which suggests that the protein component of the inactivating system functions by complexing with the ammonia-lyase.

Once enzyme activity has decayed after induction, PAL can be reactivated by exposing the tissue to 4°C for 24 hr. The reappearance of PAL activity after the tissue is transferred back to 25°C does not require light nor is it sensitive to cycloheximide. Consequently Engelsma has suggested that the original inactivation of PAL involved combination with a protein inhibitor. Treatment at 4°C weakens the complex, and active PAL is subsequently released when the tissue is returned to warmer temperature. Recently Engelsma (50) reported that a protein inhibitor of PAL is found in the diffusate of excised gherkin hypocotyls. In tissue such as red cabbage which does not synthesize an inactivating system, cold temperature treatment has no effect on the level of PAL activity (49).

During rapid inactivation of PAL in darkness, radioactivity from ^{14}C-amino acids incorporated into the enzyme prior to inactivation is lost (204). The loss of both activity and radioactivity from the labeled PAL fraction in *Xanthium* leaf discs suggests that inactivation in this case may involve degradation or disaggregation of the enzyme, although complexing with an inhibitor protein cannot be ruled out as a mechanism of inactivation. If degradation of the enzyme did occur, it would require a specific inactivation because turnover of the enzyme in darkness is much faster than that of the bulk of the soluble leaf protein (206). PAL could be made particularly susceptible to proteolysis by combining with small effector molecules such as the phenolic compounds described by Engelsma (45). Specific proteolytic enzymes that attack PAL preferentially seem a less likely possibility. However, proteases specific for apo pyridine nucleotide enzymes (110) or apo pyridoxal enzymes (111) have recently been isolated from rat intestine. These surprisingly specific proteases were induced by niacin and B_6 deficiencies respectively.

The rapid decay in PAL activity observed in darkness has been explained by assuming that light-induced synthesis of the enzyme stops in darkness, leaving an inactivating system to remove the enzyme from the tissue (195, 204). However, direct measurement of PAL synthesis in *Xanthium* leaf discs made by following incorporation of radioactive amino acids into the enzyme has demonstrated that this hypothesis is not tenable (205, 206). Incorporation of ^{14}C-amino acids into PAL was as readily demonstrated in discs in the dark as in the light (205). Hence, synthesis of the enzyme does not stop in darkness. To account for the rapid decay in enzyme activity in the presence of continuing synthesis, we must assume that inactivation of PAL increases greatly in the dark. Pulse labeling experiments with ^{14}C-leucine and ^{14}C-arginine have verified this assumption (206). Enzyme in discs pulse labeled in the light lost little radioactivity during an ensuing chase period if the tissue was maintained in the light. However, a significant loss of radioactivity from the enzyme fraction occurred if discs were transferred to darkness during the chase. These experiments demonstrate that photoregulation of PAL synthesis does not occur, at least once the induction begins. Rather, the cycloheximide-sensitive inactivating system is affected by conditions of illumination, inactivation being low or absent in the light but increasing greatly in darkness. A reasonable explanation for these observations is that synthesis of the inactivator protein is repressed by a photosynthetic product. This type of repressor would account for the participation of chlorophyll as the photoreceptor in this light-sensitive system (204). As pointed out in the section on the chymotrypsin protein inhibitor below, repression of synthesis of specific proteins by photosynthate may be a general mechanism of photoregulation. Light appears to repress PAL synthesis itself in soybean callus cultures (158).

The importance of inactivation of induced enzymes in plant tissues has been reemphasized by Marcus (129) in a recent review of enzyme induction. Unlike bacteria, plants cannot divide rapidly and outgrow their induced enzyme activities but must destroy them. Although direct measurements of synthesis in complex induction systems of plant tissues are difficult to obtain and interpret, they are vital to an understanding of the photoregulation of enzymes. One light-sensitive induction system in plants that has afforded data of this type is described in the next section.

Light-Induced Synthesis of Protein Inhibitors of Chymotrypsin

Specific proteins which bind to and inhibit chymotrypsin are synthesized in various tissues of the potato plant and in other members of the Solanaceae (166). The chymotrypsin inhibitor (Inhibitor I) crystallized from juice of potato tubers is a small protein (mol wt = 39,000). Antibodies prepared against crystalline inhibitor I can be used to measure quantitatively the amount of immunologically detectable inhibitor I protein formed in potato, tomato, and tobacco leaves (166). Ryan and his colleagues have demonstrated unequivocally that a net increase in soluble chymotrypsin inhibitor protein occurs in potato (161, 162, 164) and tomato (162, 164) leaves when they are excised and placed in the light. If the excised leaves are subjected to a 12 hour light-dark cycle (162) or are treated with DCMU in continuous light (164), the amount of protein inhibitor extracted is significantly less. Leaves not excised from the plant show very little synthesis of inhibitor proteins. These characteristics of induction of specific inhibitor proteins, i.e. requirements for leaf excision and light and sensitivity to DCMU, are similar to those of the PAL induction system in mature green leaves (3, 33, 204).

The rapid disappearance of chymotrypsin inhibitor protein from meristematic tissues under certain conditions of growth (166) suggests that inactivation of the proteinase inhibitor occurs. Intact and excised tobacco leaves show a dramatic change in immunologically detected inhibitor protein when conditions of illumination are altered (163). Synthesis of the protein in this tissue is repressed by some product of photosynthesis (163, 165). Net increase in the content of soluble inhibitor protein can be demonstrated only in the white tissue of variegated leaves of a mutant strain (165) or in etiolated tissue of normally green plants placed in darkenss for at least 24 hours (163). If the darkened leaves are returned to light, inhibitor protein disappears rapidly from the tissue (163).

Comparison of photoregulation of proteinase inhibitor 1 and some enzymes.—Experiments with incorporation of labeled amino acids into inhibitor I protein under conditions of induction indicated that it represented 12% or more of the total newly synthesized protein in either potato (162) or tomato (164) leaves. Similar calculations for the light-induced synthesis of RUDP carboxylase in greening barley leaves indicate that it represents 70% of the newly synthesized protein (120). Estimates of PAL synthesis in *Xanthium* leaves suggest that it accounts for about 5% of the newly formed protein synthesized under conditions of induction (206). In each of these cases light induction appears to affect specifically the synthesis of a few individual proteins. The nonspecific stimulation of polysome formation by light (29, 185), and the resultant increase in protein synthesis in general, would not in itself be sufficient to account for these specific light effects.

Although the light requirements for initiation of rapid loss of a specific proteinase inhibitor from tobacco and of PAL from *Xanthium* leaves are opposite, a similar mechanism may be involved, i.e. repression of protein synthesis by a diffusible product of photosynthesis. In tobacco some photosynthate appears to repress the synthesis of the inhibitor protein itself (165), while in *Xanthium* repression of the synthesis of a PAL inactivating system in the light by photosynthate has been proposed to explain a rapid increase in rate of PAL inactivation on transfer to darkness (205, 206). An unidentified photosynthate has also been proposed as a repressor of RUDP-carboxylase in light synchronized cultures of *Chlorella* (138). These studies provide no information on the nature of the repression of the specific enzymes involved, but they do suggest a potentially important mechanism whereby cell development can be regulated by light.

INDUCTION OF NITRATE REDUCTASE

Nitrate reductase (NR) is an enzyme essential for the utilization of nitrate and hence fertilizers by plants (10). The level of NR in the growing plant is highly correlated with yield (23). Synthesis of the enzyme, which can be induced by nitrate, is light dependent in many green tissues (10). The light requirement appears to be one for photosynthesis because NR does not increase in leaves exposed to light in a CO_2-free atmosphere or in leaves treated with inhibitors of photosynthesis (108). The photosynthetic effect has been attributed to a light stimulation of uptake of nitrate and to a photosynthetic generation of reducing power (10). However, conditions of illumination have been found to affect the synthesis of NR during a subsequent period of darkness (27, 187). Since the level of nitrate is not affected by preillumination and since reducing power is obviously not generated photosynthetically in the dark, neither of these hypotheses can satisfactorily account for the light dependence of NR synthesis.

Polysome formation in leaves in known to show a diurnal fluctuation; the polysomes disaggregate in darkness (29). Induction of NR in corn leaves is closely correlated with light-induced development of cytoplasmic polysomes and is related to the maintenance of polysome structure with energy derived from photophorylation (185). Diurnal fluctuations in the level of NR (157) may reflect such a nonspecific effect of light on protein synthesis, particularly where the enzyme shows a rapid turnover in darkness (10). Opposing reactions of synthesis and inactivation may also be involved in the recently described circadian oscillation in NR activity which occurs in leaves of *Chenopodium rubrum* either in continuous light or darkness (30).

The rapid rate of loss of NR activity in corn seedlings transferred from light to darkness or in tissue treated with relatively low concentrations of cycloheximide indicates that the enzyme has a rapid turnover. In barley tissue the rapid inactivation of the enzyme can be inhibited by cycloheximide (186). Thus inactivation of NR, like that of PAL (43, 203), requires synthesis of a protein inactivator. NR induction systems are also similar to the PAL induction system in *Xanthium* leaf discs (204) in that synthesis of NR in tobacco cell culture continues under conditions where the enzyme is being inactivated rapidly (201). It is possible that a specific photoregulation of NR inactivation may yet be found.

PHOTOPERIODIC REGULATION AND CIRCADIAN RHYTHMS

Photoperiodic control of enzyme synthesis provides examples of day-length effects that do not involve phytochrome. The level of PAL in *Xanthium* leaves (204) and of proteinase inhibitor I in potato leaves (162) is regulated photoperiodically by photosynthetic systems that do not involve phytochrome as the photoreceptor. A simulation of photoperiodic control of PAL synthesis involving a blue light receptor has also been demonstrated in gherkin seedlings (47). Changes in isozyme composition of enzymes such as esterases, leucine aminopeptidases and peroxidases are affected by photoperiodic treatment in citrus shoots (194).

The interaction of photoperiod, circadian rhythms, and phytochrome is illustrated in the complex photocontrol of PEP-carboxylase and malic enzyme in the crassulacean genus *Kalanchoe* (152). The level of both of these enzymes involved in crassulacean acid metabolism increases substantially during repeated exposure of the plants to long nights (photoperiodic effect). However, photoregulation is not the same for both enzymes. A brief flash of red light during the middle of the long night inhibits the rise in level of PEP-carboxylase (phytochrome control) but has no effect on the increase in malic enzyme (phytochrome insensitive). Thus a photoperiodic effect can involve phytochrome but does not require it as the only controlling photoreceptor. Since only changes in activity have been measured in these studies, it is not yet possible to tell whether photocontrol involves synthesis or activation of the enzymes. Light activation of PEP-carboxylase has already been described (176).

Conclusion

Many photomorphogenic effects in plants involve the tetrapyrrole photoreceptor molecules, phytochrome, protochlorophyllide holochrome, and, in a sense, chlorophyll. The first two of these pigment-protein complexes, at least, appear to affect biochemical change by virtue of their ability to undergo photoinduced conformational changes. These photoreceptors can regulate the synthesis and turnover or inactivation of a wide variety of enzymes in plants. Some of the photoregulated enzymes are associated with chloroplast structure and function, some with phenolic biosynthesis, while others have obvious biochemical relationships. Many, but not all, of the photoregulated enzymes show a high rate of turnover or inactivation. In fact, photocontrol can involve formation of inactivating systems as well as induction of enzyme synthesis. Few other general patterns of photoreceptor mechanisms or biochemical responses are yet discernible. Whether additional unifying characteristics remain to be uncovered is open to question. Powerful biological and evolutionary forces can operate to impose species specific patterns on reactions of fundamental importance (101).

Acknowledgments

I would like to thank my colleagues at Washington State University, Drs. R. J. Foster, L. A. Hadwiger, C. A. Ryan, and J. Scheibe, who allowed me to borrow from their manuscripts before publication. The literature search for this review was begun at the Connecticut Agricultural Experiment Station.

LITERATURE CITED

1. Amrhein, N., Zenk, M. H. 1968. *Naturwissenschaften* 55:394–95
2. Ibid 1970. 57:312–13
3. Amrhein, N., Zenk, M. H. 1971. *Z. Pflanzenphysiol.* 64:145–68
4. Andersen, W. R., Wildner, G. F., Criddle, R. S. 1970. *Arch. Biochem. Biophys.* 137:84–90
5. Armstrong, J. J., Surzycki, S. J., Moll, B., Levine, R. P. 1971. *Biochemistry* 10:692–701
6. Attridge, T. H., Smith, H. 1967. *Biochim. Biophys. Acta* 148:805–7
7. Barz, W., Hosel, W. 1971. *Phytochemistry* 10:335–41
8. Barz, W., Hosel, W., Adamek, C. 1971. *Phytochemistry* 10:343–49
9. Beale, S. I. 1971. *Plant Physiol.* 48:316–19
10. Beevers, L., Hageman, R. H. 1969. *Ann. Rev. Plant Physiol.* 20:495–522
11. Bellini, E., Hillman, W. S. 1971. *Plant Physiol.* 47:668–71
12. Bellini, E., Van Poucke, M. 1970. *Planta* 93:60–70
13. Ben-Shaul, Y., Ophir, I. 1969. *Can. J. Bot.* 48:929–34
14. Berry, D. R., Smith, H. 1971. *J. Cell Res.* 8:185–200
15. Bezanson, G. S., Desaty, D., Emes, A. V., Vining, L. 1970. *Can. J. Microbiol.* 16:147–51
16. Biehn, W. L., Kuc, J., Williams, B. 1968. *Phytopathology* 58:1255–60
17. Bitensky, M. W., Gorman, R. E., Miller, W. H. 1971. *Proc. Nat. Acad. Sci.* 68:561–62
18. Björkman, O. 1968. *Physiol. Plant.* 21:1–10
19. Bogorad, L. 1967. In *Biochemistry of Chloroplasts*, ed. T. W. Goodwin, 615–31. New York: Academic
20. Borthwick, H. A., Hendricks, S. B., Schneider, M. J., Taylorson, R. B., Toole, V. K. 1969. *Proc. Nat. Acad. Sci.* 64:479–86
21. Bottomley, W. 1970. *Plant Physiol.* 45:608–11
22. Boudet, A., Ranjeva, R., Gadal, P. 1971. *Phytochemistry* 10:997–1005
23. Bowerman, A., Goodman, P. J. 1971. *Ann. Bot.* 35:353–66

24. Butler, L. G., Bennett, V. 1969. *Plant Physiol.* 44:1285–90
25. Camm, E. L., Towers, G. H. N. 1969. *Phytochemistry* 8:1407–13
26. Chen, D., Sarid, S., Katchalski, E. 1968. *Proc. Nat. Acad. Sci.* 60:902–9
27. Chen, T. M., Ries, S. K. 1968. *Can. J. Bot.* 47:341–43
28. Cheng, C. K.-C., Marsh, H. V. Jr. 1968. *Plant Physiol.* 43:1755–59
29. Clark, M. F., Matthews, R. E. F., Ralph, R. K. 1968. *Biochim. Biophys. Acta* 155:183–87
30. Cohen, A. S. 1971. *Abstr. Can. Soc. Plant Physiol. Meet.*
31. Cohen, S. S. 1970. *Am. Sci.* 58:281–89
32. Creasy, L. L. 1968. *Phytochemistry* 7:441–46
33. Ibid, 1743–49
34. Cruickshank, I. A. M. 1963. *Ann. Rev. Phytopathol.* 1:351–74
35. Davis, L. C., Brox, L. W., Gracy, R. W., Ribereau-Gayon, G., Horecker, B. L. 1970. *Arch. Biochem. Biophys.* 140:215–22
36. De Greef, J., Butler, W. L., Roth, T. F. 1971. *Plant Physiol.* 47:457–64
37. Draffan, A. G., Russell, G. K., Lyman, H., Ledbetter, M. 1971. *Abstr. Am. Soc. Plant Physiol., N.E. Sect. Meet., New Haven, Conn.*
38. Drumm, H., Moller, J., Mohr, H. 1971. *Naturwissenschaften* 58:97–98
39. Durst, F., Duranton, H. 1970. *C. R. Acad. Sci.* 270:2940–42
40. Durst, F., Mohr, H. 1966. *Naturwissenschaften* 53:707
41. Eisinger, J. A., Lamola, A. A., Longworth, J. W., Gratzer, W. B. 1970. *Nature* 226:113–18
42. Engelsma, G. 1967. *Naturwissenschaften* 12:319–20
43. Engelsma, G. 1967. *Planta* 75:207–19
44. Engelsma, G. 1968. *Acta Bot. Neer.* 17:499–505
45. Engelsma, G. 1968. *Planta* 82:355–68
46. Engelsma, G. 1969. *Naturwissenschaften* 56:503
47. Engelsma, G. 1970. *Planta* 90:133–41

48. Ibid. 91:246–54
49. Engelsma, G. 1970. *Acta Bot. Neer.* 19:403–14
50. Engelsma, G., Van Bruggen, J. M. H. 1971. *Plant Physiol.* 48:94–96
51. Epel, B. I., Butler, W. L. 1970. *Plant Physiol.* 45:728–34
52. Everett, M. S., Briggs, W. R. 1970. *Plant Physiol.* 45:679–83
53. Feierabend, J., Pirson, A. 1966. *Z. Pflanzenphysiol.* 55:235–45
54. Filner, B., Klein, A. O. 1968. *Plant Physiol.* 43:1587–96
55. Fondeville, J. C., Borthwick, H. A., Hendricks, S. B. 1966. *Planta* 69:359–64
56. Foster, R. J. et al 1971. *Proc. 1st Eur. Biophys. Congr. Vienna,* 137–49
57. Friend, J., Reynolds, S. B., Aveyard, M. A. 1971. *Biochem. J.* 124:29P
58. Fritig, B., Hirth, L., Ourisson, G. 1970. *Phytochemistry* 9:1963–75
59. Fujita, Y., Hattori, A. 1962. *Plant Cell Physiol.* 3:209–20
60. Gassman, M., Bogorad, L. 1967. *Plant Physiol.* 42:774–80
61. Gee, R., Kylin, A., Saltman, P. 1970. *Biochem. Biophys. Res. Commun.* 40:642–48
62. Graham, D., Grieve, A. M., Smillie, R. M. 1968. *Nature* 218: 89–90
63. Graham, D., Hatch, M. D., Slack, C. R., Smillie, R. M. 1970. *Phytochemistry* 9:521–32
64. Hadwiger, L. A. 1968. *Neth. J. Plant Pathol.* 74:163–69
65. Hadwiger, L. A. 1972. *Plant Physiol.* In press
66. Hadwiger, L. A., Schwochau, M. E. 1967. *Phytopathology* 59: 223–27
67. Hadwiger, L. A., Schwochau, M. E. 1971. *Plant Physiol.* 47: 346–51
68. Ibid, 588–90
69. Hahlbrock, K., Grisebach, H. 1970. *FEBS Lett.* 11:62–64
70. Hahlbrock, K., Sutter, A., Wellmann, E., Ortmann, R., Grisebach, H. 1970. *Phytochemistry* 10:109–16
71. Hahlbrock, K., Wellmann, E. 1970. *Planta* 94:236–39
72. Hahlbrock, K., Wong, E., Schill, L., Grisebach, H. 1970. *Phytochemistry* 9:949–58
73. Hanson, K. R., Havir, E. A. 1970. *Arch. Biochem. Biophys.* 141:1–17
74. Harper, D. B., Austin, D. J., Smith, H. 1970. *Phytochemistry* 9:497–505
75. Hatch, M. D., Slack, C. R. 1970. *Biochem. J.* 112:549–58
76. Hatch, M. D., Slack, C. R., Bull, T. A. 1969. *Phytochemistry* 8: 697–706
77. Hatch, M. D., Turner, J. F. 1960. *Biochem. J.* 76:556–62
78. Haupt, W. 1970. *Physiol. Veg.* 8: 551–63
79. Havir, E. A., Hanson, K. R. 1968. *Biochemistry* 7:1896–1903
80. Ibid, 1904–14
81. Hendricks, S. B., Borthwick, H. A. 1967. *Proc. Nat. Acad. Sci.* 58: 2125–30
82. Henningsen, K. W. 1970. *J. Cell Sci.* 7:587–621
83. Henningsen, K. W., Boynton, J. E. 1969. *J. Cell Sci.* 5:757–93
84. Henningsen, K. W., Kahn, A. 1971. *Plant Physiol.* 47:685–90
85. Hess, S. L., Hadwiger, L. A., Schwochau, M. 1971. *Phytopathology* 61: 79–82
86. Higuchi, T. 1966. *Agr. Biol. Chem.* 30:667–73
87. Higuchi, T., Barnoud, F. 1964. *Symp. Int. Chim. Biochim. Lignine, Cellul., Hemicellul. Grenoble,* 255–74
88. Hillis, W. E., Ishikura, N. 1970. *Phytochemistry* 9:1517–28
89. Hodgins, D. S. 1971. *J. Biol. Chem.* 246:2977–85
90. Holowinsky, A. W., Schiff, J. A. 1970. *Plant Physiol.* 45:339–47
91. Holzer, H., Duntze, W. 1971. *Ann. Rev. Biochem.* 40:345–74
92. Hopkins, D. W., Butler, W. L. 1970. *Plant Physiol.* 45:567–70
93. Hopkins, W. G., Orkwiszewski, J. A. J. 1971. *Can. J. Bot.* 49: 129–35
94. Huault, C. 1970. *Physiol. Veg.* 8: 532
95. Hug, D., Roth, D. 1971. *Biochemistry* 10:1397–1401
96. Hyodo, H., Yang, S. F. 1971. *Plant Physiol.* 47:765–70
97. Hyodo, H., Yang, S. F. 1971. *Arch. Biochem. Biophys.* 143: 338–39
98. Imaseki, H., Uchiyama, M., Uritani, I. 1968. *Agr. Biol. Chem.* 32:387–89
99. Ivanova, T. M., Rubin, B. A., Da-

vydova, M. A. 1970. *Dokl. Biochem.* 190:1–4

100. Jaffe, M. J. 1970. *Plant Physiol.* 46:768–77

101. Janzen, D. H. 1969. *Evolution* 23: 1–27

102. Jaynes, T. A., Nelson, O. E. 1971. *Plant Physiol.* 47:629–34

103. Johnson, H. S., Hatch, M. D. 1970. *Biochem. J.* 119:273–80

104. Kahn, A. 1968. *Plant Physiol.* 43: 1769–80

105. Ibid, 1781–85

106. Kahn, A., Boardman, N. K., Thorne, S. W. 1969. *J. Mol. Biol.* 48:85–101

107. Kannangara, C. G. 1969. *Plant Physiol.* 44:1533–37

108. Kannangara, C. G., Woolhouse, H. W. 1967. *New Phytol.* 66: 553–61

109. Karkow, H., Mohr, H. 1968. *Naturwissenschaften* 56:94

110. Katunuma, N., Kito, K., Kominami, E. 1971. *Biochem. Biophys. Res. Commun.* 45:76–81

111. Katunuma, N., Kominami, E., Kominami, S. 1971. *Biochem. Biophys. Res. Commun.* 45:70–75

112. Kawashima, N. 1970. *Biochem. Biophys. Res. Commun.* 38: 119–24

113. Khavkin, E. E., Perelyaeva, A. I. 1970. *Dokl. Biochem.* 193:198–201

114. Kirk, J. T. O. 1970. *Ann. Rev. Plant Physiol.* 21:11–42

115. Kirk, J. T. O. 1971. *Ann. Rev. Biochem.* 40:161–96

116. Kirk, J. T. O., Tilney-Bassett, R. A. E. 1967. *The Plastids.* San Francisco: Freeman

117. Klee, C. 1970. *J. Biol. Chem.* 245: 3143–52

118. Klein, A. O. 1969. *Plant Physiol.* 44:897–902

119. Klein, S., Schiff, J. A., Holowinsky, A. W. 1971. *Abstr. Am. Soc. Plant Physiol., N.E. Sect. Meet., New Haven, Conn.*

120. Kleinkopf, G. E., Huffaker, R. C., Matheson, A. 1970. *Plant Physiol.* 46:416–18

121. Koukol, J., Conn, E. 1961. *J. Biol. Chem.* 236:2692–98

122. Kroes, H. H. 1970. *Physiol. Veg.* 8:533–49

123. Latzko, E., Garnier, R. V., Gibbs, M. 1970. *Biochem. Biophys. Res. Commun.* 39:1140–44

124. Latzko, E., Gibbs, M. 1969. *Progr. Photosyn. Res.* 3:1624–30

125. Lee, D., Sargent, D. F., Taylor, C. P. S. 1971. *Can. J. Bot.* 49: 651–55

126. Liedtke, M. P., Ohmann, E. 1969. *Eur. J. Biochem.* 10:539–48

127. Loomis, R. S., Williams, W. A., Hall, A. E. 1971. *Ann. Rev. Plant Physiol.* 22:431–68

128. Maier, V. P., Hasegawa, S. 1970. *Phytochemistry* 9:139–44

129. Marcus, A. 1971. *Ann. Rev. Plant. Physiol.* 22:313–36

130. Marsh, H. V., Jr., Havir, E. A., Hanson, K. R. 1968. *Biochemistry* 7:1915–18

131. Miller, W. H., Gorman, R. E., Bitensky, N. W. 1971. *Science* 174:295–97

132. Minamikawa, T., Uritani, I. 1964. *Arch. Biochem. Biophys.* 108: 573–74

133. Minamikawa, T., Uritani, I. 1965. *J. Biochem. Tokyo* 57:678–88

134. Mohr, H. 1966. *Photochem. Photobiol.* 5:469–83

135. Mohr, H. 1969. In *An Introduction to Photobiology,* ed. C. P. Swanson, 99–141. Englewood Cliffs, N.J.: Prentice-Hall

136. Mohr, H. et al 1968. *Planta* 83: 267–75

137. Moeller, J., Van Poucke, M. 1970. *Phytochemistry* 9:1803–5

138. Molloy, G. R., Schmidt, R. R. 1970. *Biochem. Biophys. Res. Commun.* 40:1125–33

139. Nadler, K., Granick, S. 1970. *Plant Physiol.* 46:240–46

140. Newman, I. A., Briggs, W. R. 1971. *Plant Physiol.* 47:S-1

141. Nitsch, C., Nitsch, J. P. 1966. *C. R. Acad. Sci.* 262:1102–5

142. Oelze-Karow, H., Schopfer, P., Mohr, H. 1970. *Proc. Nat. Acad. Sci.* 65:51–57

143. Ogawa, M., Uritani, I. 1969. *Radiat. Res.* 39:117–25

144. Okoloko, G., Lewis, L. N., Reid, B. R. 1970. *Plant Physiol.* 46:660–65

145. O'Neal, D., Keller, C. J. 1970. *Phytochemistry* 9:1373–83

146. Paleg, L. G., Aspinall, D. 1970. *Nature* 228:970–73

147. Park, R. B., Sane, P. V. 1971. *Ann. Rev. Plant Physiol.* 22: 395–430

148. Patterson, B. D., Smillie, R. M. 1971. *Plant Physiol.* 47:196–98

149. Pratt, L. H., Butler, W. L. 1970.

Photochem. Photobiol. 11:503–9

150. Preiss, J., Kosuge, T. 1970. *Ann. Rev. Plant Physiol.* 21:433–66
151. Pressey, R. 1967. *Plant Physiol.* 42:1780–86
152. Queiroz, O. 1969. *Phytochemistry* 8:1655–63
153. Reid, P. D., Marsh, H. V. 1969. *Z. Pflanzenphysiol.* 61:170–72
154. Riov, J., Monselise, S. P., Kahan, R. S. 1968. *Radiat. Bot.* 8:463–66
155. Riov, J., Monselise, S. P., Kahan, R. S. 1969. *Plant Physiol.* 44:631–35
156. Rissland, I., Mohr, H. 1967. *Planta* 77:239–49
157. Roth-Bejerano, N., Lips, S. H. 1970. *Physiol. Plant.* 23:530–35
158. Rubery, P. H., Fosket, D. E. 1969. *Planta* 87:54–62
159. Rubery, P. H., Northcote, D. H. 1968. *Nature* 219:1230–34
160. Russell, D. W. 1971. *J. Biol. Chem.* 246:3870–78
161. Ryan, C. A. 1968. *Plant Physiol.* 43:1859–65
162. Ibid, 1880–81
163. Ryan, C. A., DeMoura, J., Kuo, T. 1971. *Plant Physiol.* 57:S-48
164. Ryan, C. A., Huisman, W. 1970. *Plant Physiol.* 45:484–89
165. Ryan, C. A., Shumway, L. K. 1970. *Plant Physiol.* 45:512–14
166. Ryan, C. A., Shumway, L. K. 1971. *Proc. Int. Res. Conf. Protein Inhibitors, Munich,* 175–88
167. Satter, R. L., Marinoff, P., Galston, A. W. 1970. *Am. J. Bot.* 57:916–26
168. Scheibe, J. 1972. In preparation
169. Scherf, H., Zenk, M. H. 1967. *Z. Pflanzenphysiol.* 56:203–6
170. Ibid. 57:401–18
171. Schiff, J. A. 1970. *Symp. Soc. Exp. Biol.* 24:277–302
172. Schmid, G. H. 1971. *Phytochemistry* 10:2041–42
173. Schopfer, P., Siegelman, H. W. 1968. *Plant Physiol.* 43:990–96
174. Schwartzbach, S. D., Schiff, J. 1971. *Plant Physiol.* 47:S-45
175. Simmons, S., Butler, L. G. 1969. *Biochim. Biophys. Acta* 172:150–57
176. Slack, C. R. 1968. *Biochem. Biophys. Res. Commun.* 30:483–88
177. Smillie, R. M., Scott, N. S. 1969.

Progr. Mol. Submol. Biol. 1:136–202

178. Smith, H. 1970. *Nature* 227:665–68
179. Smith, H., Attridge, T. H. 1970. *Phytochemistry* 9:487–95
180. Steer, B. T., Gibbs, M. 1969. *Plant Physiol.* 44:775–80
181. Ibid, 781–83
182. Subba Rao, P. V., Moore, K., Towers, G. H. N. 1967. *Can. J. Biochem.* 45:1863–72
183. Suzer, S., Sauer, K. 1971. *Plant Physiol.* 48:60–63
184. Taylor, A. O., Bonner, B. A. 1967. *Plant Physiol.* 42:762–66
185. Travis, R. L., Huffaker, R. C., Key, J. L. 1970. *Plant Physiol.* 46:800–5
186. Travis, R. L., Jordan, W. R., Huffaker, R. C. 1969. *Plant Physiol.* 44:1150–56
187. Travis, R. L., Jordan, W. R., Huffaker, R. C. 1970. *Physiol. Plant.* 23:678–85
188. Treharne, K. J., Stoddart, J. L., Pughe, J., Paranjothy, K., Wareing, P. F. 1970. *Nature* 228:129–31
189. van Poucke, M., Cerff, R., Barthe, F., Mohr, H. 1970. *Naturwissenschaften* 57:132–33
190. Walton, D. C. 1968. *Plant Physiol.* 43:1120–24
191. Walton, D. C., Sondheimer, E. 1968. *Plant Physiol.* 43:467–69
192. Walton, E., Butt, V. S. 1970. *J. Exp. Bot.* 21:887–91
193. Walton, E., Butt, V. S. 1971. *Phytochemistry* 10:295–304
194. Warner, R. M., Upadhya, M. D. 1968. *Physiol. Plant.* 21:941–48
195. Weidner, M., Rissland, I., Lohmann, L., Huault, C., Mohr, H. 1969. *Planta* 86:33–41
196. Wettstein, D. von, Henningsen, K. W., Boynton, J. E., Kannangara, G. C., Nielsen, O. F. 1971. In *Autonomy and Biogenesis of Mitochondria and Chloroplasts,* ed. N. K. Boardman, A. W. Linnane, R. M. Smillie, 205–23. Amsterdam: North Holland
197. Wildner, G. F., Criddle, R. S. 1969. *Biochem. Biophys. Res. Commun.* 37:952–60
198. Yoshida, S. 1969. *Ann. Rev. Plant Physiol.* 20:41–62
199. Yoshida, S., Shimokoriyama, M. 1965. *Bot. Mag. Tokyo* 78:14–19

200. Ziegler, H., Ziegler, I., Schmidt-Clausen, H. J., Mueller, B., Doerr, I. 1969. *Progr. Photosyn. Res.* 3:1636–44
201. Zielke, H. R., Filner, P. 1971. *J. Biol. Chem.* 246:1772–79
202. Zucker, M. 1965. *Plant Physiol.* 40:779–84
203. Ibid 1968. 43:365–74
204. Ibid 1969. 44:912–22
205. Zucker, M. 1970. *Biochim. Biophys. Acta* 208:331–33
206. Zucker, M. 1971. *Plant Physiol.* 47:442–44
207. Zucker, M., Hankin, L. 1970. *Ann. Bot.* 34:1047–62

Ann. Rev. Plant Physiol. 1972. 23:157–72

WATER TRANSPORT ACROSS MEMBRANES[1]   7527

P. J. C. KUIPER

Laboratory of Plant Physiological Research
Agricultural University, Wageningen, The Netherlands

CONTENTS

A discussion on water transport across membranes involves the physical properties of the transported water, the physical and chemical properties of the respective membranes, and the energetical aspects of this transport phenomenon as dependent on metabolism and growth. All these aspects are interrelated with each other.

PHYSICAL PROPERTIES OF WATER

Structure of liquid water in bulk.—Research on the structure of water has been reviewed recently by Frank (27). Raman scattering studies clearly show that liquid water cannot be a mixture of different components in the

---

[1] Communication 307 of the Laboratory of Plant Physiological Research, Agricultural University, Wageningen, The Netherlands.

ordinary sense (132). The structure of liquid water most likely consists of hydrogen-bonded, four-coordinated frameworks with an occasional single molecule in the cavities of the framework (132). This framework is rather regular at lower temperatures and becomes more random as the temperature of the water increases. Formation of hydrogen bonds between neighboring water molecules should be accompanied by charge transfer which leads to mutual reinforcement of hydrogen bonds in series. This resonance effect in the hydrogen-bonded structure results in the formation of clusters of about 50 to 100 molecules with an average lifetime of 10^{-11} sec (28, 95, 97, 118). The cluster size depends on temperature, and it decreases from 91 molecules at 0°C to 8 at 40°C (97, 117). Anomalies of the temperature dependance of several physical properties of water such as specific heat, density, and surface tension indicate that with an increase in temperature a critical point is reached at which a particular H-bonded framework, possibly of a specific cluster size, collapses (72).

Structure of water near solutes and interfaces.—The interaction of a hydrogen-bonded cluster with a polar or hydrogen-bonding group of a membrane surface increases the resonance between the water molecules of a cluster. This results in an increased size of the cluster. If the surface has a sufficient density of polar and hydrogen-bonding groups, an almost continuous clustered sheath of water will be formed on the surface. Measurements of water transport through cellulose acetate desalination membranes indicated that the effective pore size for water transport was about 21.3 Å and that the average cluster size was 162 at room temperature (118, 119). This is twice the cluster size of liquid water in bulk at room temperature.

Nonpolar molecules also induce hydrogen bonding in water, called hydrophobic hydration. The number of hydrogen bonds of the water surrounding the nonpolar groups increases because the water cannot form hydrogen bonds with the nonpolar group of the solute molecule itself (29, 96). This highly hydrogen-bonded and ordered water around the nonpolar group of the solute forms a partial cage of water molecules near the solute molecule. Its structure is called "iceberg" or "icelike structure" or clathrate structure (53, 105, 117). It also occurs near nonpolar interfaces of biomembranes, and it has a lowered chemical potential compared with water in the bulk phase (96, 117). Solutes with an aliphatic chain induce a higher degree of icelikeness of the water molecules surrounding the solute than solute molecules of an aromatic nature (96, 117).

When aliphatic chains of two neighboring solute molecules in water approach each other within their van der Waals radii, the number of water molecules in contact with it will decrease and the icelike structure will partially melt. Formation of hydrophobic bonds between nonpolar groups thus involves a van der Waals interaction and a change in the water structure. It was computed that the contribution of the van der Waals forces was about 45%

of the total free energy of formation of a hydrophobic bond and the remaining 55% was due to a change in the water structure (96, 98). Such hydrophobic bonds can stabilize conformations of polypeptide chains. Interaction between nonpolar groups of polypeptides can also add further stabilization by induction of hydrogen bonds between the polar groups of the molecule and by induction of hydrogen bonds between the polar group and its surrounding water phase. In general, such hydrophobic interactions will enforce the strength of the internal hydrogen bonds (98) and the hydrogen bonding with water around the polar groups (93, 98).

PHYSICAL AND CHEMICAL PROPERTIES OF MEMBRANE COMPONENTS

These aspects will be discussed only insofar as they are related to the mechanism of water transport. Most membranes consist of lipids and protein, although sometimes also polysaccharides are present. Together with water and ions these compounds determine the structure and properties of biomembranes, and they may interact with the water phase surrounding the membrane in specific ways.

Lipids and water transport.—In water the hydrocarbon chain of the lipid will induce extensive formation of icelike water structure enforcing hydrogen bonding between the polar groups and neighboring water. The electrical resistance of bimolecular lipid films or "black films" is high, viz 10^6 to 10^8 $\Omega \cdot cm^2$ because of the hydrocarbon part of the film (35, 47). The effect of electrolyte concentration on the resistance indicates that the structure of the water surrounding the polar heads of the lipid molecules might partly determine the electrical resistance (107). Water permeability of black lipid films has been determined for osmotic water transport and diffusion (isotope exchange). The osmotic permeability coefficient (Lp) ranges from 5 to 100×10^{-4} cm/sec, depending on the lipid composition of the film (45, 46, 54, 55, 92). The isotope exchange permeability coefficient was found to be lower, about 2×10^{-4} cm/sec. This lower value could be attributed to unstirred water layers around the film. By superimposing an osmotic gradient or with improved stirring it could be raised to about 12×10^{-4} cm/sec, in agreement with the observed value of the osmotic permeability coefficient of the same film (47, 49).

Addition of cholesterol to egg lecithin lowered the osmotic permeability coefficient from 4.2×10^{-4} cm/sec (no cholesterol added) to about 1×10^{-4} cm/sec (lecithin/cholesterol ratio ¼), and a similar observation was made for brain phospholipid/tocopherol mixtures (23). It was concluded that addition of lipophilic material increases the stability of the membrane and at the same time reduces the osmotic water permeability coefficient. On the other hand, it may be expected that addition of more hydrophilic lipids as glycolipid, the plant sulfolipid, or lysolecithin will reduce the stability of the black film, and possibly this might result in an increased osmotic water permeability of the film. No data are reported so far. Variation of the solvent for preparation of the membrane (n-decane and n-tetradecane) did not affect water permeability of the membrane (35).

The activation energy of osmotic water transport through egg lecithin/ cholesterol films varied between 12 and 15 kcal/mole (49, 109, 111). Because egg lecithin mainly consists of the stearoyl and oleoyl conjugates, it is of interest to compare egg lecithin with plant lecithin which usually consists of the linoleoyl and linolenoyl conjugates (3, 67). This comparison would allow further experimental estimation of the contribution of the hydrocarbon chains to the potential energy barrier for water transport through black films. No differences in activation energy of water evaporation through monolayers of stearic acid/oleic acid mixtures were observed (63). Addition of oleic acid to stearic acid monolayers resulted in an increased water conductivity of the monolayer, possibly because of increased free pores for water vapor transport through the film.

Water permeability of lipid membranes can also be studied as osmotic swelling of lipid micelles (liposomes) in different salt solutions. The liposomes behave as ideal osmometers (5). Osmotic swelling of synthetic lecithin micelles increased with the degree of unsaturation of the acyl chains (20). Osmotic swelling of dilinoleoyl lecithin liposomes was nearly twice as fast as swelling of dioleoyl lecithin liposomes. Also a strong temperature effect on swelling was observed. The permeability-temperature curves ran parallel, and no significant differences in activation energy between the different lecithin micelles were noted. Liposomes of a mixture of lecithin and cholesterol show a decrease in water permeability which is proportional to the amount of cholesterol in the mixture (20). Polyene antibiotics increase the water permeability of membranes composed of phospholipids and cholesterol because these antibiotics make complexes with the membrane sterol (4).

Sugars and water transport through lipid films.—Osmotic water transport through black lipid films due to a glucose gradient differs from that induced by NaCl. Water transport by glucose or sucrose decreases with time while the effect of a superimposed NaCl gradient is greatly reduced by the presence of sugar in the bathing solution. Evidently sugar interferes with the polar groups of the membrane lipids, resulting in an increased hydrogen bonding at the surface and a reduced water permeability of the film (12). It has been suggested that under such conditions electro-osmotic water transport is significant (124). The influence of sugar on water transport through a liquid membrane containing lipids has also been studied. In this system water was transported from one aqueous compartment to another through a nonaqueous containing phospholipid under a concentration gradient (1). Addition of the polysaccharide heparin to one of the aqueous compartments blocked osmotic water transport, while addition of chondroitin, an acetylated polysaccharide, greatly enhanced the water flux. This experiment suggested that the degree of acetylation of the polysaccharide attached to the lipid molecules in the liquid membrane determined the water flux.

Melting behavior of lipids.—It is thought that for formation of lipid membranes the lipids should have a liquid crystalline structure (14). Small amounts of water affect the melting behavior of lipids (15). While the actual melting point of phospholipids is about 200°C, in differential thermal analysis absorption of heat is observed at transition temperatures, often within the physiological range (13, 16). The heat absorption at the transition temperature indicates a partial "melting" of the lipid and specifically a "melting" of the hydrocarbon part of the molecule. The transition temperature drops with added water till the ionic groups of the lipids are fully hydrated. This occurs at a water content of about 20%, and this bound water is thought to be essential for the organization of the lipid film (15).

The unsaturated phospholipids show much lower transition temperatures than the fully saturated ones. The transition temperature of mixtures of phospholipids covers a wider range than that of single lipids.

Lipid-protein interactions.—As stated before, lipids and proteins make up biomembranes. Several studies indicate that biomembranes are in a dynamic state between two phases: a lamellar phase characterized by a bimolecular lipid layer according to the Davson-Danielli-Robertson-model, and a globular phase consisting of protein-lipid subunits (79, 80, 134). In the lamellar phase lipid-lipid interaction consists of hydrophobic bonding, while the lipid-protein interaction is mainly ionic in nature. The protein structure is characterized by hydrogen bonding and mutual hydrophobic interaction inside the tertiary structure of the protein. Several membrane proteins are known to possess an α-helix structure (61, 69). In the globular phase lipid molecules are intimately associated with protein molecules by hydrophobic bonding between the acyl groups of the lipid and the nonpolar groups of the amino acids of the protein (38, 39, 131). Ionic and electrostatic interactions play a less important role in such lipoprotein complexes, though surely newly synthesized lipid molecules will be attached to the membrane surface by ionic bonding till the lipid molecule approaches the protein molecule closely enough (within their van der Waals radii) to establish hydrophobic bonding. In addition, zwitterionic groups may induce a buffering capacity of the lipoprotein complex.

Several observations point to the importance of lipid-protein interactions of a hydrophobic nature in membranes. Plasmamembranes of *Mycoplasma laidlawi* can be solubilized with the surface-active compound sodium dodecylsulphate. Removal of this compound in the presence of Mg^{2+} leads to reconstituted membrane-like particles (21, 22). Similar results were obtained in mitochondrial particles with bile salts, detergents, and aqueous acetone (39, 40). Studies with chloroplasts showed that the association of lipids with the membrane proteins was independent of the charge of hydrophilic end-group and was mainly determined by the hydrocarbon composition of the lipid (6, 57).

Studies of the digestion of lecithin films by phospholipase demonstrated the importance of an electrical double layer surrounding the head groups of the lipid film for the interaction between lipid and lipase (19).

When human erythrocyte membranes were exposed to phospholipase C, 70% of the total membrane phospholipid was lost upon treatment. Neither changes of the membrane protein conformation as determined by circular dichroism nor microscopically visible alterations were observed (70). The experiment indicated that the endgroup of the phospholipid was on the outer side of the membrane accessible to the action of phospholipase C and that the membrane structure was stabilized by hydrophobic interactions between lipids and helical membrane protein.

Lipid-protein-water interactions.—The mechanism of water transport through a membrane will depend on the membrane phase, lamellar or globular. When water molecules pass the bimolecular lipid layer of a lamellar membrane, the hydrocarbon tails of the lipids will separate to allow passage of water. Also hydrophobic hydration of the newly formed pore will occur. Obviously a relatively large quantity of energy is required to form such a pore for water transport through the hydrocarbon region of the membrane, and the lipid layers of the membrane will present a large potential energy barrier for water transport. Another possibility for water transport through such a lamellar membrane would be associated with tilting of the lipid molecules so that the polar heads of the lipids would line the formed pore for water transport. In this way water would be bound to the sheath of the pore by dipole forces and hydrogen bonding (134).

Let us consider a membrane which will go through a transition from the lamellar phase to the globular phase (43, 78). Such membranes with globular lipoproteins will show less hydrophobic hydration water because of the hydrophobic bonding between lipid and protein in the globular phase. Close association between protein and lipid within their van der Waals radii will result in partial melting of the icelike structured water surrounding the hydrophobic groups, and the membrane structure itself will be more hydrophilic in nature. This can be demonstrated by the following experiment. Defatting membrane particles with aqueous acetone results in membrane proteins which are insoluble in a water environment because of the many exposed hydrophobic endgroups of amino acids of the protein. Such defatted protein can often be solubilized again with sonicated lipid, and the lipid-protein complex behaves hydrophilic in solution.

In summary, a lamellar membrane system will be characterized by large hydrophobic areas where water will be present in the icelike water structure. The more intimate lipid-protein interaction, characteristic for the globular phase, will have more water bound by dipole forces or hydrogen bonding to the polar groups of lipid and protein ("polarized water"). The ratio of the two types of water structure present in the membrane—hydrophobic hydration water versus hydrogen-bonded and dipole bonded water—will be affected by several parameters such as the protein/lipid ratio, the hydrophobic surface area of the lipids as dependent on their degree of unsaturation, and the degree of association between lipid and protein. It is also clear that minor changes in the conformation of the membrane protein may result in a shift of the balance between hydrophobic and hydrophilic groups of the membrane, with consequent changes in the contents of hydrophobic hydration water and polarized water. Experimental evidence for these two fractions of differently structured water in frog muscles has been obtained recently (73). Finally, the important question of which type of membrane presents the highest potential energy barrier for water transport will be discussed later with respect to actual measurements on living membranes. At the moment it is sufficient to say that membranes with icelike structured water, with a high degree of internal hydrogen-bonding between the water molecules, present a larger barrier for formation of flickering clusters than water bound by hydrophilic groups. Because of the mutual interaction between the two structural water types it is difficult to make a general statement in this respect, and several individual cases will be discussed later.

In conclusion, small changes in the protein conformation can shift the balance between hydrophobic and hydrophilic groups and thus affect the membrane interaction with water. Metabolic factors delivering energy for the required conformational changes of the membrane may be important in this respect.

Membrane enzyme function in relation to the lipid-protein-water interaction.—The activity of several membrane enzymes depends on lipids, that is, activity is lost when lipid is removed, while activity is at least partially restored after addition of lipid. Thus lipid is essential for enzyme activity as well as for reconstitution of the membrane structure (38–40, 131). The lipid requirement may be nonspecific (38, 40) or specific depending on the lipid species and/or the degree of saturation of the acyl groups. Possibly the nonspecific lipid requirement reflects the necessity of a hydrophobic area for enzyme activity, while the more specific lipid requirement may reflect the need for appropiate charged groups. For a better understanding of the relation between water transport and membrane enzyme activity, the latter will be briefly considered.

Most attention has been given to ion-stimulated membrane adenosine triphosphatases (41, 42, 48, 68, 108, 125). ATPase of erythrocytes stimulated by ($Na^+ + K^+$) required phosphatidyl serine (114), while the Ca^{2+}-stimulated ATPase of muscle microsomes required binding to the protein of lecithin or lysolecithin (85). ATPases from plant tissue required lecithin and sulfolipid. The differences in permeability of Na^+ and K^+ and the often observed asymmetrical distribution of Na^+, K^+, Mg^{2+}, Ca^{2+} and other ions, sugars, and amino acids lead to the conclusion that the required energy to establish the asymmetrical distribution of the mentioned ions exceeds the total metabolic energy delivered per unit time (71). Considering the structure of water in the lipid-protein unit, specific sites for ion fixation will be present, explaining the differences in ionic distribution at both sides of the membrane. Metabolic energy mediated via ATPase could possibly induce the required conformational changes for ion-specific absorption sites, while accompanying changes of the water structure could restrict transport of other ions across the membrane, according to this association-induction hypothesis (71). Similarly, water transport and water permeability of a membrane could be affected by conformational changes of the lipoprotein structure of the membrane mediated by membrane ATPases.

Physiological Aspects of Water Transport Across Membranes

The data, ideas, and explorations so far presented on water structure and on the interactions between water, lipids, and membrane protein in the broadest sense served as an introduction to the physiological aspects of water transport across biomembranes. They will be applied in the following discussion.

Effect of surface-active chemicals and anaesthetics on water permeability of membranes.—Changes in water permeability due to added chemicals have been reviewed recently (126), and attention will be restricted to surface-active compounds which are known to increase water permeability (66). To distinguish between an increase in water permeability due to surface-active reagents and straight injury to the membrane, however, is difficult. Exposure of bean roots to decenylsuccinic acid ($10^{-3}M$) greatly increased permeability of the roots, but the roots were dead after 5 days (99). Lower concentrations were less effective in induction of a change in permeability and also less toxic.[2] The effective concentration range for water permeability of this compound proved to be narrow, however, and wider ranges were observed for its monoamides. The more zwitterionic the nature of the surfactant the wider the range of activity of this group of compounds (66). Surface-active chemicals probably affect permeability of the plasmamembrane by incorporation of the molecules into the membrane, while the hydrocarbon tail of the molecule interacts with the hydrophobic region. It is interesting that the addition of a CH_2 group to the molecule resulted in a 1.6 to 2.1-fold increase of water permeability of the bean root cell membranes. A similar effect of long chain alcohols on the osmotic permeability of the frog bladder was noted (72).

Acetylated glycerol, glucose, and sucrose as surfactants also increased water permeability of the roots while the corresponding sugars decreased water transport at the same concentrations (66). In this connection it should be

[2] The effect of decenylsuccinic acid on respiration of isolated corn mitochondria was ascribed to physical disruption of the membrane (Koeppe, D. E., Miller, R. J. 1971. *Plant Physiol.* 48:659–62).

mentioned that chondroitin, an acetylated polysaccharide, increased water permeability of liquid films (1). Evidently partial solubilization of membranes by surface-active chemicals is possible, though the distinction between injurous and noninjurous effects still remains vague. Surface-active chemicals interfere with the hydrophobic regions of the membrane. This effect might be similar to the solubilizing action of lipids on defatted membrane protein. The result is a more hydrophilic character of the membrane with increased water permeability. Excessive quantities of surfactant reduce the hydrophobicity of the membrane too much, resulting in a structural collapse of the membrane protein.

Anaesthetics increase the hydraulic permeability of erythrocyte plasma-membranes (122). They expand the membrane area by about 5% (120, 121), solubilizing ("fluidizing") the membrane (59, 88). Ethanol and other alcohols increased L_p up to 55%, chlorpromazine increased it 54%, and stearic acid also increased L_p, as in bean root cells (65). Like the surface-active compounds, the above anaesthetics increase the mobility of the water in the membrane (88), probably again by "melting" part of the hydrophobic hydration water of the membrane.

On the other hand, anaesthetic gases like xenon (7) and cyclopropane (8) reduce water transport across membranes of the gut as measured by self-diffusion. The presence of these gases in the tissue induces a cage-like iceberg surrounding the anaesthetic atoms and molecules (105) which hinders transport of water in the tissue.

Effect of temperature on water permeability.—Water uptake of roots is greatly affected by temperature (33, 65). In general, different temperature effects can be distinguished with low and high activation energy for hydraulic and diffusive transport. Mostly a temperature range of strong dependance upon temperature with Q_{10} values between 3 and 8 is followed at higher temperatures by a range with a considerably smaller effect of temperature with corresponding Q_{10} values between 1.2 and 2.0. A range of low activation energy for water transport also has been observed at very low temperature (33, 112). The range with the observed high activation energy indicates a high potential energy barrier for water transport, which decreases rather sharply above a critical temperature. This critical temperature strongly depends on environmental conditions during growth. It shifts to lower values with decreasing root temperature during growth. This shift in the water transport versus temperature curve may be correlated with several changes in membrane structure. Possibly at the critical temperature "hydrophobic melting" of the lipid molecules starts, while the increased mobility of the hydrocarbon chains would allow a transition in the membrane from the lamellar phase, with its lipid-lipid interaction, to the globular phase with its lipid-protein interaction. At the same time the icelike water structure could collapse at the critical temperature and be replaced by the polarized water structure around the charged groups of the lipid-protein complex of the globular phase. The measured activation energy indicates that icelike structured water in the membrane presents a greater barrier than the polarized water in the membrane. A similar temperature curve has been reported for the distribution of sucrose between the medium and the water in frog sartorius muscles. The temperature shifts are also explained here by changes of the water structure present in the muscle protein (72).

Discontinuous temperature response curves of membrane ATPase activity have been reported in animal tissue (41, 62) with a high activation energy in the low temperature range and a much lower activation energy in the high temperature range. The change in temperature response has been attributed to changes in enzyme conformation (41, 42), while transport of ATP to the active site of the enzyme was supposed to be rate-limiting below the critical temperature (62).

A shift in the membrane phase from lamellar to globular also could be mediated by metabolic energy supply. Water flow through corn roots was also considered to be mediated metabolically (32, 33). The discontinuous temperature curve of respiration of mitochondria from chilling-sensitive plants was also explained as a phase shift of the membrane lipids (83).

Effect of hydrostatic pressure on water permeability.—Experiments on the effect of hydrostatic pressure on water movement suggest that elastic properties of the membrane influence the hydrostatic permeability coefficient. The effect of pressure is different when proceeding from low to high pressure or reverse (64). In general, permeability of root systems increases with pressure up to a value of several atmospheres. Above this value no further increase in permeability is observed (11, 64, 77, 86).

The osmotic permeability coefficient for erythrocytes was inversely dependent on the osmotic potential of the bathing solution. It was independent of the direction of water flow or the nature of the solution used in the medium. As a consequence a constant water balance of the erythrocyte was maintained (113). When in *Chara* cells the osmotic pressure difference between cell sap and environment was increased, the hydraulic water permeability coefficient (L_p) decreased. The value of this coefficient was affected by ions: K^+ and Na^+ increased L_p, while Ca^{2+} decreased L_p (130).

The hydraulic permeability coefficient of cells from tobacco leaves was also inversely related to the water potential of the tissue, again enabling a regulation of the water balance of the leaf by tight coupling of permeability and potential (37).

Changes in L_p in relation to the water potential can be related to changes in the conformation of the lipoprotein complex of the membrane and the structure of the membrane water, whether icelike or polarized. Exposure of root cells to pressure possibly results in an increase of exposed hydrophilic groups (or a decrease in hydrophobicity of the membrane), while the reverse could be true for cells with a negative feedback mechanism for maintenance of the water status of the cell (red cells, *Chara* cells, tobacco leaf cells).

Effect of hormones on water transport.—The feedback mechanism for maintenance of the water balance of the leaf seems to be operated by the availability of two regulators which exert opposite effects:cytokinin and abscisic acid (76). Water stress exerts a profound effect on the activity of these hormones; it decreases cytokinin activity while activity of abscisic acid is increased. Cytokinin promotes stomatal opening and consequently transpiration in many plants in the range from 10^{-5} to 5×10^{-8} M (56, 75, 81, 82, 87, 90, 104). Abscisic acid reduces stomatal opening and transpiration (56, 58, 74, 89, 90).

Because of the tight coupling between L_P and water potential in leaf cells it is difficult to distinguish between the effects of these hormones on permeability of the guard cells and on the difference in water potential between the guard cells and neighboring cells. Treatment of excised oat leaves with kinetin reduced the value of the water potential and of the turgor pressure of leaf cells. The increase in difference in turgor pressure between neighbor cells and guard cells allows a wider opening of the stomata. The action of kinetin was thought to be maintenance of the protein level or stimulation of protein synthesis (104). Also, guard cell ATPase could be involved in the kinetic response (76). Light-stimulated transport of K^+ ions to the guard cells mediated by membrane ATPase could account for an increase in water potential of guard cells with a subsequent increase in turgor pressure of the guard cells and opening of the slit (24, 26, 30, 116, 134). Obviously conformational changes of the cytoplasm of the guard cells could change the chemical potential of the water present in the cell as well as the permeability of water, K^+, and Na^+. As stated earlier, Na^+ and K^+ increase L_P of *Chara* internodal cells, while Ca^{2+} decreases it. In experiments with isolated epidermis, Na^+ and K^+ stimulates opening of the stomata (24–26, 116, 133), while Ca^{2+} induces a closing reaction (103). If the observed mutual dependence of L_P and potential also holds true for the stomatal guard cells, ions could affect stomatal opening because of their effects on L_P, water potential, and turgor potential of the guard cells.

The two above hormones could be directly involved in the conformational change, e.g. by regulation of suitable binding sites for K^+ in the guard cell cytoplasm, or more indirectly by metabolic regulation of the conformational change via ATP or cyclic adenosine monophosphate.

The water balance of animals seems to be controlled by vasopressin. This hormone greatly increases the hydraulic permeability coefficient, while its effect on the diffusion permeability coefficient is much less (106). Also, the activation energy of water transport decreases from about 10 kcal/mol to 5 kcal/mol (10, 36, 63, 102). Vasopressin acts via stimulation of adenyl cyclase, and this stimulation requires the presence of tyramine (129). The vasopressin-induced water loss is further increased by addition of acidic phospholipids, while the zwitterionic lipids are ineffective (9). The hormonal effect is thus mediated via cyclic adenosine monophosphate, which might interact with the membrane lipids. The compound specifically increases the water flow across frog skin and toad bladder (10, 36). Possible functions of plant cyclic adenosine monophosphate in relation to transport are unknown so far.

The hormonal effect mediated by cyclic AMP could be by widening of the membrane pores for water transport (50, 63, 106) or by hindering the osmotic solutes to penetrate the membrane. Osmotic water flow will be larger the more effective the exclusion of the osmotic solute (18, 52, 110). At low concentrations of hormone and of cyclic AMP, penetration of the bladder membrane by solutes will be high and L_p comparatively low. Further experimentation will show if similar mechanisms of hormonal control of L_p exist in plants. It would unify, for example, the data on K^+ transport, L_p, osmotic and turgor potential, and stomatal opening in such a way that kinetin could stimulate stomatal opening by a decreased L_p (and increased osmotic and turgor potential) through deeper penetration of K^+ ions from outside into the guard cell cytoplasm, while abscisic acid would show opposite effects. Tight coupling between L_p and osmotic and turgor potential would be a prerequisite for turgor regulation of the guard cells.[3]

Effects of environmental adaptations and growth on water transport across membranes.—Several studies show the interrelations between environmental adaptation and water transport across membranes. Mitochondria from chilling-resistant crops contain a higher amount of polyunsaturated fatty acids and have a higher flexibility than mitochondria from chilling-sensitive plants (84). Fatty acid composition of the roots of alfalfa changed during cold hardening, and an accumulation of polyunsaturated fatty acids was observed (31). At the same time the hardened root cells show a higher osmotic permeability coefficient (60). Water permeability of the apple flower pistil cells is directly related to the hardiness of the pistil (2). In microorganisms a similar relation between lipid and fatty acid composition and membrane permeability exists (17). A higher degree of unsaturation of the lipids may result in the following events: a lower partial melting point of the lipid, a higher mobility of the lipid molecules, a shift in the membrane phase from lamellar to globular, less hydrophobic hydration water and more polarized bound water in the membrane, with a concomittant increase in L_P and decrease of the activation energy required for water transport.

Adaptation of wheat plants to drought also induced a greater flexibility of the membrane molecules (44). At the same time, mobility of the water in the cell was decreased and the water was more firmly bound to the cytoplasm. Conformational changes in the membrane could account for the observed changes in water mobility and permeability (115). An additional factor could be stimulated protein synthesis, because ^{15}N incorporation into protein proceeds at a faster rate in drought resistant plants (51).

[3] Abscisic acid raises the osmotic permeability as well as the diffusional permeability of carrot root cells (Linke, Z. G., Reinhold, L. 1971. *Plant Physiol.* 48: 103–5).

Discussions on active water transport should also take into consideration the possibility of growth (34, 91, 101). Careful measurements on water up-take of the "sleeves" of corn roots clearly demonstrated that a minor part of the total water uptake of the roots, thermodynamically speaking, was actively transported. It was reasoned that water upon reaching the endodermis was discharged into the xylem vessels by shrinkage of the cytoplasmic strands in the plasmodesmata (34). Protoplasmic flow could also be part of the active water uptake system. Very likely, however, synthesis of protein could develop new areas for polarized water binding, while growth aspects such as aging, cell division, and cell enlargement also might influence the water absorption behavior of an intact plant organ as a root.

Further evidence that water absorption of pea and corn roots might de-pend on protein synthesis was obtained from experiments with chloramphenicol. In the presence of indolylacetic acid, application of this compound de-creased the protein content of the roots with a concommitant decrease of the water absorption capacity (100). Growth of carrot tissue was stimulated by coconut milk, and the absorption of water increased in exactly the same ratio as the protein synthesis (128). Steward (127) also clearly pointed out that the essential problem of water transport to the top of a *Sequoia* tree was not that it had to happen in one big "swoop," but that the only requirement was to remove the water from the old wood within to the new growth in each season of growth. Thus in this much more complicated organism water trans-port also would be guided by growth.

Conclusions.—Going from the physical chemistry of water-lipid-protein interactions to the integrated physiological responses of water transport across membranes in cells and in growing tissues leaves several points still to be clarified. Summing up, experimental evidence for the distinction of differ-ently structured water, hydrophobic and polarized, is required on the plant cellular and plant tissue level. Biochemical aspects of the mechanism of the tight coupling between L_P and potential in leaf cells need further study to determine, for example, if it is affected by ions, lipids, and hormones. Finally, the role of protein synthesis in water transport process is very challenging since it adds a new dimension to the functional aspects of membranes in growing tissue.

LITERATURE CITED

1. Agostini, A. M., Schultz, J. H. 1965. *Surface activity and the microbial cell. Soc. Chem. Ind. London Monogr.* 19:37–58
2. Akabane, N. 1961. *Hokkaido Agr. Exp. Sta. (Japan) Rep. 9*
3. Allen, C. F., Good, P., Davis, H. F., Chisum, P., Fowler, S. D. 1966. *J. Am. Oil Chem. Soc.* 43:223–31
4. Andreoli, T. E., Dennis, V. W., Weigl, A. M. 1969. *J. Gen. Physiol.* 53:133–56
5. Bangham, A. D., de Gier, J., Gre-ville, G. D. 1967. *Chem. Phys. Lipids* 1:225–29
6. Benson, A. A. 1968. *Membrane Models and the Formation of Biological Membranes,* ed. L. Bolis, B. A. Pethica. New York: Wiley
7. Berger, E. J., Peckiyan, F. R.,

Kanzaki, G. 1968. *J. Gen. Physiol.* 52:876–86
8. Berger, E. J., Peckiyan, E. R., Kanzaki, G. 1969. *Am. J. Physiol.* 217:411–18
9. Borowitz, J. L. 1970. *Biochem. Pharmacol.* 19:515–24
10. Bourgoignè, J., Guggenheim, S., Kipnis, D. M., Klahr, S. 1969. *Science* 165:1362–63
11. Brouwer, R. 1954. *Proc. Kon. Ned. Akad. Wetensch.* 57:68–80
12. Cass, A., Finkelstein, A. 1967. *J. Gen. Physiol.* 50:1765–84
13. Chapman, D., Collin, D. T. 1965. *Nature* 206:189
14. Chapman, D., Fluck, D. J. 1966. *J. Cell Biol.* 30:1–11
15. Chapman, D., Wallach, D. F. H. 1968. *Biological Membranes, Physical Fact and Function,* ed. D. Chapman, 125–202. New York: Academic
16. Chapman, D., Williams, R. M., Ladbrooke, B. D. 1967. *Chem. Phys. Lipids* 1:445–75
17. Christophersen, J. 1967. *Molecular Mechanisms of Temperature Adaptations,* ed. C. Ladd Prosser, 327–48. Washington, D.C.: Am. Assoc. Advan. Sci.
18. Dainty, J. 1965. *Symp. Soc. Exp. Biol.* 19:75–85
19. Dawson, R. M. C. 1968. See Ref. 15, 203–32
20. de Gier, J., Mandersloot, J. G., van Deenen, L. L. M. 1968. *Biochim. Biophys. Acta* 150:666–75
21. Engelman, D. M., Morowitz, H. J. 1968. *Biochim. Biophys. Acta* 150:376–84
22. Ibid, 385–96
23. Finkelstein, A., Cass, A. 1967. *Nature* 216:717–18
24. Fischer, R. A. 1968. *Plant Physiol.* 43:1947–52
25. Ibid 1971. 47:555–58
26. Fischer, R. A., Hsiao, T. C. 1968. *Plant Physiol.* 43:1953–58
27. Frank, H. S. 1970. *Science* 169:635–41
28. Frank, H. S., Wen, W. Y. 1957. *Discuss. Faraday Soc.* 24:133–40
29. Franks, F. 1965. *Ann. NY Acad. Sci.* 125:277–89
30. Fujino, M. 1967. *Sci. Bull. Fac. Educ. Nagasaki Univ.* 18:1–47
31. Gerloff, E. D., Richardson, T.,

Stahmann, M. A. 1966. *Plant Physiol.* 41:1280–84
32. Ginsburg, H., Ginzburg, B. Z. 1970. *J. Exp. Bot.* 21:580–92
33. Ibid 1971. 22:337–45
34. Ginsburg, H., Ginzburg, B. Z. 1971. *J. Membrane Biol.* In press
35. Goldup, A., Ohki, S., Danielli, J. F. 1970. *Recent Progr. Surface Sci.* 3:193–260
36. Grantham, J. J., Burg, M. B. 1966. *Am. J. Physiol.* 211:255–59
37. Graziani, Y., Livne, A. 1971. *Plant Physiol.* 48:575–79
38. Green, D. E., Fleischer, S. 1963. *Biochim. Biophys. Acta* 70:554–82
39. Green, D. E., Tzagoloff, A. 1966. *J. Lipid Res.* 7:587–602
40. Green, D. E. et al 1967. *Arch. Biochem. Biophys.* 119:312–35
41. Gruener, N., Avi Dor, Y. 1966. *Biochem. J.* 100:762–67
42. Gruener, N., Avi Dor, Y. 1967. *Isr. J. Med. Sci.* 3:143–48
43. Gulik-Krzywicki, T., Shechter, E., Luzzati, V., Faure, M. 1969. *Nature* 223:1116–21
44. Gusev, N. A., Khokhlova, L. P., Gordon, L. Kh., Sedykh, N. V. 1969. *Bot. Zh.* 54:53–66
45. Hanai, T., Haydon, D. A. 1966. *J. Theor. Biol.* 11:370–82
46. Hanai, T., Haydon, D. A., Redwood, W. R. 1966. *Ann. NY Acad. Sci.* 137:731–39
47. Hanai, T., Haydon, D. A., Taylor, J. 1964. *Proc. Roy. Soc. Ser. A* 281:377–91
48. Hansson, G., Kylin, A. 1969. *Z. Pflanzenphysiol.* 60:270–75
49. Haydon, D. A. 1968. *J. Am. Oil Chem. Soc.* 45:230–40
50. Hays, R. M., Leaf, A. 1962. *J. Gen. Physiol.* 45:905–19
51. Henckel, P. A. 1970. *Can. J. Bot.* 48:1235–41
52. Hill, A. 1970. *Publ. 104 Soils and Fertilizers Lab., Technion, Israel.* 14 pp.
53. Horne, R. A. 1970. *Science* 168:151
54. Huang, C., Thompson, T. E. 1966. *J. Mol. Biol.* 15:539–54
55. Ibid 1966. 16:576
56. Imber, D., Tal, M. 1971. *Science* 169:592–93
57. Ji, T. H., Benson, A. A. 1968. *Biochim. Biophys. Acta* 150:686–93

58. Jones, R. J., Mansfield, T. A. 1970. *J. Exp. Bot.* 21:714–19
59. Johnson, S. M., Bangham, A. D. 1969. *Biochim. Biophys. Acta* 193:92–104
60. Jung, G. A., Smith, D. 1961. *Agron. J.* 53:359–66
61. Ké, B. 1965. *Arch. Biochem. Biophys.* 112:554–61
62. Kemp, A., Groot, G. S. P., Reitsma, H. J. 1969. *Biochim. Biophys. Acta* 180:28–34
63. Koefoed-Johnson, V., Ussing, H. H. 1953. *Acta Physiol. Scand.* 28:60–76
64. Kuiper, P. J. C. 1963. *Stomata and water relations in plants.* *Conn. Agr. Exp. Sta. Bull.* 664: 59–68
65. Kuiper, P. J. C. 1964. *Meded. Landbouwhogesch. Wageningen* 64:1–11
66. Ibid 1967. 67:1–23
67. Kuiper, P. J. C. 1970. *Plant Physiol.* 45:684–86
68. Kylin, A., Gee, R. 1970. *Plant Physiol.* 45:169–72
69. Lenard, J., Singer, S. J. 1960. *Proc. Nat. Acad. Sci. USA* 56: 1828–35
70. Lenard, J., Singer, S. J. 1968. *Science* 159:738–39
71. Ling, G. N. 1965. *Ann. NY Acad. Sci.* 125:401–17
72. Ling, G. N. 1967. *Thermobiology,* ed. A. H. Rose, 5–24. New York: Academic
73. Ling, G. N., Negendank, W. 1970. *Physiol. Chem. Phys.* 2:15–33
74. Little, C. H. A., Eidt, D. C. 1968. *Nature* 220:498–99
75. Livne, A., Vaadia, Y. 1965. *Physiol. Plant.* 18:658–64
76. Livne, A., Vaadia, Y. 1971. *Water Deficits and Plant Growth,* Vol. 3, ed. T. T. Kozlowski. In press
77. Lopushinsky, W. 1964. *Plant Physiol.* 39:494–501
78. Lucy, J. A. 1964. *J. Theor. Biol.* 7:360–73
79. Lucy, J. A. 1967. *4th Conference on Cellular Dynamics,* ed. L. D. Peachy. NY Acad. Sci. Interdisciplinary Comm. Progr. 147
80. Lucy, J. A. 1968. See Ref. 15, 233–88
81. Luke, H. H., Freeman, T. E. 1967. *Nature* 215:874–75
82. Ibid 1969. 217:873–74
83. Lyons, J. M., Raison, J. K. 1970. *Plant Physiol.* 45:386–89
84. Lyons, J. M., Wheaton, T. A.,

85. Martonosi, A., Donley, J., Halpin, R. A. 1968. *J. Biol. Chem.* 243: 61–70
86. Mees, G. C., Weatherley, P. E. 1957. *Proc. Roy. Soc. Ser. B* 147:381–91
87. Meidner, H. 1967. *J. Exp. Bot.* 18: 556–61
88. Metcalfe, J. C., Seeman, P., Burgen, A. S. V. 1968. *Mol. Pharmacol.* 4:87–95
89. Mittelheuser, C. J., van Steveninck, R. F. M. 1969. *Nature* 221:281–82
90. Mizrahi, A., Blumenfeld, A., Richmond, A. E. 1970. *Plant Physiol.* 46:169–71
91. Mozhaeva, L. V., Pil'shchikova, N. V. 1969. *Dokl. Mosk. Sel' skokhoz Akad. K.A.T.* 154: 183–90
92. Mueller, P., Rudin, D. O., Tien, H. T., Wescott, W. C. 1964. *Recent Progr. Surface Sci.* 1:379– 93
93. Nash, T. 1965. *Surface activity and the microbial cell.* *Soc. Chem. Ind. London Monogr.* 19:122–35
94. Natochin, J. V. 1968. *Dokl. Akad. Nauk SSSR* 182:1237–40
95. Némethy, G., Scheraga, H. A. 1962. *J. Chem. Phys.* 36:3382– 3400
96. Ibid, 3401–17
97. Ibid 1964. 41:680–89
98. Némethy, G., Steinberg, J. Z., Scheraga, H. A. 1963. *Biopolymers* 1:43–69
99. Newman, E. J., Kramer, P. J. 1966. *Plant Physiol.* 41:606–9
100. Nizná, E. 1968. *Biol. Bratislava* 23:508–22
101. Novakova, A. 1967. *Cesk. Fysiol.* 16:500–1
102. Orloff, J., Handler, J. S., Preston, A. 1962. *J. Clin. Invest.* 41: 702–9
103. Pallaghy, C. K. 1970. *Z. Pflanzenphysiol.* 62:58–62
104. Pallas, J. E., Box, J. E. 1970. *Nature* 227:87–89
105. Pauling, L. 1961. *Science* 134:15– 21
106. Persson, E. 1970. *Acta Physiol. Scand.* 78:364–75
107. Petkau, A., Chelack, W. S. 1967. *Biochim. Biophys. Acta* 135: 812–24
108. Post, R. L., Albright, C. D. 1960.

Pratt, H. K. 1964. *Plant Physiol.* 39:262–68

Membrane Transport and Metabolism, ed. A. Kleinzeller, A. Kotyk, 219–27

109. Price, H. D., Thompson, T. E. 1969. *J. Mol. Biol.* 41:443–57
110. Ray, P. M. 1960. *Plant Physiol.* 35:783–801
111. Redwood, W. R., Haydon, D. A. 1969. *J. Theor. Biol.* 22:1–8
112. Reinhold, L. 1969. *Proc. 11th Int. Bot. Congr.*
113. Rich, G. T., Sha'afi, R. I., Romualdez, A., Solomon, A. K. 1968. *J. Gen. Physiol.* 52:941–54
114. Roelofsen, B. 1968. *Some studies of the extractibility of lipids and the ATPase activity of the erythrocyte membrane.* Thesis. Univ. Utrecht. 67 pp.
115. Samuilov, F. D., Sedykh, N. V. 1969. *Dokl. Akad. Nauk SSSR* 184:489–92
116. Sawhney, B. L., Zelitch, I. 1969. *Plant Physiol.* 44:1350–54
117. Scheraga, H. A. 1965. *Ann. NY Acad. Sci.* 125:253–76
118. Schultz, R. D., Asunmaa, S. K. 1970. *Recent Progr. Surface Sci.* 3:291–332
119. Schultz, R. D., Asunmaa, S. K., Guter, G. A., Littman, F. E. 1970. Paper No. 10247 McDonnel Douglas Corp., Newport Beach, Calif.
120. Seeman, P., Kwant, W. O., Sauks, T. 1969. *Biochim. Biophys. Acta* 183:499–511

121. Seeman, P., Kwant, W. O., Sauks, T., Argent, W. 1969. *Biochim. Biophys. Acta* 183:490–98
122. Seeman, P., Sha'afi, R. J., Galey, W. R., Solomon, A. K. 1970. *Biochim. Biophys. Acta* 211:365–68
123. Shiratori, M., Mizana, H. 1968. *Bull. Jap. Soc. Sci. Fish.* 34:408–10
124. Shiratori, M., Mizuno, H., Tabata, Y., Okamoto, M. 1968. *Bull. Jap. Soc. Sci. Fish.* 34:404–7
125. Skou, J. C. See Ref. 108, 228–36
126. Stadelmann, E. J. 1969. *Ann. Rev. Plant Physiol.* 20:585–606
127. Steward, F. C. 1968. *Growth and Organization in Plants.* Reading, Mass.: Addison-Wesley
128. Steward, F. C., Bidwell, R. G. S., Yemm, E. W. 1956. *Nature* 178:789–92
129. Strauch, B. S., Langdon, R. G. 1969. *Arch. Biochem. Biophys.* 129:277–82
130. Tazawa, M., Kamiya, N. 1965. *Ann. Rep. Biol. Works Fac. Sci. Osaka Univ.* 13:123–57
131. Triggle, D. J. 1971. *Recent Progr. Surface Sci.* 3:273–90
132. Wall, T. T., Hornig, D. F. 1965. *J. Chem. Phys.* 43:2079–87
133. Willmer, C. M., Mansfield, T. A. 1970. *Z. Pflanzenphysiol.* 62:398–400
134. Wolman, M. 1970. *Recent Progr. Surface Sci.* 3:262–67

Ann. Rev. Plant Physiol. 1972. 23:173–96

TRANSFER CELLS   7528

J. S. Pate and B. E. S. Gunning

Department of Botany, Queen's University
Belfast, Northern Ireland

CONTENTS

The purpose of this review is to draw the attention of plant physiologists to a highly specialized cellular adaptation in which ingrowths of cell wall material increase the surface area of the plasma membrane (Figure 1). We will refer to cells carrying this adaptation as *transfer cells* and will advance the hypothesis that they represent a versatile apparatus facilitating transmembrane flux of solutes in a wide variety of anatomical situations in plants.

Neither the observation of cells with wall ingrowths nor the idea that they function in solute transport is new. In early descriptions of the sporophyte-gametophyte junction in bryophytes (5, 64, 65), they are given a treatment that could scarcely be bettered today, and there are a number of other less well documented cases in still earlier literature (57, 62, 74). Electron microscopy brought further examples to light: in gland cells (97), in the epidermis

173

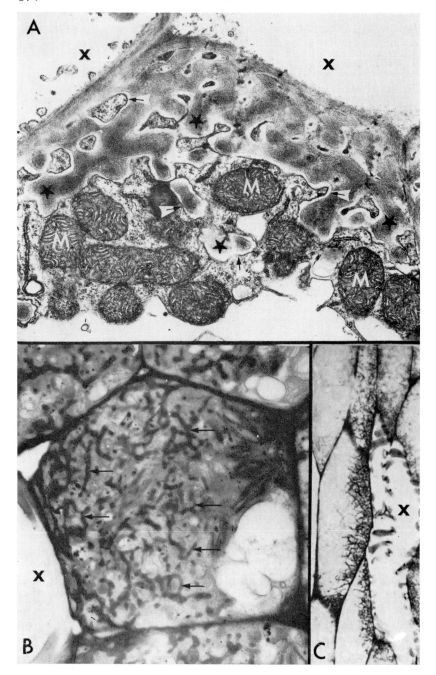

of submerged leaves of water plants (31, 110), in megagametophyte-sporophyte junctions in flowering plants (52), in companion cells in pea stem phloem (122), in xylem parenchyma in pine seedlings (125), and in phloem parenchyma of bean leaf minor veins (129). The past few years have seen the publication of many other instances, and observations of wall ingrowths by scanning (7, 50) and high voltage (39, 44; Figure 1 B) electron microscopy, but the main development has been the realization that the cells are exceedingly common and may have a common function (40).

As our own awareness of this development grew, we extended our original use of the term transfer cell—first applied to minor vein phloem parenchyma (43)—to include all of the numerous examples of cells with wall ingrowths known or suspected to function in short-distance transport of solutes. It should be emphasized that coining a new name was not intended to imply the discovery of a totally new type of cell, but rather to show that transfer cells represent specialized versions of old, familiar types of cell. For complete clarity one can speak of xylem parenchyma transfer cells, of pericycle transfer cells, of epidermal transfer cells, and so on. The nomenclature is unconventional in that it cuts across the anatomical classification of cell types, in recognition of the fact that the characteristic *wall-membrane apparatus* of transfer cells is an attribute that transcends anatomical differences.

THE WALL-MEMBRANE APPARATUS OF THE TRANSFER CELL[1]

The wall ingrowths of transfer cells develop relatively late in the life of the cell (42, 71). They are in effect a specialized form of secondary wall, deposited on the inner face of an initially perfectly ordinary cell wall. Two broad categories of wall ingrowth are recognizable: one is a type that can in its simplest form be described as papillate, as filiform if in a more elongated state, and as a labyrinth if (as is frequent) the ingrowths branch and interconnect (Figure 1); the other is a system of much more robust flanges or trabeculae, sometimes not unlike thickenings of xylem elements, and found

[1] We have made use of a number of our unpublished observations in preparing this review, but for the sake of brevity we have omitted citations for these in the text.

FIGURE 1. Structural features of transfer cells. A: The wall-membrane apparatus. Ultrathin section showing plasma membrane (arrows) involuted around wall ingrowths (asterisks) and abundant mitochondria (M) and endoplasmic reticulum ($\times$ 16,500). B: Extensive labyrinth of wall ingrowths (arrows) in 0.5 μm section viewed by high voltage electron microscopy ($\times$ 6,300). C: Polarized distribution of wall labyrinths, seen in optical micrograph ($\times$ 1,000). All show xylem parenchyma transfer cells in cotyledonary nodes. ($\times$—xylem elements) A—*Lactuca*, B—*Lamium*, C—*Ammobium*.

commonly in the grasses (44, 76, 127, 128) as well as in certain other taxa (81).

Within limits, the wall ingrowths in each species have a characteristic morphology and in extreme cases are easily recognizable [e.g. *Fumaria* (44)]. It follows that species-specific morphogenetic programs are being expressed during their formation. There is currently no cytological or biochemical information on how the cells produce ingrowths, let alone on how they produce ingrowths of specific shape. Where a plant possesses transfer cells in more than one anatomical location, marked tissue-specific differences in ingrowth morphology may be superimposed on the species specificity. Two very clear examples have been described (44, 128), and it is evident that a single genome can carry more than one specification for wall ingrowth morphology.

Regulation of wall ingrowth development is spatial as well as morphological, and three patterns have been detected thus far. (1) Wall ingrowths develop all around the cell wall, avoiding only pit fields (42, 81). (2) Wall ingrowths develop only on those areas of the cell wall bordering a neighboring compartment such as the external medium, an intercellular space, the lumen of a xylem element, or an adjacent sieve element and companion cell. Where more than one such compartment backs on to a transfer cell, the cell may produce a zone of ingrowths in relation to each compartment. (3) Wall ingrowths develop along a strip of wall girdling the cell [e.g. epidermal cells of submerged leaves of the water plant *Ranunculus fluitans* (40)]. It appears that transfer cells can regulate where their wall ingrowths are to develop, perhaps in association with stimuli emanating from neighboring compartments. The majority of transfer cells exhibit some degree of polarity in the distribution of their wall ingrowths.

One feature common to all transfer cells is that the plasma membrane follows the contours of the wall ingrowths, however irregular and labyrinthine they may be. Different morphologies amplify the surface area of the plasma membrane to different extents. In polarized transfer cells where amplification factors are only meaningful if calculated for that part of the wall bearing ingrowths, values up to more than 20 have been recorded (44).

The hypothesis that the larger the surface area of the plasma membrane, the greater the potential flux across it is simple and has been confirmed in the case of microvilli (19). The greater the expanse of membrane, the greater the diffusive flux, and the more "carriers" that can be accommodated. The relationship between total flux and total area, however, may be complicated by a number of factors. A large proportion of the plasma membrane of transfer cells is around the ingrowths, so that in traveling to or from it solutes must penetrate the fabric of the ingrowths themselves. It is therefore necessary to consider their permeability, their chemistry, and whether or not they represent unstirred compartments in which the lengthy diffusion path to or from the plasma membrane might limit the transport capacity of the whole wall-membrane apparatus.

Two studies, one of papillate (40) and one of flange-type (127) wall ingrowths, show that they are permeable to lanthanum nitrate tracer, indicating (127) the existence of open channels at least 2 nm in diameter. Presumably, therefore, the physical restrictions to passage of low molecular weight solutes will be slight. In other more conventional cell walls, the microfibrils and the matrix are minor obstacles to the movement of water and ions (115). Histochemical observations on hand-cut sections (wall labyrinths may be seen easily in e.g. a *Tradescantia* node) or material fixed in aldehydes and embedded in glycol methacrylate (32) indicate the presence of free carboxyl groups. Thus wall ingrowths normally stain metachromatically above pH 3.0; this staining is reduced by methylation (42), and they are periodic acid-Schiff positive (43). A brief rinse in dilute acid is sometimes needed to obtain intense metachromasy, suggesting that the acid groups were blocked by counter-ions at the time of fixation.

If, as seems highly likely, all wall ingrowths possess open channels in a polyanionic matrix, electro-osmosis may play a significant part in transport to the plasma membrane. The existence of potential differences between the cytoplasm and the apoplast of suitable magnitude to drive the coupled flow of ions and water, and their possible influence on short-distance transport, have been considered recently (116). A volume flow generated in the wall ingrowths would reduce limitations imposed by unstirred layers within them.

Another phenomenon for which some of the longer, labyrinthine wall ingrowths appear to be suitable is the formation of standing osmotic gradients (42). Here solute pumps located in or on the plasma membrane would either enrich or deplete (in secretory or absorptive transfer cells respectively) the solute content of the inward extremity of the ingrowth, relative to the external region. An osmotic gradient would exist at equilibrium, with a mass flow of water and solute passing in outward (secretory) or inward (absorptive) directions. The flow of water would be generated and maintained by water potential differences created between the lumen of the ingrowth and the cytoplasm by the outward or inward solute pumping activity; the flow of solutes along the ingrowths would arise both by diffusion along their own concentration gradients and by mass flow.

The standing gradient osmotic flow hypothesis (15) was developed by animal physiologists to account for the transport of isotonic or greater than isotonic fluids by absorptive epithelia possessing microvilli, surface infoldings, or intercellular clefts. These plications, like wall ingrowths, produce narrow fluid-filled compartments, joined and closed at the inner end by the plasma membrane. The hypothesis has since been extended to secretory situations and to coupled absorption and secretion (77) in animals, as well as to the transport of xylem sap in plant roots (2). The structures described recently for *Spartina* salt glands are remarkably like some of the zoological counterparts for which the hypothesis was developed (63). It should be noted that the direction in which the gradient is established depends entirely upon the vectorial properties of the solute pumps; the same morphology could therefore suit secretion or absorption.

It would be premature to apply the hypothesis to transfer cells with any degree of confidence. It is not known that there are in fact solute pumps on the plasma membrane lining the ingrowths, though the abundance of mitochondria (e.g. Figure 1 A) in the cells is in accord with active transport processes. Electron enzyme-histochemical procedures should be applied to search for ATPase activity, present in zoological parallels such as Malpighian tubules (77) and avian salt glands (1, 28). In ultrathin sections the plasma membrane looks much the same at wall ingrowths as it does elsewhere in the cell, but this does not rule out the possibility of profound biochemical differences. Perhaps selective staining or freeze-etching would reveal specialization of the membrane in the region of the hypothetical solute pumps, as, for instance, in the highly involuted electrolyte transporting membranes of pseudobranch glands and chloride cells of the marine fish *Fundulus,* where 2.5 nm repeating particles are present (90).

It is unlikely that all wall ingrowths have a sufficiently favorable geometry for formation of a standing gradient, but, on the other hand, some are many microns in length (e.g. Figure 1 B). A recent theoretical treatment (108) suggests that given a favorable perimeter: cross-sectional area ratio, even quite short channels are suitable, and the solute pumps do not even have to be restricted to the closed end of the channels.

Another feature of transfer cells is their possession of plasmodesmata, which may or may not be restricted to specific parts of the transfer cell wall. Assuming that plasmodesmata do create a symplast (91, 117), the wall-membrane apparatus of transfer cells may be regarded as the beginning or the end of transport pathways; in absorptive situations the apparatus serves whole tracts of symplast, and in secretory situations it draws upon a symplastic hinterland.

The structural features and functional potentialities described above picture the transfer cell as a module especially well equipped to handle the transport of solutes between the symplast of the plant and its extracellular environment. In the sections that follow, we attempt to show how consistent this picture is despite great diversity in the anatomical location and biochemical function of transfer cells. Specific case histories will be considered, some in fundamental elements of the plant framework, others in highly specialized organs of somewhat exotic function. Incomplete though information may be in many of the examples, viewed collectively they provide some general indication of the overall strategy and versatility of transfer cell functioning.

TRANSFER CELLS AND THE PLANT GROWTH CYCLE

Cotyledons.—The first transfer cells to appear after a seed has germinated are usually those differentiating from vascular parenchyma in the fine veins of the cotyledons. Phloem parenchyma transfer cells develop in many species exhibiting epigeal germination, while hypogeal germinators also have xylem parenchyma transfer cells (8).

Cotyledonary node.—Just after the cotyledon vein transfer cells have formed, a second, entirely distinct set of transfer cells develops in the seedling axis at the cotyledonary node. Detailed examination of lettuce (*Lactuca*) and groundsel (*Senecio*) and surveys of very many other species (44) have revealed that it is the xylem parenchyma of the cotyledonary traces of the hypocotyl that becomes transformed into transfer cells, and that, at a very early stage relative to the differentiation of its vascular elements, the vascular network of the plumule may also become endowed with phloem parenchyma transfer cells. This "collar" of transfer cells at the cotyledonary node is thought to function in the early nutrition of the young plumule, particularly before it develops proper vascular connections with the primary vascular system already connecting cotyledons and root (83). As a structure only weakly active in transpiration, the plumule is badly equipped to sequester solutes absorbed by or generated in the seedling root and supplied to the shoot in the transpiration stream. The investment of xylem parenchyma transfer cells is strategically positioned to abstract materials moving towards the cotyledons in the xylem, and so to generate a symplastic supply for the plumule. Xylem sap bleeding from groundsel seedlings decapitated above the zone of hypocotyl transfer cells contains a much lower concentration of the major nitrogenous component nitrate than is present in sap collected below the zone of transfer cells (83), so that at least in the relatively stagnant conditions in the xylem of a decapitated seedling the transfer cells appear to rob the xylem of this compound. These transfer cells might then pass on nitrate via the symplast to cells of the plumule, which in groundsel are especially active in nitrate reduction. [3]H leucine fed through the cut base of the hypocotyl of actively transpiring seedlings of groundsel or lettuce labels the xylem transfer cells and all cells of the plumule, while in contrast upper regions of the cotyledons above the gauntlet of transfer cells acquire only low levels of labeling (83).

In small-seeded epigeal seedlings, illumination and access to carbon dioxide is needed before wall ingrowths form in the presumptive transfer cells. Removing or darkening one of the cotyledons prevents proper development of xylem transfer cells in the relevant trace, but does not prevent normal transfer cells forming in strands serving the intact, undeprived sister cotyledon (83). In these small seedlings it would appear that the translocation of newly absorbed or photosynthesized solutes is necessary before a normal pattern of transfer cell differentiation can take place; possibly these solutes provide both the stimuli and the nutrients for wall ingrowth development.

Leaf minor veins.—As the shoot grows, sets of transfer cells equivalent in type and possibly in function to those of its cotyledons may develop in the minor veins of each of its leaves. Somewhat surprisingly, minor vein transfer cells are almost entirely restricted to dicotyledons, and within this group are restricted mostly to herbaceous species. An article has been published (81) dealing with the four types that occur in leaf veins and taxonomic aspects of their distribution. One, the "A cell" (Figure 32.1), is a modified companion cell; another, the "B cell," a phloem parenchyma cell (Figure 32.2); and the others, "C" and "D" cells, modifications of xylem parenchyma and bundle sheath respectively (Figure 32.3, 4).

We have suggested that these minor vein transfer cells collectively facilitate retrieval and transfer of solutes to the sieve elements, whether these solutes are supplied to the leaf in the transpiration stream or are generated by photosynthetic activity (81). The initiation of the wall ingrowths coincides with or slightly precedes the commencement of export from the leaf, and their further outgrowth parallels buildup in export activity (42, 43). The fact that ingrowths are less well developed in the white than in the green regions of variegated leaves (40) supports the idea that an exportable surplus of photosynthate, as well as of transpirationally delivered solutes, may have to be present in a leaf before extensive labyrinths of wall material can develop. We imagine minor vein transfer cells to function in any of the ways suggested in Figure 2, providing an apparatus for scavenging solutes from the free space of the leaf (Figure 2A, B), while at the same time allowing a symplastic stream of solutes to flow to or from the mesophyll and from the transfer cells themselves to the sieve elements (Figure 2C). One might argue that species possessing transfer cells in their leaves are especially well adapted for intraveinal retrieval of solutes delivered in the xylem, and indeed it has been shown that the hydrophilic and open matrix of their wall ingrowths allows access to the transfer cell plasma membranes of transpirationally fed tracer (lanthanum nitrate) (40). The same wall-membrane apparatus could facilitate absorption of inorganic solutes and photosynthetically produced materi-

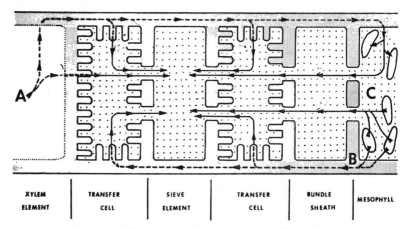

| XYLEM | TRANSFER | SIEVE | TRANSFER | BUNDLE | MESOPHYLL |
| ELEMENT | CELL | ELEMENT | CELL | SHEATH | |

FIGURE 2. Transfer cells in leaf minor veins—possible pathways for solutes. Cell walls grey, cytoplasm stippled, ——→ ——→— symplastic routes, — → — → — apoplastic routes. A—solutes arriving in transpiration stream; B—solutes leaking from mesophyll to free space of leaf; C—symplastic route from mesophyll to sieve element.

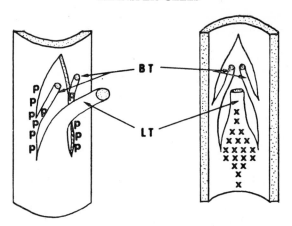

FIGURE 3. Diagrammatic representation of the distribution of xylem transfer cells (X) and phloem transfer cells (P) at the node of a stem. BT—axillary branch traces; LT—leaf trace.

als leaking from the mesophyll to the leaf free space. Such leakage has been demonstrated (3, 59, 60), though not in species which we know to possess vein transfer cells.

Nodes.—Just as equivalent sets of transfer cells occur in the minor veins of epigeal cotyledons, leaves, and indeed other photosynthetic structures such as stipules, phyllodes, cladodes, and bracts, so the counterparts of cotyledonary node transfer cells are to be encountered at every node subsequently developed on the shoot. In fact, it is here that some of the most spectacular of all wall labyrinths can be found. The habit of bearing these nodal transfer cells extends well beyond the dicotyledons, being known in all major taxa except the club mosses (44). In every instance it is the xylem parenchyma cells of the departing foliar traces (Figure 3 LT) which become transformed into transfer cells (Figure 3X), while in some species, phloem parenchyma flanking the leaf gap may be similarly modified (Figure 3 P).

Arguing solely on anatomical grounds, one possibility is that the xylem transfer cells at the node abstract solutes passing up to the leaf in the xylem, to the benefit of the axillary meristem or branch traces (Figure 3 BT), This idea is similar to that invoked for the cotyledonary node of dicotyledons in respect of the nutrition of the young plumule. Another possibility is mediation of cross-traffic of solutes between bundles supplying different classes and orthostichies of organs. This might apply especially in the complex nodes of monocotyledons where individual bundles may bear transfer cells only in their xylem, only in their phloem, or in both xylem and phloem (44, 76), and where strands of varying origin and destination may run in close proximity, with conducting elements of their xylem and/or phloem ensheathed with transfer cells.

The development of flowers and fruits can have a strong modifying influence on already existing transfer cells and may also signal the development of further tracts of vascular transfer cells. A specific example will illustrate these points. In the pea (*Pisum*) the presence of a developing fruit in the axil of a node encourages further amplification of the labyrinths of wall ingrowths in the xylem parenchyma of the node, each transfer cell eventually developing a considerably more extensive wall apparatus than in equivalent cells at nodes bearing an unfertilized flower, or where the flower has been removed early in the life of the blossom node. A hormonal influence in this phenomenon cannot be discounted. Then, as the pod enlarges, a complement of transfer cells develops in its phloem, and as its seeds mature the vascular supply to the funicle and integuments also develops transfer cells. Viewed as a whole, these successive sets of transfer cells may overcome rate-limiting barriers as the axillary flowering shoot, flowers, fruits, and finally seeds fill with assimilates mobilized from other regions of the plant. In wheat the terminal vascular supply to the spikelets possesses extensive sets of xylem parenchyma transfer cells and in common with the legume mentioned above these may aid the movement of assimilates to the flowers and fruits (127).

Reproductive tissues.—A final obstacle to transport of solutes in the plant growth cycle relates to the nutrition of the new sporophyte generation in the reproductive structures. This new generation is more or less parasitic on its parent, and since cytoplasmic contact is almost invariably lacking (105), it comes as no surprise to find the transfer cell module at virtually every site of cytoplasmic discontinuity in the reproductive organs (Table 1).

Figures 4–10 diagram the positioning of the wall apparatus with respect to donor and receptor compartments and illustrate the general theme of surface area amplification within the layered and confined reproductive structures. In some cases, e.g. the moss placenta shown in Figure 4 [one of the few transfer cell situations where physiological information is available (6, 79)] and the seeds of some angiosperms (Figures 6 and 7), zones of transfer cells face each other across a common apoplastic boundary, their combined effect interpretable as a coupled "push-pull" secretory-absorptive system from one symplast to another. Numerous differences exist between species and are readily seen in comparisons between seeds storing their main reserves in cotyledons (Figures 7 and 8), in perisperm (Figure 9), or in endosperm (Figure 10). Even within these categories further differences may be encountered. For example, some legumes develop transfer cells mainly in their suspensor (Figure 8), while others develop one or more or even all of the range of transfer cell investments depicted in Figures 6 and 7. It is necessary to find out whether these zones of transfer cells are quantitatively important entry ports for solutes contributing to cotyledon storage, and if they are, whether their distribution and ontogeny are spatially and temporally related to the vascularization of the stalk and jacket of the ovule and to the buildup of storage reserves in different regions of the cotyledons.

These reproductive situations provide some of the most convincing evi-

TABLE 1. Transfer Cells and Plant Reproduction

Organ or tissue	Location	Text figure	Distribution as presently known
Placenta	Lining cells of gametophyte and/or sporophyte	4	Bryophytes and Pteridophytes (5, 29, 40–42, 53, 64, 65, 69)
Embryo sac	Synergids, antipodals, endosperm	5	Wide variety of Angiosperms (16–18, 52, 88, 103, 119–121, 124)
Embryo	(i) Suspensor cells	6–9	Several Angiosperms (11, 42, 70, 102, 104)
	(ii) Cotyledon-epidermis	7	Certain Leguminosae (38, 40, 42)
Endosperm	(i) Outer and inner faces	6	*Capsella* (105), certain Leguminosae (42, 70)
	(ii) Adjoining perisperm	9	*Mesembryanthemum* (42)
	(iii) Aleurone at placenta	10	Caryopses of Gramineae (54, 94, 95, 128)
Integumentary endothelium	Facing embryo	7	*Vicia, Lathyrus* (42)
Pollen sac	Tapetum	—	*Paeonia* (72)
Microgametophyte	Pollen tube wall in compatible pistil	—	*Lilium* (92, 123)
Stigmatoid cells of ovary	(i) Style in contact with pollen tube	—	*Muscari* (42)
	(ii) "Transmitting tissue"	—	*Lilium* (93), *Fritillaria* (42)

dence of transfer cell action in solute transport. In at least three of the examples (Figures 4, 7, and 10) it is an anatomical necessity that all material flowing from the parent must pass through these specialized cells.

TRANSFER CELLS AND SOLUTE EXCHANGE WITH THE ENVIRONMENT
EXTERNAL TO THE PLANT

At least three functions for the wall-membrane apparatus in the exchange of solutes between the symplast and the external environment can be envis-

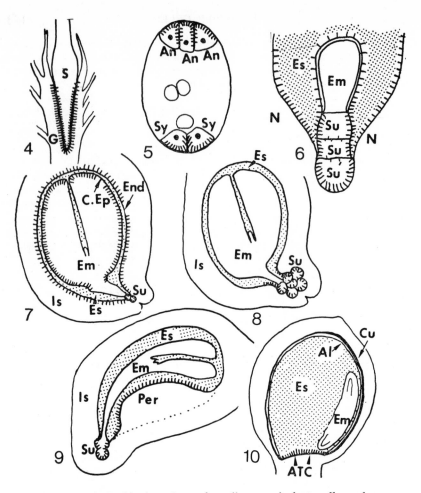

FIGURES 4–10. Positioning of transfer cells or equivalent wall-membrane apparatus in reproductive structures of plants.

Figure 4: transfer cells lining placental junction of gametophyte (G) and sporophyte (S) in lower plants.

Figure 5: wall ingrowths in synergids (Sy) or antipodals (An) in embryo sac.

Figure 6: wall ingrowths on outer and inner layers of endosperm (Es) and in suspensor (Su).

Figure 7: transfer cells of suspensor (Su), integumentary endothelium (End), and abaxial faces of cotyledonary epidermis (C.Ep). Endosperm wall ingrowths not shown.

Figure 8: giant suspensor transfer cells (Su) in *Phaseolus* seed.

Figure 9: endosperm wall ingrowths facing perisperm (Per) of *Mesembryan-themum* seed.

Figure 10: aleurone transfer cells (ATC) of grass seed.

Other symbols: N-nucellus, Em-embryo, Is-integuments, AL-aleurone layer, Cu-cuticle. See Table 1 for further details.

aged. These are absorption from and secretion to the exterior, and selective resorption of solutes from among those expelled from the plant on a nonselective basis. The three are illustrated in the following sections devoted to transfer cells in various specialized plant organs.

Haustoria.—The feeding haustoria of angiosperm parasites develop wall ingrowths when in contact with a rather peculiar external environment, namely the vascular tissues of their host plant. Examples are *Orobanche* haustoria on and in xylem elements[2] (Figure 32.10) and *Cuscuta* haustoria against sieve elements (20–22, 55) (Figure 32.9). The former probably absorb solutes from the xylem sap in the manner of xylem parenchyma transfer cells, while the latter may have a more subtle effect, first causing the sieve elements to leak (51) and then retrieving the solutes. Other structures that perhaps should be included here are the plasma membrane invaginations (without wall ingrowths) of fungal hyphae in lichens (9, 86, 87).

Submerged leaves of water plants.—It seems reasonable to attribute absorptive functions to the wall labyrinths found on the outer walls of epidermal cells of submerged leaves of certain water plants. This was one of the first anatomical locations for the apparatus to be discovered by electron microscopy, and the original observations on *Elodea* (31, 110) have now been extended (40) to other aquatics, including *Lagarosiphon* and *Vallisneria,* also extensively used in experiments on solute uptake (107, 112). In these genera the outermost abaxial wall bears the ingrowths (Figure 11), but this is not the only possibility. In species of *Potamogeton* with narrow strap-shaped leaves the ingrowths extend around the other walls of the epidermal layer and the walls of the single mesophyll layer, thus enhancing the area of plasma membrane open to the leaf apoplast as well as to the external medium (Figure 12). In batrachian species of *Ranunculus,* with cylindrical leaves (Figure 13a), the apparatus occupies a paradermal girdle in each epidermal cell (Figure 13b), in contact with the external medium via intercellular clefts (40, 73). Possession of this apparatus is evidently one facet of leaf heterophylly, for in *R. peltatus* it fails to develop in the epidermal cells of intermediate, floating, or aerial leaves.

Well-developed wall ingrowths have been found in two other situations that are equivalent in physiological terms to the epidermis of a submerged leaf. One mentioned already is the epidermis of cotyledons of developing embryos (Figure 7), the other is the outward facing wall of cell aggregates formed in liquid cultures of sycamore tissue (*Acer*) (12). Absorptive functions can scarcely be doubted in either case.

Hydropotes.—Restriction of solute uptake to specialized regions of the epidermis termed *Hydropoten* (hydropotes, "water drinkers") has been noted

[2] R. Kollmann & I. Dörr, personal communication.

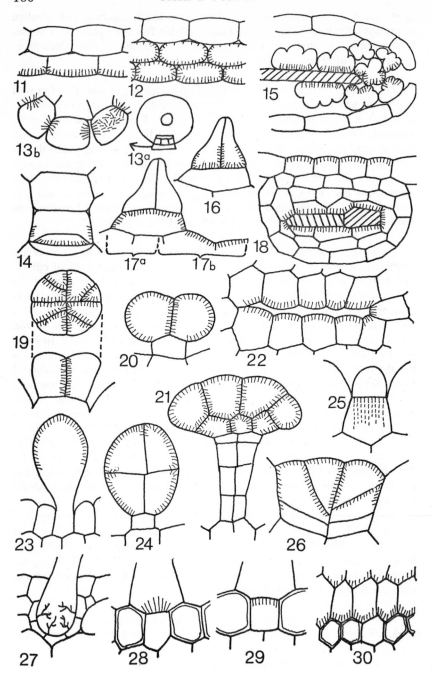

by many workers (see 73, 107). The water lilies *Nuphar* (56) and *Nymphaea* (67, 68; Figure 14) provide examples of wall ingrowth formation in hydropotes, but these are somewhat unusual in having many structural characteristics of glands (67). Lüttge and co-workers (68) have shown that *Nymphaea* hydropotes can absorb sulphate, and they are about twice as effective as the surrounding epidermis in this respect when uptake is expressed per unit area of leaf surface. It can be estimated from the published electron micrographs (67, 68) that the plasma membrane surface area amplification brought about by the wall ingrowths varies within the range of two- to five-fold.

Hydathodes.—Many structures possess the property of releasing aqueous solutions from the plant. The simplest of these hydathodes guttate xylem sap at a pore in the epidermis overlying a blindly ending strand of xylem. Epithem hydathodes are more complex, with specialized, usually lobed, parenchyma cells (epithem) between the xylem and the pore, and recent electron microscope work shows that this tissue develops wall ingrowths in a number of plants (85; Figure 15). Our own work shows that the ingrowths are well developed where the epithem is in contact with xylem elements, that is to say, where it may be regarded as a form of xylem parenchyma. The likely function is retrieval of solutes from the escaping xylem exudate.

In other hydathodes the aqueous fluid must pass through the symplast to the exterior, good examples being trichome hydathodes (peltate glands, *glandes en bouclier, Schildrüsen*) in those holo- and hemi-parasites which are considered to have evolved them as devices by which the xylem sap diverted from the host at a haustorium is expelled from the plant, its solutes having been retrieved en route. One investigation has shown that the solutes are re-

---

←⫷⫷

FIGURES 11–30. Positioning of wall ingrowths in specialized plant organs. See text for details.

Figures 11–13: water plants, submerged leaves
Figure 14: hydropote
Figure 15: epithem hydathode
Figures 16 and 17: trichome hydathodes
Figure 18: epidermal hydathode
Figures 19–21: insectivorous plant glands
Figure 22: septal nectary
Figure 23: trichome nectary
Figure 24: capitate gland with external wall labyrinths
Figures 25 and 26: salt glands
Figure 27: root hair *Stratiotes*
Figures 28–30: root hypodermis

trieved in large measure between the xylem and the exudate from the hydathodes (37), and a possible site for this is the zone of wall ingrowths in the trichome itself. The trichome consists of a stalk cell with cutin-impregnated anticlinal walls, surmounted by shield-shaped cap cells. In *Odontites, Rhinanthus,* and *Lathraea* the cap cells develop wall labyrinths (Figure 16), but the hydathodes of *Castilleja, Euphrasia,* and *Pedicularis* have labyrinths in the large stalk cell (Figure 17a). In some *Pedicularis* species the wall labyrinths extend to neighboring cells (Figure 17b).

Yet another type of hydathode is exemplified by the circular areas of specialized epidermal cells lying above a vein termination in the leaf of the fern *Nephrolepis* (45, 47, 111). We find that here too wall ingrowths develop in the epidermal cells in the same manner as in a submerged leaf (cf Figures 11 and 18), and also in the xylem parenchyma at the vein termination in the same manner as in epithem (cf Figures 15 and 18).

Glands.—There is a structural and functional gradation between hydathodes and nectaries (66), and wall ingrowths are as much a feature of the one as they are of the other. They occur in trichome nectaries (e.g. floral nectary of *Lonicera* (30; Figure 23), extra-floral nectary of *Vicia* (33, 34, 40, 126; Figure 24), in epithelial nectaries (e.g. septal nectaries (99, 101; Figure 22) and in the more bulky types where internal tissues participate (e.g. *Vicia* floral nectary).

Other types of glands incorporate the wall membrane apparatus. It occurs in both stalked and sessile glands of insectivorous plants; in *Drosera* (Figures 20 and 21) and *Drosophyllum* the ingrowths develop on all walls (98, 101), while in *Pinguicula* (48; Figure 19) and *Dionaea* (96, 106) they tend to lie on the internal walls. Histochemical work on *Pinguicula* indicates a role for the ingrowths as a route whereby esterase, ribonuclease, and acid phosphatase are released to the exterior from the cytoplasmic embayments between the ingrowths during the secretion phase of activity (48). Functions in solute uptake during the subsequent absorption phase are also probable (106). The apparatus also occurs commonly in salt glands (40, 109, 113, 114, 130) (e.g. *Frankenia,* Figure 26), and in one remarkable case (*Spartina,* Figure 25) plasma membrane invaginations extend far beyond the inner terminations of the wall ingrowths (63). There are yet other glands, of unknown function, such as the capitate glands of *Lathraea* (89, 100), *Rhinanthus, Euphrasia,* and *Vicia,* which possess spectacular wall labyrinths (Figure 24).

Clearly, the functional potentialities of the wall-membrane apparatus have been exploited many times in the course of evolution of eccrine glandular systems, absorptive systems, and back-resorption systems. Neither the nature of the system nor its specificity with respect to solutes, however, can be determined by the morphology of the cells and their wall ingrowths. Nectaries can look like absorptive epidermes; hydropotes can look like hydathodes or salt glands; glands very different in function can be strikingly similar in structure.

Transfer Cells in Roots

The vast majority of roots that have been examined lack transfer cells. One exception is the diminutive root of the water fern *Azolla,* where each of the six pericycle cells develops ingrowths backing on to the proto- and metaxylem elements. Two other hydrophytes have ingrowths in the bulbous base of their root hair cells (46, 57, 61, 118; Figure 27). The hypodermis, lying between the inner and outer cortex, deserves to be investigated in more detail than heretofore, for wall labyrinths occur on its outer face in a number of monocotyledons such as *Aspidistra* (57), and other work (13, 14, 75) suggests that this may be a widespread phenomenon in Liliaceae, Amaryllidaceae, and Araceae (Figure 30). The outer tangential wall of passage cells (36; Figure 29), and the inner tangential walls of the outer cortical cell immediately outside them (Figure 28), are other sites where the early microscopists described what may well be wall labyrinths (47, 62, 74).

Giant cells.—Another instance of wall ingrowth formation in roots is seen in the response of plants to invasion by parasitic nematodes. Those of the family Heteroderidae induce the formation of "giant cells" in the root of the host, usually in the stele (23, 25). Wall ingrowths are known to develop in giant cells of *Cucumis sativus* (49, 78), *Vicia faba* (49), *Ipomoea batatas* (58), *Lycopersicon esculentum* (4, 10, 78, 84), *Lolium perenne,* and *Solanum tuberosum.* Various species and strains of *Meloidogyne* and *Heterodera* bring about the response, but oddly enough, neither of these genera seems to induce wall ingrowth formation in infected soybean roots, though giant cells do develop (24, 26, 35). In our own studies of tomato infected with *Meloidogyne* and potato infected with *Heterodera,* we have noted that the zones of wall ingrowths in the giant cells tend to back on to the xylem and the phloem (Figure 32.8). The cells are very large, multinucleate, densely cytoplasmic, and must undoubtedly represent a substantial sink for transportable solutes. It may be that the wall labyrinths against the xylem support an absorptive membrane (as in epithem or stem xylem parenchyma transfer cells) while those against the phloem, which resemble the haustoria of the parasite *Cuscuta* in the disposition of the ingrowths relative to the sieve elements, enhance absorption of phloem-mobile material. Together they would provide a balanced diet for giant cell development and ultimately for the nematode which feeds on the giant cells.

Root nodules.—The root nodules of certain species of legumes provide the final example of root-located transfer cells (82). The vascular network of the nodule is peripheral to a central core of nitrogen-fixing bacterial tissue, and each vascular strand is bounded by an endodermis with well developed Casparian thickenings. Immediately within this is a concentric layer of transfer cells whose wall ingrowths amplify the symplast: apoplast interface of the nodule bundle, being especially prominent adjacent to the xylem elements (Figure 32.7). Figure 31 depicts a model of nodule function. The transfer cells have a dual role, ferrying sugars outward from the sieve elements across the bundle symplast, and secreting selectively to the bundle apoplast amides and certain amino acids produced in the bacterial tissue and diffusing inwards from the bacterial tissue. The level of the nitrogenous solutes is in fact somewhat lower in the outer rind of the nodule where the bundles are located than in the donor bacterial tissue. The amino compounds secreted by the transfer cells lower the water potential of the bundle apoplast, leading to an influx of water across the endodermis, flushing the solutes out to the rest of the plant in the xylem. Evidence of glandular activity can be obtained if a nodule is detached and placed with its apical end in water. Immediately, and over the next few hours, there is copious bleeding from the cut xylem at the distal end of the nodule, and this sap contains amides and certain amino acids at concentrations up to ten times those in the donor nitrogen-fixing tissues of the nodule. The sap, however, does not contain sugars, showing that the membranes of the transfer cells, although greatly enlarged, do not lose detectable amounts of the carbohydrates which they carry from the phloem (82). Of course, the levels of amino compounds in the nodule xylem may be considerably lower in a nodule attached to a transpiring plant than in a detached nodule, but judging from rates of water uptake by attached nodules, and by whole roots, it may well be that under certain conditions, e.g. at night or in high humidity or water stress, the transfer cells may maintain an "uphill" gradient of these compounds from their protoplasts into the bundle apoplast,

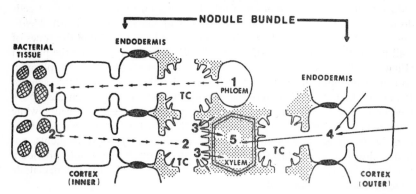

FIGURE 31. Diagrammatic representation of solute transport in the root nodule of a legume. 1—symplast gradient for sugars utilized in nitrogen fixation; 2—symplast gradient for amino compounds produced in bacterial tissue and destined for export; 3—selective secretion by transfer cells (TC) of amino compounds to the bundle apoplast (stippled); 4—entry of water into bundle across endodermis (Casparian strip shown in black); 5—discharge of concentrated solution of amino compounds in xylem.

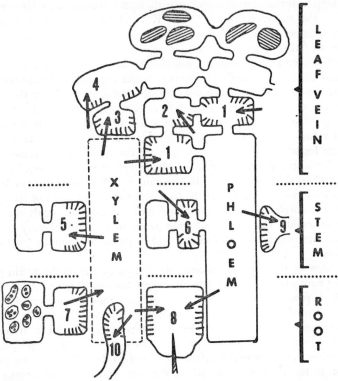

FIGURE 32. Diagram summarizing the spatial relationships of the various types of vascular transfer cells to the conducting elements of xylem and phloem. The orientation of the wall ingrowths and the presumed direction of solute flux is shown. Relevant regions of the symplast are enclosed by full lines. Legend for transfer cells: 1–4—A, B, C, and D cells of minor vein; 5—xylem parenchyma of foliar trace; 6—phloem parenchyma of leaf gap margins and branch trace; 7—pericycle of legume root nodule; 8—giant cell of nematode root gall; 9—haustorium of *Cuscuta;* 10—haustorium of *Orobanche.*

providing a most effective means for transporting nitrogen with a minimum loss of water from the nodule.

TRANSFER CELL DEVELOPMENT

In every situation where observations of transfer cell development have been made, it appears that wall ingrowth formation coincides with the onset of intensive solute transport. This applies to glands (30, 66), nodules (82), and minor veins (40, 42). Cause and effect cannot be distinguished in this relationship, but a plausible explanation is that the growth of the wall-membrane apparatus is a response to the presence of solutes, in a manner akin to the adaptive formation of membrane transport proteins in microorganisms (80), or the augmentation of ($Na^+ - K^+$) ATPase activity associated with plasma membrane surface area amplification during salt-stressing of avian salt glands (27). There is a need for methods which will provide quantitative data on surface area amplification in transfer cells in relation to solute fluxes and their changes during development or under experimental conditions.

The stimuli that elicit transfer cell development, whatever their nature and origin, are only effective in certain cells, tissues, and organisms, and one of the greatest mysteries in the subject is the wide variability in the distribution of transfer cells. Plants range from those in which no transfer cells have yet been found to others in which they are very prolific. The vetches (*Vicia* spp.), for instance, have wall ingrowths in xylem parenchyma of departing foliar and cotyledonary traces, in companion cells of minor veins in leaves and various vascular strands in the fruit, in the pericycle cells of the nodules, in both the floral and the extra-floral nectaries, in capitate glands in the pods, in the synergids, suspensor and endosperm, in the cotyledonary epidermis and the integumentary endothelium, and in nematode-induced giant cells in the roots.

In view of this superabundance, the failure of *Vicia* minor vein phloem packing parenchyma cells to produce ingrowths and become "B-type" transfer cells is puzzling. Their genetic competence cannot be in question, so why is the potential for ingrowth formation not realized? Likewise in the water plant *Ranunculus fluitans* the epidermis expresses its potential for transfer cell formation but the hydathode epithem does not. In the tree *Acer* the vein phloem does not, the nodal xylem does, and so do cell aggregates in tissue culture. In tomato and potato, where the veins and nodes lack transfer cells, the stimulus provided by nematode infection releases the potential.

The occurrence of transfer cells in sporophyte-gametophyte junctions of ferns, horsetails, and bryophytes indicates that plants became genetically competent to produce the wall-membrane apparatus very early in the course of evolution. This suggests that those higher plant taxa which possess transfer cells have merely realized a basic and fundamental potential, which may be present but latent in other taxa, though as far as we are aware no single higher plant has been investigated sufficiently to warrant the statement that the cells are completely lacking. This view of the uneven taxonomic distribution of the various classes of transfer cell may be controversial, for it implies that if an appropriate physiological selection pressure related to eccrine solute transport between symplast and apoplast is exerted, any higher plant could produce transfer cells.

EPILOGUE

Despite the overwhelming circumstantial evidence implicating transfer cells in short-distance transport of solutes, it is extremely depressing to note how few items can be rescued from the literature giving concrete information on the physiology of their action in the intact plant. We are still woefully ignorant of the function, if any, of many of the internally located types of transfer cells, there is still no clearly defined biophysical or biochemical prescription of the mode of functioning of the wall-membrane apparatus, and, worst of all, the issue still has to be settled as to whether the transfer cell does indeed function in a superlative manner in exchanging solutes with its extracellular environment. The time is now ripe to embark on an examination in depth of the function of transfer cells. If we are to progress beyond teleological platitudes and blind faith that plants would not produce them if they did not intend to use them, cytologists, physiologists, and biophysicists must combine forces in a concerted attack on the problems and challenges posed by their very existence.

LITERATURE CITED

1. Abel, J. H. 1969. *J. Histochem. Cytochem.* 17:570–84
2. Anderson, W. P., Aikman, D. P., Meiri, A. 1970. *Proc. Roy. Soc. London, Ser. B* 174:445–58
3. Bernstein, L. 1971. *Plant Physiol. Lancaster* 47:361–65
4. Bird, A. F. 1961. *J. Biophys. Biochem. Cytol.* 11:701–15
5. Blaikley, N. M. 1933. *Trans. Roy. Soc. Edinburgh* 57:699–709
6. Bopp, M., Weniger, H. P. 1971. *Z. Pflanzenphysiol.* 64:190–98
7. Briarty, L. G. 1971. *J. Microsc.* 94:181–84
8. Briarty, L. G., Coult, D. A., Boulter, D. 1970. *J. Exp. Bot.* 21:513–24
9. Brown, R. M., Wilson, R. 1968. *J. Phycol.* 4:230–40
10. Christie, J. R. 1936. *Phytopathology* 26:1–22
11. Clutter, M. E., Sussex, I. M. 1968. *J. Cell Biol.* 39:26a
12. Davey, M. R., Street, H. E. 1971. *J. Exp. Bot.* 22:90–95
13. Deshpande, B. D. 1955. *Sci. Cult.* 21:41–42
14. Ibid 1956. 21:686–87
15. Diamond, J. M., Bossert, W. H. 1967. *J. Gen. Physiol.* 50:2061–83
16. Diboll, A. G. 1968. *Caryologia* 21:91–95
17. Diboll, A. G. 1968. *Am. J. Bot.* 55:787–806
18. Diboll, A. G., Larson, D. A. 1966. *Am. J. Bot.* 53:391–402
19. Dick, E. G., Dick, D. A. T., Bradbury, S. 1970. *J. Cell Sci.* 6:451–76
20. Dörr, I. 1967. *Naturwissenschaften* 54:474–75
21. Dörr, I. 1968. *Protoplasma* 65:435–48
22. Dörr, I. 1968. *Vortr. Gesamtgeb. Bot. N.F.* 2:24–26
23. Dropkin, V. H. 1969. *Ann. Rev. Phytopathol.* 7:101–22
24. Dropkin, V. H., Nelson, P. E. 1960. *Phytopathology* 50:442–47
25. Endo, B. Y. 1971. In *Plant Parasitic Nematodes,* 2:91–117. New York, London: Academic
26. Endo, B. Y., Veech, J. A. 1970. *Phytopathology* 61:347–53
27. Ernst, S. A., Ellis, R. A. 1969. *J. Cell Biol.* 40:305–21

28. Ernst, S. A., Philpott, C. W. 1969. *J. Histochem. Cytochem.* 18:251–62
29. Eymé, J., Suire, C. 1967. *C.R. Acad. Sci. Paris* 265:1788–91
30. Fahn, A., Rachmilevitz, T. 1970. In *New Research in Plant Anatomy* [*J. Linn. Soc. (Bot.)* 63 Suppl.] London: Academic
31. Falk, H., Sitte, P. 1963. *Protoplasma* 57:290–303
32. Feder, N., O'Brien, T. P. 1968. *Am. J. Bot.* 55:123–44
33. Figier, J. 1968. *Planta* 83:60–79
34. Ibid 1971. 98:31–49
35. Gipson, I., Kim, K. S., Riggs, R. D. 1971. *Phytopathology* 61:347–53
36. Goebel, K. 1922. *Flora, Jena* 115:1–26
37. Govier, R. N., Brown J. G. S., Pate, J. S. 1968. *New Phytol.* 67:963–72
38. Graham, T. A., Gunning, B. E. S. 1970. *Nature London* 228:81–82
39. Gunning, B. E. S. 1970. *Roy. Microsc. Soc. Proc.* 5:202–3
40. Gunning, B. E. S., Pate, J. S. 1969. *Protoplasma* 68:107–33
41. Gunning, B. E. S., Pate, J. S. 1969. *Planta* 87:271–74
42. Gunning, B. E. S., Pate, J. S. 1972. Transfer cells. In *Dynamic Aspects of Plant Ultrastructure,* ed. A. W. Robards. Maidenhead, England: McGraw-Hill. In press
43. Gunning, B. E. S., Pate, J. S., Briarty, L. G. 1968. *J. Cell Biol.* 37:C7–12
44. Gunning, B. E. S., Pate, J. S., Green, L. W. 1970. *Protoplasma* 71:147–71
45. Guttenberg, H. von 1943. *Encycl. Plant Anat.* 5:1–216
46. Ibid 1968. 8(5):1–472
47. Haberlandt, G. 1914. *Physiological Plant Anatomy* (transl.) London: Macmillan. 4th ed.
48. Heslop-Harrison, Y., Knox, R. B. 1971. *Planta* 96:183–211
49. Huang, C. S., Maggenti, A. R. 1969. *Phytopathology* 59:931–37
50. Idle, D. B. 1971. *J. Microsc.* 93:77–79
51. Jacob, F., Neumann, S. 1968. *Flora, Jena* 159:191–203

52. Jensen, W. A. 1965. *Am. J. Bot.* 52:238–56
53. Kelley, C. 1969. *J. Cell Biol.* 41: 910–14
54. Kiesselbach, T. A., Walker, E. R. 1952. *Am. J. Bot* 39:561–69
55. Kollmann, R., Dörr, I. 1969. *Ber. Deut. Bot. Ges.* 82:415–25
56. Kristen, V. von 1969. *Flora, Jena* 159:536–56
57. Kroemer, K. 1903. *Bibl. Bot.* 59: 1–147
58. Krusberg, L. R., Nielsen, L. W. 1958. *Phytopathology* 48:30–39
59. Kursanov, A. L. 1968. In *Transport and Distribution of Matter in Cells of Higher Plants.* Berlin: Akademie-Verlag
60. Kursanov, A. L., Brovchenko, B. 1969. *Fiziol. Rast.* 16:965–72
61. Küster, E. 1956. *Die Pflanzenzelle,* 3. Jena: Fischer
62. Leitgeb, H. 1864. *Sb. Akad. Wiss. Wien, math. nat. Klasse* 49:1
63. Levering, C. A., Thomson, W. W. 1971. *Planta* 97:183–96
64. Lorch, W. 1925. *Ber. Deut. Bot. Ges.* 43:120–26
65. Lorch, W. 1931. Anatomie der Laubmoose. In *Handbuch der Pflanzenanatomie,* 7:1. Berlin: Borntraeger
66. Lüttge, U. 1971. *Ann. Rev. Plant Physiol.* 22:23–44
67. Lüttge, U., Krapf, G. 1969. *Cytobiologie* 1:121–31
68. Lüttge, U., Pallaghy, C. K., Willert, K. von 1971. *J. Membrane Biol.* 4:395–407
69. Maier, K. 1967. *Planta* 77:108–26
70. Marinos, N. G. 1970. *Protoplasma* 70:261–79
71. Marks, I. 1971. *Ultrastructural studies on transfer cells in minor veins of leaves.* PhD thesis. Univ. Belfast
72. Marquardt, H., Barth, O. M., Rahden, U. von 1968. *Protoplasma* 65:407–21
73. Mayr, F. 1915. *Beih. Bot. Zentralbl.* 32:278–371
74. Meinecke, E. P. 1894. *Flora, Jena* 78:133–203
75. Mulay, B. N., Deshpande, B. D. 1959. *J. Indian Bot. Soc.* 38: 383–97
76. O'Brien, T. P., Zee, S. Y. 1971. *Aust. J. Biol. Sci.* 24:207–17
77. Oschman, J. L., Berridge, M. J. 1971. *Fed. Proc.* 30:49–56

78. Owens, R. G., Specht, H. N. 1964. *Contrib. Boyce Thompson Inst.* 22:471–89
79. Paolillo, D. J., Bazzaz, F. A. 1968. *The Bryologist* 71:335–43
80. Pardee, A. B. 1968. *Science* 162: 632–37
81. Pate, J. S., Gunning, B.E.S. 1969. *Protoplasma* 68:135–56
82. Pate, J. S., Gunning, B. E. S., Briarty, L. G. 1968. *Planta* 85: 11–34
83. Pate, J. S., Gunning, B. E. S., Milliken, F. F. 1970. *Protoplasma* 71:313–34
84. Paulson, R. E., Webster, J. M. 1969. *Can. J. Bot.* 48:271–76
85. Perrin, A. 1971. *Z. Pflanzenphysiol.* 65:39–51
86. Peveling, E. 1968. *Naturwissenschaften* 55:452
87. Peveling, E. 1969. *Z. Pflanzenphysiol.* 61:151–64
88. Pluijm, J. E. van der 1964. In *Pollen Physiology and Fertilization,* 6-16. Amsterdam: North-Holland
89. Renaudin, S. 1966. *Bull. Soc. Bot. Fr.* 113:379–85
90. Ritch, R., Philpott, C. W. 1967. *Exp. Cell Res.* 55:17–24
91. Robards, A. W. 1971. *Protoplasma* 72:315–23
92. Rosen, W. G., Gawlik, S. R. 1966. In *Electron Microscopy. Proc. 6th Int. Microsc. Symp. Kyoto.* Tokyo: Maruzen
93. Rosen, W. G., Thomas, H. R. 1970. *Am. J. Bot.* 57:1108–14
94. Rost, T. L. 1970. *Am. J. Bot.* 57: 738
95. Rost, T. L., Lersten, N. R. 1971. *Protoplasma* 71:403–8
96. Scala, J., Schwab, D., Simmons, E. 1968. *Am. J. Bot.* 55:649–57
97. Schnepf, E. 1960. *Planta* 54:671–74
98. Schnepf, E. 1961. *Flora, Jena* 151: 73–87
99. Schnepf, E. 1964. *Protoplasma* 58: 137–71
100. Schnepf, E. 1964. *Planta* 60:473–82
101. Schnepf, E. 1969. *Protoplasmatologia* 8(8):1–177
102. Schnepf, E., Nägl, W. 1970. *Protoplasma* 69:133–43
103. Schulz, S. R., Jensen, W. A. 1968. *Am. J. Bot.* 55:541–52

104. Schulz, S. R., Jensen, W. A. 1969. *Protoplasma* 67:139–63
105. Schulz, S. R., Jensen, W. A. 1971. *J. Cell Sci.* 8:201–27
106. Schwab, D. W., Simmons, E., Scala, J. 1969. *Am. J. Bot.* 56:88–100
107. Sculthorpe, C. D. 1967. *The Biology of Aquatic Vascular Plants.* London: Arnold
108. Segel, L. A. 1970. *J. Theor. Biol.* 29:233–50
109. Shimony, C., Fahn, A. 1968. *J. Linn. Soc. Bot. London* 60:283–87
110. Sitte, P. 1963. *Protoplasma* 57:304–33
111. Sperlich, A. 1939. *Encycl. Plant Anat.* 4:1–184
112. Steward, F. C., Sutcliffe, J. F. 1959. Plants in relation to inorganic salts. In *Plant Physiology,* Vol. 2, ed. F. C. Steward. New York: Academic
113. Thomson, W. W., Berry, W. L., Liu, L. L. 1969. *Proc. Nat. Acad. Sci. USA* 63:310–17
114. Thomson, W. W., Liu, L. L. 1967. *Planta* 73:201–20
115. Tyree, M. T. 1968. *Can. J. Bot.* 46:317–27
116. Tyree, M. T. 1969. *J. Exp. Bot.* 20:341–49
117. Tyree, M. T. 1970. *J. Theor. Biol.* 26:181–214
118. Uphof, J. C., Hummel, K. 1962. *Encycl. Plant Anat.* 4:1–292
119. Vazart, B., Vazart, J. 1966. *Rev. Cytol. Biol. Veg.* 24:251–66
120. Vazart, J. 1968. *C. R. Acad. Sci.* 266:211–13
121. Vazart, J. 1969. *Rev. Cytol. Biol. Veg.* 32:227–40
122. Wark, M. C. 1965. *Aust. J. Biol. Sci.* 13:185–94
123. Welk, M., Millington, W. F., Rosen, W. G. 1965. *Am. J. Bot.* 52:774–81
124. Went, J.van, Linskens, H. F. 1967. *Genet. Breed. Res.* 37:51–56
125. Wooding, F.B.P., Northcote, D. H. 1965. *J. Ultrastruct. Res.* 12:463–72
126. Wrischer, M. 1962. *Acta Bot. Croatica* 21:75–94
127. Zee, S. Y., O'Brien, T. P. 1971. *Aust. J. Biol. Sci.* 24:35–49
128. Ibid, 391–95
129. Ziegler, H. 1965. *Ber. Deut. Bot. Ges.* 78:466–77
130. Ziegler, H., Lüttge, U. 1966. *Planta* 70:193–206

Ann. Rev. Plant Physiol. 1972. 23:197–218

TRANSLOCATION OF INORGANIC SOLUTES[1]   7529

ANDRÉ LÄUCHLI

Department of Plant Sciences, Texas A&M University
College Station, Texas

CONTENTS

INTRODUCTION

Growth, differentiation, and reproduction in higher plants can only proceed normally if the acquisition of all the essential elements is accomplished. In land plants this is achieved through absorption of inorganic nutrients from soil by the roots and photosynthetic fixation of CO_2 from the atmosphere which takes place mainly in the leaves. The sites of utilization of the nutrients, however, are often remote from those of absorption, and one or several translocation steps linking the initial entry with the final consumption are necessary for the life processes of a plant to occur in a regulated and orderly manner. The translocation of inorganic and organic solutes therefore constitutes a prominent part of whole-plant physiology.

Some 30 years ago, three distinct features of the translocation of inorganic solutes, which in this review will be referred to as ions, had been revealed in classical studies. First, Crafts & Broyer (39) advanced a hypothesis concerning the migration of ions from the root surface into the xylem which

---

[1] A contribution of the Texas Agricultural Experiment Station.

197

stimulated numerous experimental studies and made the problem of how ions are transported laterally in roots one of the most fascinating topics in the field of translocation research. This hypothesis will be discussed later. Second, Stout & Hoagland (150) demonstrated by utilizing radioisotopes of the elements K, Na, P, and Br that the path of upward translocation of ions is through the xylem, and furthermore that there may be a rapid lateral transfer of ions from the xylem to the phloem in the stem. Third, Biddulph & Markle (17) used ^{32}P to show that P is exported from leaves in the phloem and may cross over to the xylem.

Much of the older literature on movement in the two principal channels of longitudinal translocation, i.e. the xylem and the phloem, has been reviewed by Bollard (23) and Zimmerman (169) respectively, while the entire subject of this review has been covered periodically (15, 61, 107, 153, 160). Tiffin (153) recently contributed a symposium paper on the translocation of micronutrients, whereas certain aspects of the movement of Na^+ in plants are incorporated in a review by Rains (137) elsewhere in this volume. The present review addresses itself to some principles and problems relative to the pathways and mechanisms of ion translocation in the vegetative organs of higher plants, the emphasis being on the macronutrients. A corresponding treatise of ion translocation to fruits was published by Bollard (24).

Methods of localization.—In any consideration of the translocation of solutes in plants, one feature is central, i.e. numerous types of cells and tissues are implicated. Consequently, it is of great help in unraveling mechanisms of translocation if one is able to establish the pattern of distribution of a particular solute in a tissue or organ. To determine the pathways of translocation, a thorough knowledge of such distribution patterns is a necessity. This requires methods of localization that are specific and sensitive and permit a precise correlation of solute distribution with structures. Only recently have such methods become available.

Two methods are available for localization of ions on the light-microscopic level: microautoradiography for detecting radioactively labeled ions, and electron probe analysis for making an essentially complete chemical analysis in microscopic regions of tissue sections. In other words, electron probe analysis provides information on chemical amounts, whereas microautoradiography yields data on radioactivities. An in-depth treatise of the principles and perspectives of these two methods is given in a book (109) to which the reader is referred for detailed descriptions. One may point out that the ultimate spatial resolution of electron probe analysis is projected to be about 0.1 μm with an improved technique of specimen preparation (103). This will permit one to analyze cell compartments. Moreover, it is now conceivable to measure specific activities of labeled ions in situ, when both microautoradiography and electron probe analysis are combined experimentally (100, 109). In studies on absorption of radioactively labeled ions by plant tissues, the experimental material often is subjected to chemical analysis and radioassay to obtain an estimate of the specific activity of the ions in the whole tissue. Such an approach is also useful in translocation experiments, mainly when the detection of the specific activity can be carried out at the cellular or subcellular level.

Some ions may be localized and detected intracellularly with an electron microscope, if one is able to precipitate them specifically and quantitatively, such as Ca^{++}, Cl^-, and SO_4^{--} as Ca-oxalate, AgCl and $BaSO_4$ respectively (109). With one exception, this method has not been employed in studies of ion translocation. Ziegler & Lüttge (167) examined effectively the pathway of Cl^- in leaves using a combination of microautoradiographic detection of ^{36}Cl with electron microscopic localization of the AgCl precipitate. The localization with the electron microscope as a means for elucidating the pathways of translocation of ions merits further investigation.

An attempt was also made to apply ion-sensitive microelectrodes for measuring activities of ions such as K^+ in cell compartments (48, 59). Technical difficulties in manipulating a microelectrode intracellularly, coupled with the lack of absolute specificity of, for instance, the K^+–sensitive electrode, hampered the widespread application of this method. With additional technical improvements in the microelectrodes this may become a useful tool, mainly because such electrodes sense ionic activities which in some cases seem physiologically more meaningful than concentrations of the element.

LATERAL TRANSPORT INTO THE XYLEM OF ROOTS

Structural aspects.—There is general agreement that the main transfer of ions into the xylem occurs approximately within the terminal 10 cm of the roots, where the cells of the cortex are fully vacuolated and the endodermis has formed the Casparian strip in its radial cell walls (4, 20, 23, 35, 91, 95, 107, 113). An inspection of a transverse section through a root taken about one to several centimeters from the apex reveals that an ion moving from the surface into the xylem has to traverse several distinct cells with very diverse structures. The surge of electron microscopy as a significant tool in the plant sciences has led to a renewed interest in anatomical and cytological investigations of roots as related to lateral transport of ions into the xylem. On the one side, the studies centered on examining the structural basis of the theory of symplasmic transport in roots (cf 6, 39); on the other, the emphasis was on gathering more detailed information about the ultrastructures of those cells that are thought to play some role in lateral ion transport, i.e. the endodermal and xylem parenchyma cells and the xylem vessels themselves.

Kollmann & Dörr (89), in a review on the structural aspects of intercellular solute transport, conclude that the plasmodesmata do provide a symplasmic pathway in a tissue, though they admit that this has never really been proved. In respect to transfer across roots, two problems must be settled before symplasmic movement can be accepted unequivocally: (*a*) are all the tissues across which lateral transport occurs connected with one another through plasmodesmata; and (*b*) are the plasmodesmata indeed plasmatic in nature with the cytoplasm of the various cells forming a continuum. As for the symplasmic continuum in roots, a mutual symplasmic contact between the cells of the cortex, endodermis, and pericycle has been demonstrated (25, 75). No plasmodesmata, however, were observed in the walls of the stelar parenchyma cells (4). This isolated report has yet to be confirmed, as its implication is that the symplasm extends only into the pericycle and not all the way to the xylem vessels. Such a conclusion would not be compatible with the general pattern of ion distribution in roots as will be outlined below.

The ultrastructure of the plasmodesmata is not fully understood, though most investigators interpret the experimental findings as indicative of a cytoplasmic connection across a plasmodesma. There is agreement as to the plasmalemma being continuous and lining the canal (64, 75, 104, 125, 141). More complex is the question of whether or not the endoplasmic reticulum extends through the canal. Robards (141) suggested that the individual plasmodesmatal canals are plugged by a desmotubule (cf 125). In a subsequent paper (142), Robards interpreted these desmotubules to be in open continuity with the endoplasmic reticulum of the neighboring cells, thereby confirming other reports to the effect that the endoplasmic reticulum does extend through the canal (64, 104). Thus, the endoplasmic reticulum may be synonymous with the symplasm (142). In summary, symplasmic transport of ions appears to occur in plants according to numerous physiological data (cf 74), although the structural basis underlying this transport concept is not entirely established as yet.

The prominent feature of the cells of the endodermis is the Casparian strip which forms a complete band in the primary radial walls; it is composed of lignin, suberin, or both and appears to block the submicroscopic capillaries in the wall (56, 132). Thus, the Casparian strip has long been thought to interrupt the free space of the root, creating a barrier to passive movement of water and ions between cortex and stele (39, 93, 107). According to this view, ions can only pass across the endodermis by way of the symplasm through the plasmodesmata (cf 75). Ultrastructural studies revealed that the endodermal cells do not appear to be specialized for any active secretion of solutes into the stele (25) and are in line with their function as a barrier for transport in the free space. Nevertheless, structural evidence alone is not sufficient to determine the efficiency of the Casparian strip as a barrier. In primary roots with branch roots not yet developed, the Casparian strip does form a complete band in the radial walls of each endodermal cell (47, 132). As branch roots develop and elongate, there is a period when the formation of the Casparian strip lags behind the division of the endodermal cells so that there may be a temporary pathway for passive movement of solutes from the cortex into the stele (47). The extent to which such passive transport in the free space would contribute to the total transfer of ions into the xylem has not been examined. Yet, some lateral movement of ions may always proceed by mass flow (91), which could occur at the sites of emergence of branch roots.

In many monocotyledonous plants, the cell walls of the endodermis undergo secondary thickenings, thus increasing their resistance to the flow of water and solutes; only the passage cells remain in the primary stage with the Casparian strip present (56, 168). The movement of ions across the endodermis in its secondary or tertiary stage has gained some attention, but is probably not very important in the nutritional economy of a higher plant. Nevertheless, phosphate ions can move in barley roots through the tertiary endodermis into the stele (38). Such transport apparently may proceed through the entire endodermis, since plasmodesmata are abundant even in the cells of the tertiary stage (36). On the other hand, Ziegler et al (168) obtained microautoradiographic evidence with $^{35}SO_4^{--}$ that the transport across the tertiary endodermis in the opposite direction, i.e. from the stele into the cortex, occurs only through the passage cells.

Another intriguing problem is what role the xylem parenchyma cells exercise in lateral ion transfer. The cells of the xylem parenchyma often abut upon the smaller metaxylem vessels that are oriented towards the periphery of the stele. These outer metaxylem vessels appear to be the major functional xylem elements for upward transport to the shoot (35). The xylem parenchyma cells are small when compared to those of the cortex; they contain dense cytoplasm and are not highly vacuolated (20, 102, 108). Membrane systems in their cytoplasm are abundant (102) and mitochondria are present (4). The general structure of these cells, coupled with their association with the conducting xylem vessels, appears to qualify them as cells with a secretory function reminiscent of the function of transfer cells. However, transfer cells in roots have not been found as yet (71). A thorough study of the ultrastructures of xylem parenchyma cells is needed, the emphasis being on a search for the occurrence of plasmodesmata and on unraveling the nature of the cytoplasmic membrane system. This would facilitate the assignment of the probable function to the xylem parenchyma cells.

The development of the outer metaxylem vessels in corn roots, which are functional in upward transport (35), has been studied by Anderson & House (4). They showed that these conducting vessels exhibit a distinct variation in their structure with distance from the root apex in that the cytoplasm persists until after the secondary walls have been laid down (approximately 6 cm from the tip in 6 to 7-day-old roots). In addition, the ability of a root to transport ions into the xylem decreased from the apex to the base in a manner quantitatively correlated with the diminution in the number of vessels that contain cytoplasm. The implications of these findings will be discussed in connection with the possible mechanisms of lateral transport of ions in roots. A great deal of additional information on the ultrastructure of roots is necessary before the anatomical and cytological basis of the mechanisms of ion translocation in roots is fully elucidated.

Excised roots vs intact plants.—Ion absorption by roots and lateral passage of ions into the xylem has often been studied with excised roots of young seedlings. Yet, the common practice of severing the roots from the shoots may lead to experimental findings on rates of absorption and translocation of ions that are different from the actual rates in the intact plant. In spite of this potential risk, the utilization of excised roots in such work bears several experimental advantages over working with intact plants.

One of the principal effects of shoot excision which has troubled many investigators is the fact that the rate of ion transport into the xylem exudate is often declining gradually with time (9, 31). This undesirable situation may be attributed to a change in the supply of energy to the root system, since the decrease in the rate of transport depends upon the sugar status of the root (31) and may be overcome by adding glucose to the root medium (9). Bange (9) reasoned that the downward flow of carbohydrates from the leaves is necessary for maintaining the supply of energy to some physiological event connected with the transfer of ions from the root tissue to the conducting vessels. Experimental evidence in support of Bange's hypothesis was subsequently furnished by Bowling (27, 28).

Another feature which is manifest in experiments with excised roots of low salt status is that upon exposure of such a root to a salt solution the rate of ion translocation into the xylem exudate rises at first and becomes constant only after some time. The initial rise is due to a predominant sequestration of the absorbed ions by the root vacuoles. Once the vacuolar concentration of ions arrives at a certain level, the root is preloaded, and ions which are absorbed by the root thereafter essentially bypass the vacuoles and move into the xylem vessels at a constant rate. In other words, steady-state conditions are obtained (see 51 for a review of the older literature, plus references 41, 73, 98, 99, 102). Läuchli & Epstein (98, 99) worked out a technique for study of lateral transport of ions into the xylem exudate of single roots in short-term experiments under steady-state conditions. Preloading of the root vacuoles was accomplished while the seedlings were still intact. Thus, the time during which the root was without import of respiratory substrates from the shoot was minimized, and the transport of K^+ and Cl^- into the vessels of corn roots proceeded at constant rates for at least 100 min after preloading was complete. When the experimental periods chosen are longer, steady-state conditions may be upset, as the rate of ion efflux from the vacuoles of preloaded roots tends to increase progressively (77). Anderson & Allen (3) clearly demonstrated that such rates of translocation obtained with excised roots under steady-state conditions resemble closely the rates of transport to the shoot in the intact plant. Their findings bear out the previous conclusion that excised roots may be used in studies of the lateral passage of ions across the root into the conducting vessels, provided that those sources of error, which were discussed above, are taken into account in the experimental regime employed.

Patterns of distribution in roots.—In the last 10 years, several attempts have been made to determine the distribution of ions in those regions of the root that are the main sites of lateral transfer into the xylem, i.e. about 1 to 10 cm from the apex. By means of microautoradiography, the distribution of ^{45}Ca (20, 37, 67), ^{35}S (20, 65, 156), ^{32}P (42, 43), and ^{59}Fe (30) was examined, whereas the elements K (95, 96, 102, 103) and P (95, 96) were localized in roots by electron probe analysis. Determinations of K^+ activities in the vacuoles of the various root cells with K^+-sensitive microelectrodes (48) supplemented the electron probe analyses of the general distribution of potassium. In some of the studies, the resolution of the methods employed was high enough to permit one to distinguish between all the types of cells that an ion has to pass on its way from the surface of the root to a xylem vessel. The results of these studies are compiled in Table 1. The experimental regimes as indicated in Table 1 were such that steady-state conditions for the transfer of ions to the root xylem were prevailing. Weigl & Lüttge (156) were the first to unravel the distribution of $^{35}SO_4^{--}$ in corn roots. Table 1 shows that there was some accumulation of sulfate in the epidermis. The concentrations in cortex and endodermis, however, were low whereas certain stelar tissues, above all the cells of the xylem parenchyma and the outer metaxylem vessels themselves, contained the highest concentrations of this divalent anion. When a similar experiment was done in the presence of the inhibitor azide, no accumulation of sulfate in the stele was found. The authors interpreted their experiments to indicate that the endodermis may actively pump ions into the stele. A comparable study on *Vicia faba* (65) essentially confirmed these results. In the region of a bean root where the endodermis featured a complete Casparian strip, the concentration of $^{35}SO_4^{--}$ was greatest in the parenchymatous tissues at the xylem poles—in other words, in those cells of the pericycle and xylem parenchyma that are in the vicinity or associated with the xylem vessels (20). This was taken by Biddulph (20) to mean that the cells of the xylem parenchyma are possibly implicated in the absorption and transfer of this ion to the vessels. Experiments by Crossett (42, 43) and Läuchli (95, 96) on the lateral distribution of phosphate in corn roots revealed an accumulation in the epidermis as well as in the xylem and phloem regions of the stele.

Before Table 1 will be discussed further, the reader's attention is called to two factors that must be taken into account if mechanisms of translocation are to be deduced from patterns of sulfate and phosphate distribution. First, such patterns of distribution do not necessarily reflect solely the distribution of the ions SO_4^{--} and PO_4^{3-} because they may be incorporated to a certain extent into organic sulfur and phosphorus compounds while they are moving laterally in the root (106, 127). Second, microbial contamination may affect the results of studies on the distribution of phosphate if the experiments are performed under nonsterile conditions (11–13, 26). Particularly, the high concentrations of P in the epidermis as demonstrated for nonsterile roots (42, 95) appear to be associated with bacteria (13) and do not represent a true phosphate accumulation in the higher plant tissue.

Recently, Clarkson & Sanderson (37) made an extensive study of the dis-

TABLE 1. Lateral Distribution of Ions in Roots

Plant Material	Ion	Uptake Period	Method of Localization
Zea mays, primary root of intact seedling	$^{35}SO_4^{--}$ (20 mM)	2.5 hr	microautoradiography[a]
Hordeum vulgare, root of intact seedling (1 cm from apex)	$^{45}Ca^{++}$ (0.1 mM)	2 hr	microautoradiography[a]
Zea mays, root of intact seedling pretreated, excised thereafter (1 cm from apex)	K^+ (0.2 mM)	3 hr (intact) +1 hr (excised)	electron probe analysis: K($K\alpha$)
Zea mays, root of intact seedling (1.5–2 cm from apex)	K^+ (1 mM)	24 hr	detection with K^+-sensitive microelectrode[b]

[a] Values for entire cell (except for Xy).
[b] Values for vacuoles.

tribution of ^{45}Ca in barley roots. Table 1 exhibits that at a distance of 1 cm from the apex the pattern was as follows: The epidermis and especially the cortex were low in their Ca content. In contrast, the endodermis, pericycle, and xylem parenchyma contained on the order of three times as much Ca as did the cortex and the center of an outer metaxylem vessel. Further unpublished data by the same authors revealed that the sites of most extensive Ca accumulation were usually arranged in a ring surrounding the conducting vessels and coincided with the protoxylem poles and xylem parenchyma. Moreover, when ^{45}Ca-labeled roots of intact seedlings were transferred to a solution containing unlabeled Ca, ^{45}Ca previously accumulated in the xylem parenchyma was removed in a slow process requiring several hours. This indicates that Ca^{++} is transportable from the xylem parenchyma in the root through the xylem to the shoot, although by a slow process. Earlier reports on the distribution of Ca in roots (67, 95) were indicative of some accumulation in the epidermis and the root hairs. When ^{45}Ca was fed to roots of bean plants (20), a large fraction of the absorbed Ca was trapped as Ca-oxalate crystals present in the cortex in the vicinity of the endodermis. The activity due to ^{45}Ca was also high in the vessels and their allied xylem parenchyma. Accumulation of Ca in the cells of the xylem parenchyma, as demonstrated in roots of barley (37) and of bean plants (20), suggests that this tissue is implicated in the transfer of Ca^{++} from the root into the conducting elements of the xylem.

Electron probe analysis was used by Läuchli et al (95, 96, 102, 103) to

Unit	Ep	Co	Lateral Distribution[e] En	Pe	Pa	XyP	Xy	Reference
relative grain density	6.9	1.5	1.5	3.5	11.8	14.5	18.5	Weigl & Lüttge (156, Fig. 1)
relative grain density (average)	0.96	0.73	1.62	2.02	—	2.28	0.75	Clarkson & Sanderson (37)
relative x-ray intensity (average)	16	6	10	18	17	39	4	Läuchli et al (102, 103, and unpublished data)
activity (mM)	100	112	115	107	108	111	—	Dunlop & Bowling (48)

[e] Ep: epidermis; Co: mid-cortex; En: endodermis; Pe: pericycle; Pa: unspecified stelar parenchyma; XyP: xylem parenchyma; Xy: center of outer metaxylem vessel.

localize the element K in transverse sections of corn roots. The experiments were done under nonsterile conditions, a procedure which appears to be justified for K determinations according to a report by Epstein (53). All the data collected at the University of California at Davis (102, 103, and unpublished data) were averaged to reveal the pattern of lateral distribution of K in roots (Table 1). It appears that the K concentration in the epidermis was fairly high compared to the low level in the cortex, but the xylem parenchyma in the stele accumulated K to a much greater extent than any other tissue. An attempt to detect K separately in the cytoplasmic phase of the xylem parenchyma cells and in their vacuoles proved to be a difficult task because of protoplasmic strands traversing the vacuoles. Nevertheless, no significant differences in K concentration between cytoplasm and vacuole were apparent (102). In this context, an experimental study by Dunlop & Bowling (48) of the K^+ activities in the cell vacuoles of corn roots is of utmost interest. As is shown in Table 1, there were no significant differences in K^+ activities of the vacuoles, not even between cortex and xylem parenchyma. Hence, does this indicate that the high K concentrations in the cells of the xylem parenchyma, as detected by electron probe analysis, are due mainly to an accumulation of this ion in the cytoplasm? A definitive answer to this question cannot be given until more experimental results become available.

    In summary, Table 1 shows that the cells of the xylem parenchyma associated with the outer metaxylem vessels in roots were identified as accumulators of three important inorganic nutrients, i.e. the cations K^+ and Ca^{++} and the anion SO_4^{--}. Whether these high ionic concentrations are distributed equally over the cell or whether it is the cytoplasm that is the accumulatory compartment of a xylem parenchyma cell awaits further investigation.

Mechanisms of lateral transport into the xylem.—Some of the major mechanisms of lateral ion transport that have been advanced will now be discussed, keeping in mind the pattern of ion distribution in roots. The relationship between the transfer of water and ions has been adequately reviewed (31, 91, 144) and will be excluded from this discussion. One of the most stimulating hypotheses is the one put forth by Crafts & Broyer (39). In brief, they proposed that ions are absorbed and accumulated within the cytoplasm of cortical cells through an active transport process operating at the plasmalemma of these cells. The accumulated ions then travel in the symplasm along a concentration gradient and finally leak out passively from the symplasm into the xylem vessels. This hypothesis implies several critical features which will be examined in the light of new experimental evidence.

First of all, the Casparian strip in the endodermis must form a permeability barrier between the free spaces of cortex and stele if the Crafts-Broyer model is to operate. Most likely, this prerequisite appears to be fulfilled on structural grounds (see previous discussion) and as a consequence of some physiological evidence showing that there is probably a barrier for mass flow of phosphate between the cortex and the stele of barley roots (81) and that the Casparian strip effectively interrupts the free space in roots of corn (92).

Active transport across the plasmalemma of cortical cells is another essential feature. It will be examined here only in relation to lateral ion transport; a full account of ion uptake by roots is presented by Anderson (2) elsewhere in this volume. In accord with the Crafts-Broyer hypothesis, Brouwer (31) concluded in a critical review that the cytoplasm of root cells is unlikely to be passively accessible to ions and that ion transfer to the xylem vessels is under metabolic control. This conclusion was subsequently challenged by Laties' group (93, 110, 111), who claimed that the dual mechanisms of ion transport described by Epstein (52) operate in series with mechanism 1 at the plasmalemma and mechanism 2 at the tonoplast respectively. Furthermore, they considered the plasmalemma to be sufficiently permeable to ions at concentrations in the medium greater than 1 mM—where mechanism 2 is in evidence (52)—so that considerable diffusive movement across the plasmalemma is superimposed on the activity of mechanism 1 and absorption via mechanism 1 is not rate limiting (93). Laties (93) reviewed the evidence that is in accordance with his view. Moreover, Edwards (50) contended that the isotherm for translocation of phosphate in the concentration range of mechanism 2 is linear. Transport of phosphate to the shoot, however, depends upon its prior incorporation into organic compounds in the root (106); consequently, an event other than transport across the plasmalemma may have been rate limiting.

Several kinds of experimental findings speak against the concept of Laties outlined above. First, at moderately high concentrations of potassium or chloride in the medium (5 to 10 mM), the concentration of these ions in the xylem exudate exceeded that in the medium considerably (5, 49, 99, 158), and the electrochemical potential differences for K$^+$ and Cl$^-$ were positive (49) indicating that both ions were pumped actively into the xylem. Second, the transfer of Cl$^-$ into the xylem exudate at external concentrations in the range of mechanism 2 was inhibited by Br$^-$ in a manner implicating competition (99) and was depressed severely by various metabolic inhibitors (99, 155). Such evidence is not in line with a large diffusive movement of ions across the plasmalemma of the cortex cells. In fact, it corroborates the view of the parallel operation of the dual mechanisms in the plasmalemma (158) and can be reconciled with Crafts & Broyer's original hypothesis (cf 39).

Following active transport across the plasmalemma, ions move in the symplasm to the stele and eventually into the vessels (84, 99). The nature of symplasmic transport is still a matter of speculation. It is more rapid than by diffusion, the rate of transport being one to several cm/hr (7). The transport of Cl$^-$ across a corn root into the xylem and basipetally therein was found to be fast, on the order of 75–250 cm/hr (102). Epstein (54 and unpublished data) measured separately the lateral and the longitudinal components of this ion flux. The rate of lateral transport into the xylem was about 1 cm/hr, which is in good agreement with earlier observations on the rates of symplasmic transport (7) and confirms that the symplasm is the main pathway for lateral transfer of ions in roots. The basipetal flow along the axis of the root, however, was faster by about two orders of magnitude, emphasizing the role of metabolic transport of ions across the root. The question of where in the stele the symplasm ends has received some attention recently. The inner end of the symplasm must be bound by a membrane, as there is certainly a structural separation between the cytoplasm of the live stelar cells and the free space of the conducting vessels. Anderson and co-workers (4, 5; cf 78) proposed that the plasmalemmae or tonoplasts of the xylem vessels themselves constitute the final membranes through which ions are transported from the symplasm into the xylem exudate. They arrived at this conclusion on the grounds that the extent of ion transport into the xylem decreased basipetally along the root as did the number of vessels that had retained the cytoplasm. In contrast, the great ability of the xylem parenchyma cells to accumulate ions (see Table 1) emphasizes the plasmalemma facing the common wall between the vessel and the xylem parenchyma cell as the inner boundary of the symplasm.

It was originally proposed by Crafts & Broyer (39) that the ions in the symplasm of the stele leak into the vessels. That is, the passage of ions through the membrane, which separates the symplasm from the free space of the vessels, would proceed passively. Experiments with isolated steles (72, 94, 112) revealed that upon excision they did not have the ability to accumulate ions. Such evidence appears to support the Crafts-Broyer view. On the other hand, Yu & Kramer (162, 163) demonstrated that steles of intact roots do accumulate ions, a result that casts doubt on the validity of using isolated steles in studies of the mechanism of ion transfer into the xylem. On the contrary, Yu & Kramer's findings are in accordance with the pattern of ion distribution in roots established from the data in Table 1. It was therefore proposed (102) that two active mechanisms are involved in lateral ion transfer, namely active absorption across the plasmalemma into the cytoplasm of cortex cells and active secretion from the cytoplasm of the xylem parenchyma cells into the vessels. Other independent evidence also supports the concept of two active steps in lateral transport. The almost instantaneous response of exudation (8) and lateral ion transfer (99) to inhibitors suggests that some process other than ion absorption in the cortex is implicated. Also, the transfer of K^+ into the xylem is selective against Na^+ (10, 133, 134), which can be explained by a carrier-mediated transport of K^+ across the plasmalemma into the vessels. Moreover, studies of the movement of Ca^{++} in roots showed that the Ca concentration in the exudate can greatly exceed that in the medium (105, 120) and that transport of Ca^{++} to the shoot is under metabolic control (46, 60).

In summary, accumulating experimental evidence indicates that there is a secretory process implicated in the transport of ions from the symplasm into the xylem vessels. This secretion is located in the xylem parenchyma cells adjoining the vessels and is probably driven across the plasmalemma by a carrier-mediated transport, as proposed by Läuchli et al (102).

Genetic control.—Most experimental work has been done with few plant species, a restriction which may turn out to be unfortunate. Boyer (29) found that the roots of soybean form a considerably higher resistance to transport of water than those of sunflower and bean. Similar differences between species may become known with regard to ion transport. More to the point is the approach of using mutants and varieties in studies of transport mechanisms, as the selective transport of ions appears to be under genetic control (55). Whether or not a carrier-mediated ion transfer from the symplasm to the vessels occurs may be sorted out by screening single-locus mutants for efficiency and selectivity of ion transport to the xylem of roots. Experimental studies in which this approach was used are scarce. The work by Munns et al (121–123) represents a first step in this direction. Their experiments showed that two commercial varieties of *Avena fatua,* namely Sun II and Algerian, which are not single-locus mutants, differed consistently in the concentration of Mn in the shoots, the differences being heritable. Furthermore, the part of absorbed Mn that is translocated to the shoot was found to pass through a labile fraction in the root. The varietal differences in shoot-Mn were attributed to variations in size and turnover rate of the labile fraction. One is tempted to speculate that the labile Mn fraction is located in a cytoplasmic compartment of the xylem parenchyma cells.

There are varieties of celery, *Apium graveolens,* varying in their Mg nutrition in which the utilization of Mg is conditioned by a single gene (135).

The mutant Utah 10 B is prone to Mg-deficiency chlorosis in the leaves. It would be interesting to determine whether the low Mg content in the leaves of the mutant Utah 10 B is due to an inefficient absorption of Mg^{++} by the roots or due to a poorly developed mechanism of Mg^{++} transport into the xylem.

Still the best example of the genetic approach in testing transport mechanisms is the study by Brown (32) on the absorption and translocation of Fe by the varieties Hawkeye and PI of soybean *Glycine max,* whose variation in the utilization of Fe is governed by a single gene (157). Brown's conclusion (32) that two mechanisms are implicated in the absorption-translocation of Fe by the Hawkeye mutant, of which the translocation mechanism is probably missing in the PI mutant, makes these single-locus mutants most suitable for testing the hypothesis of the carrier-mediated secretion of ions into the xylem as advanced in this review.

TRANSPORT IN THE XYLEM

The bulk of the essential elements moves in the xylem vessels as inorganic ions except for N, which is translocated mainly in organic form (23), and for Fe occurring in the xylem sap as an organic complex in the anionic form, most probably as citrate complex (152). Some organic phosphorus and sulfur compounds such as phosphorylcholine (114) and methionine (127) were identified in xylem exudates, but the elements P and S also appear to be largely translocated in the xylem as inorganic ions.

The classical concept of mass flow in xylem vessels, by which the solutes would move with the same velocity as water, no longer holds for Ca^{++} and probably not for other divalent and trivalent cations. Results from Biddulph's laboratory (14, 18) showed that the conducting vessels in bean stems may be looked upon as negatively charged exchange columns for Ca^{++}. The binding of Ca^{++} to the vessel walls was a reversible exchange process, the bound Ca^{++} being exchangeable by Ca^{++}, Sr^{++}, or Mg^{++} supplied by the root but not by K^+ (14). Jacoby (82, 83), who worked with derooted bean plants, came to a similar conclusion. The interactions between the vessel walls and the ions in the transpiration stream deserve further investigation because electron probe analyses of transverse sections of corn roots revealed intensive K peaks over the walls of the xylem vessels (95).

Lateral movement of ions from the xylem vessels into the surrounding tissues has been known to occur in stems ever since its discovery by Stout & Hoagland (150). Recent measurements of the osmotic potential of xylem sap at various heights in tomato and pepper plants disclosed that salt is gradually removed from the xylem sap as it passes through the petiole and the leaf (87). In stems, various ions may cross over into the phloem. For instance, in willow some accumulation of such ions as K^+, Na^+, SO_4^{--}, and PO_4^{3-} takes place at first in tissues between the vessels of the xylem and the sieve tubes of the phloem before they enter the sieve tubes (76, 129). Calcium is also exported from the xylem vessels of stems and may in part be accumulated irreversibly as Ca-oxalate (18), or moves toward the periphery of the stem where it is deposited in the epidermis and in hypodermal tissues (95, 143). It is not known what tissues constitute the pathway of lateral ion transfer between the xylem and the phloem. Steward (149) and Biddulph (15) stressed the significance of the cambium, if present. The rays may also function in mediating transport of ions between the two channels of longitudinal translocation.

There may be an import of solutes to the xylem from tissues that are not located in the absorbing region of the root, as was demonstrated for K^+ (115). Transport of ions from the sieve tubes to the vessels of stems occurs as well (130, 166). In the stem of tree seedlings, $^{35}SO_4{}^-$ probably is translocated from the phloem to the xylem in rays (166), but very little is known about the pathways and possible mechanisms of translocation between phloem and xylem and vice versa. This appears to be a challenging area in translocation research.

TRANSPORT IN LEAVES

Patterns of distribution in leaves.—The pathways of translocation and the actual distribution of ions in leaves has not been examined as much as in roots. One may envision that ions are transported along with water in the leaf free space which appears confined to the cell walls (44). While being conveyed in the free space in direction of the transpiring water, ions can be transported across the plasmalemma into the inner compartments of the leaf cells by means of carrier mechanisms identical to the ones demonstrated in roots (147). Then they eventually move in the symplasm to other parts of the leaf including the phloem. A certain fraction of the ions may not be exported from the free space at all, but follows the lateral pathway of water to the main regions of water loss, i.e. the stomata and other epidermal cells (44). The loss of salt through salt glands to the surface of the leaves was reviewed by Lüttge (108) and will not be covered here. Part of the ions in the xylem are translocated up to the endings of the xylem elements, and from there they can enter other cells or are lost with the water by guttation. Bollard (23) has already pointed out that more knowledge should be acquired concerning the role of the vein endings in distributing ions in the leaves. He mentioned that the endings of the veins are surrounded frequently by a parenchymatous bundle sheath whose cell walls stain similarly to the Casparian strip in the endodermis of the root. The presence of a suberized layer in the radial walls of bundle-sheath cells has been confirmed for several grasses by O'Brien & Carr (124). They raised the question of whether a bundle sheath in the leaf may function as does the endodermis in the root. If this were the case, then ions could only proceed from the vascular bundle to the mesophyll through a symplasmic pathway. Another interesting structural feature in the leaves of most monocotyledonous plants is the absence of transfer cells (128), the significance of which in respect to ion transfer awaits elucidation.

The distribution of the elements K, Ca, and P in leaves of corn was examined by means of electron probe analysis (95, 101). In the blade of the third leaf of plants grown in solution culture for 4 weeks, these elements were mainly accumulated in the vascular bundles of the midrib arranged at the lower side of the leaf. Comparatively little accumulation was found in the mesophyll. Within the vascular bundles they were concentrated in the cell walls of the bundle sheath and the sclerenchyma, which is located between the bundle sheath and the lower epidermis. Such a pattern of distribution is in accordance with the concept that there is considerable export of ions from the xylem vessels in a lateral direction. It is too early, however, to draw any conclusions as to the possible role of the suberized layer in the bundle-sheath cell wall in ion transport across the bundle sheath to the sclerenchyma. Electron probe analyses revealed further that the guard cells contain a high K concentration when the stomata are open (80, 146). This paramount physiological feature will be discussed later.

Some attention has been paid to the pattern of Ca distribution in leaves. Millikan & Hanger (116) demonstrated with autoradiographs that ^{45}Ca in leaves of subterranean clover was initially accumulated in the veins, a situation resembling that in corn leaves (95). Under conditions of adequate Ca supply, the sites of initial accumulation were saturated rapidly and ^{45}Ca then appeared in the interveinal tissues (116). Moreover, microautoradiographic studies (67, 138) showed that ^{45}Ca is capable of moving laterally from the vascular bundles through the mesophyll and may accumulate in particular cells of both epidermal layers of oat leaves, namely in the guard and accessory cells of the stomata and in the bulliform cells (138).

Some evidence exists to the effect that transport of ions through the mesophyll of leaves may utilize both the free space and the symplasm as pathways. Ziegler & Lüttge (167) obtained microautoradiographs from leaves of *Limonium vulgare* that had been exposed to $^{36}Cl^-$, in which the labeled ion was located in the cell walls and the cytoplasm of the mesophyll cells, but not in their vacuoles. Corresponding experiments with other plant species and for other ions are virtually lacking.

Control mechanisms.—Environmental and internal factors undoubtedly exert a profound influence on the complex physiological phenomena of ion transport in leaves. What are the factors that control these processes in leaves? Unfortunately, these factors have not been clearly identified as yet, and so far as their mode of action is concerned, our knowledge is still in its infancy. Yet the environmental factor light has been demonstrated convincingly to affect ionic movements in leaves. Rains (136) showed that light enhanced the rate of K^+ absorption by corn leaf tissue and advanced the hypothesis that the role of light may be to provide the energy for the enhanced uptake of K^+ through synthesis of ATP in cyclic photophosphorylation.

Light may cause the stomata to open. In experiments with epidermal strips from leaves of *Vicia faba*, Fischer & Hsiao (63) obtained evidence that absorption of K^+ may be the primary mechanism of light-stimulated stomatal opening. Through using electron probe analysis, this process was causally linked to a specific transport of K^+ into the guard cells (80, 146). Moreover, the K^+ influx into and the K^+ efflux from the guard cells paralleled stomatal opening in the light and closing in the dark respectively (79). From the same publication it appeared that light with wavelength greater than 700 nm was as effective as white light. This led the authors to suggest that the role of light is to activate photosystem I in photosynthesis, which could furnish the energy necessary for K^+ influx through cyclic photophosphorylation. Studies concerned with the effect of metabolic inhibitors on K^+ flux and stomatal opening led to the same conclusion (79) and were in line with earlier observations by Rains (136) on the source of energy for light-enhanced K^+ absorption by leaf tissue. The K^+ specific process of light-stimulated stomatal opening may require Ca^{++} (126), which is present in the guard and the accessory cells of the stomata in ample amounts (138).

With regard to movement of ions in leaves, the intriguing phenomenon of the specific K^+ influx into the guard cells in the light exposes several problems which need further study. Is the K concentration of the free space great enough in the vicinity of the stomata to supply the potassium required for the stomata to open, or is K^+ furnished by other leaf cells through a net loss from their inner compartments? Rains (136) ascertained little or no efflux of K^+ previously accumulated in leaf tissue in the light or dark. Thus K^+ influx into the stomata in the light and K^+ efflux in the dark might be matched by movements of K^+ in the free space of the leaf. This problem is still to be solved, as is the question of how K^+ moves out of the guard cells in the dark.

Light not only has a controlling influence on ion transport in leaves by providing the energy for active transport across cellular membranes, but it also affects ionic movements through activation of the phytochrome system. The form of phytochrome absorbing at 730 nm (P_{fr}), which is considered to be the active form of the pigment, may exert its control by regulating membrane permeability (66). Köhler (88) observed that the phytochrome system affected ion transport directly and not via an alteration of growth. Specifically, red light transforming the inactive phytochrome P_r to the active form P_{fr} enhanced transport of K^+ into the plumule of pea seedlings. The effect of red light on K^+ transport preceded the red-light induced promotion of growth and was abolished by far-red light applied immediately after red light. Another case of phytochrome-controlled ion translocation in leaves was discovered by Satter et al (145). The movement of leaflets in *Albizzia julibrissin* is regulated by phytochrome through turgor changes in the motor cells of the pulvinule. Leaflet pairs exposed to red light close within 30 to 90 min after transfer to darkness. Interruption of darkness by far-red light inhibits closing of the leaflet. Electron probe analyses revealed that the phytochrome-controlled leaflet closing is caused by the efflux of K^+ from the ventral and influx of K^+ into the dorsal motor cells of the pulvinule (145). The investigators assumed that opening of the leaflets is brought about by a net transport of K^+ from the dorsal into the ventral motor cells, and they suggested that the K^+ efflux from the ventral cells during closing requires the active phytochrome P_{fr} located in the plasma membrane of these cells. The phytochrome-controlled K^+ flux in leaves represents the first known instance of an ionic movement in a plant organ that is regulated by an endogenous pigment system.

It has been argued many times that the transport of solutes in plants may be under hormonal control. However, as Milthorpe & Moorby (119) clearly stated in a review, most studies fail to distinguish between direct effects on transport and those arising from altered growth. Unless new evidence to the contrary is produced, it appears that most effects of hormones on transport could arise from the stimulation of metabolic sinks.

Import and export.—Leaves not only import ions through the xylem, but they are also capable of exporting ions through the phloem. The balance between import and export in a particular leaf depends greatly on its physiological age. In addition, only those ions that are phloem-mobile are exported from leaves to any large extent. From early studies (17, 34, 161) it appeared that K^+, Cl^-, and P were exported readily from leaves, whereas Ca^{++} was considered to be essentially phloem-immobile. Sulfur and some of the micronutrients were intermediate so far as their export was concerned. The highly mobile K^+ ions are imported mainly into young leaves, while there is very little export until the leaf reaches maturity and export matches import (70). Likewise, the stage of leaf development controls the export of P through the phloem in such a way that the extent of P export increases with the age of the leaf (1, 69, 90, 154). Magnesium is exported from leaves to a lesser degree than K^+ and P (33, 148), though it is not considered phloem-immobile.

There has been a revived interest in studying any possible retranslocation of Ca^{++} from leaves. Biddulph et al (16) confirmed that very little ^{45}Ca was exported from bean leaves over a period of up to 4 days after foliar application. When the roots of bean plants were allowed to absorb ^{45}Ca and were then transferred to solutions containing unlabeled Ca^{++}, Greene & Bukovac (68) found that ^{45}Ca first accumulated in the primary leaves. In a period of 19 days, however, over half of the ^{45}Ca accumulated in the primary leaves was lost and redistributed to newly formed tissues. This indicates that Ca in leaves becomes partly exchangeable when the sites of Ca accumulation are saturated. The exchangeable fraction of Ca may then be exported, possibly through the phloem. Such a conclusion is in agreement with results by Millikan & Hanger (117, 118) and is supported by Ringoet et al (139, 140), who demonstrated that ^{45}Ca applied to oat leaves at a low concentration moved only in an acropetal direction. However, when higher ^{45}Ca levels were applied, there was also some basipetal transport which probably took place in the phloem (140). Hence, one can argue that the limited redistribution of Ca^{++} from leaves is not due to its immobility in the phloem but rather to the great capacity of Ca accumulation by other leaf tissues. Once this capacity is saturated, Ca^{++} is slowly exported from a leaf through the phloem.

TRANSPORT IN THE PHLOEM

No attempt will be made to treat the structure of the phloem, nor will the mechanism of phloem transport be discussed. These aspects of phloem transport are covered in a comprehensive monograph by Crafts & Crisp (40). What will be discussed are the questions of how ions are taken up into the sieve tubes and what ions are translocated in these longitudinal channels. Obviously, the latter process depends directly upon uptake.

Crafts & Crisp (40) look upon the sieve-tube system as a functional phase of the symplasm. Consequently, the plasmalemma of the sieve tubes has to be a functional membrane and should form a permeability barrier to the flow of solutes. One criterion for testing the function of the plasmalemma as a membrane is the successful plasmolysis of the particular cell concerned. Currier et al (45) showed that mature sieve tubes can be plasmolyzed and deplasmolyzed and concluded that the sieve-tube protoplast exhibits a differential permeability. Most researchers agree that sugars are transported actively into the sieve tubes (57, 164), a process commonly referred to as phloem loading. This raises the possibility of ions also being transported across the plasmalemma of the sieve tubes by some active process. Attempts to elucidate the existence and nature of active ion transport into the sieve tubes have encountered experimental difficulties. One needs to demonstrate that an ion is pumped into the sieve tube, leading to a higher ion concentration within the sieve tube than outside of the membrane. Ionic concentrations of sieve-tube saps can be determined accurately by means of the aphid technique (131, 165). The quantitative estimation of ions exterior to the sieve-tube plasmalemma is more difficult to achieve and will have to be approached by some method of quantitative localization, as outlined earlier in this review. Furthermore, it is not understood whether ions are supplied to the plasmalemma of sieve tubes by way of the free space or through the symplasm, or whether transfer cells, which occur in the phloem of minor veins in many species (71), are implicated in the delivery of ions to the sieve-tube system.

In experiments with excised phloem tissues from several species, Bieleski (21, 22) intended to unravel the mechanisms of phosphate and sulfate uptake by sieve tubes. The phloem tissues absorbed both ions against a concentration gradient; the accumulation by the phloem was many times greater than by parenchymatous tissue. Microautoradiographs revealed that the sieve tubes were among the most actively accumulating cells of the phloem tissue. It was concluded that translocating sieve tubes are capable of accumulating sulfate and phosphate and must have a semipermeable membrane possessing specific transport mechanisms. The microautoradiographic evidence of sulfate and phosphate localization within sieve tubes (22) was in accordance with earlier observations by Biddulph (19). Accumulation of other ions in sieve tubes has not been demonstrated as yet. Fischer (62), however, proposed a model of ion absorption by sieve tubes which accounts for some features of ion export from leaves. According to Fischer's model, the plasmalemma of sieve tubes would contain the same selective mechanisms of absorption as that of cells in the root. Consequently, the ions K^+, NO_3^-, and PO_4^{3-} would be transported into the sieve tubes selectively and rapidly by an active process, whereas Ca^{++} would move passively across the membrane. Yet Johanson & Joham (85) obtained evidence in favor of a carrier-mediated uptake of Ca^{++} by cotton roots which leaves the nature of Ca^{++} transport into the sieve tubes open for further investigation.

Some information is available about the presence or absence of ions in the sap of sieve tubes. Analyses of sieve-tube sap with the aphid technique (131, 165) which yields essentially uncontaminated samples, revealed high K concentrations ranging from about 0.1% up to 2%. Inorganic phosphate generally ranked second in concentration (165), but it was not found in samples from willow (131). Nor was phosphate detected (86) in sieve-tube sap obtained by phloem incision. Yet it was present in the phloem exudate from inflorescences of *Yucca flaccida* at a concentration of about 100 ppm or roughly 30% of the total P in the exudate (151). Magnesium was detectable at a concentration similar to that of phosphate (165), whereas 10 to 20 ppm Ca were found (151, 165). All the reports showed clearly the absence of nitrate and sulfate in sieve-tube sap; apparently, the elements N and S are translocated in the phloem as organic compounds. Moreover, only small amounts of Cl were detectable. This is in contradiction to its ability to be exported readily from leaves (161).

The presence of Ca in sieve-tube sap, though at a relatively low concentration (151, 165), confirms our previous conclusion that Ca^{++} may be translocated in the phloem. Calcium was also localized directly in sieve tubes of the fruit stalk of peas (97) and was shown to be translocated in the phloem of *Yucca flaccida* when the supply of Ca was fairly high (159). Eventually, Ca may be deposited on the sieve plates during the formation of callose, possibly as a Ca-phosphate precipitate (58).

ACKNOWLEDGMENTS

The author is indebted to several colleagues for permission to use data in advance of publication and to E. Epstein, Department of Soils and Plant Nutrition, University of California at Davis, and several colleagues in the Department of Plant Sciences, Texas A&M University, for critically discussing the manuscript.

LITERATURE CITED

1. Ahlgren, G. E., Sudia, T. W. 1964. *Bot. Gaz.* 125:204–7
2. Anderson, W. P. 1972. *Ann. Rev. Plant Physiol.* 28:51–72
3. Anderson, W. P., Allen, E. 1970. *Planta* 93:227–32
4. Anderson, W. P., House, C. R. 1967. *J. Exp. Bot.* 18:544–55
5. Anderson, W. P., Reilly, E. J. 1968. *J. Exp. Bot.* 19:19–30
6. Arisz, W. H. 1956. *Protoplasma* 46:5–62
7. Arisz, W. H., Wiersema, E. P. 1966. *Proc. Kon. Ned. Akad. Wetensch. Ser. C* 69:223–41
8. Baker, D. A. 1968. *Planta* 83:390–92
9. Bange, G. G. J. 1965. *Plant Soil* 22:280–306
10. Bange, G. G. J., van Vliet, E. 1961. *Plant Soil* 15:312–28
11. Barber, D. A. 1968. *Ann. Rev. Plant Physiol.* 19:71–88
12. Barber, D. A., Loughman, B. C.

1967. *J. Exp. Bot.* 18:170–76
13. Barber, D. A., Sanderson, J., Russell, R. S. 1968. *Nature* 217:644
14. Bell, C. W., Biddulph, O. 1963. *Plant Physiol.* 38:610–14
15. Biddulph, O. 1959. *Plant Physiology: A Treatise*, ed. F. C. Steward, 2:553–603. New York: Academic. 758 pp.
16. Biddulph, O., Cory, R., Biddulph, S. F. 1959. *Plant Physiol.* 34: 512–19
17. Biddulph, O., Markle, J. 1944. *Am. J. Bot.* 31:65–71
18. Biddulph, O., Nakayama, F. S., Cory, R. 1961. *Plant Physiol.* 36:429–36
19. Biddulph, S. F. 1956. *Am. J. Bot.* 43:143–48
20. Biddulph, S. F. 1967. *Planta* 74: 350–67
21. Bieleski, R. L. 1966. *Plant Physiol.* 41:447–54
22. Ibid, 455–66
23. Bollard, E. G. 1960. *Ann. Rev. Plant Physiol.* 11:141–66
24. Bollard, E. G. 1970. *The Biochemistry of Fruits and Their Products*, ed. A. C. Hulme, 1:387–425. New York: Academic. 620 pp.
25. Bonnett, H. T. Jr. 1968. *J. Cell Biol.* 37:199–205
26. Bowen, G. D., Rovira, A. D. 1966. *Nature* 211:665–66
27. Bowling, D. J. F. 1965. *Nature* 206:317–18
28. Bowling, D. J. F. 1968. *J. Exp. Bot.* 19:381–88
29. Boyer, J. S. 1971. *Crop Sci.* 11: 403–7
30. Branton, D., Jacobson, L. 1962. *Plant Physiol.* 37:546–51
31. Brouwer, R. 1965. *Ann. Rev. Plant Physiol.* 16:241–66
32. Brown, J. C. 1963. *Soil Sci.* 96: 387–94
33. Bukovac, M. J., Teubner, F. G., Wittwer, S. H. 1960. *Proc. Am. Soc. Hort. Sci.* 75:429–34
34. Bukovac, M. J., Wittwer, S. H. 1957. *Plant Physiol.* 32:428–35
35. Burley, J. W. A., Nwoke, F. I. O., Leister, G. L., Popham, R. A. 1970. *Am. J. Bot.* 57:504–11
36. Clarkson, D. T., Robards, A. W., Sanderson, J. 1971. *Planta* 96: 292–305
37. Clarkson, D. T., Sanderson, J. In preparation
38. Clarkson, D. T., Sanderson, J.,

Russell, R. S. 1968. *Nature* 220:805–6
39. Crafts, A. S., Broyer, T. C. 1938. *Am. J. Bot.* 25:529–35
40. Crafts, A. S., Crisp, C. E. 1971. *Phloem Transport in Plants.* San Francisco: Freeman. 481 pp.
41. Crossett, R. N. 1966. *New Phytol.* 65:443–58
42. Crossett, R. N. 1967. *Nature* 213: 312–13
43. Crossett, R. N. 1968. *Aust. J. Biol. Sci.* 21:1063–67
44. Crowdy, S. H., Tanton, T. W. 1970. *J. Exp. Bot.* 21:102–11
45. Currier, H. B., Esau, K., Cheadle, V. I. 1955. *Am. J. Bot.* 42:68–81
46. Drew, M. C., Biddulph, O. 1971. *Plant Physiol.* 48:426–32
47. Dumbroff, E. B., Peirson, D. R. 1971. *Can. J. Bot.* 49:35–38
48. Dunlop, J., Bowling, D. J. F. 1971. *J. Exp. Bot.* 22:434–44
49. Ibid, 445–52
50. Edwards, D. G. 1970. *Aust. J. Biol. Sci.* 23:255–64
51. Epstein, E. 1960. *Am. J. Bot.* 47: 393–99
52. Epstein, E. 1966. *Nature* 212: 1324–27
53. Epstein, E. 1968. *Experientia* 24: 616
54. Epstein, E. 1971. *Plant Physiol. Suppl.* 47:26
55. Epstein, E. 1972. *Mineral Nutrition of Plants: Principles and Perspectives*, 325–44. New York: Wiley. 412 pp.
56. Esau, K. 1965. *Plant Anatomy*, 489–93. New York: Wiley. 2nd ed. 767 pp.
57. Eschrich, W. 1970. *Ann. Rev. Plant Physiol.* 21:193–214
58. Eschrich, W., Eschrich, B., Currier, H. B. 1964. *Planta* 63:146–54
59. Etherton, B. 1968. *Plant Physiol.* 43:838–40
60. Evans, E. C. III. 1964. *Science* 144:174–77
61. Fischer, H. 1967. *Encyclopedia of Plant Physiology*, ed. W. Ruhland, 13:200–68. Berlin: Springer
62. Fischer, H. 1967. *Z. Pflanzenernaehr. Bodenk.* 118:100–11
63. Fischer, R. A., Hsiao, T. C. 1968. *Plant Physiol.* 43:1953–58
64. Frey-Wyssling, A., Mühlethaler, K. 1965. *Ultrastructural Plant Cy-*

tology, 279. Amsterdam: Elsevier. 377 pp.
65. Gahan, P. B., Rajan, A. K. 1965. *Exp. Cell Res.* 38:204–7
66. Galston, A. W., Davies, P. J. 1970. *Control Mechanisms in Plant Development,* 1–39. Englewood Cliffs: Prentice-Hall. 184 pp.
67. Gielink, A. J., Sauer, G., Ringoet, A. 1966. *Stain Technol.* 41: 281–86
68. Greene, D. W., Bukovac, M. J. 1968. *Proc. Am. Soc. Hort. Sci.* 93:368–78
69. Greenway, H., Gunn, A. 1966. *Planta* 71:43–67
70. Greenway, H., Pitman, M. G. 1965. *Aust. J. Biol. Sci.* 18: 235–47
71. Gunning, B. E. S., Pate, J. S. 1969. *Protoplasma* 68:107–33
72. Hall, J. L., Sexton, R., Baker, D. A. 1971. *Planta* 96:54–61
73. Helder, R. J. 1964. *Acta Bot. Neer.* 13:488–506
74. Helder, R. J. See Ref. 61, 20–43
75. Helder, R. J., Boerma, J. 1969. *Acta Bot. Neer.* 18:99–107
76. Hoad, G. V., Peel, A. J. 1965. *J. Exp. Bot.* 16:742–58
77. Hodges, T. K., Vaadia, Y. 1964. *Plant Physiol.* 39:104–8
78. House, C. R., Findlay, N. 1966. *J. Exp. Bot.* 17:627–40
79. Humble, G. D., Hsiao, T. C. 1970. *Plant Physiol.* 46:483–87
80. Humble, G. D., Raschke, K. 1971. *Plant Physiol.* 48:447–53
81. Jacobson, L., Hannapel, R. J., Moore, D. P. 1958. *Plant Physiol.* 33:278–82
82. Jacoby, B. 1966. *Nature* 211:212
83. Jacoby, B. 1967. *Ann. Bot. London* 31:725–30
84. Jarvis, P., House, C. R. 1970. *J. Exp. Bot.* 21:83–90
85. Johanson, L., Joham, H. E. 1971. *Plant Soil* 34:331–39
86. Kimmel, Ch. 1962. *Über das Vorkommen anorganischer Ionen in Siebröhrensäften und den Transport von Salzen im Phloem.* Dissertation. Tech. Hochschule, Darmstadt
87. Klepper, B., Kaufmann, M. R. 1966. *Plant Physiol.* 41:1743–47
88. Köhler, D. 1969. *Planta* 84:158–65
89. Kollmann, R., Dörr, I. 1969. *Ber. Deut. Bot. Ges.* 82:415–25
90. Koontz, H., Biddulph, O. 1957. *Plant Physiol.* 32:463–70
91. Kramer, P. J. 1969. *Plant and Soil Water Relationships: A Modern Synthesis,* 214–57. New York: McGraw-Hill. 482 pp.
92. Krichbaum, R., Lüttge, U., Weigl, J. 1967. *Ber. Deut. Bot. Ges.* 80:167–76
93. Laties, G. G. 1969. *Ann. Rev. Plant Physiol.* 20:89–116
94. Laties, G. G., Budd, K. 1964. *Proc. Nat. Acad. Sci. USA* 52: 462–69
95. Läuchli, A. 1967. *Planta* 75:185–206
96. Läuchli, A. 1968. *Vortr. Gesamtgeb. Bot.* 2:58–65
97. Läuchli, A. 1968. *Planta* 83:137–49
98. Läuchli, A., Epstein, E. 1970. *Plant Physiol.* 45:639–41
99. Ibid 1971. 48:111–17
100. Läuchli, A., Lüttge, U. 1968. *Planta* 83:80–98
101. Läuchli, A., Schwander, H. 1966. *Experientia* 22:503–5
102. Läuchli, A., Spurr, A. R., Epstein, E. 1971. *Plant Physiol.* 48:118–24
103. Läuchli, A., Spurr, A. R., Wittkopp, R. W. 1970. *Planta* 95: 341–50
104. Lopez-Saez, J. F., Gimenez-Martin, G., Risueno, M. C. 1966. *Protoplasma* 61:81–84
105. Lopushinsky, W. 1964. *Nature* 201:518–19
106. Loughman, B. C. 1966. *New Phytol.* 65:388–97
107. Lüttge, U. 1969. *Protoplasmatologia* VIII, 7b:91–114
108. Lüttge, U. 1971. *Ann. Rev. Plant Physiol.* 22:23–44
109. Lüttge, U., Ed. 1972. *Methods of Microautoradiography and Electron Probe Analysis.* Berlin: Springer. In press
110. Lüttge, U., Laties, G. G. 1966. *Plant Physiol.* 41:1531–39
111. Ibid 1967. 42:181–85
112. Lüttge, U., Laties, G. G. 1967. *Planta* 74:173–87
113. Lüttge, U., Weigl, J. 1962. *Planta* 58:113–26
114. Maizel, J. V., Benson, A. A., Tolbert, N. E. 1956. *Plant Physiol.* 31:407–8
115. Meiri, A., Anderson, W. P. 1970. *J. Exp. Bot.* 21:908–14

116. Millikan, C. R., Hanger, B. C. 1964. *Aust. J. Biol. Sci.* 17: 823–44
117. Ibid 1965. 18:211–26
118. Ibid 1966. 19:1–14
119. Milthorpe, F. L., Moorby, J. 1969. *Ann. Rev. Plant Physiol.* 20: 117–38
120. Moore, D. P., Mason, B. J., Maas, E. V. 1965. *Plant Physiol.* 40: 641–44
121. Munns, D. N., Jacobson, L., Johnson, C. M. 1963. *Plant Soil* 19: 193–204
122. Munns, D. N., Johnson, C. M., Jacobson, L. 1963. *Plant Soil* 19: 115–26
123. Ibid, 285–95
124. O'Brien, T. P., Carr, D. J. 1970. *Aust. J. Biol. Sci.* 23:275–87
125. O'Brien, T. P., McCully, M. E. 1969. *Plant Structure and Development,* 5. London: Macmillan. 114 pp.
126. Pallaghy, C. K. 1970. *Z. Pflanzenphysiol.* 62:58–62
127. Pate, J. S. 1965. *Science* 149:547–48
128. Pate, J. S., Gunning, B. E. S. 1969. *Protoplasma* 68:135–56
129. Peel, A. J. 1963. *J. Exp. Bot.* 14: 438–47
130. Ibid 1967. 18:600–6
131. Peel, A. J., Weatherley, P. E. 1959. *Nature* 184:1955–56
132. Peirson, D. R., Dumbroff, E. B. 1969. *Can. J. Bot.* 47:1869–71
133. Pitman, M. G. 1965. *Aust. J. Biol. Sci.* 18:10–24
134. Ibid 1966. 19:257–69
135. Pope, D. T., Munger, H. M. 1953. *Proc. Am. Soc. Hort. Sci.* 61: 472–80
136. Rains, D. W. 1968. *Plant Physiol.* 43:394–400
137. Rains, D. W. 1972. *Ann. Rev. Plant Physiol.* 23:367–88
138. Ringoet, A., Rechenmann, R. V., Gielink, A. J. 1971. *Z. Pflanzenphysiol.* 64:60–64
139. Ringoet, A., Rechenmann, R. V., Veen, H. 1967. *Radiat. Bot.* 7: 81–90
140. Ringoet, A., Sauer, G., Gielink, A. J. 1968. *Planta* 80:15–20
141. Robards, A. W. 1968. *Planta* 82: 200–10
142. Robards, A. W. 1971. *Protoplasma* 72:315–23
143. Roland, J. C., Bessoles, M. 1968.
144. Russell, R. S., Barber, D. A. 1960. *Ann. Rev. Plant Physiol.* 11: 127–40
145. Satter, R. L., Marinoff, P., Galston, A. W. 1970. *Am. J. Bot.* 57:916–26
146. Sawhney, B. L., Zelitch, I. 1969. *Plant Physiol.* 44:1350–54
147. Smith, R. C., Epstein, E. 1964. *Plant Physiol.* 39:992–96
148. Steucek, G. L., Koontz, H. V. 1970. *Plant Physiol.* 46:50–52
149. Steward, F. C. 1954. *Symp. Soc. Exp. Biol., 8th, 1953* 8:393–406
150. Stout, P. R., Hoagland, D. R. 1939. *Am. J. Bot.* 26:320–24
151. Tammes, P. M. L., van Die, J. 1964. *Acta Bot. Neer.* 13:76–83
152. Tiffin, L. O. 1967. *Plant Physiol.* 42:1427–32
153. Tiffin, L. O. 1972. *Symp. Micronutrients in Agriculture.* In press
154. Wanner, H., Bachofen, R. 1961. *Planta* 57:531–42
155. Weigl, J. 1969. *Planta* 84:311–23
156. Weigl, J., Lüttge, U. 1962. *Planta* 59:15–28
157. Weiss, M. G. 1943. *Genetics* 28: 253–68
158. Welch, R. M., Epstein, E. 1968. *Proc. Nat. Acad. Sci. USA* 61: 447–53
159. Wiersum, L. K., Vonk, C. A., Tammes, P. M. L. 1971. *Naturwissenschaften* 58:99
160. Willenbrink, J. See Ref. 61, 178–99
161. Woolley, J. T., Broyer, T. C., Johnson, G. V. 1958. *Plant Physiol.* 33:1–7
162. Yu, G. H., Kramer, P. J. 1967. *Plant Physiol.* 42:985–90
163. Ibid 1969. 44:1095–1100
164. Ziegler, H. 1956. *Planta* 47:447–500
165. Ziegler, H. 1962. *Verh. 11th Int. Kongr. Entomol.* 2:537–40
166. Ziegler, H. 1965. *Isotopes and Radiation in Soil-Plant Studies,* 361–70. Wien: International Atomic Energy Agency
167. Ziegler, H., Lüttge, U. 1967. *Planta* 74:1–17
168. Ziegler, H., Weigl, J., Lüttge, U. 1963. *Protoplasma* 56:362–70
169. Zimmermann, M. H. 1960. *Ann. Rev. Plant Physiol.* 11:167–90

C. R. Acad. Sci. Ser. D 267: 589–92

Ann. Rev. Plant Physiol. 1972. 23:219–34

BLUE LIGHT AND CARBON METABOLISM[1]   7530

NATALIA P. VOSKRESENSKAYA

*K. A. Timiryasev Institute of Plant Physiology
USSR Academy of Science, Moscow*

CONTENTS

INTRODUCTION

This review is an attempt to consider blue light as a factor in the self-regulation of photosynthetic metabolism of carbon.

Over 30 years ago, Lubimenko (74) pointed out the possibility of regulation of photosynthetic carbon metabolism by means of the "secondary" photochemical reactions induced by blue light. Later, experimental evidence was obtained for higher plants (127, 131) and algae (61, 97) which established that among newly formed organic substances the amount of carbohydrate decreases and the amount of protein increases under blue light as compared with red light. Enhancement of protein formation by blue light is now an accepted fact (see 80, 97, 131). The regulatory role of blue light on metabolism is ascribed to absorption of the light by flavins or carotenoids (65, 127).

[1] The following abbreviations are used: FAD (flavin-adenine-dinucleotide); FMN (flavin-mononucleotide); GPDH (glyceraldehyde 3-phosphate dehydrogenase); PEP (phosphoenolpyruvate); 3-PGA (3-phosphoglycerate); OAA (oxalacetate); RuDP (ribulose 1,5-diphosphate).

A low energy level of red light is also known to regulate the biochemical activities of plants, and its function usually is explained by phototransformation of phytochrome (44, 80).

In principle the regulatory role of both blue and red light occurs in the absence of photosynthesis (62, 71, 80). However, it is in illuminated green plants that an interaction between the regulatory light reactions and photosynthesis is most likely to occur. Warburg, Krippahl & Schröder (147) first showed that catalytic amounts of blue light might be necessary to obtain high quantum yields of photosynthesis. Later this regulatory effect of blue light was observed for stomata movement (54, 76), the orientation of chloroplasts in the cell (38, 148) and their conformational changes (149), the photochemistry of the chlorophyll (66), the energy efficiency of photosynthesis (5, 6), and for the CO_2 and O_2 gas exchange (31, 131, 148).

In this review we will consider mainly short-term (seconds, minutes) action of blue light compared with red light on carbon metabolism in green plants previously grown under white light. Considerable attention will be given to the possibility of regulation by blue light in terms of withdrawal of carbon from the photosynthetic carbon cycle at the level of 3-PGA or the glycolate pathway reactions. Reports concerning the restoration of photosynthesis by blue light in plants grown under red light alone will also be discussed.

Recent reviews have covered the principal pathways of photosynthetic carbon metabolism [although without taking into account the effects of blue light (4, 29, 37, 53, 123)], as well as the energy and substrate interaction of the chloroplast and of the entire cell (4, 146) and the phenomenon of photorespiration (49). In most cases we believe it is sufficient to refer the reader to these papers.

Principal Regulatory Effects of Blue Light and Conditions of their Manifestation

The regulatory role of blue light in carbon metabolism usually is demonstrated by comparing the effects of blue and red light under conditions identical for the photochemical act of photosynthesis, i.e. when the light fluxes are equal with respect to the quanta absorbed. However, due to the activation of respiration (49, 60, 65, 129, 144) the efficiency of blue light for photosynthesis under such conditions, and especially at low light intensities, is much lower than that of red light. Therefore, in order to separate the specific effect of the quality of light on carbon metabolism from the possible effect of low light intensity, blue light is equalized to red light with respect to the rate of CO_2 assimilation (131). In this case a difference in the distribution of ^{14}C is revealed under red and blue light. Blue light activates [within a wide range of CO_2 concentrations (39, 55, 68, 86, 104) and light intensities (55, 86, 131, 144)] the incorporation of ^{14}C into certain amino acids (2, 39, 55, 86, 104, 128, 131) and also on occasion into intermediates of the tricarboxylic acid cycle (2, 39, 104) and protein, according to Keerberg (132). On the other hand, it inhibits carbon incorporation into various carbohydrates (28, 39, 86, 104, 131).

This effect of blue light on photosynthetic carbon metabolism has been observed for green algae and higher plants assimilating CO_2 either by the Calvin cycle or by the C_4-dicarboxylic acid pathway (131). However, it has not been reported in the photosynthetic bacteria (18, 39) and in the blue-green alga *Anacystis nidulans* (15).

The maximum increase in the incorporation of radioactivity from $^{14}CO_2$ into aspartate, glutamate, and malate, and sometimes in serine and glycine (28) concomitant with a decrease in the radioactivity of sucrose and sugar phosphates, was found under blue light over the wavelength range 458-480 nm compared to red light of 670–680 nm (28, 86, 104). For amino acids the effect is diminished considerably at 505 nm, and for sugars it disappears altogether (104). Since the maximum increase in the radioactivity of amino acids coincides with the maximum in the action spectrum of activation of protein synthesis by blue light (62), a connection is thought to exist between the two phenomena. The similar enhancement of ^{14}C incorporation into amino acids and organic acids is found not only on exposure to blue light but also on adding small amounts of blue light to red light (40), or under white light as compared with red light (23). The aftereffect of blue light, furthermore, is observed under red light (41). These observations point to: (*a*) a low light saturation of the effects; (*b*) its manifestations in the presence of red light; and (*c*) a prolonged action of blue light effects on metabolism which only gradually disappears in the absence of this light.

EFFECT OF BLUE LIGHT IN THE ABSENCE OF PHOTOSYNTHESIS

The regulatory nature of blue light on metabolism is radically independent of photosynthesis. This was established by experiments with *Chlorella* cells whose photosynthesis was inhibited by DCMU (78, 86). While in the absence of photosynthesis the incorporation of ^{14}C into aspartate, glutamate, and malate under red light remains the same as in the dark, weak ($\sim$300 erg /cm^2 sec) blue light substantially increases the radioactivity of these compounds. In accounting for these results it is not improbable that blue light increases the reaction of glycolysis and in turn the formation of PEP. As a result of the activation of the PEP carboxylation under blue light, OAA is formed which is partially reduced to malate while the bulk of it is converted to aspartate. The localization of ^{14}C in the carboxyl carbon of aspartate supports this pathway (78, 86). Oxalacetate can also serve as a source for the continuous regeneration of the Krebs cycle reactions leading to the synthesis of glutamate.

In higher plants and algae (both green and nongreen) activation of respiration by blue light has been reported (49, 63, 65, 130, 144). This phenomenon has the same action maximum and low light saturation as the blue light activation of ^{14}C incorporation in amino and carboxylic acids and in protein formation (60, 71). Therefore, the possibility exists that blue light action on carbon metabolism and respiration are interdependent. This suggestion is supported by experiments with colorless *Chlorella* mutant. In this organism Miyachi (personal communication) observed that both respiration and $^{14}CO_2$ fixation were enhanced by blue light. However, in a yellow *Chlorella* mutant the respiration is activated by blue light but the carbon metabolism is not altered (104). It is possible that blue light in green plants increases the accessibility of the substrate for the PEP carboxylation outside of chloroplasts. In this connection it is interesting to note that in the absence of photosynthesis the protein synthesis in *Chlorella* at the expense of endogenous sugars occurs only under blue light. Upon addition of sugars, the activation of protein synthesis has been detected both under blue light and to a smaller extent under red light. It is assumed, therefore, that the effect of blue light in the former case is due to its action on the permeability of chloroplast membranes and an increase in the accessibility of substrates to protein synthesis outside of the chloroplasts (71, 96).

EFFECT OF WEAK BLUE LIGHT IN PRESENCE OF PHOTOSYNTHESIS

Under photosynthetic conditions, the enhancement of ^{14}C incorporation in amino acids and organic acids formed at the expense of PGA, as well as the decrease in radioactivity of carbohydrates, is found at all levels of light intensity (2, 28, 39, 86, 104, 131, 143). At the same time, the light intensity probably determines the origin of the substrates and the time-dependence of blue light effects. Thus at low light intensity (~500 erg/cm^2 sec) the kinetics of tracer incorporation into aspartate, malate, and glutamate are the same during the first 5 minutes of exposure to red and blue light. Subsequently the activity of these compounds progressively increases under blue but not under red light. On the contrary, the radioactivity incorporated into the sugar phosphates, which is the same for blue and red light during the first 5 minutes, subsequently becomes lower under blue light (86). Weak blue light (as distinct from red light) with or without the presence of photosynthesis seems to provide possibilities for the continuous inflow of substrates for the formation of PEP and its carboxylation. The time of appearance of the effect of blue light on carbon metabolism (after 5 minutes of illumination) coincides with the time of blue light activation of respiration (60).

Respiratory substrates may include, among others, sugars newly formed during photosynthesis (86, 104). The activation of ^{14}C incorporation in aspartate, malate, and glutamate at low blue light intensity either in the presence or absence of photosynthesis probably is connected with the increased possibility of anaplerotic mechanisms associated with the activation of Krebs cycle reactions. Thus the effects of low blue light during photosynthesis are revealed mainly outside the chloroplast. However, the causal relationship of blue light effects on respiration and carbon metabolism is not yet clear. For instance, Pickett (95) did not detect a correlation between the effect of blue light flashes on respiration and protein synthesis.

Effects of Strong Blue Light in the Presence of Photosynthesis

It is well known that alanine, malate, and glutamate can be formed as early products of photosynthesis (4, 29). The distribution of the tracer in alanine and aspartate under both strong blue and red light points to their origin from 3-PGA of the Calvin cycle (17, 86). Activation by strong blue light of ^{14}C incorporation in amino acids, as distinct from low-intensity blue light, is revealed after a few seconds (17, 28, 128). The differential effect of blue light is particularly reflected in the specific activity of amino acids (17). At the end of 30 seconds the incorporation of the tracer both in alanine and aspartate may be 40 to 50% higher under blue light than under red light (128). In contrast to weak light, the incorporation of the tracer in these amino acids under strong light is a process definitely stable in time both under red and blue light (28, 55, 86, 128), but it is more active under the latter. Thus one can conclude that at high intensities of either blue or red light 3-PGA originated from the Calvin cycle is a main source of PEP formation.

The radioactivity of alanine (39, 55, 86, 104, 131) is often found to increase under blue light. This effect may take place at the same time (55, 104, 128, 131) or later than for aspartate (86). Occasionally the activating effect of blue light on ^{14}C incorporation in amino acids is limited to alanine (55). In any case, the activation of tracer incorporation in alanine or aspartate under blue light apparently is due to the greater amount of substrate than under red light and the possibility of transformation of 3-PGA to PEP and its carboxylation. Therefore, accumulation of tracer in alanine is not likely to be a property of blue light per se (120).

Blue Light and the Activity of Phosphoenolpyruvate Carboxylase

To account for the effects of blue light involving oxalacetate, and subsequently aspartate and malate, the postulate that in green plants two types of PEP carboxylase are present may provide an explanation (92, 114, 119). It is further postulated that one of them is activated by light and the only PEP carboxylase plays the key role in the photosynthetic metabolism of carbon (114, 119). The PEP carboxylation and the formation of C_4-dicarboxylic acids is the main path for CO_2 assimilation in a large group of plants (37, 53). It has been shown that PEP carboxylase of sugar-cane chloroplasts is activated by light (3). But even for species of plants whose primary carboxylation reaction is catalyzed by RuDP carboxylase; β-carboxylation may account for 3 to 12% (145) or more (26, 81) of the total assimilation of CO_2. Miyachi (77) reported that preliminary illumination of the *Chlorella* cells by blue light does not influence the activity of RuDP carboxylase but enhances the activity of PEP carboxylase by a factor of 2 to 3 compared with red light.

In the dark following the white light, the activity of the carboxylation reaction involving phosphoenolpyruvate gradually decreases (88). Weak blue but not red light restores the enzyme activity after the darkness. Since chloramphenicol and cycloheximide inhibit this restoration, it is assumed that blue light activates the synthesis of the enzyme. This finding with poisons also explains the fact that after darkness (39, 68, 88, 104) or after a preliminary exposure to red light (71) all effects of blue light on metabolism are especially pronounced. However, immediately after illumination with white light (in whose integral flux blue rays are present, providing optimal conditions for manifestation of the activity of PEP carboxylase) the difference in the radioactivity of aspartate for blue and red light is smaller (143) or quite negligible (55).

Since an enhanced ^{14}C incorporation in aspartate by blue light (143) may precede activation of ^{14}C incorporation into protein (132), one can assume that perhaps some other mechanism also may be available for the regulation of PEP carboxylase, the allosteric enzyme, by blue light. Acetyl-coenzyme A has been shown to affect PEP carboxylation in bacteria (113) but not in the green alga *Chlamydomonas* (50). The reaction product OAA inhibits PEP carboxylase (73), but the role this property of the enzyme plays in blue light effects has not been determined. An increased flow of 3-PGA into PEP may be the result of an inhibition by blue light of 3-PGA reduction to glyceraldehyde 3-phosphate. In spite of the fact that pools of 3-PGA in whole cells are similar under blue and red light (28), their transformation into carbohydrates or amino acids is determined by light quality. Schürmann reported [in contrast to Galmiche (28)] that in isolated chloroplasts blue light favors the fixation of $^{14}CO_2$ into 3-PGA and red light into sugar phosphates (110). The reduction of 3-PGA is dependent upon the availability of ATP and reduced pyridine nucleotide (4).

The enzyme responsible for the reduction of PGA to triose phosphates is GPDH. Its activity has the same dependence on light intensity as photosynthesis. NADPH and ATP are considered to be allosteric effectors of the enzyme (101). The inhibition of 3-PGA transformation in Calvin cycle to triose phosphates can be determined either by altering the GDPH activity or by formation of NADPH and ATP. Enzyme activity as well as O_2 evolution in whole cells of *Lemna gibba* have been reported decreased by blue light even in the presence of red light, as compared with red light alone (108), because of reduced pyridine nucleotide deficiency caused by the increase of respiration under this light (60, 130). On the other hand, mitochondrial respiration is suppressed under high intensities (49) of both red (47, 131) and blue light (131). Therefore, the inhibition of GPDH by blue light may be an effect of this light on the electron transport of photosynthesis.

Thus it has been shown that in green plants light induces the uptake of oxygen as a function of light intensity (47, 131); this is observed even when photosynthesis is light-saturated (14). This uptake activated by blue light ($\lambda = 400 - 580$ mn) (131, 134) may not be accompanied by the evolution of CO_2 (49) and the acceleration of glycolate pathway reactions (14). It is stable to sodium azide (131) and cyanide (47) and completely disappears in the dark (129, 131). It is likely that this kind of oxygen uptake causes the reoxidation of reduced ferredoxin (121) or carotenoid (32, 118) taking part in photosynthesis electron transfer (27). If the more active oxygen uptake under blue light increases the rate of reduced components of electron transfer chain oxidation, a decrease in the formation of NADPH could be expected. Indeed, the blue light as compared with red light inhibits the activity of the Hill reaction (32, 35, 36).

In this connection it is interesting to note that even under strong blue light the intensities of photosynthesis in higher plants as measured in terms of oxygen exchange are smaller (131) than those measured in terms of CO_2 exchange (136). Apparently this effect is caused not by the activation of photorespiration by blue light (136, 143) but by extra interaction of oxygen with photosynthetically generated electrons or reductant.

Data currently available on the blue light action on photophosphorylation do not permit any specific conclusions about the problem (10, 11, 75, 85, 87). One attractive suggestion is that the low activity of cyclic photophosphorylation is the cause of a decrease in carbohydrate formation (especially in that of starch) under blue light (117, 143). Thus the inhibition of cyclic photophosphorylation in isolated chloroplasts simultaneously leads to the decrease in sugar phosphate formation and to the increase in the amount of 3-PGA (111). The oxygen consumption activated by blue light seems to be the main reason for cyclic photophosphorylation inhibition caused by light in vivo (142).

Similarity of the Effect of Blue Light and Nitrogen on Carbon Metabolism

The responses of photosynthetic carbon metabolism to short-term action of NH_4^+ and of blue light are similar in many respects. In both cases incorporation of ^{14}C into sucrose is reduced, whereas incorporation of pyruvate and alanine and also of aspartate, glutamate, and organic acids of the Krebs cycle is increased (46, 51, 52). The similarity does not exclude the possibility that the effect of blue light is associated at least partially with the increase of NH_4^+ concentration in the cell. This is indicated by reduction of the effects of blue light at a high nitrogen concentration in the medium (104, 124, 133). Blue light, as compared with red light, may increase the level of NH_4^+ due to an enhanced reduction of oxidized forms of nitrogen by this light (115, 135). An increase in the level of NH_4^+ in the chloroplast and the cell under blue light due to change in the membrane permeability may also be a factor (104). At the same time the similarity of the effect of blue light and that of the ammonium ion evidently is not absolute because the activation of protein synthesis by blue light occurs both at high and low nitrogen concentrations (133).

Factors Limiting Blue Light Effects

The intensive removal of carbon from the Calvin cycle and the active PEP carboxylation under blue light are attributed to the utilization of the formed substrates in active protein biosynthesis, since in the presence of inhibitors of protein synthesis the difference disappears between the rate of ^{14}C incorporation in aspartate, malate, or carbohydrates (sucrose) under red and blue light (88, 104).

Under anaerobic conditions the radioactivity of aspartate, glutamate, and Krebs cycle acids decreases (79, 143). The glycolate formation also drastically decreases simultaneously with the increase in the label of alanine and serine both under red and blue light. However, some difference in favor of blue light in the labeling of all amino acids except alanine remains in nitrogen (79, 143). The low level of PEP formation seems to be the cause of the inhibition of PEP carboxylation, so the PEP addition to the *Chlorella* cells in nitrogen stimulates the formation of malate and aspartate (Miyachi, personal communication). Since the removal of PGA from the Calvin cycle in nitrogen is intensified under both blue and red light (143), the decrease in the carboxylation of PEP in both cases cannot be caused by the deficiency of PGA.

It is possible that the inhibition of PEP formation in nitrogen is due to the activation of pyruvate kinase, since alanine formation in nitrogen is greatly enhanced in both cases of illumination, but especially in red light (79, 143). The radioactivity of starch, which is much lower in air under blue light, becomes in nitrogen even higher than under red light (143). It is not clear whether all these phenomena are accompanied by inhibition of enhancement of protein biosynthesis in nitrogen under blue light or by other causes. Most of the blue light effects examined seem to be indirect in nature. They occur as a result of blue light action on respiration or on the photosynthetic electron transfer chain.

EFFECT OF BLUE LIGHT ON FLAVOENZYMES AND ON REACTIONS OF THE GLYCOLATE PATHWAY

A direct effect of blue light on carbon metabolism may be its influence upon enzymes containing chromophore groups such as flavin (FMN or FAD) coenzymes. Enzymes containing flavin are widespread in the plant kingdom (4). They have a high acceptor specificity (24). Light absorption by the flavin coenzyme, and its excitation and transition to a triplet state (4), may determine the activity of the enzyme. In model experiments in the presence of FMN, photoinactivation of such flavoenzymes as D-amino-oxidase is revealed (94, 106).

In the presence of FMN, blue but not red light inhibits the activity of L-lactic dehydrogenase, xanthine oxidase, uric oxidase (106), and glycolate oxidase (105). At the same time, a colorless strain of *Chlorella* is found to contain the deaminating oxidase of glycine (107), whose activity increases under blue light. The possibility of inhibition of the oxidases of D-amino acids with blue light, as well as the occurrence of these enzymes in green plants, needs investigating. It would be interesting to study this question in order to shed some light on the causes of preferential and stable incorporation of ^{14}C in amino acids under blue light, which persists for most amino acids.

According to Schmid (105), the activity of preparations of glycolate oxidase isolated from tobacco leaves is inhibited by weak blue light and does not
change under red light of any intensity. After preincubation of the enzyme
with FMN under blue light without its substrate glycolate, the inhibition increases. This is assumed to be the result of an intramolecular interaction of
the triplet of FMN with the apoenzyme. In this case the apoenzyme plays the
role of a coenzyme reducer instead of a substrate. Inhibition by blue light of
glycolate oxidase, including its FMN-containing analog in algae (64, 83),
should be extremely important for the regulation of the glycolate pathway,
which is widespread both in higher plants and algae (56, 58, 64, 90, 91,
123). However, on short-term exposures in vivo the limitation of this pathway by blue light seems to take place at an earlier stage in the formation of
glycolate itself, because the pools not only of radioactive glycine and serine
but also of glycolate in stationary photosynthesis can be smaller under blue
light than under red light (55, 143).

The absence of glycolate excretion into the medium in algal suspensions
briefly exposed to blue light is also explained by the inhibition of glycolate
formation (72). It has been postulated that the "active glycolaldehyde"
formed either in a transketolase reaction involving sugar phosphates (30, 98,
123) or from the C_2 fragment of RuDP is the precursor of glycolate (89).
Recently Osmond has demonstrated that labeled ribose-5-phosphate, fructose-
6-phosphate, and fructose 1,6-diphosphate metabolized to glycolate with equal
rates (91). Plaut & Gibbs (98) found that the oxidation of two carbon fragments occurs in the chloroplast apparently because of the hydrogen peroxide
obtained through the interaction of atmospheric oxygen with reduced ferredoxin, or by a photo-oxidant generated in photosystem II of the electron transport pathway. In the chloroplast suspension glycolate is formed more actively
from fructose diphosphate. Addition of FMN or FAD accelerates the reaction
(12). If the enzyme taking part in the oxidation of active glycolaldehyde interacts with the flavin, it is possible that blue light can inhibit the glycolate
formation as well as its oxidation.

In accord with the inhibition of the glycolate pathway reactions, a particular activation of photorespiration by blue light as compared with red light is
not observed. Photosynthesis decreases equally in 21% of O_2 as compared
with 2% O_2 under red and blue light at nonsaturating and sometimes at saturating (136, 143) light intensities and above the compensation point (13,
136). On the other hand, Poskuta (99) reported that at saturating light intensities photorespiration was activated by blue light. One might expect that
if high intensities of blue light quench the excited state of flavin coenzymes,
then this can restore the activity of the enzymes taking part in the formation
and breakdown of glycolate (105).

In some cases the blue light can activate the glycolate pathway reactions,
as observed in yellow *Chlorella* mutant in an anaerobic atmosphere (64) and
in *Ankistrodesmus braunii* (25). An increase in the label of glycine is observed in bean plants subjected to high intensities of light in contrast to low

intensities (143). In pulse-chase experiments with tomato leaves, Galmiche (28) has also found that the increase in radioactivity of glycine and serine is faster under blue light than under red. Distribution of the label in these amino acids at saturating light intensities in *Chlorella* cells indicates that they are formed by the glycolate pathway only under blue and not under red light (17). The glycolate is synthesized in chloroplasts, and its formation is dependent on oxygen concentration in the organelle (30, 98). Therefore, the fine mechanism of blue light effects on the glycolate pathway reactions needs further investigation at the chloroplast level.

Photoreceptors and the Primary Mechanism of Blue Light Action

Blue light effects are explained as a result of light absorption either by flavins or carotenoids (62, 127). The absorption of light by the latter can be accompanied by transformation of carotenoids from the nonactive *trans* form into the active *cis* form (65) as is postulated for phytochrome (44). The influence of blue light on respiration and carbon metabolism is reported for the colorless and carotenoidless *Chlorella* mutant (Miyachi, personal communication). The mutant contains a small quantity of flavoenzymes including glycine oxidase (107). However, it is not quite clear whether the effects of blue light on carbon metabolism in green plants are associated with the absorption of light by flavins solely (32, 35, 36).

Of special interest is the fact that in model systems enzymes which are not photosensitive as such may change their activity under light in the presence of chromophore molecules (9). In the presence of FMN, blue light inhibits the activity of the nonphotosensitive enzyme uric oxidase (106). Therefore, one cannot deny the possibility of photoregulation of nonphotosensitive enzymes in living cells with the intervention of chromophore molecules absorbing light selectively in different regions of the spectrum. The excited flavins can alter the activity of some allosteric enzymes of photosynthesis by means of interaction with their SH group (82). In this case light serves as one of several factors in the dynamic regulation of enzymes. The specific feature of regulation by light is that the reaction results from light absorption by the photoeffector molecule (132). Here the selective effect of light of different quality on metabolism should be determined by the nature of the apoenzyme (24, 43, 105), by the absorption spectrum of the cofactor, and by the accessibility of the chromophore group to the interaction with protein. In the case of allosteric enzymes, the noncovalent binding of the photosensitive effector by the regulatory (allosteric) center of the enzyme may influence the cooperative structural rearrangement of all subunits of the macromolecule (126) and, in the final analysis, change the activity of the enzyme. If, as is assumed now, many of the enzymes and also flavin enzymes (70) are built into membranes (43, 59, 112), one might expect that not only multienzymes but also the membrane of the cell could be subjected to fine functional regulation by blue light (126). The correlation between conformational changes and molecular reconstruction of photosynthetic electron transfer chain reactions excited by blue light has been shown recently by Harnischfeger (33, 34, 36).

PHOTOSYNTHESIS UNDER PROLONGED ACTION OF BLUE OR RED LIGHT

It is known that in both higher plants and in algae the prolonged action of blue light provokes morphological effects different from red light effects (57, 80, 97, 131). Simultaneously the blue light activates the biosynthesis of protein (15, 19, 80, 93, 97, 103, 131, 139), ribonucleic acids (16, 80, 97, 103, 139), and chlorophyll, principally chlorophyll b (45, 80, 116, 131), and inhibits carbohydrate synthesis (16, 80, 114, 117, 131), especially that of starch (21, 117). Under prolonged action the effects are more stable in time. They are accompanied by an enlargement of chloroplasts (7, 80, 131) which is not observed in the presence of chloramphenicol (8). Therefore, we do not exclude the possibility that blue light controls the state of a protein synthesizing system of chloroplasts.

Such a conclusion may be based on the experiments with fern gameto-phytes cited above (8) and also on the work with unnucleated *Acetabularia* (21). Blue light as compared with red accelerates the formation of a fine structure of chloroplasts (125), and like kinetin (69) it also restores the structure of aged barley leaves grown under red light (140). After transfer of "blue light plants" to red light, the high level of protein, nucleic acids, and chlorophyll progressively decreases (within 1 or 2 days) to the level peculiar to the plant grown under red light, but the effect is completely reversible (80, 139). The spectral composition of red light usually used in experiments (21, 45, 100, 122, 131) provides excitation of both photochemical systems of photosynthesis. However, in contrast to the blue light, the red light is less effective for photosynthesis under prolonged illumination (21, 93, 141). Restoration of photosynthesis may be obtained by blue light treatment in the course of 2 to 3 days (21). The higher level of photosynthesis in plants grown under blue light is accompanied by a corresponding enhancement of the Hill reaction (20, 141), noncyclic, and particularly pseudocyclic photo-phosphorylation (141). However, the enhancement of photosynthesis under prolonged blue light may be provoked not only by activation of photosynthetic electron transfer chain reactions, but by high activity of ribulose-1,5-diphosphate carboxylase. The blue light does not seem to be involved in allosteric and substrate regulation of the enzyme (77). At the same time, it activates the de novo biosynthesis of this carboxylase. For horse bean plants (100) and the flagellate *Chlorogonium elongatum* (114) grown under red light, exposure to blue light for 8-10 hours increased the carboxydismutase activity. Similar results were obtained with GPDH (114). Incorporation of ^{14}C into protein begins to increase in the first hours of illumination by blue light, and within 8 hours it exceeds the control value by a factor of 5. On the other hand, the level of photosynthesis remains constant during the first 8 hours of illumination (21).

Then enhancement of photosynthesis and alteration in carbon metabolism

peculiar to blue light was observed. This phenomenon seems to be connected with extra protein synthesis (21). In plants grown under blue light,
reactions of the glycolate pathway are more active—in algae the incorporation of ^{14}C from $^{14}CO_2$ into glycolate is enhanced (45), and in higher plants
the glycolate content is markedly increased in the presence of bisulphite
(138). After only 16 hours of illumination by blue light, glycolate oxidase
and glyoxylate transformation are activated (137). Thus prolonged absence
of blue light (i.e. growth under a single red light) disturbs the normal course
of photosynthesis at the expense of limiting the synthesis of those enzymes
which play an essential role in the photosynthetic carbon metabolism both
inside and outside the chloroplasts.

Conclusion

The short-term action of blue light as compared with red markedly alters
the photosynthetic carbon metabolism in plants grown under white light.
Since white light contains blue rays also, it can ensure extra protein synthesis.
Therefore, the short-term action of blue light after white light may be considered most likely for substrate and allosteric control of the enzymes available.
The regulatory effect of blue light on photosynthetic metabolism in this case
may manifest itself in different ways such as the following: 1. Direct action
—i.e. photoregulation of enzymes containing a chromophore group and taking part in carbon metabolism both in the chloroplasts and in the entire cell.
2. Indirect action—(a) photoregulation of photosynthetic and oxidative electron transport and energy accumulating reactions; and (b) photoregulation
of the conformal state and permeability membranes of the cell.

In any case, the nature of metabolites formed under blue light is such that
one may expect the enhancement of substrate and energetic interaction of
chloroplasts and the cell (4, 42, 67, 146). The more pronounced effects of
blue light are observed after preliminary exposure of plants under the red
light or in the dark. It appears that the alterations in metabolism provoked by
blue light do not disappear immediately after the transfer from blue light into
other conditions of illumination.

The prolonged exposure of plants to red light eliminates the possibility of
the enhancement of protein biosynthesis by blue light. A very early (but evidently not the primary) step in the chain of responses to blue light in this
case is the enhancement of protein biosynthesis, probably at the expense of
blue light action on this process at the level of transcription or translation of
genetic information (80). It is evident that under white light the most optimal conditions for photosynthesis exist when all the photoreactions are in
interaction. However, transfer or growth of plants under the light of different
spectral composition represents not only the means for studying the action
spectra of different photoreactions but also the mode of control by light for
both photosynthetic metabolism of the plant and integration of all life processes (44, 80, 84, 102). We must emphasize the fact that the alterations in
metabolism provoked by blue light are reversible. At best they are restricted

by the life cycle of plant individuum, i.e. they have modificative nature. However, one cannot exclude the fact that in natural environments the selectivity of plant is genetically linked to the light of different spectral composition (74, 109).

LITERATURE CITED

1. Aerov, I. L., Guliaev, B., Manuiskii, V. 1969. *Dokl. Akad. Nauk SSSR* 187:1194–97
2. Ahmed, A. M. M., Ries, E. 1969. In *Progress in Photosynthesis Research*, ed. H. Metzner, 3:1662–68. Munich: Goldmann-Verlag
3. Baldry, C. W., Bucke, C., Coombs, J. 1969. *Biochem. Biophys. Res. Commun.* 37:828–32
4. Bassham, J. A., Kirk, M. 1968. In *Comparative Biochemistry and Biophysics of Photosynthesis*, ed. K. Shibata, A. Takamiya, A. T. Jagendorf, R. C. Fuller, 365–78. Univ. Tokyo Press. 445 pp.
5. Bell, L. N., Shuvalova, N. P. 1971. *Photosynthetica* 5(2): 113–23
6. Bell, L. N., Shuvalova, N. P., Mironova, G. S., Nichiporovich, A. A. 1968. *Dokl. Akad. Nauk SSSR* 182:1439–42
7. Bergfeld, R. 1964. *Z. Naturforsch. B* 19:1076–78
8. Bergfeld, R. 1968. *Planta* 81:274–79
9. Bieth, J., Wassermann, N., Vratsanos, S. M., Erlanger, B. F. 1970. *Proc. Nat. Acad. Sci. USA* 66: 850–54
10. Black, C. C., Fewson, C. A., Gibbs, M., Gordon, S. A. 1963. *J. Biol. Chem.* 238:3802–5
11. Black, C. C., Turner, J. F., Gibbs, M., Krogmann, D. W., Gordon, S. A., Ellanger, P. 1962. *J. Biol. Chem.* 237:580–83
12. Bradbeer, J. W., Anderson, C. M. A. 1967. In *Biochemistry of Chloroplasts*, ed. T. W. Goodwin, 2:175–79. New York: Academic. 776 pp.
13. Bulley, N. R., Nelson, C. D., Tregunna, E. B. 1969. *Plant Physiol.* 44:678–84
14. Bunt, J. S., Heeb, M. A. 1971. *Biochim. Biophys. Acta* 226: 354–59
15. Buschbaum, A. 1968. *Vergleich-

end-physiologische Untersuchungen an Algen aus Blau und Rotlichtkulturen.* PhD thesis. Georg-August Univ., Göttingen
16. Casper, R. 1963. *Wirkungen von blauen und rotem Licht auf die Zusammensetzung der Blätter von Kalanchoe rotundifolia.* PhD thesis. Georg-August Univ., Göttingen
17. Cayle, T., Emerson, R. 1957. *Nature* 179:89–90
18. Champigny, M. L. 1959. *Rev. Cytol. Biol. Veg.* 21:3–43
19. Clauss, H. 1968. *Protoplasma* 65: 49–80
20. Clauss, H. 1970. *Planta* 91:32–37
21. Clauss, H. 1970. In *Biology of Acetabularia*, ed. J. Brachet, S. Bonotto, 177–91. New York: Academic
22. Cornish, A. S. 1971. *Arch. Biochem. Biophys.* 142:584–90
23. Das, V., Raju, P. 1965. *Indian J. Plant Physiol.* 8:1–4
24. Dixon, M. 1971. *Biochim. Biophys. Acta* 226:269–84
25. Döhler, G., Braun, F. 1971. *Planta* 98:357–61
26. Döhler, G., Wegmann, R. 1969. *Planta* 89:266–74
27. Fork, D. C. See Ref. 2, 2:800–10
28. Galmiche, J. M. 1971. In *Progress in Photosynthesis Research. Proc. 11th Int. Congr. Photosyn. Res., Stresa, Italy.* In press
29. Gibbs, M. 1967. *Ann. Rev. Biochem.* 36:757–84
30. Gibbs, M. 1969. *Productivity of Photosynthetic Systems. Proc. Int. Symp. Photosyn. Res., Moscow,* part 2. In press
31. Gold, V. M. 1969. *Fiziol. Rast.* 16:594–602
32. Gross, J. A., Whitfield, M. D. 1970. *Biochem. Biophys. Res. Commun.* 40:1216–23
33. Harnischfeger, G. 1970. *Planta* 92:164–77
34. Harnischfeger, G. See Ref. 28
35. Harnischfeger, G., Gaffron, H. 1969. *Planta* 89:385–88
36. Ibid 1970. 93:89–105

37. Hatch, M. D., Slack, C. R. 1970. *Ann. Rev. Plant Physiol.* 21: 141–62
38. Haupt, E. 1970. *Ber. Deut. Bot. Ges.* 83:201–3
39. Hauschild, A. H. W., Nelson, C. D., Krotkov, G. 1962. *Can. J. Bot.* 40:1619–30
40. Hauschild, A. H. W., Nelson, C. D., Krotkov, G. 1964. *Naturwissenschaften* 51:274–75
41. Ibid 1965. 52:435–36
42. Heber, U., Santarius, K. A. 1970. *Z. Naturforsch. B* 25:718–28
43. Hemmerich, P., Nagelschneider, G., Veeger, C. 1970. *FEBS Lett.* 8:69–83
44. Hendricks, S. B., Borthwick, H. A. 1967. *Proc. Nat. Acad. Sci. USA* 58:2125–30
45. Hess, J. L., Tolbert, N. E. 1967. *Plant Physiol.* 42:1123–30
46. Hiller, R. G. 1970. *J. Exp. Bot.* 21:628–38
47. Hoch, G., Owens, O. v.H., Kok, B. 1963. *Arch. Biochem. Biophys.* 101:171–80
48. Horváth, I., Fehér, I. V. 1965. *Acta Bot. Hung.* 11:159–64
49. Jackson, W. A., Volk, R. J. 1970. *Ann. Rev. Plant Physiol.* 21: 385–432
50. Jones, R. F., Chen, J. H. 1970. *Plant Physiol.* 46:761–62
51. Kanazawa, T., Kanazawa, K., Kirk, M., Bassham, J. A. 1970. *Plant Cell Physiol.* 11:445–52
52. Kanazawa, T., Kirk, M., Bassham, J. A. 1970. *Biochim. Biophys. Acta* 205:401–8
53. Karpilov, Yu. S. 1970. *Kooperativnyi Fotosintez kserofitov Mold. nauchn. Inst. orosh. zemledelia Ovoshevod.*, 3–17. Kishinev
54. Keerberg, H., Keerberg, O., Parnik, T., Viil, J., Vark, E. 1971. *Photosynthetica* 5:99–106
55. Keerberg, H., Vark, E., Keerberg, O., Parnik, T. 1971. *Izv. Akad. Nauk Est. SSR Biol.* 20:350–52
56. Keerberg, O., Vark, E., Keerberg, H., Parnik, T. 1970. *Dokl. Akad. Nauk SSSR* 195:238–41
57. Kleschnin, A. F. 1954. *Rastenie i svet.* Moskwa: Nauka. 456 pp.
58. Kolesnikov, P. A. 1968. *Usp. Sovrem. Biol.* 65:20–33
59. Konev, S. V., Aksentsev, S. L., Chernitskii, E. A. 1970. *Kooperativnii perehodi belkov v kletke*

60. Kowallik, U., Kowallik, W. 1969. *Izd.* Minsk: Nauka i tehnika. 202 pp.
60. Kowallik, U., Kowallik, W. 1969. *Planta* 84:141–57
61. Kowallik, W. 1962. *Planta* 58: 337–65
62. Ibid 1966. 69:292–95
63. Kowallik, W. 1967. *Brookhaven Symp. Biol.* 19:467–77
64. Kowallik, W. 1969. *Planta* 87: 372–84
65. Kowallik, W., Gaffron, H. 1967. *Nature* 215:1038–40
66. Krasnovskii, A. A., Michailova, E. S. 1970. *Dokl. Akad. Nauk SSSR* 194:953–56
67. Krause, G. H. 1971. *Z. Pflanzenphysiol.* 65:13–23
68. Krotkov, G. 1964. *Trans. Roy. Soc. Can.* 2:205–15
69. Kursanov, A. L. et al 1964. *Fiziol. Rast.* 11:838–48
70. Lardy, H. A., Ferguson, S. M. 1969. *Ann. Rev. Biochem.* 38: 991–1034
71. Laudenbach, B., Pirson, A. 1969. *Arch. Mikrobiol.* 67:226–42
72. Lord, J. M., Codd, G. A., Merrett, M. J. 1970. *Plant Physiol.* 46: 855–56
73. Lowe, I., Slack, C. R. 1971. *Biochim. Biophys. Acta* 235:207–9
74. Lubimenko, V. N. 1935. *Fotosintez i chemosintez v rastitelnom mire.* Leningrad: Selchozgiz. 321 pp.
75. Lundegårdh, H. 1964. *Biochim. Biophys. Acta* 88:37–56
76. Meidner, H. 1968. *J. Exp. Bot.* 19: 146–51
77. Miyachi, S. 1969. *Proc. 11th Int. Bot. Congr. Seattle*, 149
78. Miyachi, S. See Ref. 28
79. Miyachi, S., Hogetsu, D. 1970. *Can. J. Bot.* 48:1203–7
80. Mohr, H. 1968. *Lehrbuch der Pflanzenphysiologie.* Berlin: Springer-Verlag. 408 pp.
81. Mokronosov, A. T., Nekrasova, G. F. 1966. *Fiziol. Rast.* 13: 385–97
82. Neims, A. H., Hellerman, L. 1970. *Ann. Rev. Biochem.* 39: 867–88
83. Nelson, E. B., Tolbert, N. E. 1969. *Biochim. Biophys. Acta* 184: 263–70
84. Nichiporovich, A. A., Voskresenskaya, N. P., Butenko, R. G. 1959. *Proc. Int. Bot. Congr. Montreal*, 11:1039–46

85. Nobel, P. S. 1968. *Biochim. Biophys. Acta* 153:170–82
86. Ogasawara, N., Miyachi, S. 1970. *Plant Cell Physiol.* 11:1–14
87. Ibid., 411–16
88. Ogasawara, N., Miyachi, S. *Plant Cell Physiol.* 12:675–82
89. Ogren, W. L., Bowes, G. 1971. *Nature* 230:159–60
90. Osmond, C. B. 1971. In *Photosynthesis and Photorespiration,* ed. M. D. Hatch, C. B. Osmond, R. O. Slatyer, 472–82. New York: Wiley/Interscience
91. Osmond, C. B., Harris, B. 1971. *Biochim. Biophys. Acta* 234: 270–82
92. Pau, D., Waygood, E. R. 1971. *Can. J. Bot.* 49:631–44
93. Payer, H. D. 1969. *Planta* 86: 103–15
94. Penzer, G. R. 1970. *Biochem. J.* 116:733–43
95. Pickett, J. M. 1971. *Plant Physiol.* 47:226–29
96. Pirson, A. See Ref. 30
97. Pirson, A., Kowallik, W. 1964. *Photochem. Photobiol.* 4:489–97
98. Plaut, Z., Gibbs, M. 1970. *Plant Physiol.* 45:470–74
99. Poskuta, J. 1968. *Experientia* 24: 796–97
100. Poyarkova, N. M., Drozdova, I. S., Voskresenskaya, N. P. 1971. *Fiziol. Rast.* 18:683–90
101. Preiss, J., Kosuge, T. 1970. *Ann. Rev. Plant Physiol.* 21:433–66
102. Protasova, N. N., Kefeli, V. I., Kof, E. M., Chastukhina, E. A. See Ref. 30
103. Raghavan, V. 1969. *Am. J. Bot.* 56:871–80
104. Ries, E. 1970. *Beziehungen zwischen Gaswechsel und ¹⁴C-Einbau monochromatisch bestrahlter Algenzellen.* PhD thesis. Eberhard-Karls Univ., Tübingen
105. Schmid, G. H. 1969. *Z. Physiol. Chem.* 350:1035–46
106. Ibid 1970. 351:575–78
107. Schmid, G. H., Schwarze, P. 1969. *Z. Physiol. Chem.* 350:1513–20
108. Schmidt-Clausen, H. J., Ziegler, J. See Ref. 2, 1646–52
109. Schulgin, I. A. 1969. *Nauch. Dokl. Vyssh. Shk. Biol. Nauki* 9:7–25
110. Schürmann, P. 1968. *Verh. Schweiz. Naturforsch. Ges.* 148: 121–22
111. Schürmann, P., Buchanan, B. B., Arnon, D. I. See Ref. 2
112. Severin, S. E. 1969. *Izv. Akad. Nauk SSSR Ser. Biol.* 6:797–99
113. Smith, T. E. 1970. *Arch. Biochem. Biophys.* 137:512–22
114. Stabenau, H. 1969. *Der Einfluss von Licht und Acetat auf Chlorogonium elongatum Dangeard.* PhD. thesis. Georg-August Univ., Göttingen
115. Stoy, V. 1955. *Physiol. Plant.* 8: 963-86
116. Szász, K., Horváth, I., Szász-Barsi, E., Garay, A. S. 1969. *Acta Bot. Hung.* 15:167–70
117. Szász, K., Szász-Barsi, E. 1971. *Photosynthetica* 5:71–73
118. Takeguchi, C. A., Yamamoto, H. Y. 1968. *Biochim. Biophys. Acta* 153:459–65
119. Tamas, I. A., Yemm, E. W., Bidwell, R. G. S. 1970. *Can. J. Bot.* 48:2313–17
120. Tarchevsky, I. A. 1964. *Fotosintez i zasucha.* Izd. Kaz. Univ. 198 pp.
121. Telfer, A., Cammack, R., Evans, M. C. W. 1970. *FEBS Lett.* 10: 21–24
122. Terborgh, J. 1966. *Plant Physiol.* 41:1401–10
123. Tolbert, N. E., Yamazaki, R. K. 1969. *Ann. NY Acad. Sci.* 168: 325–41
124. Truhin, N. V. 1967. *Tr. Inst. Biol. Vnutr. Vod. Akad. Nauk SSSR* 14(17):10–21
125. Vlasova, M. P., Drozdova, I. S., Voskresenskaya, N. P. 1971. *Fiziol. Rast.* 18:5–11
126. Volkenshtein, M. V. 1970. *Biofizika* 25:215–24
127. Voskresenskaya, N. P. 1953. *Dokl. Akad. Nauk SSSR* 93: 911–14
128. Voskresenskaya, N. P. 1956. *Fiziol. Rast.* 3:49–57
129. Voskresenskaya, N. P. 1961. In *Progress in Photobiology,* ed. B. C. Christensen, B. Buchmann, 149–53. New York: Elsevier
130. Voskresenskaya, N. P. 1965. In *Biochimiya i Biofizika Fotosinteza,* 219–35. Moskwa: Nauka
131. Voskresenskaya, N. P. 1965. *Fotosintez i spektral'nyi sostav sveta.* Moskva: Izd. Akad. Nauk SSSR. 276 pp.
132. Voskresenskaya, N. P. See Ref. 30
133. Voskresenskaya, N. P., Grishina,

G. S. 1958. *Fiziol. Rast.* 5:147–55
134. Ibid 1961. 8:726–33
135. Ibid 1962. 9:7–15
136. Voskresenskaya, N. P., Grishina, G. S., Chmora, S. N., Poyarkova, N. M. 1970. *Can. J. Bot.* 48:1251–57
137. Voskresenskaya, N. P., Grishina, G. S., Sechenska, M., Drozdova, I. S. 1970. *Fiziol. Rast.* 17:1028–36
138. Voskresenskaya, N. P., Hodjiev, K. A. 1972. *Dokl. Akad. Nauk Tadzh. SSR.* In press
139. Voskresenskaya, N. P., Nechayeva, E. P. 1967. *Fiziol. Rast.* 14:299–309
140. Voskresenskaya, N. P., Nechayeva, E. P., Vlasova, M. P., Nichiporovich, A. A. 1968. *Fiziol. Rast.* 15:890–98
141. Voskresenskaya, N. P., Oshmarova, I. S. See Ref. 2, 1669–74
142. Voskresenskaya, N. P., Viil, Y. A. 1966. *Fiziol. Rast.* 13:762–68
143. Voskresenskaya, N. P., Viil, Y. A., Grishina, G. S., Parnik, T. P. 1971. *Fiziol. Rast.* 18:488–93
144. Voskresenskaya, N. P., Zak, E. G. 1957. *Dokl. Acad. Nauk SSSR* 114:375–78
145. Walker, D. A. 1962. *Biol. Rev. Cambridge Phil. Soc.* 37:215–56
146. Walker, D. A., Crofts, A. R. 1970. *Ann. Rev. Biochem.* 39:389–428
147. Warburg, O., Krippahl, G., Schröder, W. 1954. *Z. Naturforsch. B* 9:667–75
148. Zurzycki, J. 1967. In *Photochemistry and Photobiology in Plant Physiology*. Eur. Photobiol. Symp. book of abstr., 131–34, Hvar, Yugoslavia
149. Zurzycki, J. See Ref. 12, 609–12

Ann. Rev. Plant Physiol. 1972. 23:235–58

AUXINS AND ROOTS 7531

Tom K. Scott

Department of Botany, University of North Carolina
Chapel Hill, North Carolina

CONTENTS

During the past 10 years there have been several reviews dealing with the physiology of root growth and development (51, 208, 209, 212, 222). (see also the proceedings of the symposium titled *Root Growth*, 233). The subject of this review has been restricted to consideration of only one growth regulator, auxin, in an attempt to help bridge that "curious gap" in our understanding of hormonal control mechanisms in root growth and development (216). Clearly, auxin can act to regulate root growth both in promotive and inhibitory capacities. It is also clear that whereas auxin may be a limiting factor in root growth, it is not necessarily so at the exclusion of other substances which are also present in the root at different times and at different tissue and cellular locations. The growth of roots results from a multiple of interactions, but such interactions apparently do not operate in the absence of auxin (208).

By the same token, though auxin is necessary to, it is not the exclusive controlling agent of, normal differentiation of different root tissues (222–224).

The two words of the title of this review prove difficult to define restrictively. They will usually be used hereafter to include the following and sometimes wide-ranging meanings: 1. An auxin is construed to be any substance, by virtue of its chemical characteristics and/or its biological activity, which resembles indole-3-acetic acid (IAA) (154). 2. A root is considered to be any organ possessing those organizational and anatomical characteristics which have been outlined by Torrey (222). On the other hand, the two words as used in the studies cited below are virtually always restricted, in the operational sense, to include: 1. an auxin as an exogenous or endogenous compound (of known or suspected character) which promotes, or inhibits, or otherwise affects growth or some other developmental process in roots; 2. a root as a primary or lateral organ which has been observed or treated either while intact, in culture, or as an isolated segment. Two other terms requiring operational definitions are growth and differentiation. Growth is expressed as a measure of elongation (cell or organ), cell division, wet and dry weight, or protein, cell wall, DNA, or RNA synthesis. Differentiation is generally intended to relate to those steps which occur in which different cells acquire different morphological and chemical characters.

The following topics will be dealt with: (a) the occurrence and nature of native auxin(s) in roots; (b) auxin metabolism in roots; (c) the role of auxin in root growth; (d) the role of auxin in differentiation in roots.

THE OCCURRENCE AND NATURE OF NATIVE AUXIN IN ROOTS

Indole acetic acid.—Reports of the naturally occurring auxin, indole-3-acetic acid, or indol-3yl-(acetic acid), (IAA), in roots are numerous and have persisted since the first paper chromatographic isolation of auxin made by Bennet-Clark et al in 1952 (22). Conventional ether or alcoholic extraction of tissues followed by ether, ethyl acetate, or acid partitioning has given evidence for the presence of IAA, among other promoters and inhibitors, in such various roots as wheat (Lexander 131), sunflower (Audus & Thresh 16, Phillips 159), corn (Kefford 115), *Lens* (Pilet 162, 163), water hyacinth (Sircar & Ray 200), sugar cane (Cutler & Vlitos 62), pea (Gasparikova 83), *Vicia* (Lahiri & Audus 127), tomato (Thurman & Street 217), chicory (Vardjan & Nitsch 230), bean (Scharf & Günther 186), lupine (Dullaart 72) and *Cycus* (Dullaart 73). Other reports indicate that IAA exists in roots, but that its presence may either be masked by other substances (39), or that conversion or loss of it may take place during isolation and purification procedures (217, 241). However, since the time of Bentley's review (24), notes of caution or disclaimers have appeared in increasing numbers regarding the previously held generalization that IAA as such is universally present in root tissue. Bennet-Clark et al (23), for instance, using starch column chromatography, showed that the active auxin in *Vicia* roots is not necessarily IAA. Further, as Audus and co-workers have pointed out (14, 15, 38, 39, 127), water-soluble auxins of roots, probably indoles, have as much or more activity than IAA, but they are probably not identical with IAA. Finally, Street et al (210, 211) and Woodruffe et al (241) have shown that carefully isolated and purified ether soluble indoles from wheat and tomato roots did not include IAA. In fact, the four indoles isolated by Woodruffe et al (241), using DEAE cellulose chromatography, did not correspond to any of the 26 known natural or synthetic indoles which were co-tested. These reports also show that the major auxin activity is associated with water-soluble fractions (210, 211, 241). Bohling (26), Bayer (19), and Schneider (188) have isolated IAA-like substances directly from the diffusate of *Vicia* and corn roots.

Other indole auxins.—Without going into details, it may be said that starting from the time of the earliest work (Bennet-Clark & Kefford 21, Lexander 131) which reported the isolation and identification of IAA (plus one other promoter and an inhibitor), and going through to the present, virtually all known or suspected indole auxins or their precursors have been reported as isolated and identified from extracts of root tissue (see references in the paragraph above). However, once again the results of Woodruffe et al (241) emphasize the difficulties associated with the technology involved in these chemical isolations. Following scrupulous purification and separation procedures applied to wheat root extracts (involving paper, thin-layer, DEAE cellulose chromatography and paper electrophoresis), they showed that the four as yet unknown indoles (run against the 26 known indoles tested) were the only ones present in wheat roots. It is apparent then that previous reports of IAA and associated indole isolations must be treated with caution (especially those involving ether or ethyl acetate partitioning) and that certainly more attention needs to be given to the characterization and identification of water-soluble auxins (Burnett et al 39).

Persistent reports of the isolation of tryptophan from root extracts are worthy of note (39, 74, 210, 211, 217, 230, 241). The relatively large amounts of this indole, presumably without auxin activity, would nevertheless seem to favor the biosynthesis of at least some IAA (39, 164). However, Elliott (74) has identified α-N-malonyl-D-tryptophan from wheat root extracts, and he has shown it to be the tryptophan isolate of *Vicia* and tomato roots previously reported by Street et al (210). He argues persuasively that the D-isomer of malonyl tryptophan is not likely to be involved with either growth or auxin metabolism in the wheat root. Nevertheless, the fact that autoclaved tryptophan has been found to promote growth in cultured wheat roots grown in the dark suggests a biosynthetic involvement of the compound (Sutton et al 213).

Recently, Hofinger et al (101) have identified 3-(3-indolyl)-acrylic acid as the major auxin in *Lens* root extracts. Its presence had previously been indicated in root extracts of *Lens* by Collet et al (61) and *Vicia* by Burnett et al (39), but the report of its identification did not appear until 1969 (Hofinger 99). Indole acrylic acid behaves like an auxin (indeed IAA) in that it inhibits *Lens* root growth at high concentrations (99) and antagonizes kinetin inhibition of *Lens* root growth (Darimont et al 63). That this compound may be an artifact of extraction has recently been suggested by Hofinger (100).

The presence of nonindolic auxins in roots remains speculative. Burnett et al (39) have suggested that the "citrus auxin" (a naphthol), characterized by Khalifah et al (118), and 3,4-dihydroxyphenylalanine occur in *Vicia* root extracts and that each may possess auxin activity. Also, it is possible that auxin complexes or compounds are growth active (Street et al 210). However no such complex has been shown with certainty to exist in root tissue.

AUXIN METABOLISM IN ROOTS

Auxin oxidation.—The greatest preponderance of literature relating to the metabolism of auxin in roots is concerned with oxidative destruction. Auxin catabolism has presumably received such wide attention since it was suggested long ago that the control of the level of auxin in roots may be positively correlated with the control of growth rate (Pilet & Galston 172). This area of research is particularly interesting since various feedback mechanisms are suggested.

The difficulties in making clear-cut distinctions between enzymes reflecting IAA oxidase, peroxidase, and polyphenol oxidase activities continues. Janssen (110) found identical elution patterns for the three carefully separated and analyzed enzyme activities from crude pea root extracts when recovered from Sephadex G-100 columns. Likewise, Pilet and associates (174–176) found that the activities from extracts of *Lens* roots, first gel chromatographed and then eluted from Sephadex G-25 and G-100, appeared in the same fractions. It would seem possible, as Janssen (110) suggests, that the auxin destruction system specific for IAA in the pea root may be due to different activity centers on one enzyme. Pilet et al (175), on the other hand, suggest the possibility that IAA oxidase, IAA-destroying peroxidase, and IAA-destroying polyphenol oxidase may be a combination of enzymes or of isoenzyme systems. Van der Mast (142), using polyacrylamide electrophoretic separation methods, was able to show that pea root homogenates contain one protein complex capable of degrading IAA, but the complex consists of at least two enzymes: a peroxidase and, by inference, a polyphenol oxidase (141). This result is consistent with the findings of others who based their similar conclusions on experiments which utilized inhibitors, various separation procedures, and molecular weight determinations (Sequeira & Mineo 194, Janssen 109). Van der Mast (142, 143) raises the interesting possibility that the active IAA-destroying peroxidase is active when membrane bound and that nondestroying peroxidases may participate in IAA destruction by serving as traps for IAA free radicals (79). Mace (139) has shown that banana roots contain enzymic IAA-destroying activity which may not necessarily be restricted to one type of peroxidase. Penon et al (157) have shown that three peroxidases (one acid and two basic) are associated with the ribosomes of *Lens* roots and that IAA treatment of the root tissue contributes to the de novo biosynthesis of the two basic forms.

It is necessary to point out that the auxin oxidase system may not be specific to IAA. As mentioned above, there is reason to challenge the premise that IAA as such is universally present in roots. Further, there are reports of the presence of an IAA-oxidase system in roots in which the presence of IAA has never been claimed. For example, Altman et al (4), Goldschmidt et al (85), Goren & Goldschmidt (86), and Goren & Tomer (87) have demonstrated oxidase activity from citrus root extracts which is active in decreasing both endogenous auxin and synthetic IAA levels. Though IAA may be present in citrus roots, the real identity of "citrus auxin" has not been specified with certainty and thus should be considered unknown (Dr. Lowell Lewis, personal communication). In addition, Hofinger et al (101) have shown that indolacrylic acid is destroyed along similar patterns as IAA in the presence of a purified *Lens* root peroxidase preparation. However, as stated previously, this substance may not occur naturally in *Lens* roots (Hofinger 100).

Two final points regarding auxin destruction are worthy of mention. Much of the work which has been done on or in connection with root "IAA-oxidase" systems has been compared and equated with the "IAA-oxidase" system of shoots (Hare 89). Niemann (152), while investigating the effects of phenylamine on the IAA-oxidase activity of cucumber, showed that the oxidase system isolated from the cucumber root may differ very considerably from that of the shoot. Lastly, there is the nagging worry as to what is the general physiological significance of such isolated extract or homogenate activity (177). Further, the absence of any reports of the in vivo oxidation products, hydroxymethyloxindole and methyleneoxindole (227), of IAA warrants continued caution with respect to the implications of such studies. However, Morris et al (149) reported the presence of large amounts of indolealdehyde from extracts and diffusates of pea roots following injection of ^{14}C-IAA into the intact shoot, which does indicate the presence of degradative activity.

Nonoxidative fates of auxin.—In addition to undergoing oxidative decarboxylation, it is known that auxin may conjugate with aspartic acid in intact roots and root segments (1, 9, 10, 103, 116, 149, 218). This alternate inactivating metabolic process is of wide occurrence and has been construed as a growth-regulating mechanism in roots. Andreae has proposed that the conjugate remains in the tissue as a relatively growth-inactive substance and thus as a detoxification product (6). He demonstrated clearly (5) that the uptake of IAA from the external medium and its conjugation with aspartic acid followed closely the kinetics of the adaptation of pea root tips to IAA and also to the recovery of growth inhibition following the removal of IAA. Later experiments (6) showed that such a recovery could be demonstrated and correlated with both IAA and naphthalene acetic acid (NAA) being metabolized to aspartate conjugates. Significantly, no growth recovery took place when the herbicide 2,4-D was applied at inhibitory levels and then removed. This substance accumulates in the root tissue but it is neither decarboxylated nor otherwise metabolized (5, 6).

Andreae & Collet (7) noted that in the experiments cited above, which involved IAA, some decarboxylation was reported. However, by adding phenols (known to decrease or increase IAA-induced growth through inversely increasing or decreasing the activity of the IAA-oxidase system) to the external medium and by following the evolution of carbon dioxide, they were able to show that the effect of the oxidase took place in the external solution and not in the tissue. This was due apparently to release of oxidizing enzyme from the pea root segments into the external solution. They concluded: "oxidase activity cannot control the auxin level of applied IAA at growth active sites." These findings are in basic agreement with Fang et al (78), who showed that GA-induced increased growth in corn and pea roots could be ascribed to a large decrease in the formation of indoleacetylaspartate and only a relatively small amount of decarboxylation. Root segments of rice, peas, and corn have been shown to take up IAA from the external solution in three separate phases (Igari 102, 103). Accordingly, it should be mentioned that the formation of the conjugate may be a complex process.

Other reported products of IAA metabolism (103), which were isolated from "feeding" experiments, do not seem to correspond to any known growth regulator.

THE ROLE OF AUXIN IN ROOT GROWTH

The question relating to the role of auxin in root growth has been rather extensively reviewed (2, 43–45, 205, 207, 208, 221, 222, 239). Over the last decade, doubt has vanished that auxin is an essential factor in the promotion of growth in roots. The following reasons are offered as evidence (206): (a) IAA at various concentrations can significantly increase the elongation of intact and isolated root segments (for example: Audus & Thresh 16, Brauner & Diemer 31, Burström 47, Davidson et al 68, Konings 121 and 124, Naylor & Rappaport 151, Pilet 166, and Pilet et al 173). (b) Auxin increases the growth of excised roots maintained in culture (for example: Butcher & Street 50, Robbins & Harvey 180, Roberts & Street 181, Weston & Street 232). (c) Isolated or intact roots whose growth has been inhibited by synthetic and natural inhibitors may undergo growth recovery as a result of auxin application (for example: Burström 45, Davidson et al 68, Libbert 132, Pilet 166), or conversely, auxin-inhibited root growth may be reversed and stimulated by the addition of auxin antagonists or anti-auxins (for example: Burström 41, 42).

The relationship of auxin to root growth will be dealt with at three different levels: (a) auxin as a limiting factor in linear growth and geotropism; (b) the mechanism of auxin action in growth; and (c) the interaction of auxin with other substances in growth promotion.

Auxin as a Limiting Factor in Root Growth

It now seems clear that auxin is active in promoting growth of roots while at the same time serving as the agent which brings about its cessation (40, 44, 47). Burström (47), in reaffirming his original two-phase theory, has shown that NAA treatment of intact wheat roots at concentrations between 3 $\times$ 10^{-8} M and 3 $\times$ 10^{-7} M increased the rate of elongation over the control at all concentrations tested ("first effect"). However, the first effect may be masked since the duration of the promotion of elongation was progressively shortened with increasing auxin concentration ("second effect"). Further, the different kinetics of growth, at the concentrations applied, make it clear that cells at different stages of elongation respond in the two ways (hereafter referred to as "step one" and "step two") as a consequence of their chronological age. Thus, the two effects are independent. Accordingly, although the system is by no means uncomplicated, it does appear that the limiting action of auxin is *not* at the promotional phase ("step one")—since the experiments were performed on intact roots presumably containing their own endogenous supply of auxin—but rather on the duration phase ("step two"). Roots, according to Burström, may differ from shoots in that "in roots auxin exhibits a second, inhibitory action at the end of the elongation period." Net growth promotion then is the sum of varying "step one" and "step two" responses throughout the root axis.

Burström's hypothesis goes a long way in explaining reports of the effects of auxin on root growth which are reflected as a two-phase response, as a transient response, or as a recovery-reactivation response. Pilet (161), for instance, showed that auxin first stimulated the growth of intact *Lens* roots and then depressed it. Burström (43, 47) showed clearly that the extension of root cells took several hours but was followed by a sharp or abrupt cessation. These two phases were presumed due to (*a*) increased deformability of the cell wall (elongation, "step one") and (*b*) synthesis of new wall material (cessation, "step two"). Morré & Bonner (148), upon careful analysis of these precepts and by interpretation of their growth data with corn roots, concluded that increased deformability is associated with increased growth; that cell wall synthesis may be accelerated during IAA-induced growth; and that growth inhibition induced by auxin is due to a shortening of the period of elongation. Díez et al (70, 71) have very clearly demonstrated that cells in two regions of intact onion roots respond differently to added auxin. Cells which were in their initial stage of growth had their elongation phase of growth strongly enhanced by IAA, while those which had stopped growing were not affected. List (135), by using the streak photographic method of Erickson & Sax (76) and by making precise computer analyses of growth, has shown that IAA-induced reduction of growth rate of corn roots consists of a series of dampened oscillations which reflect serial promotions and inhibitions in response to a pulse application of auxin. Thus apparently "step one" and "step two" responses to auxin (47) can occur at the same time in the intact root, but one response may appear to have the upper hand over the other at any one time. Such transient growth enhancement activities of corn roots due to applied IAA have also been demonstrated by Barlow (18).

As stated above, the cessation of elongation growth is also an effect of auxin (47, 148). Furthermore, protracted exposure to auxin or exposure to high concentrations of auxin can speed cessation of growth to a point which, in sum, results in an overall inhibition of growth (Burström 47, List 135). Hejnowicz & Erickson (93, 94) have demonstrated an apparent paradox with respect to this phenomenon. They have shown that when *tips* of roots of corn, pea, sunflower, and onion were subjected to an inhibitory treatment of auxin for a short period (4–60 min), the lag period before elongation commenced again was protracted (one hour), while if the treatment itself lasted longer than an hour, the lag period before elongation commenced again was reduced very considerably. They attributed this difference in reaction time to a possible radial gradient of auxin in the tissue. They reasoned that a positive gradient (concentration decreasing with distance from the center) would obtain after a prolonged exposure to auxin and that such a gradient favors growth. A negative gradient (concentration increasing with distance from the center), on the other hand, would obtain after a short exposure to auxin and impose an inhibition on growth until the gradient was reversed. This interpretation fits nicely with the kinetics of IAA accumulation shown by Andreae & van Ysselstein (10) and with the conclusions of Andreae (6), Hejnowicz (91, 92), List (135), and Manos (140) that complete growth inhibition is a rapidly reversible process and that it ceases when the auxin concentration on the outside of the root is less than that inside of isolated pea root tips. The interesting implications of a positive radial gradient in the growing region and its relation to a linear auxin gradient, auxin transport, and growth, are discussed below.

Andreae (6) has shown that growth inhibition due to excess auxin very likely takes place before auxin enters the cytoplasm, and he thus concludes that growth inhibition due to auxin occurs at sites in the cell wall or the membrane. The cell wall itself is known to be the location of a great deal of relatively autonomous metabolic activity, and accordingly it may respond directly to auxin (128, 179). Burström (46) has shown that calcium may extend or enhance the auxin-stimulated elongation phase of growth ("step one"), and thus he rejects the hypothesis that calcium alone may stiffen the wall and cause elongation to cease. Auxin-induced growth cessation or inhibition when auxin is present in nontoxic amounts requires alternative explanation(s) which will certainly be related to the properties of the cell wall.

Growth cessation has also been accounted for on the basis of endogenous inhibitors. Libbert (132) proposed that an auxin-induced inhibitor in roots (Hoepfner 98), perhaps "bound" auxin itself, is responsible for growth inhibition. He has further shown that inhibition in root growth may be reversed by leaching out the inhibitor. Other inhibitors which have been suggested and which might directly interact with or antagonize auxin-induced root growth are: an auxin-associated inhibitor of an unknown character (Fransson 80); β inhibitor (Björn et al 25, Masuda 144); the oxindole derivatives of auxin-oxidase activity (Tuli & Moyed 227); epiphytic bacterial-produced inhibitors (Burström et al 48); and abscisic acid (ABA) (Belhanafi & Collet 20, Pilet 168). It now appears likely that ABA, the naturally occurring inhibitor of shoots (3), is present in roots (Tietz 219). Accordingly, it seems likely that previous reports of inhibitors, such as β inhibitor isolated from roots, may be attributed to the presence of ABA in the extracts (Addicott & Lyon 3). Thus, correlations which have been described between differential growth responses and differences in inhibitor levels (Björn et al 25, Masuda 144) and to distributional changes of inhibitors (Gibbons & Wilkins 84) may, at least in part, be due to ABA. Finally, auxin-induced synthesis of ethylene has been proposed as the agent responsible for the inhibition of root growth (Burg & Burg 37, Chadwick & Burg 55, Sankhla & Shukla 185). It has long been known that ethylene inhibits root growth (Zimmerman & Wilcoxon 249). However, whereas auxin-induced inhibition is readily reversible, ethylene-induced inhibition under some circumstances does not appear to be so (8). Furthermore, differing kinetics and magnitude of inhibition, differing calcium and pH effects, and differing responses to ethylene concentration-solution volume ratios led Andreae et al (8) to conclude that ethylene evolution does not account in large measure for the inhibition of pea root growth induced by IAA. Chadwick & Burg (56) accounted for some of the discrepancies between themselves (55) and Andreae et al (8), but concluded that not all the inhibition due to IAA application which they record involves ethylene. Thus it would appear that ethylene is not the sole inhibitory factor and that its role in the regulation of root growth requires further elucidation.

Auxin distribution in roots.—It is pertinent that auxin-induced growth promotion and inhibition be discussed in relation to auxin distribution in the root. There is considerable indirect evidence that endogenous auxin is synthesized by the tip of the root. This evidence rests on observations that there is usually more auxin activity found in the tip than basal to it (Pilet 161, Thimann 215) and also on experiments which showed that auxin may be obtained from root tips by the agar-diffusion method (Boysen-Jensen 30, Hawker 90, and van Raalte 178). More recent experiments using colchicine (Davidson 65), barban (Burström 46), and X irradiation (Davidson 64, Woodham & Bedford 240) have been interpreted to indicate that IAA may be synthesized in the actively dividing meristem.

The significance of these types of observations must now be questioned on grounds that an acropetally oriented polar auxin transport system has been demonstrated in the tips of a variety of roots. The presence of this system, which has been brought to light by applying ^{14}C radioactively labeled IAA to subapical root segments using the classical donor-receiver agar block method (231), has been reported by: Aasheim & Iversen (1); Hillman & Phillips

(97); Iversen & Aasheim (104); Kirk & Jacobs (120); Pilet (165); Scott & Wilkins (192, 193); Wilkins & Scott (237, 238); Zaerr (248). As in shoots, auxin transport in roots is an active metabolically driven process (Wilkins & Scott 237, 238) which is both light (Scott & Wilkins 193) and temperature (Wilkins & Cane 236, Wilkins & Scott 238) sensitive. Cane & Wilkins (52) using corn and Hillman & Phillips using peas (97) demonstrated that the transport of IAA in the base to apex direction in short-term experiments is most pronounced at the tip of the root, and thus acropetal transport takes place in the region of the root which is growing. Cane & Wilkins (52) also showed evidence that the acropetal flux decreases to a diminishingly small amount behind this region (this observation has been confirmed verbatim by E. Merrell and T. K. Scott, unpublished). A small basipetal transport component, thought to be accounted for by virtue of physical diffusion (Wilkins & Scott 237, 238), is present over the entire length of the corn root (Cane & Wilkins 52).

The presence of the strongly acropetal system at the tip of the root seems so general that reported exceptions require comment. Faber (77) using several species, Hertel & Leopold (96) using corn, Iversen & Aasheim (104) using sunflower and cabbage, and Pilet (165) using *Lens,* found evidence of movement of auxin in both directions, while Yeomans & Audus (246) detected little movement of IAA in *Vicia* (although, like Scott & Wilkins 192, they showed evidence of an accumulation of auxin in the tissue of the root tip rather than the base). It would seem that these results may be readily accounted for on the basis of variable recoveries due to: the duration of the transport period (192); the length of the root segment and the region of the root from which the segment was selected (52); or cut-surface auxin inactivation (1, 104, 177) and further auxin metabolism (97, 177). Inactivation of IAA does not appear, however, to be a factor in recovery from segments taken from corn roots (Scott & Batra 189). Bonnett & Torrey (28) were only able to detect acropetal movement of IAA in regions outside the growing zone in excised, cultured *Convolvulus* roots. The absence of transport in the growing zone may have resulted from developments deriving from the culture conditions.

Reports of basipetal auxin transport, previously taken to be the normal direction of movement in intact roots, require further explanation. Nagao & Ohwaki (150) used the inhibition of growth as the index for auxin transport, and they were able to show that in *Vicia* roots growth was inhibited basal to the point of IAA application. However, they also reported control-level growth promotion proximal or apical to the point of application. This latter point may in fact serve as evidence for the acropetal movement of auxin. Hejnowicz (92) has shown that basally applied IAA at concentrations of 10^{-7} M and 10^{-6} M did not inhibit growth, and that when growth inhibition first appeared ($10^{-5}M$) it was rapidly reversed by removing the auxin. Konings (124) found a similar stimulating effect (as reflected by a positive geotropic response) by adding auxin at various distances basal to the region where the response was observed.

The last two reports may be taken as direct evidence for acropetal movement of auxin in intact roots. It is likely that Nagao & Ohwaki (150), like Hejnowicz & Erickson (94), induced growth inhibition when the tip of the root was treated in such a way as to impose a negative centripetal gradient of auxin. Thus by indirection it may be concluded that the promotive, positive radial gradient of auxin is imposed by the acropetal transport system. Further, it should be noted that if transport takes place in the center of the root in or near the stele and the protostele (Hejnowicz 92), then the positive centripetal gradient of auxin would be the expected condition.

If roots can be made to respond positively using concentrations of auxin which promote the growth of shoots (Burström 47, Hejnowicz 91, Libbert 132) and if transported or transportable auxin—the hormonal entity—is the form which is mostly strictly correlated with growth (Hertel et al 95, Jacobs 107, Scott & Briggs 190, Steeves & Briggs 204, and confirmed for roots by Hejnowicz 92, and Konings 124), then a source of the auxin which is moving toward the tip must reside basally. The possible alternative sources suggested below are speculative.

Sources of auxin in roots.—The aerial portion of the seedling or the shoot was first proposed as the source of the root growth-promoting hormone "rhizocaline" (Bouillenne & Went 29). Thimann (215) also pointed out that auxin found in the root could come from the shoot. Long-term experiments with intact seedlings indicate that diffusible auxin may move from the shoot to the root (Altman et al 4, Iversen et al 105, Morris et al 149, Phillips 158, 159). However, Morris et al (149) have shown that radioactive IAA tended to accumulate in the developing lateral roots, while there was no evidence that IAA reached the primary root. Indirect evidence that auxin does not reach the root from the shoot in younger seedlings, in which the primary root is the predominant organ, is provided in cases in which there does not appear to be auxin transport through the shoot-root transition zone (Jacobs 107, Scott & Briggs 190 and 191, Thimann 215), and in cases in which there is very little if any base to tip auxin transport demonstrable in the base of the primary root (52). Thus, an alternative auxin source for the growing zone in the tip of the primary root would exist within the root itself or be present immediately outside (or both).

Sheldrake (195) has commented that the most probable auxin source within the root is differentiating xylem cells. Sheldrake & Northcote (196–198) were the first to suggest that cell death is, in and of itself, an auxin production mechanism and that xylogenesis can produce hormonal auxin (196). Thus there is the paradox of auxin being required for vascular differentiation (Jacobs 108) but produced as a byproduct of it if differentiation is fully determinant to the point where cells undergo complete autolysis. Spatially this would seem an appropriate source of auxin for the acropetal transport system in roots. Senescing root hair cells and perhaps sloughed root cap cells would also seem possible candidates as auxin sources.

A complication related to the above proposal would appear to be that there are two endogenous auxin systems, both acropetal, which occur in the root: one which serves the elongation zone of the primary root and which is more or less restricted to it; and one which may derive from the shoot and involve differentiative processes other than growth. Buis (35) has strongly inferred that such a situation indeed may be the case. As a result of his analysis of growth and development of lupine roots, there is strong evidence for more than one active "factor" being present in different places and at different times in the root during its development. Buis considered both the shoot and root to be sources of the factors.

The last alternative source of auxin would be external to the root. Such a source has been widely noted and is thought to come from epiphytes associated with the rhizosphere. Many bacteria (133) and fungi (182) are known to produce auxins (including IAA), and auxin has been isolated from soils in which microorganisms have been known to occur (88, 156). Furthermore, the interaction between the root and associated microflora appears to affect root growth. The interaction is reflected as the sum of the activities of bacteria and fungi which can be both stimulatory and inhibitory to root growth (Sobieszczański 203). Those microorganisms which have been identified and demonstrated to be associated with the root surface can be distinguished from and are separate from other soil microorganisms (Lim 134). Thus the rhizosphere is a biological realm in which the higher plant and its associated microflora may be assumed to benefit mutually from each other's activities.

Libbert et al (133) have shown that plants in the presence of epiphytic bacteria have more endogenous auxin than those which are maintained in a sterile condition. Thus exogenously produced auxin apparently gets into the tissue. Burström et al (48) have shown bacteria to be present in the mucilaginous layer of the root, and their metabolites may be rapidly taken up by the epidermis. Substrates for bacterial synthesis of auxin would appear to be plentiful in the soil, due in part to the presence of the root itself. Exudates from a variety of sterile intact roots have been shown to contain a large number of proteins (Juo & Stotzky 114). Indolic substances have been found in "staled" media in which cultured roots have been allowed to grow for protracted periods (Winter & Street 239, Woodruffe et al 242). One such compound was reported to be a tryptophan-containing peptide (239) which itself might well be utilized by bacteria (in the soil under normal conditions) to synthesize indolic auxin(s) (Erdmann & Schiewer 75). Since there appear to be exoenzymes (Chang & Bandurski 57) of root and bacterial origin present outside the root which are capable of participating in both auxin synthesis and its subsequent metabolism (Andreae 6, Pilet & Chalvignac 171, Wichner 234), it would appear that sufficient metabolic apparatus, necessary for auxin synthesis, exists on or immediate to the root surface. Further, it has been amply demonstrated that roots have the capacity to take up a variety of organic compounds, including indoles, and that the region of uptake is most probably the zone of differentiation or where root hairs are most abundant (Bolli 27, Khavkin et al 119, Scheffer et al 187). It should be noted in this connection that IAA transported to the root tip is apparently not secreted by the tip (Scott & Wilkins 192) and thus probably is not available to be recycled as such. Thus there is the suggestion of a second auxin source available to the primary root which is not of shoot origin.

Geotropism in roots.—It would seem appropriate to review tropistic behavior of roots in view of the preceding discussion regarding the stimulation and inhibition of root growth and the movement, distribution, and possible sources of auxin. Since research on phototropism in roots is so scant and preliminary, the subject will not be treated here.

The topic of geotropism of roots has recently been thoroughly reviewed by Audus (11), Ball (17), Rufelt (183), and Wilkins (235). The classical theory of Cholodny (58) and Went (231), when applied to the positive geotropic response of roots, can no longer be sustained as valid. As mentioned above and as pointed out by Ball (17): (*a*) auxin production by the root tip has not been satisfactorily demonstrated; and (*b*) the presence of a supraoptimal concentration of auxin in the root tip has not been reliably verified over the years. Both points are minimal requirements and fundamental to the Cholodny-Went hypothesis, which when rephrased states: that auxin in the tip is present in growth inhibiting amounts; that consequence of gravity is to further inhibit growth of the lower half of a horizontally displaced root by causing an increase of auxin in the lower half. The increased amount of auxin was presumed to be due to lateral transport.

There is good evidence for unequal distribution of ^{14}C-labeled IAA in horizontal pea roots, with more below than above (124). However, Konings (123) has challenged that differential growth is in response to excess or supraoptimal amounts of auxin. He cited his own evidence and that of others which showed that removal of the tip of the root did not cause an increase of growth (release from inhibition). In other experiments, Iversen et al (105) and Konings (124) showed that IAA applied to the base of intact *Phaseolus* and pea roots, respectively, actually enhanced the geotropic growth curvature at the tip. This result led Konings to the conclusion that auxin was present in suboptimal amounts in the geosensitive region of the pea root (124). In experiments of a similar nature, Audus & Brownbridge (12) were also able to increase as well as decrease both linear growth and the positive geotropic response in pea roots by adding different concentrations of auxin. In addition, they found that under circumstances in which curvature was altered, there was no *differential* effect on the growth rate of the upper and the lower sides. The above results, taken together, indicate that positive curvature of roots induced by gravity is a result of differential growth, but the difference is not due to supraoptimal amounts of auxin in any region of the root.

Burström (47) has questioned the usefulness of the terms supra and suboptimal amounts of auxin when using them in reference to root growth. As an example to amplify the point, he has suggested that the effect of gravity on roots of wheat maintained vertically may be an increased auxin supply to the growing region, which in turn would result in greater linear growth of the root. He has shown that vertical roots grow at a faster rate than those growing horizontally on a clinostat. It may also be concluded, in the case of the horizontal roots, that a differential in the sequence of "step one" and "step two" growth responses, due to unequal amounts of auxin in the upper and lower halves, may result in a positive downward curvature. Whereas these two conclusions may seem contradictory, one need only add that root cells are highly sensitive to a relatively narrow concentration range of auxin. Indeed, Zinke (250) has shown that in the course of an overall downward bending of the pea root, there are both negative and positive reactions manifested at different times. His observations would seem to underscore the sensitivity of the responding cells on the upper and lower halves during transient changes in auxin ratios between the two.

Scott & Batra (189) have confirmed that vertical orientation of corn roots, before testing for transport, intensifies the acropetal movement of auxin in segments without altering the basipetal component. Thus an effect of gravity may well result in the delivery of more auxin to the growing region. Scott & Wilkins (192) did not detect a gravitational influence on auxin transport during the course of their transport experiments. Presumably the orientation effect was not apparent due to insufficient elapsed time during the experiment. That more auxin can be transported to the growing zone seems to be indicated by the experiments of Hejnowicz (92). He has shown that IAA, at concentrations of 10^{-7} and 10^{-6} M applied to the basal region of corn roots, did not affect growth at the tip, whereas a concentration of 10^{-5} M was inhibitory. These reports provide evidence that it is possible to increase the amount of auxin transported and that this can, though need not necessarily, result in growth changes. Whether or not such changes can affect the geotropic response is not known. However, there is the suggestion that there is a promotion of acropetal transport of IAA in geotropically stimulated corn roots (Scott & Wilkins 193). Also Konings (124) and Parker (155) were able to inhibit the geotropic response by applying 2,3,5-triiodobenzoic acid (TIBA) to roots. The presence of this inhibitor was presumed to interfere with auxin movement associated with the acropetal transport system (124).

It may be postulated from the above investigations that the promotive phase in the positive geotropic growth response may be due to both linear and transverse auxin movement. The differential growth causing the curvature is a consequence of unequal amounts of auxin which may result in sequential differences in the two "step" phases of growth (47) because of promotive and inhibitory auxin gradients on the upper and lower sides (94).

If auxin coming from the base of the root participates in the control of growth and geotropic behavior, then what part does the root cap play in the gravity response of roots? Juniper et al (113) and Konings (123) have provided indisputable evidence that the root cap must be present in order for a positive geotropic response to occur (113, 123) or for the lateral migration of ^{14}C lableled IAA to take place near the tip (123). The loss of geotropic sensitivity, which is not attended by a loss in linear growth, is regained with the regeneration of the root cap (Juniper et al 113, Cercek 54). The role of the root cap as the geoperception organ is no longer disputed (112, 130).

Three possible ways in which the root cap might mediate the curvature response in geotropism suggest themselves. First, it may be that the cap synthesizes an auxin which could run "countercurrent" in the tip of the root and alter the growth response (Burström 47). As already mentioned, the evidence of this occurring does not appear convincing (Juniper et al 113). Secondly, the root cap may also be the site of synthesis of compounds which participate in the metabolism of auxin. The reports of the unequal distribution of auxin on the upper and lower sides of horizontally displaced roots may thus reflect unequal metabolism of auxin. Pilet (167), for instance, has demonstrated high IAA oxidase activity in the root cap of *Lens*, which implicates the cap as the site of its production. Konings (122, 123) has shown that auxin oxidase-inhibiting phenols applied to the root cap inhibit geotropic curvature and also reduce the differential in ^{14}C-labeled auxin between upper and lower halves of pea roots. This may be taken as evidence that there is an unequal distribution of IAA oxidase in these two regions. Differentials in other metabolic activities involving auxin may also account for the apparent difference in auxin content between upper and lower tissues (Iversen et al 105, Lahiri 126). Finally, Gibbons & Wilkins (84) have suggested that the root cap is the production site of a growth inhibitor which itself could be unequally distributed either because of unequal synthesis or because of its lateral migration. The presence of a growth-inhibiting substance participating in the root georesponse has also been suggested by Audus & Brownbridge (12, 13). Burg & Burg (36, 37) and Chadwick & Burg (56) have suggested that ethylene acts as an inhibitor and, in part, causes the geotropic responses in roots and shoots. However, for reasons mentioned earlier, the role of ethylene in root growth remains obscure. Obviously, until the nature of or the identification of the inhibitor is established and the sensitivity of growing tissue to it is known, the relationship of the root cap in gravi-perception and the geotropic response remains an open question.

The Mechanism of Auxin Action in Root Growth

The dilemmas which have arisen concerning the mechanism of auxin action in shoot growth have intrigued and haunted investigators for over 40 years (see Ray 179). Clearly, the same problems apply to roots. No single molecular, cytological, or tissue site has been identified with certainty as the location of the primary action of auxin. Many have been suggested, but only the cell wall persists as the most likely place where auxin has its first effect.

As pointed out by Ray and his co-workers (179), determining the timing of the initiation of the growth response by auxin is critical to the interpretation of its mechanism of action. As is the case in shoots, auxin can evoke almost immediate growth responses in roots. Indirect evidence of this is seen from the enhanced cytoplasmic streaming (Jackson 106, Sweeney 214) and growth (Jackson 106) in root hairs which took place within a few minutes following application of IAA. Hejnowicz (92) and Hejnowicz & Erickson (94) have recorded very rapid responses of roots to auxin. Burström et al (49) have shown, by analyzing resonance frequency differences and applying the principle of Young's modulus, that auxin-induced elastic extensibility could be detected within 15 minutes in pea roots. Furthermore, they were able to demonstrate that the change in the elastic modulus was restricted to the region where IAA was thought to promote cell elongation. The overall growth enhancement and growth inhibition (and recovery from inhibition) so frequently reported as a consequence of auxin application appear to be the sum(s) of many extremely rapid dampened oscillations in growth rate (List 135). Thus it would appear from this evidence that both "first step" and "second step" auxin effects may take place very rapidly (43, 47).

The evidence is convincing that the action of auxin on cell elongation in roots, at least initially, is on the mechanical properties of the cell wall (Burström et al 49), and perhaps the plasma membrane (Cocking 60). This rapid action is very likely respiration independent (Gasparikova 82) and may be considered general for higher plant tissues (59). It is also generally assumed that the time course of the initiation of elongation is such as to preclude direct involvement of the nucleus. Thus an immediate and direct effect of auxin on gene expression in this connection seems highly doubtful (49, 179). The primary mechanism of action remains obscure and a subject of considerable controversy. However, it does seem clear that reports of auxin-influenced nucleic acid syntheses, metabolite syntheses, and enzyme syntheses and activations (all processes which take place after the initiation of elongation and perhaps as a consequence of it) play further roles in root development (see below).

The Interaction of Auxin with other Endogenous Growth Regulators

It has been demonstrated positively that at least two other major native growth-promoting substances, gibberellins and cytokinins, occur in roots. Thus, it is entirely possible that reports of interactions between auxin and these two substances have physiological significance.

Gibberellins.—The presence of gibberellins in roots has been shown by Carr et al (53), Jones & Phillips (111), Phillips & Jones (160), Skene (201), and others. Furthermore, there is good evidence that gibberellins are synthesized by the root and exported to the shoot (33, 111, 129).

The majority of the reports describing the effect of gibberellins on roots show that inhibition of growth of intact roots results from its application whether in the presence or the absence of applied auxin (for example: Lacoppe & Gaspar 125, Manos 140, Tognoni et al 220). However, such is not always the case (for example: Brown & Gifford 32, Devlin & Brown 69, Fang et al 78), and thus the picture is not a clear one.

The best analysis of a gibberellin-auxin interaction in the regulation of root growth has been provided by Mertz (145, 146). By using both a gibberellin-less dwarf mutant and a normal corn variety, he was able to demonstrate that the lack of gibberellin in the former resulted in reduced root growth because of a decrease in the level of endogenous auxin. This conclusion was drawn from experiments in which root growth in the normal variety was reduced as a consequence of the application of inhibitors of gibberellin biosynthesis. Significantly, this induced inhibition in the normal plant was reversed by application of either gibberellin or IAA. What part gibberellin might play in auxin synthesis or increased auxin availability is not known. Reports of both Odhnoff (153) and Lacoppe & Gaspar (125) suggest that gibberellin has a sparing effect on auxin in roots.

Cytokinins.—Cytokinin activity has been found in root extracts (Bui-Dang-Ha & Nitsch 34) and from root exudates on numerous occasions (Kende 117, Loeffler & van Overbeek 136, Vaadia & Itai 228, Yoshida et al 247, and others). It is thought that cytokinins are also synthesized in the root (Kende 117).

Like gibberellins, the cytokinins (such as kinetin or benzyladenine), when applied alone or together with auxin, can cause a net growth inhibition in roots (for example: Gaspar & Xhaufflaire 81, Lacoppe & Gaspar 125, Manos 140, Tognoni et al 220, Yang & Dodson 245). There are no satisfactory explanations for this effect on root growth. However, since the activity of cytokinins is so strongly associated with RNA metabolism (Skoog & Armstrong 202), it would appear that growth effects are most likely to occur at the level of cell division and events closely related to genetic expression.

THE ROLE OF AUXIN IN DIFFERENTIATION IN ROOTS

Torrey comprehensively detailed the involvement of auxin with various developmental processes of roots in 1965 (222). Since the publication of that review there have been large numbers of reports giving evidence of auxin-dependent processes at all levels of root development. Unfortunately, it is not possible to treat the subject in detail here. However, there follows a discussion of observations relating to differentiation in roots which correlates with the recent evidence relating to the presence and distribution of auxin in roots.

Cell division.—IAA is known to stimulate cell division in the tip of the primary root and in lateral roots (Davidson & MacLeod 66, 67) and the vascular cambium (Torrey & Loomis 225, 226). The results of experiments in which cell division was inhibited by colchicine in *Vicia* root tips (66–68) indicate that reactivation and reorganization of a new meristem is stimulated by application of IAA. The significance of this point is that auxin is influential in the induction of meristematic activity in the root but is *not* synthesized as a result of it. Thus one implication of these results is that auxin supplied from the base of the root is active in promoting cell division. It should be noted, in addition, that colchicine-induced inhibition is strictly correlated with alterations in the polarity of root cells (MacLeod 137). It follows, therefore, that cell polarity changes may alter the polarity and functioning of the auxin transport system, which in turn may account for the inhibition of cell division and loss of tissue organization.

It is highly likely that cytokinins are also involved in the cell division process in the root tip (MacLeod 138, Van't Hof 229). It would seem reasonable that significant interactions between these two substances go on in this region and that if availability of auxin is altered in some way so as to create an imbalance, an alteration in growth and development would be expected.

RNA and Protein Metabolism.—Since the subject was last reviewed 5 years ago (Street 208), there has been a spate of reports which show a connection between auxin distribution and auxin treatment with the synthesis and activation of RNA and various enzyme proteins in roots.

It is interesting to note that as early as 1960 Woodstock & Skoog showed a concomitant increase in RNA content (but not DNA) with growth rate in corn roots (243). They further described evidence which indicated that the role of RNA, synthesized in the apex of the root, was associated with the control of the rate of growth but not its duration (244). They concluded that the determining influence of RNA, while implicated in determining the final size of the root, occurs at the time of cell division rather than during elongation. This conclusion is strongly supported by the studies of Torrey et al (224) which show that a series of unalterable events are set into motion at the time the cell "originates" and cannot be stopped short of full cellular differentiation. Also, Burström's hypothesis (47, 49) that the rapidly induced promoting and duration-limiting effects of auxin in root elongation are RNA-independent processes seems consistent with the findings of Woodstock & Skoog (244).

Other investigations indicate that auxin may be responsible for the synthesis of RNA in the root tip (Miassod et al 147, Pilet & Braun 169, 170). This suggestion derives from pulse-labeling experiments (147) and from the positive correlation between auxin levels and RNA levels in various tissues of *Lens* root (170). An alternative hypothesis is that auxin decreases RNA catabolism through an inhibition of RNase activity in the root tip (170). Furthermore, IAA treatment of intact *Lens* roots has been reported to cause an increase in endogenous auxin as well as RNA levels, thus suggesting an auxin-auxin feedback mechanism. Finally, there is the long-standing observation that there is a negative correlation between levels of endogenous auxin and the auxin oxidase system (Pilet 161, Pilet & Galston 172). The oxidase system, if induced by the presence of auxin (172), could alter an auxin-auxin feedback mechanism (Pilet & Braun 170) in a variety of ways.

The unchanging picture emerging from the studies of Pilet and his coworkers is that the strict correlations (positive and negative) between auxin, RNA, and enzymes explain the cause and effect relationships in root growth and development. It should be pointed out, however, that all the experiments cited above, in which there was the feeding or application of substances, involved long incubation periods. Thus it is very difficult to separate primary, secondary, or tertiary effects on even such processes as growth. These studies do show that the role of the gene is expressed through RNA synthesis and subsequent enzyme synthesis. Unfortunately, these studies do not provide information regarding the sequence of developmental events in the normal, intact root.

The importance of the time and place of the initial cell division to *all* developmentally connected processes which follow has been properly and emphatically pointed out by Torrey et al (224). Thus, the cell is "programmed" and it will respond variously to such hormonal stimuli as auxin and a cytokinin as time progresses and as it is exposed to different amounts of hormones. Of course, it is critical to know in this connection what the sources of the auxin and the cytokinin are and in what direction they are moving. From the available evidence, these two hormones may move and thus operate in the "countercurrent" fashion suggested by Burström (47).

Whereas both hormones are clearly necessary for the final disposition of the genetic and metabolic interactions which occur during root development, there are several examples of the sequence and polarity in which cell differentiation takes place which emphasize the importance of the direction from which auxin comes. For example: xylem regeneration, when stimulated directly by auxin application to the root, takes place in an acropetal direction (Simon 199, Sachs 184); auxin-induced lateral root initiation and development in cultured *Convolvulus* and wheat roots takes place from the base to the apex (Bonnett & Torrey 28, Street 209); and auxin-dependent vascular cambium activation, in radish roots maintained in culture, progresses from the base toward the apex (Torrey & Loomis 225, 226).

In conclusion, it should be repeated that the preferential movement of auxin is consistently from the morphological top to the bottom of the plant, and normally in accordance with the gravitational vector, and that features of auxin regulation of root growth and development appear to be manifested as a consequence of this movement.

CONCLUSIONS

The role of auxin in the development of roots includes the following essential features:

1. Auxin is necessary for the elongation of root cells, and it may be stimulatory at concentrations which promote shoot growth.

2. Auxin also regulates root growth by inhibiting the duration of cell elongation.

3. IAA may not be present as such in the root, but it may exist as a complex or as a member of a group of other promoters and inhibitors.

4. Auxin moves predominantly in a base to apex direction in the root tip in an active polar transport system.

5. Auxin metabolism in the root is complex, and some of its many features may be related directly to cell growth and differentiation.

ACKNOWLEDGMENTS

I wish to thank Miss Betty Fulton for her extensive help in the literature search and for her conscientious and careful typing of the manuscript. I am also greatly indebted to numerous colleagues on both sides of the Atlantic who provided me information, preprints, and valuable advice and criticism during the preparation of this review.

LITERATURE CITED

1. Aasheim, T., Iversen, T-H. 1971. *Physiol. Plant.* 24:325–29
2. Åberg, B. 1957. *Ann. Rev. Plant Physiol.* 8:153–80
3. Addicott, F. T., Lyon, J. L. 1969. *Ann. Rev. Plant Physiol.* 20:139–64
4. Altman, A., Monselis, S. P., Mendel, K. 1966. *J. Hort. Sci.* 41:215–24
5. Andreae, W. A. 1964. *Régulateurs Naturels de la Croissance Végétale,* ed. J. P. Nitsch, 559–73. Paris: CNRS
6. Andreae, W. A. 1967. *Can. J. Bot.* 45:737–53
7. Andreae, W. A., Collet, G. 1968. *Biochemistry and Physiology of Plant Growth Substances. Proc. 6th Int. Conf. Plant Growth Substances,* ed. F. Wightman, G. Setterfield, 553–61. Ottawa: Runge
8. Andreae, W. A., Venis, M. A., Jursic, F., Dumas T. 1968. *Plant Physiol.* 43:1375–79
9. Andreae, W. A., Ysselstein, M. W. H. van 1956. *Plant Physiol.* 31:235–40
10. Ibid 1960. 35:225–32
11. Audus, L. J. 1969. *The Physiology of Plant Growth and Development,* ed. M. B. Wilkins, 205–42. London: McGraw-Hill
12. Audus, L. J., Brownbridge, M. E. 1957. *J. Exp. Bot.* 8:105–24
13. Ibid, 235–49
14. Audus, L. J., Gunning, B. E. S. 1958. *Physiol. Plant.* 11:685–97
15. Audus, L. J., Lahiri, A. N. 1961. *J. Exp. Bot.* 12:75–84
16. Audus, L. J., Thresh, R. 1953. *Physiol. Plant.* 6:451–65
17. Ball, N. G. 1969. *Plant Physiology: A Treatise. Analysis of Growth: Behavior of Plants and their Organs,* ed. F. C. Steward, 5A:119–228. New York: Academic
18. Barlow, P. W. 1969. *Planta* 88:215–23
19. Bayer, M. 1961. *Planta* 57:215–34
20. Belhanafi, A., Collet, G. F. 1970. *Physiol. Plant.* 23:859–70
21. Bennet-Clark, T. A., Kefford, N. P. 1953. *Nature* 171:645–49
22. Bennet-Clark, T. A., Tambiah, M. S., Kefford, N. P. 1952. *Nature* 169:452–53
23. Bennet-Clark, T. A., Younis, A. F., Esnault, R. 1959. *J. Exp. Bot.* 10:69–86
24. Bentley, J. A. 1958. *Ann. Rev. Plant Physiol.* 9:47–80
25. Björn, L. O., Suzuki, Y., Nilsson, J. 1963. *Physiol. Plant.* 16:132–41
26. Bohling, H. 1959. *Planta* 53:69–108
27. Bolli, H. K. 1967. *Ber. Schweiz. Bot. Ges.* 77:61–102
28. Bonnett, H. T. Jr., Torrey, J. G. 1965. *Plant Physiol.* 40:813–18
29. Bouillenne, R., Went, F. W. 1933. *Ann. Gard. Bot. Buitenzorg* 43:25–202
30. Boysen-Jensen, P. 1933. *Planta* 20:688–98
31. Brauner, L., Diemer, R. 1967. *Planta* 77:1–31
32. Brown, C. L., Gifford, E. M. Jr. 1958. *Plant Physiol.* 33:57–64
33. Brown, M. E., Jackson, R. M., Burlingham, S. K. 1968. *J. Exp. Bot.* 19:544–52
34. Bui-Dang-Ha, D., Nitsch, J. P. 1970. *Planta* 95:119–26
35. Buis, R. 1970. *Physiol. Vég.* 8:1–33
36. Burg, S. P., Burg, E. A. 1967. *Plant Physiol.* 42:891–93
37. Burg, S. P., Burg, E. A. See Ref. 7, 1275–94

38. Burnett, D., Audus, L. J. 1964. *Phytochemistry* 3:395–415
39. Burnett, D., Audus, L. J., Zinsmeister, H. D. 1965. *Phytochemistry* 4:891–904
40. Burström, H. 1942. *Kgl. Lantbruks-Högsk. Ann.* 10:209–40
41. Burström, H. 1950. *Physiol. Plant.* 3:277–92
42. Ibid 1951. 4:470–85
43. Burström, H. 1953. *Ann. Rev. Plant Physiol.* 4:237–52
44. Burström, H. 1957. *Symp. Soc. Exp. Biol.* 11:44–66
45. Burström, H. G. 1968. *Physiol. Plant.* 21:1137–55
46. Burström, H. G. 1968. *Biol. Rev. Cambridge Phil. Soc.* 43:287–316
47. Burström, H. G. 1969. *Am. J. Bot.* 56:679–84
48. Burström, H. G., Persson, P. I., Stjernquist, I. 1970. *Physiol. Plant.* 23:202–8
49. Burström, H. G., Uhrström, I., Olausson, B. 1970. *Physiol. Plant.* 23:1223–33
50. Butcher, D. N., Street, H. E. 1960. *J. Exp. Bot.* 11:206–16
51. Butcher, D. N., Street, H. E. 1964. *Bot. Rev.* 30:513–86
52. Cane, A. R., Wilkins, M. B. 1970. *J. Exp. Bot.* 21:212–18
53. Carr, D. J., Reid, D. M., Skene, K. G. M. 1964. *Planta* 63:382–92
54. Cercek, L. 1970. *Int. J. Radiat. Biol.* 17:187–94
55. Chadwick, A. V., Burg, S. P. 1967. *Plant Physiol.* 42:415–20
56. Ibid 1970. 45:192–200
57. Chang, C. W., Bandurski, R. S. 1964. *Plant Physiol.* 39:60–64
58. Cholodny, N. 1926. *Jahrb. Wiss. Bot.* 65:447–59
59. Cleland, R. 1971. *Ann. Rev. Plant Physiol.* 22:197–222
60. Cocking, E. C. 1961. *Nature* 191: 780–82
61. Collet, G., Dubouchet, J., Pilet, P. E. 1964. *Physiol. Veg.* 2:157–94
62. Cutler, H. G., Vlitos, A. J. 1962. *Physiol. Plant.* 15:27–42
63. Darimont, E., Gaspar, T., Hofinger, M. 1971. *Z. Pflanzenphysiol.* 64: 232–40
64. Davidson, D. 1960. *Ann. Bot.* 24: 287–95
65. Davidson, D. 1961. *Chromosoma* 12:484–504
66. Davidson, D., MacLeod, R. D.

1966. *Chromosoma* 18:421–37
67. Davidson, D., MacLeod, R. D. 1968. *Bot. Gaz.* 129:166–71
68. Davidson, D., MacLeod, R. D., Taylor, J. 1965. *New Phytol.* 64:393–98
69. Devlin, R. M., Brown, D. P. 1969. *Physiol. Plant.* 22:759–63
70. Díez, J. L., López-Sáez, J. F., González-Bernáldez, F. 1970. *Planta* 91:87–95
71. Díez, J. L., Torre, C. de la, López-Sáez, J. F. 1971. *Planta* 97:364–66
72. Dullaart, J. 1967. *Acta Bot. Neer.* 16:222–30
73. Ibid 1968. 17:496–98
74. Elliott, M. C. 1971. *New Phytol.* 70:1005–15
75. Erdmann, N., Schiewer, U. 1971. *Planta* 97:135–41
76. Erickson, R. O., Sax, K. B. 1956. *Proc. Am. Phil. Soc.* 100:487–98
77. Faber, E. R. 1936. *Jahrb. Wiss. Bot.* 83:439–69
78. Fang, S. C., Bourke, J. P., Stevens, V. L., Butts, J. S. 1960. *Plant Physiol.* 35:251–55
79. Fox, R., Purves, W. K. 1968. *Plant Physiol.* 43:454–56
80. Fransson, P. 1960. *Physiol. Plant.* 13:398–428
81. Gaspar, T., Xhaufflaire, A. 1967. *Planta* 72:252–57
82. Gasparikova, O. 1967. *Biologica* 22:401–6
83. Ibid 1970. 25:11–17
84. Gibbons, G. S. B., Wilkins, M. B. 1970. *Nature* 226:558–59
85. Goldschmidt, E. E., Goren, R., Monselise, S. P. 1967. *Planta* 72:213–22
86. Goren, R., Goldschmidt, E. E. 1966. *Phytochemistry* 5:153–59
87. Goren, R., Tomer, E. 1971. *Plant Physiol.* 47:312–17
88. Hamence, J. H. 1946. *Analyst* 71: 111–16
89. Hare, R. C. 1964. *Bot. Rev.* 30: 129–65
90. Hawker, L. E. 1932. *New Phytol.* 31:321–28
91. Hejnowicz, Z. 1961. *Acta Soc. Bot. Pol.* 30:25–42
92. Ibid 1968. 37:451–60
93. Hejnowicz, Z., Erickson, R. O. 1967. *Wachstums regulatoren bei Pflanzen. Internationale Vortragstagung Rostock, Wissenschaftlichen Zeitschrift der Universitat Rostock Mathematisch-*

Naturewissenschaftliche Reihe, ed. E. Libbert, B. Steyer, 16: 533–34. Jena, Rostock: Fischer-Verlag

94. Hejnowicz, Z., Erickson, R. O. 1968. *Physiol. Plant.* 21:302–13

95. Hertel, R., Evans, M. L., Leopold, A. C., Sell, H. M. 1969. *Planta* 85:238–49

96. Hertel, R., Leopold, A. C. 1963. *Planta* 59:535–62

97. Hillman, S. K., Phillips, I. D. J. 1970. *J. Exp. Bot.* 21:959–67

98. Hoepfner, K-H. 1961. *Flora* 151: 398–410

99. Hofinger, M. 1969. *Arch. Int. Physiol. Biochem.* 75:225–30

100. Hofinger, M. 1970. *Biol. Plant.* 12:428–30

101. Hofinger, M., Gaspar, T., Darimont, E. 1970. *Phytochemistry* 9:1757–61

102. Igari, M. 1966. *Sci. Rep. Tohoku Univ. Ser. 4 (Biol.)* 32:35–46

103. Ibid, 47–63

104. Iversen, T-H., Aasheim, T. 1970. *Planta* 93:354–62

105. Iversen, T-H., Aasheim, T., Pedersen, K. 1971. *Physiol. Plant.* 25. In press

106. Jackson, W. T. 1960. *Physiol. Plant.* 13:36–45

107. Jacobs, W. P. 1950. *Am. J. Bot.* 37:551–55

108. Jacobs, W. P. 1969. *Int. Rev. Cytol.* 28:239–73

109. Janssen, M. G. H. 1969. *Acta Bot. Neer.* 18:429–33

110. Ibid 1970. 19:73–80

111. Jones, R. L., Phillips, I. D. J. 1966. *Plant Physiol.* 41:1381–86

112. Juniper, B. E., French, A. 1970. *Planta* 95:314–29

113. Juniper, B. E., Groves, S., Landau-Schacher, B., Audus, L. J. 1966. *Nature* 209:93–94

114. Juo, P-S., Stotzky, G. 1970. *Can. J. Bot.* 48:713–18

115. Kefford, N. P. 1955. *J. Exp. Bot.* 6:129–51

116. Kendall, F. H., Park, C. K., Mer, C. L. 1971. *Ann. Bot.* 35:565–79

117. Kende, H. 1965. *Proc. Nat. Acad. Sci. USA* 53:1302–7

118. Khalifah, R. A., Lewis, L. N., Coggins, C. W. Jr. 1963. *Science* 142:399–400

119. Khavkin, E. E., Tokareva, E. V., Obrucheva, N. V. 1967. *Fiziol.*

Rast. 14:997–1005

120. Kirk, S. C., Jacobs, W. P. 1968. *Plant Physiol.* 43:675–82

121. Konings, H. 1964. *Acta Bot. Neer.* 13:566–622

122. Ibid 1967. 16:161–76

123. Ibid 1968. 17:203–11

124. Ibid 1969. 18:528–37

125. Lacoppe, J., Gaspar, T. 1968. *Planta* 80:27–33

126. Lahiri, A. N. 1968. *Proc. Nat. Inst. Sci. India Part B* 34:21–26

127. Lahiri, A. N., Audus, L. J. 1960. *J. Exp. Bot.* 11:341–50

128. Lamport, D. T. A. 1970. *Ann. Rev. Plant Physiol.* 21:235–70

129. Lang, A. 1970. *Ann. Rev. Plant Physiol.* 21:537–70

130. Larsen, P. 1969. *Physiol. Plant.* 22:469–88

131. Lexander, K. 1953. *Physiol. Plant.* 6:406–11

132. Libbert, E. See Ref. 5, 387–405

133. Libbert, E., Wichner, S., Schiewer, U., Risch, H., Kaiser, W. 1966. *Planta* 68:327–34

134. Lim, G. 1969. *Plant Soil* 31:143–48

135. List, A. Jr. 1969. *Planta* 87:1–19

136. Loeffler, J. E., Overbeek, J. van. See Ref. 5, 77–82

137. MacLeod, R. D. 1966. *Planta* 71: 257–67

138. MacLeod, R. D. 1968. *Chromosoma* 24:177–87

139. Mace, M. E. 1967. *Can J. Bot.* 45: 945–48

140. Manos, G. E. 1961. *Physiol. Plant.* 14:697–711

141. Mast, C. A. van der 1969. *Acta Bot. Neer.* 18:620–26

142. Ibid 1970. 19:363–72

143. Ibid, 727–36

144. Masuda, Y. 1962. *Physiol. Plant.* 15:780–90

145. Mertz, D. 1966. *Plant Cell Physiol.* 7:125–35

146. Mertz, D. 1967. *Advan. Frontiers Plant Sci.* 18:89–96

147. Miassod, R., Penon, P., Teissere, M., Ricard, J., Cecchini, J. P. 1970. *Biochim. Biophys. Acta* 224:423–40

148. Morré, D. J., Bonner, J. 1965. *Physiol. Plant.* 18:635–49

149. Morris, D. A., Briant, R. E., Thomson, P. G. 1969. *Planta* 89:178–97

150. Nagao, M., Ohwaki, Y. 1968. *Bot. Mag.* 81:44–45

151. Naylor, A. W., Rappaport, B. N. 1950. *Physiol. Plant.* 3:315–33

152. Niemann, G. J. 1970. *Acta Bot. Neer.* 19:415–18
153. Odhnoff, C. 1963. *Physiol. Plant.* 16:474–83
154. Overbeek, J. van, Muir, R. M., Went, F. W., Tukey, H. B. 1954. *Plant Physiol.* 29:307–8
155. Parker, C. See Ref. 233, 402–3
156. Parker-Rhodes, A. F. 1940. *J. Agr. Sci.* 30:654–71
157. Penon, P. et al 1970. *Phytochemistry* 9:73–86
158. Phillips, I. D. J. 1964. *Ann. Bot.* 28:17–35
159. Ibid, 37–45
160. Phillips, I. D. J., Jones, R. L. 1964. *Planta* 63:269–78
161. Pilet, P. E. 1951. *Bull. Soc. Bot. Suisse* 61:410–24
162. Pilet, P. E. 1958. *Rev. Gen. Bot.* 65:605–33
163. Pilet, P. E. 1961. *Plant Growth Regulation. Int. Conf., 4th, Plant Growth Regulation,* ed. R. M. Klein, 167–79. Iowa State Univ. Press, Ames
164. Pilet, P. E. See Ref. 5, 542–58
165. Pilet, P. E. 1964. *Nature* 204: 561–62
166. Pilet, P. E. 1968. *C. R. Hebd. Seances Acad. Sci. Ser. D* 267: 1142–45
167. Pilet, P. E. 1969. *Experientia* 25: 1036
168. Pilet, P. E. 1970. *J. Exp. Bot.* 21: 446–51
169. Pilet, P. E., Braun, R. 1967. *Physiol. Plant.* 20:870–78
170. Ibid 1970. 23:245–50
171. Pilet, P. E., Chalvignac, M. A. 1970. *Ann. Inst. Pasteur* 118: 349–55
172. Pilet, P. E., Galston, A. W. 1955. *Physiol. Plant.* 8:888–98
173. Pilet, P. E., Kobr, M., Siegenthaler, P. A. 1960. *Rev. Gen. Bot.* 67:575–601
174. Pilet, P. E., Lavanchy, P. 1969. *Physiol. Veg.* 7:19–29
175. Pilet, P. E., Lavanchy, P., Sevhonkian, S. 1970. *Physiol. Plant.* 23:800–4
176. Pilet, P. E., Sevhonkian, S. 1969. *Physiol. Veg.* 7:325–33
177. Raa, J. 1971. *Physiol. Plant.* 24: 498–505
178. Raalte, M. H. van 1937. *Rec. Trav. Bot. Neer.* 34:278–332
179. Ray, P. M. 1969. *Communication in Development,* ed. A. Lang, 172–205. Develop. Biol. Suppl. 3. New York: Academic

180. Robbins, W. J., Hervey, A. 1969. *Proc. Nat. Acad. Sci. USA* 64: 495–97
181. Roberts, E. H., Street, H. E. 1955. *Physiol. Plant.* 8:238–62
182. Rovira, A. D. 1965. *Ann. Rev. Microbiol.* 19:241–66
183. Rufelt, H. See Ref. 233, 54–64
184. Sachs, T. 1968. *Ann. Bot.* 32: 391–99
185. Sankhla, N., Shukla, S. N. 1970. *Z. Pflanzenphysiol.* 63:284–87
186. Scharf, P., Günther, G. 1970. *Biochem. Physiol. Pflanz.* 161:320–29
187. Scheffer, F., Kickuth, R., Schlimme, E. 1968. *Plant Soil* 28:453–59
188. Schneider, H. 1965. *Z. Bot.* 52: 451–99
189. Scott, T. K., Batra, M. 1971. *ASB Bull.* 18:54–55
190. Scott, T. K., Briggs, W. R. 1960. *Am. J. Bot.* 47:492–99
191. Ibid 1963. 50:652–57
192. Scott, T. K., Wilkins, M. B. 1968. *Planta* 83:323–34
193. Ibid 1969. 87:249–58
194. Sequeira, L., Mineo, L. 1966. *Plant Physiol.* 41:1200–8
195. Sheldrake, A. R. 1971. *New Phytol.* 70:519–26
196. Sheldrake, A. R., Northcote, D. H. 1968. *Planta* 80:227–36
197. Sheldrake, A. R., Northcote, D. H. 1968. *New Phytol.* 67:1–13
198. Sheldrake, A. R., Northcote, D. H. 1968. *Nature* 217:195
199. Simon, S. 1908. *Ber. Deut. Bot. Ges.* 26:364–96
200. Sircar, S. M., Ray, A. 1961. *Nature* 190:1213–14
201. Skene, K. G. M. 1967. *Planta* 74: 250–62
202. Skoog, F., Armstrong, D. J. 1970. *Ann. Rev. Plant Physiol.* 21: 359–84
203. Sobieszczański, J. 1965. *Acta Microbiol. Pol.* 14:183–202
204. Steeves, T. A., Briggs, W. R. 1960. *J. Exp. Bot.* 11:45–67
205. Street, H. E. 1957. *Biol. Rev. Cambridge Phil. Soc.* 32:117–55
206. Street, H. E. 1961. *Advan. Sci.* 18: 13–18
207. Street, H. E. 1962. *Viewpoints in Biology,* ed. J. D. Carthy, C. L. Duddington, 1:1–49. London: Butterworths
208. Street, H. E. 1966. *Ann. Rev. Plant Physiol.* 17:315–44

209. Street, H. E. See Ref. 233, 20–41
210. Street, H. E., Bullen, P. M., Elliott, M. C. See Ref. 93, 407–16
211. Street, H. E., Butcher, D. N., Handoll, C., Winter, A. See Ref. 5, 329–41
212. Street, H. E., Winter, A. 1963. *Plant Tissue and Organ Culture: A Symposium*, 82–104. Int. Soc. Plant Morphol. Delhi
213. Sutton, D., Scott, E. G., Street, H. E. 1961. *Physiol. Plant.* 17:712–24
214. Sweeney, B. M. 1944. *Am. J. Bot.* 31:78–80
215. Thimann, K. V. 1934. *J. Gen. Physiol.* 18:23–34
216. Thimann, K. V. 1963. *Ann. Rev. Plant Physiol.* 14:1–18
217. Thurman, D. A., Street, H. E. 1960. *J. Exp. Bot.* 11:188–97
218. Ibid 1962. 13:369–77
219. Tietz, A. 1971. *Planta* 96:93–96
220. Tognoni, F., Halevy, A. H., Wittwer, S. H. 1967. *Planta* 72:43–52
221. Torrey, J. G. 1956. *Ann. Rev. Plant Physiol.* 7:237–66
222. Torrey, J. G. 1965. *Encyclopedia of Plant Physiology*, ed. W. Ruhland, 15/1:1256–1327. Berlin: Springer-Verlag
223. Torrey, J. G., Fosket, D. E. 1970. *Am. J. Bot.* 57:1072–80
224. Torrey, J. G., Fosket, D. E., Hepler, P. K. 1971. *Am. Sci.* 59:338–52
225. Torrey, J. G., Loomis, R. S. 1967. *Am. J. Bot.* 54:1098–1106
226. Torrey, J. G., Loomis, R. S. 1967. *Phytomorphology* 17:401–9
227. Tuli, V., Moyed, H. S. 1967. *Plant Physiol.* 42:425–30
228. Vaadia, Y., Itai, C. See Ref. 233, 65–79
229. Van't Hof, J. 1968. *Exp. Cell Res.* 51:167–76
230. Vardjan, M., Nitsch, J. P. 1961. *Bull. Soc. Bot. Fr.* 108:363–74
231. Went, F. W. 1928. *Rec. Trav. Bot. Neer.* 25:1–116
232. Weston, G. D., Street, H. E. 1968. *J. Exp. Bot.* 19:628–35
233. Whittington, W. J., Ed. 1969. *Root Growth: Proc. 15th Easter Sch. Agr. Sci., Univ. Nottingham.* London: Butterworths
234. Wichner, S. See Ref. 93, 443–44
235. Wilkins, M. B. 1971. *Gravity and the Organism*, ed. S. A. Gordon, M. J. Cohen, 107–24. Univ. Chicago Press
236. Wilkins, M. B., Cane A. R. 1970. *J. Exp. Bot.* 21:195–211
237. Wilkins, M. B., Scott, T. K. 1968. *Nature* 219:1388–89
238. Wilkins, M. B., Scott, T. K. 1968. *Planta* 83:335–46
239. Winter, A., Street, H. E. 1963. *Nature* 198:1283–88
240. Woodham, C. H., Bedford, J. S. 1970. *Int. J. Radiat. Biol.* 18:501–5
241. Woodruffe, P., Anthony, A., Street, H. E. 1970. *New Phytol.* 69:51–63
242. Woodruffe, P., Anthony, A., Street, H. E. 1970. *Physiol. Plant.* 23:488–97
243. Woodstock, L. W., Skoog, F. 1960. *Am. J. Bot.* 47:713–16
244. Ibid 1962. 49:623–33
245. Yang, D-P., Dodson, E. O. 1970. *Can. J. Bot.* 48:19–25
246. Yeomans, L. M., Audus, L. J. 1964. *Nature* 204:559–61
247. Yoshida, R., Oritani, T., Nishi, A. 1971. *Plant Cell Physiol.* 12:89–94
248. Zaerr, J. B. 1968. *Physiol. Plant.* 21:1265–69
249. Zimmerman, P. W., Wilcoxon, F. 1935. *Contrib. Boyce Thompson Inst.* 7:209–29
250. Zinke, H. 1968. *Planta* 82:50–72

Ann. Rev. Plant Physiol. 1972. 23:259–92

BIOSYNTHESIS AND MECHANISM OF ACTION OF ETHYLENE

7532

FRED B. ABELES

Plant Air Pollution Laboratory
United States Department of Agriculture, Agricultural Research Division
Plant Science Research Division, Beltsville, Maryland

CONTENTS

The list of ethylene regulated phenomenon include: breaking of dormancy, regulation of swelling and elongation, hypertrophy, induction of adventitious roots, epinasty, hook closure, inhibition of leaf expansion, control of flower induction, exudation, ripening, senescence, and abscission. Many excellent reviews have already appeared which describe these phenomenon and related topics. For general reviews see (26, 55, 80, 144, 177, 203, 204, 232, and 265). Fruit ripening, because of its economic impact, has been extensively reviewed. Appropriate references include (41–47, 49, 60, 66, 109, 111, 121, 139, 185, 197–199, 223, 227, 231, 244, 260, 264, 277, 281–283). Other special reviews including ethylene biosynthesis (135, 175, 176, 201, 202, 298), abscission (6, 19, 23, 24, 57, 85), air pollution (29, 76, 77, 84, 233, 268, 274), flowering (212), and senescence (286) have been prepared.

Research in the mechanism of action of ethylene has been active, resourceful, and stimulating. Nevertheless, the primary act of ethylene or any plant hormone still remains unknown. One immediate problem centers on the question of whether or not ethylene is a hormone. Strictly speaking, it probably is not.

A hormone is defined as a naturally produced substance which is transported from a site of production to a point of action elsewhere in the organism. It is also generally assumed that hormones are effectors and, as such, are not degraded or otherwise metabolized in the course of the activation process. However, degradative metabolism occurs subsequent to the control functions of hormones and prevents accumulation in the tissue. This definition of hormones originated with animal studies, and, while it fits what we know about insulin, the part about directed transport does not properly describe ethylene and possibly other plant hormones. Directed transport is not an essential feature of ethylene action. The gas, as far as we know, is produced to some extent in every cell of higher plants. For example: regulation does not evolve translocation of ethylene from the stem to the fruit to cause ripening. The fruit makes its own ethylene, and the ability to respond depends on changes in the rate of production and/or a change in sensitivity of the tissue to levels of the gas already being produced. To get around the present definition which legally requires the idea of transport or movement, we can either change the meaning of the word or come up with a new term for hormones in which directed transport is not a factor. One of the prerogatives of a reviewer is to stray from accustomed and normal ways of describing things and substitute his own set of ideas. With this in mind, I'd like to call ethylene a "stirone" (to stir up the cell). The definition of a stirone would be the same as that employed for a hormone, except that it can move from site of production to site of action by simple diffusion, and no activated or directed transport is required. Stirones are naturally produced organic molecules, effective in small quantities, which cause recognizable physiological phenomena and are not consumed during the course of action. While it is not assumed that stirone will become a generally accepted term to describe hormones in which directed transport is not important, it is hoped that this discussion will direct attention to this unique characteristic of gaseous hormones.

Another phenomenon that arises when gaseous hormones are being considered is that the organism does not require any detoxification or degradation mechanism to reduce levels of the substance in its tissue. Levels of ethylene are reduced simply by permitting the gas to diffuse into the surrounding atmosphere. The internal level of ethylene is controlled solely by the rate of synthesis. Why this method of regulating hormone levels was evolved for only one compound and not others is not known. As yet, gaseous regulators from other organisms have not been positively identified. If ethylene represents a successful way of controlling growth and development, it seems strange that it has not been mimicked elsewhere.

The constant production of ethylene by plants would result in the accumulation of ethylene in the atmosphere if there were no way to remove or destroy ethylene. We have estimated that the rate of ethylene production by plants in the United States is in the order of 2×10^4 tons annually (9). The major mechanisms for removal include oxidation by ozone, reaction with nitrogen oxides in the light (157), and uptake by organisms in the soil (9). The relative effectiveness of these processes is unknown, but they are apparently sufficient to keep levels of ethylene in rural areas very low in spite of the overwhelming production of ethylene by cars and other human activities (15×10^6 tons/year).

BIOCHEMISTRY OF ETHYLENE PRODUCTION

FUNGI

Even though Gane in 1935 established the fact that plants produced ethylene (98, 99), biochemical studies on the pathway of ethylene did not start until the late 1950s. Most research has centered around work with higher plants, though ethylene production by fungi (primarily *Penicillium*) has also been studied intensively. After Miller et al (200) and Biale (40) demonstrated that *Penicillium* produced ethylene, early experiments consisted of growing fungi on different carbon media and determining the relative rates of ethylene production. Fergus (94) found that the best carbon sources were D-mannitol and D-mannose, followed by D-xylose, D-galactose, D-fructose, and other sugars. For the most part, the rate of ethylene production followed growth curves of the fungus, and the greatest rates of gas production were associated with the time when mycellium growth was complete. After maximal size of the mycelia was obtained, rate of ethylene production decreased. Similar observations were made by Phan-Chon-Ton (224).

The high rate of ethylene production and the ease of introducing proposed intermediates offered distinct advantages to investigators working with fungi. However, in spite of a considerable amount of effort by a large number of workers, the biochemistry of ethylene formation in fungi remains unknown. The pathway with the greatest amount of experimental evidence in its favor is also the simplest: namely, the dehydration of ethanol. Dehydration

TABLE 1. Conversion Efficiency of Various Substrates into Ethylene by Penicillium

$$\% \text{ conversion efficiency} = \frac{\text{DPM}/\mu\text{mole ethylene formed} \times 100}{\text{DPM}/\mu\text{mole precursor added}}$$

Substrate-^{14}C	% Conversion efficiency	Reference
Glucose-U	1.32, 0.0096	225, 289
Glucose-1	0.03	289
Glucose-2	0.02	289
Glucose-3	0.052	289
CH$_2$OH–CHOH–CHOH–CHOH–CHOH–CHO		
Alanine-U	2.6	224
Alanine-1	2.0, 0.57	132, 289
Alanine-2	168, 0.91	132, 289
Alanine-3	1.5, 3.96	132, 289
CH$_3$–CHNH$_2$–COOH		
Glycine-1	0.64	289
Glycine-2	0.056, 11.6	289, 267
CH$_2$NH$_2$–COOH		
Aspartic acid-3	1.5	289
Aspartic acid-4	0.12	289
COOH–CH$_2$–CHNH$_2$–COOH		
Glutamic acid-1	0.025	289
Glutamic acid-2	0.15	289
Glutamic acid-3,4	3.1	289
Glutamic acid-5	0.04	289
COOH–CH$_2$–CH$_2$–CHNH$_2$–COOH		
Methionine-U	5.0, 0.0	132, 141
Methionine-1	0.0	132
Methionine-2	0.3	132
Methionine-methyl	2.5	132
CH$_3$–S–CH$_2$–CH$_2$–CHNH$_2$–COOH		
Serine-3	196.0	267
CH$_2$OH–CHNH$_2$–COOH		
Betaine-methyl	6.0	267
(CH$_3$)$_3$–N–CH$_2$–COOH		
Glyoxylate-1	15.1	267
HCO–COOH		
Acrylate-1	0.0	132
Acrylate-2	1.7	132
Acrylate-3	0.4	132
CH$_2$–CH–COOH		
Propionate-1	0.5	132
Propionate-2	10.0	132
Propionate-3	570.0	132
CH$_3$–CH$_2$–COOH		

TABLE 1. (*Continued*)

Substrate-^{14}C	% Conversion efficiency = $\dfrac{\text{DPM}/\mu\text{mole ethylene formed} \times 100}{\text{DPM}/\mu\text{mole precursor added}}$	Reference
	% Conversion efficiency	Reference
Ethanol-1	20.0, 0.35	224, 100
Ethanol-2	1.55	100
	CH_3–CH_2OH	
Acetate-1	0.0, 0.0, 12.0	141, 102, 132
Acetate-2	1.1, 3.8, 370.0	141, 102, 132
	CH_3–$COOH$	
Citrate-5,1	0.0	141
Citrate-2,3	4.5	141
	$COOH$–CH_2–$CHOH(COOH)$–CH_2–$COOH$	
Fumarate-1,4	0.0	132
Fumarate-2,3	84.0	132
	$COOH$–CH=CH–$COOH$	
Malate-3	0.9, 149.0	102, 132
Malate-U	0.67	224
	$COOH$–CH_2–$CHOH$–$COOH$	
Pyruvate-2	0.003	102
Pyruvate-3	0.12	102
	CH_3–CO–$COOH$	
Succinate-1,4	3.2	132
Succinate-2,3	11.0	132
	$COOH$–CH_2–CH_2–$COOH$	

of ethanol at high temperatures with H_2SO_4 in the presence of a suitable catalyst such as aluminum oxide was the method used by a large number of workers including Neljubow (208), the discoverer of the biological effects of ethylene, to prepare ethylene.

$$CH_3\text{–}CH_2OH \longrightarrow CH_2{=}CH_2 + H_2O \qquad \textbf{1.}$$

There are a number of observations that support this pathway of ethylene formation. A number of investigators have observed that feeding ethanol to *Penicillium* resulted in a greater production of ethylene (100, 105, 222) and feeding labeled ethanol to the fungus resulted in the formation of labeled ethylene (see Table 1). Unpublished inhibitor studies by some former colleagues (J. Lonski & H. E. Gahagan) demonstrated that dimedone (5,5-dimethyl-1,3-cyclohexanedione), which is an aldehyde fixing agent, had no effect on fungal growth but completely blocked ethylene formation.

According to the published results summarized in Table I, carbons 1 and 2 of glucose were not converted into ethylene as well as carbon 3. The con-

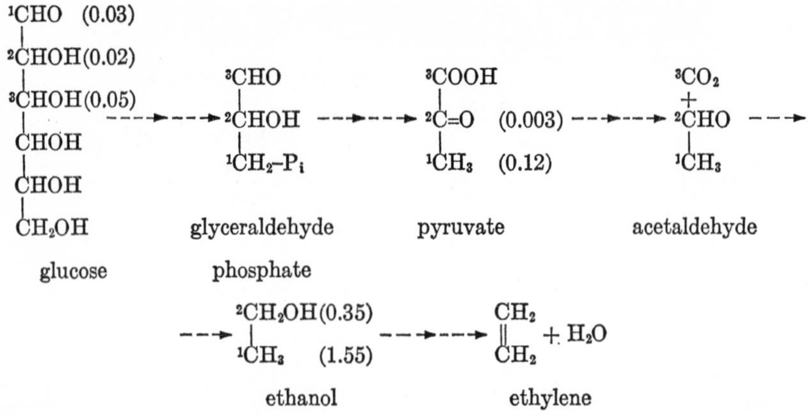

^{1}CHO (0.03)

^{2}CHOH(0.02)

^{3}CHOH(0.05) ^{3}CHO ^{3}COOH ^{3}CO$_2$
                                                                      +
CHOH $- - - \to - - \to$ ^{2}CHOH $- - \to - - \to$ ^{2}C=O (0.003) $- - \to - - \to$ ^{2}CHO $- - \to$

CHOH ^{1}CH$_2$–P$_i$ ^{1}CH$_3$ (0.12) ^{1}CH$_3$

CH$_2$OH

glucose   glyceraldehyde   pyruvate   acetaldehyde

                 phosphate

                      ^{2}CH$_2$OH(0.35)   CH$_2$
              $- - \to$ | $- - \to - - \to$ ‖ + H$_2$O
                      ^{1}CH$_3$ (1.55)   CH$_2$

                      ethanol   ethylene

FIGURE 1. A possible mechanism for ethylene production by fungi from glucose following the glycolysis pathway. Numbers in parentheses are % conversion efficiencies. The data were obtained from Table 1.

version efficiencies of various precursors of ethanol are shown in Figure 1. According to these observations, carbon 3 of glucose was converted into ethylene more effectively than other carbon atoms. However, carbon 3 is destined to form CO_2, according to the accepted glycolysis pathway.

Another confusing aspect of precursor studies is the fact that pyruvate and subsequently ethanol form ethylene in an asymmetrical fashion. If ethanol were the immediate precursor of ethylene, then both carbons should have been converted into ethylene with equal efficiency.

The possibility of amino acids as precursors of ethylene has also been examined. Unlike the pathway followed in higher plants, methionine does not appear to be an effective precursor. However, the data accumulated up to now is confusing, and an examination of the published results defies the construction of a simple scheme. Figure 2 shows that many of the compounds studied are readily interconvertible and that the conversion efficiencies are so dissimilar from one compound to the next that a logical pattern is not readily apparent. Incontrovertible evidence in favor of any one pathway for ethylene production in fungi remains to be established.

HIGHER PLANTS

A number of substances have been proposed as precursors of ethylene in higher plants. These include methionine, linolenic acid, β-alanine, propanal, ethanol, organic acids, acrylic acid, thiomalic acid, glycerol, sucrose, glucose, and acetic acid. While it is now known that methionine is the most probable precursor of ethylene, a review of the work on the other candidates will serve to show the way research in this area of ethylene physiology has progressed.

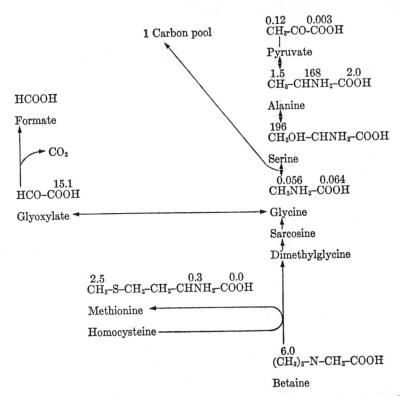

FIGURE 2. Metabolic conversions of proposed precursors of ethylene in fungi. Numbers represent % conversion efficiencies of various carbon atoms in substrates fed to *Penicillium*. Data derived from Table 1.

Acetate.—The fact that ethylene and acetate both have two carbons has tempted a number of investigators to see if ^{14}C-acetate could be converted into ethylene. Table 2 summarizes these experiments. Some workers (248) have reported a high percent conversion efficiency (8%), while others (54, 59) reported a lower percent conversion efficiency (< 0.06%). Some investigators have found that both carbons were converted into ethylene with equal facility (59, 248), while another found more rapid conversion of carbon 2 (252). In one case, the ratio depended on whether or not the tissue was healthy or diseased (248).

Linolenate.—The possibility of linolenic acid serving as a precursor of ethylene was first suggested by Lieberman & Mapson (168). Evidence in favor of this idea stems from results obtained in vitro (194) and with experiments on the production of ethylene catalyzed by apple peel extracts (239). Since oxidation is an important prerequisite for the breakdown of linolenic

TABLE 2. Conversion of Labeled Acetate into Ethylene

Substrate	Tissue	% Conversion efficiency	Ref.
Acetate-1	apple	0.015	54
Acetate-1	apple	0.057	59
Acetate-2	apple	0.064	59
Acetate-1	healthy sweet potato	8.16	248
Acetate-2	healthy sweet potato	8.01	248
Acetate-1	diseased sweet potato	0.0039	248
Acetate-2	diseased sweet potato	0.306	248
Acetate-1	apple	1.0[a]	252
Acetate-2	apple	12.0[a]	252
Acetate-2	banana	0.0174	255

[a] Data represent an estimate of the relative conversion of acetate into ethylene after assigning a value of 1 to acetate-1-^{14}C. It was not possible to calculate true percent conversion efficiency since specific activity of ethylene formed was not indicated in this reference.

acid (163) into ethylene, a number of workers pointed to an increase in peroxidase activity in ripening fruit as support for the idea that ethylene was formed from linolenic acid (195, 293).

However, it is unlikely that linolenic acid serves as the precursor for ethylene in plants. The in vitro reactions are not specific for linolenic acid and will occur with any substance containing a terminal CH_3-CH_2-CH=CH- group such as clupanodonic acid (a C-22 fatty acid from herring), 2-pentenoic acid, 3-hexenoic acid, and 3-heptene (2). The test tube reaction was inefficient. After a 20-hour incubation, only 0.02% of linolenic acid was converted into ethylene (2). Besides forming ethylene, linolenic acid breakdown also forms ethane and propane. These gases are not produced naturally in any significant quantity. Finally, the breakdown of linolenic acid should form a C-16 acid with two double bonds. This material has not been observed in plants. Conclusive evidence against the production of ethylene from linolenic acid was obtained when a number of workers demonstrated that tissue supplied with ^{14}C-linolenic acid did not produce labeled ethylene while ^{14}C-methionine did (30, 182).

Propanal.—Lieberman & Kunishi (162) subsequently suggested that propanal, a decomposition product of linolenic acid, may be the ethylene precursor (see Equation 2).

$$CH_3-CH_2-CHO \longrightarrow CH_2=CH_2 + CHOOH \qquad 2.$$

They found that of the 23 compounds tested, propanal was the most active ethylene precursor and formed ethylene to the extent of 5% of the starting material. However, other compounds could produce ethylene with some efficiency. For example: 1-propanol (1%), propyl ether (1%), and acrolein (1%). The production of ethylene was not specific since every compound tested formed at least trace amounts of ethylene ($> 0.02\%$).

Baur & Yang (30) studied the conversion of ^{14}C-propanal into ethylene. They found that C-1 labeled material did not form labeled ethylene while the C-2 label did. As a result, they concluded that the 3 and 2 carbons form ethylene as a unit, and carbon 1 was converted into formic acid and CO_2. These results show that propanal is not a precursor of ethylene in tissue, since apple slices fed with propanal-2-^{14}C did not form labeled ethylene while those fed with ^{14}C-methionine did.

Ethanol.—Since physiologists concerned with determining the source of ethylene in fungi have proposed ethanol as a probable precursor, it was natural to try to extend these findings to higher plants. This idea was supported by the observation that, to a small extent (0.6%), ethanol will form ethylene in the presence of copper and ascorbate (162). Further support stems from Phan-Chon-Ton's report (226) that petunia blossoms fed with ethanol produced ten times as much ethylene as controls. However, Biale (43) reported that he was unable to observe conversion of labeled ethanol to ethylene in cherimoya even though $^{14}CO_2$ was readily formed. On the other hand, feeding apples with ^{14}C-ethanol did give rise to labeled ethylene (59). However, Burg & Burg (59) pointed out that the conversion was not due to metabolism of the fruit but rather to a reaction between ethanol and mercuric perchlorate to form small amounts of ethylene.

Glucose.—Burg & Burg (59, 71) exposed apple tissue to labeled glucose and found that the specific activity of the ethylene produced was only 0.07% of the glucose supplied. A variety of unrelated compounds such as glycerol, propionate, pyruvate, acetate, fumarate, and CO_2 gave similar results. This suggests that these compounds were converted into other substances which eventually were converted into ethylene. Shimokawa & Kasai (252) performed similar experiments and also found a very low rate of conversion of glucose -U-^{14}C into ethylene. After an incubation period of 8 hours, only 0.0057% of the radioactivity was converted into ethylene. In similar experiments, acetate and pyruvate were better substrates. The conversion for acetate was 0.066% and for pyruvate 0.22%.

Conversion of glucose-U-^{14}C was also studied by Sakai et al (248). Unlike earlier workers, they reported that glucose was converted into ethylene with great efficiency. The specific activity of the glucose used was 5500 dpm/nmole and the specific activity of the ethylene formed was 627 dpm/nmole. However, pyruvate-^{14}C was a better precursor. Pyruvic acid-^{14}C with a specific activity of 5500 dpm/nmole gave rise to ethylene with a specific activity of 1610 dpm/nmole.

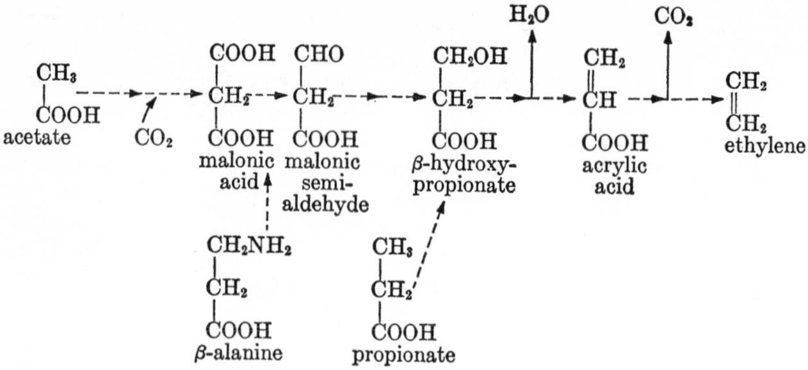

$$
\begin{array}{ccccccccc}
& & & & & & H_2O\uparrow & & CO_2\uparrow \\
& COOH & CHO & & CH_2OH & & CH_2 & & CH_2 \\
CH_3 & | & | & & | & & \| & & \| \\
| & CH_2 \rightarrow & CH_2 \rightarrow \cdots \rightarrow & CH_2 \cdots \rightarrow & CH & \cdots \rightarrow & CH_2 \\
COOH & \nearrow\ | & | & & | & & | & & \\
acetate & CO_2 & COOH & COOH & COOH & COOH & ethylene \\
& malonic & malonic & & \beta\text{-hydroxy-} & acrylic \\
& acid\uparrow & semi- & & propionate & acid \\
& & aldehyde
\end{array}
$$

(pathway diagram)

Figure 3 structures:

acetate: CH_3 / $COOH$ → CO_2 → malonic acid: $COOH$ / CH_2 / $COOH$ → malonic semialdehyde: CHO / CH_2 / $COOH$ → β-hydroxypropionate: CH_2OH / CH_2 / $COOH$ → $H_2O\uparrow$ → acrylic acid: CH_2 ‖ CH / $COOH$ → $CO_2\uparrow$ → ethylene: CH_2 ‖ CH_2

β-alanine: CH_2NH_2 / CH_2 / $COOH$

propionate: CH_3 / CH_2 / $COOH$

FIGURE 3. Proposed acrylic acid pathway according to
Shimokawa & Kasai (255–257).

Acrylic acid pathway.—Thompson & Spencer (275, 276) and Shimokawa & Kasai (255–257) have examined the possibility that acrylic acid is the precursor of ethylene. The structure of acrylic acid suggests that it could be a likely precursor of ethylene since a simple decarboxylation would result in the formation of ethylene and CO_2. A pathway for the formation of acrylic acid from acetate and β-alanine is shown in Figure 3.

Some evidence in favor of this scheme has been obtained. ^{14}C-ethylene is produced from ^{14}C-acetate, and the degree of conversion is blocked by the addition of unlabeled malonate and β-hydroxypropionate (255). Ethylene production from tissue slices was increased by the addition of β-hydroxypropionate, and ^{14}C-acetate gave rise to labeled malonate and β-hydroxypropionate (255). When tissue sections were fed ^{14}C-propionate, the rate of conversion to ethylene was decreased by the addition of unlabeled β-hydroxypropionate. Burg & Burg (59) have reported that [14]CO_2 will produce ^{14}C-ethylene. Cell-free preparations from bananas and bean cotyledons which convert acrylic acid into ethylene have also been described (257, 275). They are, however, extremely inefficient. Shimokawa & Kasai (257) found that only 0.006% of the acrylic acid added to their preparation was converted into ethylene. The observation of Lieberman & Kunishi (162) that all of the 23 compounds (including acrylic acid) they tested produced at least some ethylene under nonmetabolic conditions tends to minimize the significance of these observations. It would appear that a number of critical experiments remain to be done to substantiate the idea that acrylic acid pathway of ethylene biogenesis functions in living tissue. For example, labeled acrylic acid should give rise to labeled ethylene, and the formation of labeled ethylene from ^{14}C-labeled precursors should be reduced by unlabeled acrylic acid.

Miscellaneous.—A number of other compounds have been considered or tested at one time or another as potential precursors of ethylene. These include indoleacetic acid (48), thiomalic and thioglycolic acid (160), shikimic acid (71), Krebs cycle acids (59, 226), and ethionine (253). None of these compounds have continued to receive serious consideration as potential intermediates and, for one reason or another, have been shown not to be direct precursors.

Methionine.—It is fairly well established by now that methionine is the precursor of ethylene in plants. The first report that methionine may serve as the source of ethylene was that of Lieberman & Mapson in 1964 (168). During the course of an investigation in which ethylene production from a reaction mixture consisting of copper, ascorbate, and linolenic acid was being studied, they observed that methionine would also form ethylene. No other amino acid would function in this system, and unlike other reaction mixtures, ethylene was the only hydrocarbon produced. Support for the role of methionine in ethylene production was obtained when Lieberman & Mapson, and subsequently other investigators, reported that tissue fed with methionine produced more ethylene than controls (4, 96, 163, 167, 168, 182). However, some investigators failed to observe a promotion of ethylene production when methionine was fed to tissue (83, 154) or cell-free preparations (21).

While increased ethylene production by tissues fed with methionine would normally be considered as evidence in favor of a biochemical pathway from methionine to ethylene, a number of factors complicate the interpretation. First of all, increased ethylene production from plants can be due to wounding, stress, and a wide range of chemicals including amino acids, inorganic substances, and herbicides (34, 75, 78, 106, 129, 207, 245, 246). Secondly, ethylene production can occur nonenzymatically by means of Cu^{+2} + ascorbate, Cu^+, or Fe^{+2} + H_2O_2 catalyzed reactions (166). This means that ethylene production from methionine might arise if the tissue or homogenate contained significant amounts of heavy metal ions and a system capable of generating H_2O_2. Thirdly, it has been shown that peroxidase is capable of generating ethylene when supplied with methionine derivatives such as methional (181–183) and 2-keto-4-methylthiobutyric acid (KMBA) (151). However, Yang (296) has found that methione failed to form ethylene when incubated with peroxidase. Finally, though of little significance in tissue studies, methionine can form ethylene in the presence of flavin mononucleotide in the light (299, 300).

Additional support for the idea that methionine served as a precursor to ethylene was obtained when Lieberman et al (166) added ^{14}C-methionine labeled in positions 1, 2, 3, 4, or in the methyl group to apple slices. Only methionine labeled in positions 3 and 4 produced significant amounts of ethylene. In addition, the percent conversion efficiency was high (60%). These findings were subsequently confirmed by Burg & Clagett (68). They reported that etiolated pea stem tissue fed with methionine-U-^{14}C produced ^{14}C-ethylene when treated with a concentration of IAA capable of increasing ethylene evolution over the normal basal rate. They also demonstrated that the carboxylic acid group of methionine was converted into CO_2 and that carbons 3 and 4 were converted into ethylene. The second carbon was thought to be converted into formate and the CH_3-S- group into nonvolatile components including methionine (see Equation 3).

$$\overset{5}{CH_3}-S-\overset{4}{CH_2}-\overset{3}{CH_2}-\overset{2}{CH}NH_2-\overset{1}{COOH}\cdots\rightarrow\overset{5}{CH_3}-S-R + \overset{4}{CH_2}=\overset{3}{CH_2}$$
$$+ NH_3 + \overset{2}{HCOOH} + \overset{1}{CO_2} \quad 3.$$

The enzymatic reaction with methionine in tissue appears to favor utilization of the natural L-isomer. Baur & Yang (31) reported that unlabeled L-methionine would reduce incorporation of radioactivity from D,L-methionine into ethylene better than unlabeled D-methionine.

A number of what would appear to be logical intermediates between methionine and ethylene have been examined. However, Baur & Yang (31) concluded that in apple slices L-methionine was the most effective substrate, followed by KMBA, DL-homoserine, L-methionine sulfoxide, methional, and β-methyl-thiopropylamine.

However, Mapson et al (178) have suggested that KMBA is an intermediate in the conversion of methionine into ethylene. This idea is based upon the observation that KMBA promoted ethylene production when added to cauliflower tissue and was converted more efficiently into ethylene than methionine. In addition, unlabeled methionine increased incorporation of ^{14}C-KMBA into ethylene, while in the reverse experiment, unlabeled KMBA decreased the incorporation of ^{14}C-methionine into ethylene. This idea was questioned by Lieberman & Kunishi (164). They found that cauliflower sections would leak peroxidase into buffer solution surrounding the tissue and suggested that the ethylene production observed was due in fact to an ethylene-forming peroxidase system located outside of the cell.

The peroxidase catalyzed ethylene production is different from the ethylene production pathway in tissue in a number of respects. The peroxidase system will utilize methional as a substrate (179, 181, 273), while in intact tissue methional is a poor precursor (31). The peroxidase system will utilize KMBA 100 times more effectively than methionine (148), while plant tissue will utilize KMBA twice as fast (178) or slower than methionine (31). It is also known that monophenols or m-diphenols promote ethylene production from the peroxidase system while o-diphenols are active inhibitors (295). In contrast, ethylene production from intact tissue was not greatly influenced by the addition of either class of phenolic substances (97).

Finally, Kang et al (140) reported that there was no correlation between the amount of peroxidase in pea stem tissue and the rate of ethylene evolution. They found that ethylene production from etiolated pea plants was greater in the apical part of the seedling compared to the subapical portion.

However, they found twice as much peroxidase in the subapical part than in the apical part of the seedlings. They also reported that a high dose of IAA caused a burst of ethylene production from pea tissue. However, during the course of the dramatic increase in ethylene production there was a slight decrease in peroxidase content of the tissue.

PHYSIOLOGY OF ETHYLENE BIOSYNTHESIS
ENHANCEMENT OF ETHYLENE PRODUCTION

A number of workers have found that conversion of methionine to ethylene occurs only during periods of accelerated ethylene production. Burg & Clagett (68) reported no conversion of methionine to ethylene unless the pea stem tissue sections were first treated with IAA. Similarly, Baur et al (32) observed conversion of methionine to ethylene in the climacteric avocado but not in preclimacteric fruit. Baur et al (32) have shown that the increase in ethylene production was not due to changes in the level of methionine in tissue. They found similar amounts of methionine were present in preclimacteric and climacteric tissue even though the rate of ethylene production had increased 3000-fold. During the climacteric peak, levels of methionine had dropped by about 60%. They concluded that at high rates of ethylene production, methionine must be actively turned over since only a 3-hour supply of methionine was available in the tissue. We have found, however, that DL-methionine -U-^{14}C was converted into ethylene in normal primary leaves of bean (*Phaseolus vulgaris*) plants. In addition, when ethylene production from the leaves was enhanced by treatment with NAA, $CuSO_4$, endothal (3,6-endoxohexahydrophthalic acid), or ozone, the additional ethylene was also produced from methionine (unpublished results).

The increase in ethylene production by auxin is thought to represent the synthesis of enzymes required for the conversion of methionine to ethylene (3). This idea is based on the observation that inhibition of protein synthesis prevents or retards the development of auxin-induced ethylene production. Kang et al (140) have shown that the half-life of the biosynthetic pathway is short once auxin is removed (about 2 hours), and that the continued presence of auxin is required for a high rate of ethylene production.

The close relationship between auxin and ethylene has interested plant physiologists since Zimmerman & Wilcoxon first discovered that auxin increased ethylene production (301). A number of phenomena formerly attributed solely to auxin action are now known to involve, to some degree, increased rates of ethylene production. Some of these include enhancement of abscission (20), auxin-induced inhibition of growth of stems (61) and roots (74) [but see also Andreae et al (25) for an opposite point of view], induction of flowering in pineapples (62), inhibition of flowering in cockleburs (5), as well as the original observations of root initiation, intumescence formation, and epinasty described by the Boyce Thompson scientists (81). The fact that auxin promotes ethylene production means that ethylene production by plant tissue may give an estimate of internal auxin levels. We found that ethylene production was higher in the lower side of horizontal stem tissue than in the upper side. Similarly, more ethylene was produced by the dark side of unilaterally illuminated plants than by the light side (20). These observations agree with the Colodny-Went theory of tropistic responses which states that higher levels of auxin occur in the lower and dark side of plants exposed respectively to transverse gravitational fields and unilateral illumination. Higher rates of ethylene production in horizontal versus vertical plant tissue have also been reported (12, 87). This is in agreement with reports that there are greater quantities of auxin in horizontal tissue than in vertical tissue. Higher rates of ethylene production are also associated with apical tissue (65, 103) where auxin synthesis is presumed to occur. Decapitation of apical tissue or ringing stem tissue below the apex with triiodobenzoic acid was found to cause a reduction in ethylene production by the subtending tissue (20).

The recent work of Kang et al (140) supports this idea. They found that the rate of ethylene production by pea stems after the application of IAA closely followed the levels of free unconjugated IAA in the tissue. When a substance like 2,4-D was added, which was not subject to a similar detoxification mechanism, rates of ethylene production remained high. Curtis (83) has reported a similar phenomenon.

Inhibitor Studies

Inhibitor studies with dinitrophenol and other compounds which block the formation of ATP during respiration have shown that oxidative metabolism is required for ethylene formation (70, 167, 252, 262). However, a number of workers have found that cyanide has no effect on ethylene production (70, 252). Similarly, Lieberman et al (167) claimed that CO had no effect on ethylene production by apple slices. The role of oxidative metabolism in ethylene formation was first demonstrated by Hansen in 1942 (110). He found that ethylene production did not occur when oxygen was removed from pears. Since that time many investigators have reported similar findings (32, 69, 72, 82, 83, 156, 167, 169, 171, 180, 188, 211, 230, 261, 292). Hansen also noticed that there appeared to be a surge of ethylene production following the return of pear tissue to air. A similar phenomenon in apple tissue was also observed by Burg & Thimann (69). It is not clear whether oxygen participates directly in the conversion of methionine to ethylene or causes energy-rich metabolites to be made available as a result of respiration. Baur et al (32) have shown that the oxygen requirement occurs at some point between methionine and ethylene. A nitrogen gas phase inhibited conversion of ^{14}C-methionine to ethylene. Returning the tissue to air caused an increase in ethylene production, which again suggests the formation of some oxygen-dependent intermediate. If molecular oxygen is required for ethylene biogenesis, then there appears to be no recognition of this fact in present proposed biochemical pathways.

The formation of ethylene from methionine also appears to be sensitive to rhizobitoxine, a phytotoxin produced by certain strains of the bacterium *Rhizobium japonicum* (220). The precise structure remains to be elucidated; however, it is known to be a basic sulfur-containing amino acid which yields a derivative of homoserine upon desulfurization. Rhizobitoxin inhibits the enzyme β-cystathionase, which plays a role in methionine biosynthesis in bacteria. It also inactivates β-cystathionase isolated from spinach leaves. Owens et al (220) applied rhizobitoxin to sorghum seedlings to test the idea that methionine biosynthesis was important in ethylene formation. They found that rhizobitoxin did reduce ethylene production, and the addition of methionine partially overcame the effect of the inhibitor. The lack of complete reversal was seen as due to a second effect of rhizobitoxin on the conversion of methionine to ethylene. They found that conversion of radioactivity from ^{14}C-methionine into ethylene was blocked to the same extent as ethylene production.

Subcellular Localization

The site of ethylene production is as yet unknown. A number of investigators have experimented with the possibility that subcellular organelles contain enzymes required for ethylene biosynthesis. This idea was encouraged by the observation that ethylene production was associated with some tissue component sensitive to disruption and osmotic shock. Subdividing apples into smaller slices decreases ethylene production (70, 165). Tomatoes, on the other hand, produced more ethylene as they were sectioned (165). However, ethylene produced from both fruits stopped when they were homogenized. Burg & Thimann (70) found that the molarity of the bathing solution in which apple tissue was suspended had a profound effect on the rate of gas production. Soaking apple tissue in water reduced ethylene production by 50%, while solutions of glycerol, KCl, and other compounds prevented the reduction in ethylene evolution. Nichols (210) reported that ethylene production from flowers was halted when they were exposed to freezing temperatures. In addition to inhibiting ethylene production, these low temperatures caused a collapse of the petals and a rapid loss of water.

A number of investigators have examined the idea that subcellular fractions of cells might be the site of ethylene production. Ethylene production from chloroplasts (228, 229) and mitochondria (191–193, 235, 263, 266, 271) have been described. However, there are a number of reasons for questioning the significance of these findings. Spencer & Meheriuk (266) reported that exposing mithochondria to 0°C and 100°C failed to prevent ethylene production. Even though animals are not generally recognized as being capable of producing ethylene, mitochondria and other particulate preparations from beef heart (100, 101) and rats (161, 234) produced ethylene. Ku & Pratt (149) and Stinson & Spencer (271) failed to observe ethylene production by purified preparations of mitochondria but reported that other subcellular fractions did produce ethylene.

Systems consisting of soluble enzymes have also been studied. We (3) found that preparations from etiolated pea seedlings would produce ethylene from substrates later shown to be methionine (296) when flavinmononucleotide or Fe^{+2} ions were added. Mapson et al (181, 182) and Yang (297) have elucidated a system involving peroxidase as the enzyme responsible for ethylene production. However, for reasons discussed earlier, it is unlikely that natural ethylene production is mediated by peroxidase.

MECHANISM OF ETHYLENE ACTION
ATTACHMENT SITE OF ETHYLENE

Even though ethylene has a large number of different effects, it is likely that only a single attachment site exists. The attachment site can be characterized by: dose-response relationships, the action of homologs, and the competitive action of CO_2.

The dose-response curves for a variety of ethylene effects show a striking similarity. Normal values for threshold effects are 0.01 ppm; for half-maximal responses, 0.1 ppm; and for saturation, 10 ppm. Until oxygen becomes a limiting factor, increasing concentrations of ethylene usually have no further action, and toxic effects which mask or reverse the original observation are not seen.

The relative effectiveness of ethylene homologs, such as CO, acetylene, prophylene, and butylene, have been shown to be approximately the same from one ethylene-mediated process to another. While homolog experiments are difficult to perform because of potential ethylene contamination of the gases used, a number of investigators have shown that ethylene is the most effective compound, followed by propylene, vinyl chloride, CO, vinyl fluoride, acetylene, allene, methyl acetylene, and 1-butene (63).

Though CO_2 has long been recognized as having an effect opposite to ethylene, Burg & Burg (63) were the first to clearly state that its action was similar to that observed for competitive inhibitors of enzyme reactions. Using a Lineweaver-Burk plot similar to that employed for analysis of inhibition of enzyme activity, they demonstrated that combinations of ethylene and CO_2 applied to pea stems gave rise to curves identical to those observed when competitive inhibitors are added to enzymes. However, it should be pointed out that the mathematical model developed to equate the mechanism of enzyme action with chemical kinetics bears little relationship to complicated phenomena such as growth. Burg & Burg pointed out that CO_2 was a close structural analog of allene, a compound which mimics ethylene action. Since CO_2 has some structural resemblence to allene and could occupy the same site in the cell when levels of the gas were high

$$O=C=O \qquad\qquad H_2C=C=CH_2 \qquad\qquad H_2C=CH_2$$
$$CO_2 \qquad\qquad\qquad allene \qquad\qquad\qquad ethylene$$

(about 10% by volume), it failed to act as an effector because of a negative charge on each end.

Burg & Burg (63) have also suggested that molecular oxygen was required to oxidize the ethylene attachment site, and oxidation was required for full receptiveness or sensitivity to ethylene. While this criterion for characterizing the site of ethylene action may be valid, other investigators have failed to observe similar phenomena (13).

One question which arises concerns the nature of the forces bonding ethylene to the attachment site. The alternatives consist of covalent, hydrogen, ionic, and Van der Waals bonding. Covalent bonding would result in the incorporation of ethylene into some cellular component of the cell. While extremely unlikely, the dissociation of hydrogen would occur during hydrogen or ionic bonding, and an exchange between labeled (deuterated) ethylene and normal hydrogen of the cell should be observable.

A number of investigators have examined the possibility that ethylene can be incorporated into cellular material. Behmer (33) exposed apple tissue to 1000 ppm ^{14}C-ethylene for 19 days and reported that no incorporation was observed. Carlton et al (73) exposed carrot roots to ^{14}C-ethylene and reported no incorporation into a bitter-tasting isocoumarin derivative formed in the presence of the gas. However, no mention was made of the possibility of incorporation into other fractions. Hansen (110) measured levels of ethylene surrounding pear tissue and found they remained constant for long periods of time, which also suggests lack of gas uptake.

However, a number of workers have reported incorporation of ^{14}C- and ^3H-labeled ethylene into plant tissue. Buhler et al (53) found that ^{14}C-ethylene was incorporated in the order of 0.05% of added label into the organic acid fraction of avocado and pear tissue. On the other hand, experiments with ripe oranges, limes, papayas, green apples, tomatoes, and grapes failed to show any incorporation into fruit tissues.

Hall et al (107) found that cotton and *Coleus* plants incorporated ^{14}C-ethylene into a variety of substances. The rate of incorporation was enhanced by first absorbing ethylene on mercuric perchlorate and then releasing it. Shimokawa & Kasai (254) also studied incorporation of ^{14}C-ethylene regenerated from mercuric perchlorate. They found that the tracer was associated with DNA, RNA, and protein. On the basis of these findings, they suggested that binding of ethylene causes conformational changes in 4S RNA which, in turn, plays a role in hormonal regulation by ethylene. Further work (258) suggested that regenerated ethylene was incorporated faster in the light than in the dark, and that once it was incorporated it was not transported to other parts of the plant.

Jansen (136) also studied the difference in incorporation between fresh and mercuric perchlorate-regenerated ethylene. Compared with fresh ethylene, regenerated ethylene: (*a*) was taken up to a greater extent; (*b*) was not incorporated into volatile products; (*c*) produced radioactive CO_2 equal to 12% of incorporated label compared to 1% for fresh label; and (*d*) labeled succinic, malic, and citric acids which were not tagged when fresh ethylene was used. Obviously, ethylene released from mercuric-ethylene complex was modified to some extent to form some non-ethylene ^{14}C-labeled compounds.

Jansen (133) reported that 0.015% to 0.04% of tritium-labeled ethylene was incorporated into avocados and that 12% of the label wound up in the methyl group of toluene. When he later learned (134) that only 2% of the ^{14}C-labeled ethylene formed toluene, it became obvious that different pathways were being followed for the incorporation of different isotopes.

Radioactive ethylene is not stable. Tolbert & Lemmon (278) have reported that radiation-induced decomposition of ethylene results in the formation of polyethylene, hydrogen, and methane. In view of the known instability of tagged ethylene, the fact that only some workers report incorporation of radioactive ethylene, and then only a small percentage of the starting level, makes it reasonable to assume that the incorporation measured represents an artifact due to impurities in the ethylene used. It seems that this would be especially true in the cases where regenerated ethylene is used.

Beyer (35) and ourselves (22) have examined the physiological effect of deuterated ethylene and the possibility that exchange between the deuterium of the ethylene and cell might occur. However, there appeared to be no difference between the effectiveness of deuterated and normal ethylene as measured by its ability to inhibit the growth of pea stem and root tissue. In addition, no exchange of hydrogen between the ethylene and tissue was observed. These results indicate that ethylene was not attached to protoplasm in a way which might lead to cission of carbon-to-carbon or hydrogen-to-hydrogen bonds. While not conclusive, the data failed to support the idea that ethylene might be attached to a metal. If metal-to-ethylene binding were a part of ethylene action, then the deuterated molecule might be more effective than normal ethylene, since Atkinson (28) has shown that the affinity of silver to ethylene increased with increasing amounts of deuterium. Burg & Burg (60) proposed that a metal may play a central role in the binding site of ethylene. This idea was advanced because of the well-known affinity of olefins for metals and the fact that CO, which is an ethylene homolog, binds to a cytochrome which is a metalprotein. Support for this idea was obtained by Warner (290). He found that low concentrations of EDTA ($10^{-5}M$) reversed ethylene-induced inhibition of growth, ethylene-induced abscission, and ethylene-induced color development in tomato fruit. However, EDTA failed to repress the development of the respiratory climacteric. When chelated EDTA (FeEDTA) was used, it failed to reduce the effectiveness of ethylene.

We conclude that, since there is no evidence to the contrary, ethylene may be bound to its site of action by means of weak Van der Waals forces, and the site may contain a metal.

EFFECT OF ETHYLENE ON ISOLATED ENZYMES

A number of investigators have examined the possibility that ethylene would have a direct effect on enzyme activity. However, investigations with β-glucosidase, emulsin, salicin (92), α-amylase (93), invertase (93, 250), peroxidase (240), and adenosine triphosphatase (216) have shown there was no effect of ethylene on these enzymes. Some unpublished observations from my laboratory have also indicated that ethylene had no effect on cellulase and carbonic anhydrase. The latter enzyme was tested because it appeared as a likely candidate to show some positive response. Carbonic anhydrase contains zinc and has the ability to combine with CO_2—two features which suggest potential sensitivity to ethylene. Nelson (209) reported that ethylene increased the activity of trypsin. However, this effect was thought to be due to the removal of oxygen, since hydrogen had the same effect. Killian & Moritz (143) tested the idea that ethylene could combine with hemoglobin. Hemoglobin is known to combine readily with CO as evidenced by a change in the absorption spectra of this iron-containing protein. However, no difference between ethylene-treated and untreated hemoglobin was observed. However, Nord & Weicherz (214) reported that treating cell-free preparations of yeast with ethylene or acetylene resulted in the reduction of the viscosity of these solutions. They interpreted these results as showing a binding of ethylene to protein.

MEMBRANES

A number of investigators have tested the idea that ethylene regulated physiological phenomenon by virtue of some effect on the permeability of membranes. Because ethylene is more soluble in oil than in water and membranes contain large quantities of lipid, it seemed reasonable to examine membranes as a site of ethylene action. However, proponents of this idea failed to note that CO, an ethylene homolog, does not share the lipid solubility characteristics of ethylene.

It appears that a disruptive effect of ethylene on membranes causing a change in permeability and subsequent alteration of compartmentalization is not a valid idea. Earlier a number of investigators noted that ripening fruits exhibited obvious changes in terms of permeability and retention of soluble components. It seemed natural to suggest that ethylene caused changes in membrane permeability, which in turn led to softening and increased respiration. However, present evidence suggests that changes in the characteristics of membranes are a result of ripening rather than a cause (51, 57, 67, 247). A similar situation exists in flowers. Nichols (211) pointed out that leakage of solutes from carnation increased during senescence. Senescence and leakage were promoted by ethylene and reversed by CO_2.

Ethylene has no influence on membrane permeability of potato (209), pea (56), avocado, banana, bean, Rhoeo (247). However, von Guttenberg & Beythien (287) reported that ethylene increased the rate of deplasmolysis of Rhoeo leaves. Burg (57), however, failed to confirm their results.

There have been a number of studies with yeast, but the results are confusing. Nord and co-workers (213, 214) claimed that ethylene increased permeability of yeast cells. However, the effects were small, and Shaw (250) found no effect of ethylene on yeast cell permeability. Even if ethylene did regulate permeability, it would tell us little about its action since, as far as we know, ethylene has no effect on yeast physiology. The same criticism is applicable to red blood cell studies. Here ethylene has been shown to increase permeability (155), but the effect was not specific as it was mimicked with nitrous oxide. Similarly, early reports on the response of mitochondria to ethylene suggested some influence by high levels of ethylene on conformation (173, 215). While these effects were real, subsequent work pointed out the effects were not typical of normal ethylene action. High concentrations of ethylene were required to induce conformational changes in mitochondria, and saturated gases such as propane and ethane had similar effects (147, 190).

A number of investigators have found that ethylene increased or controlled the rate of enzyme secretion or release from cells. Jones (138) reported that ethylene increased the release of α-amylase from barley half seeds. However, the effect was not solely on secretion since ethylene inhibited the amount of α-amylase synthesized by the half seeds when no gibberellic acid was present. Unlike most ethylene-regulated phenomena, Jones reported that high concentrations of ethylene had an inhibitory effect on secretion. Normally, the dose-response curve of ethylene is asymptotic, and no additional effect is seen as the concentration is increased.

Ridge & Osborne (240) studied the regulation of peroxidase activity in pea stem tissue. They reported that ethylene inhibited leakage of peroxidase activity from apical tissue and, to a lesser extent, basal tissue.

Abscission is known to involve the enzymatic dissolution of cell walls with the result that leaves, flowers, fruit, and other parts are cast off from the plant. Horton & Osborn (120) found that the enzyme responsible for the dissolution was cellulase and that its synthesis was increased by ethylene. Later, dela Fuente & Leopold (86) pointed out that, following a short lag, ethylene reduced the breakstrength of abscission zone explants. Removal of ethylene by flushing the gas phase prevented further reduction in breakstrength. Abeles & Leather (18) examined the possibility that the strict control of breakstrength by ethylene was due to rapid synthesis of cellulase when ethylene was added and the disappearance of cellulase when the gas was removed. They found, however, that while cellulase synthesis followed the addition of ethylene, levels of cellulase remained constant after the gas was removed, and they concluded that ethylene must be having another effect in addition to the regulation of enzyme synthesis. Evidence was presented to show that this additional effect was the control of cellulase secretion and that this accounted for the maintenance of breakstrength once ethylene was removed. Ethylene not only controlled cellulase synthesis, but it also regulated the movement of cellulase from the cytoplasm to the cell wall through the membrane, and once the gas was removed, secretion of cellulase stopped.

In conclusion, the majority of the data suggest that physiological concentrations do not have a disruptive effect on the integrity of cellular membranes. However, evidence is accumulating that secretory phenomena are affected, and that ethylene can influence transport of materials through membranes.

REGULATION OF NUCLEIC ACID AND PROTEIN METABOLISM

The fact that there is a similarity between dose-response requirements, homolog activity, and competitive inhibition with CO_2 suggests that there is a single site of action which in turn regulates a wide variety of phenomena. Some of these phenomena seem to require the synthesis of new enzymes. They include: senescence (17), abscission (7, 18, 79, 159, 236), ripening (122, 170, 189), and swelling of pea stem tissue (58). It is important to emphasize that there is no reason to believe that these processes can be explained solely on the basis of new enzyme synthesis, and in fact evidence to the contrary is available in the case of abscission. Ethylene has been shown to increase aging (10) and to control cellulase secretion (18) in addition to cellulase synthesis. In the case of other ethylene effects, such as the rapid inhibition of stem elongation and epinasty, there is no evidence that enzyme synthesis is involved.

Enzyme induction.—Regeimbal & Harvey (237) were the first to report that ethylene-treated tissue contained greater quantities of a particular enzyme than controls. They found that ethylene increased the amount of protease and invertase extracted from pineapple fruits. Since that time, reports on the effect of ethylene on other enzymes have appeared. The list includes: acid phosphatase (115), ATPase (174, 270), α-amylase (115, 145, 259, 269), catalase (90, 115, 131), cellulase (7, 18, 79, 159, 236), chitinase (8), chlorophyllase (170), cinnamic acid 4-hydroxylase (124), cytochrome c reductase (137), diaphorase (137), β-1,3-glucanase (8, 11), invertase (130, 131, 152, 237, 251, 259), malic enzyme (122, 123, 137, 238), pectinesterase (115), peroxidase (27, 52, 108, 115, 125, 126, 128, 130, 131, 186, 187, 207, 217, 240, 241, 242, 249, 269), phenylalanine ammonium lyase (91, 126, 128, 243), polygalacturonase (189), polyphenol oxidase (115, 126, 128, 251, 269), protease (114, 237), and pyruvic carboxylase (122, 123, 137).

In a number of cases the induction of a particular enzyme does not depend solely on the presence of ethylene. In these cases, excising the tissue causes an induction of enzyme activity, and ethylene functions to reduce the lag or increase the rate of synthesis. Examples of enzymes whose formation does not depend strictly on ethylene include β-1,3-glucanase (11), malic enzyme (238), phenylalanine ammonium lyase (91), and peroxidase (97, 217). Excising tissue can cause wound ethylene, and it is possible that wound ethylene production plays some role in induction of enzyme activity observed in tissue slices. In the case of abscission, enzyme induction is more dependent on ethylene, and for a reasonable length of time no cellulase synthesis or abscission occurs following excision unless ethylene is added to the gas phase.

Action of cycloheximide.—Cycloheximide has been shown to prevent enzyme synthesis and the action of ethylene in a number of cases. Rhodes et al (238) found that the inhibitory effect was reversible if cycloheximide was washed out of the tissue. A number of workers (15, 86) have reported that cycloheximide will block abscission. However, cycloheximide has some secondary effects—notably, the ability to increase ethylene production (237)—and this fact has to be considered when applying cycloheximide to abscission zone explants. Fruit ripening was also blocked or delayed with cycloheximide. Frenkel et al (95) reported that cycloheximide blocked ripening and enhanced ethylene synthesis in Bartlett pears. However, it did not have an effect on the respiratory climacteric. Brady et al (51), studying the effect of cycloheximide on bananas, found that both ripening and the respiratory climacteric were inhibited. Burg et al (58) have reported that cycloheximide prevented the swelling of pea stems induced by ethylene.

The formation of a number of enzymes has been shown to be blocked by cycloheximide. The list includes β-1,3-glucanase (11), peroxidase (97, 217), cinnamic acid 4-hydroxylase (124), malate enzyme (238), cellulase (7, 236), and phenylalanine ammonium lyase (240). It would appear reasonable to conclude that the evidence points to de novo synthesis of protein. More critical evidence to establish this point was supplied by Lewis & Varner (159). They found that the deuterium of heavy water was incorporated into cellulase during abscission. This means that some existing protein was hydrolyzed and deuterium incorporated in the amino acids released. When cellulase was synthesized some of these labeled amino acids were incorporated, giving rise to cellulase which was denser than pre-existing proteins.

Action of actinomycin D.—In a number of cases, actinomycin D will block ethylene-regulated phenomenon as well as enzyme synthesis. This inhibitor of RNA synthesis will block abscission (14, 15, 117) and ethylene-induced swelling of pea stems (58). The synthesis of cellulase (7), malate enzyme (238), peroxidase (217), and cinnamic acid 4-hydroxylase (124) was also blocked by actinomycin D.

However, a number of workers have found that some enzymes whose activity was increased by ethylene, and whose synthesis was blocked by cycloheximide, were not greatly affected by actinomycin D. They include β-1,3-glucanase (11), peroxidase (97, 240), and phenylalanine ammonium lyase (243). These enzymes were also induced by slicing plant tissue. At this point the data suggest that some enzyme systems, induced by slicing tissues, can be increased by ethylene and do not appear to require RNA synthesis. The nature of this preformed RNA, its location in the cell, and its mechanism of activation remain unknown.

Protein synthesis.—Increases in protein synthesis during ripening, abscission, or after fumigation of vegetative tissue have been observed. Kidd et al (142) reported an increase in protein content in ripening pears and apples. These observations have been confirmed subsequently by others (112, 113, 121). Increased synthesis of protein in ripening fruits has been observed in bananas (51), apples (123), pears (95), and figs (184). However, Sacher &

Salminen (247) failed to observe an increase in protein synthesis when preclimacteric bananas or avocados were treated with ethylene.

Incorporation of amino acids into protein during abscission has been observed by a number of investigators (14, 15, 158). These observations have been substantiated by data obtained by histochemical staining and radioautography. Webster (291) and Stösser (272) have presented clear and vivid microphotographs showing enhanced protein synthesis in the separation layer of bean and cherry tissue, respectively. Valdovinos (285) demonstrated that endoplasmic reticulum of tobacco flower pedicels undergoing abscission was more pronounced than in intact controls.

Ethylene has also been found to increase protein levels in vegetative tissue. Elmer (90) reported that ethylene increased the protein content of potato sprouts. Soybean seedlings fumigated with ethylene were shown to contain greater quantities of protein in the basal and elongating parts but not in the apical region (118).

Ethylene accelerates the loss of protein in aging tissue (15, 16). However, these changes reflect total protein levels. Specific proteins associated with degradation are probably synthesized during aging. Cycloheximide and, to a lesser extent, actinomycin D will prevent the loss of chlorophyll, RNA, and protein in aging abscission zone explants (17).

RNA synthesis.—The first report on the promotion of RNA synthesis by ethylene is that of Turkova et al (280). They found that an increase in RNA was associated with epinasty of tomato leaves. Whether or not RNA synthesis was required for epinasty was not shown, though the idea is intriguing. Inhibitor studies with actinomycin D suggest that RNA synthesis occurred during abscission (14, 15) and was required for the process to occur. Support for this interpretation stems from the work of a number of investigators (14, 117, 218, 272, 291). It is now known that the increase in RNA synthesis precedes that of protein synthesis (14) and is localized in or near the separation layer (218, 272, 291). The increase in RNA occurred in all fractions, mRNA, rRNA, and sRNA, though the magnitude varied. Differential extraction of the nucleic acids indicated that the ethylene stimulation was confined to the fraction extracted with sodium lauryl sulfate, with the increase mainly in rRNA and mRNA. 5-Fluorouracil, which blocked 50% of the ethylene-enhanced ^{32}P incorporation, did not inhibit abscission. The greatest inhibition occurred in sRNA and rRNA fractions, indicating that not all fractions were required for abscission. Presumably, as long as sufficient mRNA was being synthesized, enough rRNA and sRNA was already available in the cell to permit abscission to occur. When all RNA synthesis was blocked with actinomycin D, then abscission stopped (117).

Ethylene has also been found to increase RNA synthesis in preclimacteric fruit (117). Marei & Romani (184) reported that, as in the case of abscission, the synthesis of all classes of RNA in fig fruit was promoted by ethylene. Hulme et al (123) found that ethylene increased RNA synthesis in apples and that the increase in RNA synthesis was followed by an increase in protein synthesis. However, Sacher & Salminen (247) reported that they failed to observe an increase in RNA synthesis when preclimacteric bananas or avocadoes were treated with ethylene.

Chromatin activity.—Holm et al (118, 119) have reported that ethylene inhibited growth of the apical part of soybean seedlings and increased it in the elongating and basal part. At the same time, RNA levels in the apical portion were reduced while they were increased in the elongating and basal portions. Chromatin from various parts of these seedlings was studied to determine its capacity for RNA synthesis. They found that activity was reduced in the apex and increased in the elongating and basal portions of the seedlings. The rate of response was rapid. The increase in chromatin activity was apparent after 3 hours and the increase in RNA after 6 hours. Nearest neighbor analysis of the RNA synthesized demonstrated that there was a qualitative difference between RNA synthesized by normal tissue and that treated with ethylene. They concluded that ethylene can regulate RNA synthesis so that a change in quantity and kind of RNA formed can occur.

DNA metabolism.—Plant growth is either promoted, inhibited, or unaffected by ethylene depending upon the kind of tissue involved. Examples of growth promotion are swelling, epinasty, hook closure, rice seedling elongation, and seed germination. Bud break is probably a special case. Here no growth takes place as long as ethylene is present. However, after the gas is removed, growth of the buds ensue. Growth inhibition is seen as arrested development of buds, leaves, or apical meristems. Mature tissue such as stems and leaves do not undergo any change in size or weight though premature senescence usually occurs.

Since growth, or the lack of it, may be associated with cytokinesis, it is of interest to learn if DNA synthesis is controlled by ethylene. Holm & Abeles (117) reported that DNA synthesis or DNA content in bean leaf tissue was not affected by a 7-hour exposure to ethylene though abscission was promoted. Later (117, 118) Holm found that soybean seedlings treated with ethylene stopped synthesizing DNA in the apex where growth was inhibited and promoted DNA synthesis in the subapical part where swelling took place. Burg et al (58) found that a similar situation existed in pea seedlings. They found that inhibition of cell division, measured as the reduction in metaphase figures, occurred within 2 hours after ethylene was added to pea seedlings. However, it is not clear if the change in DNA synthesis was the cause or result of inhibited growth. The speed at which ethylene slows growth of pea seedlings is very fast. Warner (290) found pea seedling growth slowed 6 minutes after ethylene was introduced into the gas phase and returned to normal 16 minutes after the ethylene was removed. Kinetic studies on changes in DNA or other postulated sites of action are required to establish the relationship between cause and effect. Burg et al (58) have suggested that ethylene-regulated DNA synthesis by some action in microtubule structure is essential for spindle fibre formation during mitosis. If the action of ethylene was directed toward microtubules, this might also explain the reorientation of microfibril deposition that occurs during swelling.

REGULATION OF HORMONE ACTIVITY

Enhancement of auxin action.—Van der Laan (153) observed that ethylene altered the ability of oat coleoptile sections to respond to auxin. After a 2-hour fumigation, ethylene increased the activity of auxin, and after 6 hours it decreased the activity. Similar experiments were performed by Michener (196). He found that ethylene increased the response of both oat and pea seedlings to ethylene. These phenomena have been confirmed by Yamaki (294) and Burg et al (58). Increased sensitivity to auxin action may explain the ability of ethylene to promote the growth of rice seedlings observed by Russian workers in 1937 (27, 145). Recently, Ku et al (150) confirmed this phenomenon and also observed that CO_2 instead of inhibiting ethylene action enhanced it. Imaseki & Pjon (127) have reported that ethylene enhanced the sensitivity of rice coleoptiles to auxin. Ethylene itself has no effect on isolated rice coleoptile sections. A possible explanation of ethylene action on the growth of rice seedlings may be that it enhances the action of endogenous auxin, and it has no separate effect of inhibiting elongation which is more characteristic of ethylene action on dicot stems. In general, ethylene has less effect on the growth of monocots stems than dicots.

Another example where ethylene action can be enhanced by auxin is in rooting. Krishnamoorthy (146) reported that ethrel-induced rooting of mung bean hypocotyls was improved by the addition of IAA.

Effect on levels of auxin in plant tissue.—Auxin effects are often opposite to that of ethylene. For example, auxin promotes elongation while ethylene inhibits it, and auxin delays abscission while ethylene accelerates it. Because of this, it seems reasonable to postulate that a mechanism of ethylene action includes control of auxin levels in plant tissue. In other words, ethylene might slow growth and promote abscission by reducing the levels of auxin in plant tissue. While a number of investigators have explored this possibility, the evidence is fragmentary, due for the most part to the fact that auxin assays are difficult to perform and are always subject to criticism in regard to specificity and precision. A review of the available data shows that ethylene reduces auxin levels in oat and pea tissue (50, 58, 153, 284, 288). Burg et al (58) have examined this phenomenon more closely and have found that the effect of ethylene varies according to the portion of the seedling under consideration. The largest reduction of auxin occurs in the subapex of pea seedlings, less in the hook region, and an increase in the apical region. Increases in auxin content after ethylene treatment have been observed by a number of other investigators. Dostal (88) reported increases in pea epicotyls following treatment with illuminating gas. He also (89) found that treating potato seed pieces with ethylene gave rise to plants that had higher levels of auxin in their tubers. Turkova (279) reported that ethylene increased the amount of auxin in the base of tomato petioles. Increases in the auxin content of leaf bases may explain the increase in cell size associated with leaf epinasty. The ethylene homolog, CO, has also been reported to increase auxin levels in leaf of *Mercurialis ambigua* (116).

Auxin synthesis.—Changes in auxin levels in plant tissue can be attributed to changes in synthesis, transport, or degradation. Valdovinos et al (284) have studied the effect of ethylene on the conversion of tryptophan to auxin in *Coleus.* They found that tryptophan conversion was inhibited by treating plants with ethylene and measuring the efficiency by which cell-free preparations would decarboxylate tryptophan. Cell-free preparations extracted from ethylene-treated plants were less active than controls. However, ethylene had no effect on the preparations themselves. They concluded that ethylene regulated auxin levels by controlling the activity of auxin biosynthesis.

Auxin transport.—Auxin transport is thought to play a role in abscission, growth, and epinasty. It follows that one possible mechanism of ethylene action would center around the ability to reduce auxin movement from the site of production to the site of action. For example, if auxin did not reach the separation layer, abscission would occur, or if auxin failed to reach the zone of elongation, no growth would occur. Evidence has been obtained that ethylene inhibits auxin transport and that this plays a role in the various phenomena described. However, it is important to distinguish between a direct effect on the machinery which moves auxin in plant tissue and an effect on the general metabolism resulting in reduced transport. The way to demonstrate a direct effect of ethylene on auxin transport is to conduct experiments with isolated sections of plants. All of the investigators who have done such experiments have failed to observe an effect of ethylene on auxin transport (1, 61, 64, 153, 219) in stem or petiole tissue.

However, ethylene does block auxin transport if transport is measured in sections isolated from plants pretreated with ethylene (36–39, 64, 205, 206, 219, 221, 288). The response is rapid and can be seen as soon as 1.5 hours after fumigation (37). Auxin transport can be regulated by the amount of auxin moved or by the rate of movement. Burg & Burg (64) have shown that the point of control is at the level of capacity and not velocity of movement. Beyer & Morgan (37) confirmed this finding, though a close examination of the data they presented reveals a slight reduction in velocity as well. Reduced auxin transport after ethylene treatment was shown in a variety of species.

Morgan et al (205) found that cotton and okra were highly susceptible, cowpea and English pea were intermediate in sensitivity, while tomato and sunflower showed little response after 15 hours. However, Palmer & Halsall (221) reported that they did observe an inhibition of auxin transport in ethelene-fumigated tomato plants.

The reduced capacity for transport can be due to either an effect on uptake or degradation. A number of investigators (36, 38, 61, 64, 288) have shown that ethylene does not influence the uptake of auxin through the cut surface of plant tissue. Similarly, degradation of auxin via decarboxylation does not appear to regulate transport. Burg & Burg (64) and Valdovinos (284) found that ethylene had no effect on the formation of $^{14}CO_2$ from IAA-1-^{14}C. On the other hand, Beyer & Morgan (36, 205) reported an increase in $^{14}CO_2$ production from IAA-1-^{14}C following ethylene treatment. However, in the same experiments, decarboxylation of NAA-1-^{14}C was not influenced by ethylene even though transport of this IAA homolog was reduced. Hall & Morgan (108) have suggested that an increase in auxin oxidase occurred in cotton leaves following ethylene treatment. However, Gowing & Leeper (104) failed to find an increase in auxin oxidase in pineapples after ethylene treatment.

Beyer & Morgan (38) have found that auxin was readily converted into a number of metabolites, including indoleacetyl aspartate, and that significantly more ^{14}C-IAA metabolites were recovered from ethylene-pretreated sections. However, they pointed out that accumulation of auxin metabolites might be the result, instead of the cause, of disrupted auxin transport.

Phototropic and geotropic curvature of certain plants can be blocked by ethylene. Tropistic phenomena involve, among other things, rapid lateral auxin transport, and Burg et al (58) have suggested that the cause of ethylene action may be an inhibition of lateral transport. There is good reason to believe that their interpretation is correct. Dostal (88) reported that illuminating gas prevented the lateral transport of auxin in horizontal pea tissue. This was confirmed by Burg & Burg (59), who showed that the effect was very rapid in pea tissue but did not occur to a great extent in oat coleoptile tissue. Burg et al (58) have suggested that their results agree with the effect of ethylene on geotrophic curvature in intact plants since geotropic curvature in pea tissue was effectively blocked by ethylene while it was only partially inhibited in monocot tissue. They also point out that ethylene does not prevent phototropic curvature of the fungus *Phycomyces*, which apparently has the same light receptor as higher plants but does not utilize auxin transport to mediate the preception to light. They also proposed that blockage of lateral auxin transport may explain why ethylene prevents nutation and the rapid curvature which occurs when isolated sections of pea seedlings are placed in buffer.

Epinasty is thought to involve a growth phenomenon resulting from an accumulation of auxin in the upper side of the petiole. Lyon (172) has postulated that this accumulation of auxin resulted from a disruption of lateral auxin transport in the petiole. He found that ethylene altered lateral auxin transport in petioles and that in the presence of ethylene auxin accumulated on the upper side of the leaf petiole.

LITERATURE CITED

1. Abeles, F. B. 1966. *Plant Physiol.* 41:946–48
2. Abeles, F. B. 1966. *Nature* 210: 23–25
3. Abeles, F. B. 1966. *Plant Physiol.* 41:585–88
4. Abeles, F. B. 1967. *Physiol. Plant.* 20:442–54
5. Abeles, F. B. 1967. *Plant Physiol.* 42:608–9
6. Ibid 1968. 43:1577–86
7. Ibid 1969. 44:447–52
8. Abeles, F. B., Bosshart, R. P., Forrence, L. E., Habig, W. H. 1971. *Plant Physiol.* 47:129–34
9. Abeles, F. B., Craker, L. E., Forrence, L. E., Leather, G. R. 1971. *Science* 173:914–16
10. Abeles, F. B., Craker, L. E., Leather, G. R. 1971. *Plant Physiol.* 47:7–9
11. Abeles, F. B., Forrence, L. E. 1970. *Plant Physiol.* 45:395–400
12. Abeles, F. B., Gahagan, H. E. 1968. *Life Sci.* 7:653–55
13. Abeles, F. B., Gahagan, H. E. 1968. *Plant Physiol.* 43:1255–58
14. Abeles, F. B., Holm, R. E. 1966. *Plant Physiol.* 41:1337–42
15. Abeles, F. B., Holm, R. E. 1967. *Ann. NY Acad. Sci.* 144:367–73
16. Abeles, F. B., Holm, R. E., Gahagan, H. E. 1967. *Plant Physiol.* 42:1351–56
17. Abeles, F. B., Holm, R. E., Gahagan, H. E. 1968. *Biochemistry and Physiology of Plant Growth Substances,* ed. F. Wightman, G. Setterfield, 1515–23. Ottawa, Canada: Runge
18. Abeles, F. B., Leather, G. R. 1971. *Planta* 97:87–91
19. Abeles, F. B., Leather, G. R., Forrence, L. E., Craker, L. E. 1971. *HortScience* 6:371–76
20. Abeles, F. B., Rubinstein, B. 1964. *Plant Physiol.* 39:963–69
21. Abeles, F. B., Rubinstein, B. 1964. *Biochim. Biophys. Acta* 63: 675–77
22. Abeles, F. B., Ruth, J. M., Forrence, L. E., Leather, G. R. 1971. *Plant Physiol.* 47:S–14 (Abstr.)
23. Addicott, F. T. 1968. *Plant Physiol.* 43:1471–79
24. Addicott, F. T. 1970. *Biol. Rev.* 45:485–524
25. Andreae, W. A., Venis, M. A., Jursic, F., Dumas, T. 1968. *Plant Physiol.* 43:1375–79
26. Apeland, J. 1962. *Gartneryrket* 52:1026–28, 1037
27. Asmaev, P. G. 1937. *Proc. Agr. Inst. Krasnodar* 6:49–100
28. Atkinson, J. G., Russell, A. A., Stuart, R. S. 1967. *Can. J. Chem.* 45:1963–69
29. Barth, D. S., Chairman Comm. 1970. *Air Quality Criteria for Hydrocarbons. U.S. Dep. HEW Publ. AP-64*
30. Baur, A., Yang, S. F. 1969. *Plant Physiol.* 44:189–92
31. Baur, A. H., Yang, S. F. 1969. *Plant Physiol.* 44:1347–49
32. Baur, A. H., Yang, S. F., Pratt, H. K., Biale, J. B. 1971. *Plant Physiol.* 47:696–99
33. Behmer, M. 1958. *Mitt. Klosterneuburg Ser. B: Obst Garten* 8:257–73
34. Ben-Yehoshua, S., Biggs, R. H. 1970. *Plant Physiol.* 45:604–7
35. Beyer, E. M. 1971. *Plant Physiol.* 47:S-13 (Abstr.)
36. Beyer, E. M., Morgan, P. W. 1969. *Plant Cell Physiol.* 10:787–99
37. Beyer, E. M., Morgan, P. W. 1969. *Plant Physiol.* 44:1690–94
38. Ibid 1970. 46:157–62
39. Ibid 1971. 48:208–12
40. Biale, J. B. 1940. *Science* 91:458–59
41. Biale, J. B. 1950. *Ann. Rev. Plant Physiol.* 1:183–206
42. Biale, J. B. 1960. *Handb. Pflanzenphysiol.* 12:536–92
43. Biale, J. B. 1960. *Advan. Food Res.* 10:293–354
44. Biale, J. B. 1961. *The Orange, Its Biochemistry and Physiology,* ed. W. B. Sinclair, 96–130. Univ. California Press
45. Biale, J. B. 1962. *Food Preserv. Quart.* 22:57–62
46. Biale, J. B. 1964. *Science* 146: 880–88
47. Biale, J. B., Young, R. E. 1962. *Endeavour* 21:164–74
48. Bitancourt, A. A. 1968. *Cienc. Cult. Sao Paulo* 20:400 (Abstr.)
49. Borgstrom, G. 1945. *Sver Pomol. Foren. Arsskr.* 46:202–23

50. Botjes, J. O. 1942. *Proc. Ned. Akad. Wetensch.* 45:999–1002
51. Brady, C. J., Palmer, J. K., O'Connell, P. B. H., Smillie, R. M. 1970. *Phytochemistry* 9:1037–48
52. Buchanan, D. W., Hall, C. B., Biggs, R. H., Knapp, F. W. 1969. *HortScience* 4:302–3
53. Buhler, D. R., Hansen, E., Wang, C. H. 1957. *Nature* 179:48–49
54. Burg, S. P. 1959. *Arch. Biochem. Biophys.* 84:543–46
55. Burg, S. P. 1962. *Ann. Rev. Plant Physiol.* 13:265–302
56. Burg, S. P. 1964. *Colloq. Int. Centre Nat. Rech. Sci.* 123:719–25
57. Burg, S. P. 1968. *Plant Physiol.* 43:1503–11
58. Burg, S. P., Apelbaum, A., Eisinger, W., Kang, B. G. 1971. *HortScience* 6:359–64
59. Burg, S. P., Burg, E. A. 1964. *Nature* 203:869–70
60. Burg, S. P., Burg, E. A. 1965. *Science* 148:1190–96
61. Burg, S. P., Burg, E. A. 1966. *Proc. Nat. Acad. Sci. USA* 55:262–69
62. Burg, S. P., Burg, E. A. 1966. *Science* 152:1269
63. Burg, S. P., Burg, E. A. 1967. *Plant Physiol.* 42:144–52
64. Ibid, 1224–28
65. Burg, S. P., Burg, E. A. See Ref. 17, 1275–94
66. Burg, S. P., Burg, E. A. 1969. *Qual. Plant. Mater. Veg.* 19:185–200
67. Burg, S. P., Burg, E. A., Marks, R. 1964. *Plant Physiol.* 39:185–95
68. Burg, S. P., Clagett, C. O. 1967. *Biochem. Biophys. Res. Commun.* 27:125–30
69. Burg, S. P., Thimann, K. V. 1959. *Proc. Nat. Acad. Sci.* 45:335–44
70. Burg, S. P., Thimann, K. V. 1960. *Plant Physiol.* 35:24–35
71. Burg, S. P., Thimann, K. V. 1961. *Arch. Biochem. Biophys.* 95:450–57
72. Bussel, J., Maxie, E. C. 1966. *Proc. Am. Soc. Hort. Sci.* 88:151–59
73. Carlton, B. C., Peterson, C. E., Tolbert, N. E. 1961. *Plant Physiol.* 36:550–52
74. Chadwick, A. V., Burg, S. P. 1970. *Plant Physiol.* 45:192–200
75. Chalutz, E., Stahmann, M. A. 1969. *Phytopathology* 59:1972–73
76. Clayton, G. D. 1966. Prepared by G. D. Clayton & Assoc., Detroit, Mich., for Automobile Mfr. Assoc.
77. Clayton, G. D., Platt, T. S. 1967. *Am. Ind. Hyg. Assoc. J.* 28:151–60
78. Cooper, W. C., Rasmussen, G. K., Rogers, B. J., Reece, P. C., Henry, W. H. 1968. *Plant Physiol.* 43:1560–76
79. Craker, L. E., Abeles, F. B. 1969. *Plant Physiol.* 44:1139–43
80. Crocker, W. 1932. *Proc. Am. Phil. Soc.* 71:295–98
81. Crocker, W., Hitchcock, A. E., Zimmerman, P. W. 1935. *Contrib. Boyce Thompson Inst.* 7:231–48
82. Curtis, R. W. 1969. *Plant Physiol.* 44:1368–70
83. Curtis, R. W. 1969. *Plant Cell Physiol.* 10:909–16
84. Das Gupta, S. N. 1957. *Proc. Indian Sci. Congr.* 1957:88–107
85. dela Fuente, R. K., Leopold, A. C. 1968. *Plant Physiol.* 43:1486–1502
86. Ibid 1969. 44:251–54
87. Denny, F. E. 1936. *Contrib. Boyce Thompson Inst.* 8:99–104
88. Dostal, R. 1942. *Jahrb. Wiss. Bot.* 90:199–232
89. Dostal, R. 1944. *Bodenk. Pflanzenernaehr.* 33:215–35
90. Elmer, O. H. 1936. *J. Agr. Res.* 52:609–26
91. Engelsma, G., Van Bruggen, J. M. H. 1971. *Plant Physiol.* 48:94–96
92. Engles, D. T., Dykins, F. A. 1931. *J. Am. Chem. Soc.* 53:723–26
93. Engles, D. T., Zannis, C. D. 1930. *J. Am. Chem. Soc.* 52:797–802
94. Fergus, C. L. 1954. *Mycologia* 46:543–55
95. Frenkel, C., Klein, I., Dilley, D. R. 1968. *Plant Physiol.* 43:1146–53
96. Fuchs, Y., Lieberman, M. 1968. *Plant Physiol.* 43:2029–36
97. Gahagan, H. E., Holm, R. E., Abeles, F. B. 1968. *Physiol. Plant.* 21:1270–79
98. Gane, R. 1934. *Nature* 134:1008
99. Gane, R. 1935. *Gt. Brit. Food Invest. Board Rep.* 1934:122–23
100. Gibson, M. S. 1963. *The biogenesis of ethylene.* PhD thesis. Purdue Univ., Lafayette, Ind.

101. Gibson, M. S. 1963. *Biochim. Biophys. Acta* 78:528–30
102. Gibson, M. S., Young R. E. 1966. *Nature* 210:529–30
103. Goeschl, J. D., Rappaport, L., Pratt, H. K. 1966. *Plant Physiol.* 41:877–84
104. Gowing, D. P., Leeper, R. W. 1961. *Bot Gaz.* 123:34–43
105. Hall, W. C. 1951. *Bot. Gaz.* 113:55–65
106. Ibid 1952. 113:310–22
107. Hall, W. C., Miller, C. S., Herrero, F. A. 1961. *Proc. 4th Int. Conf. Plant Growth Regul.* 4:751–78
108. Hall, W. C., Morgan, P. W. 1964. *Regulateurs naturels de la croissance vegetale,* ed. J. P. Nitsch, 727–45. Paris: C.N.R.S.
109. Haller, M. H. 1952. *U.S. Dep. Agr. Biblio. Bull. 21.* 105 pp.
110. Hansen, E. 1942. *Bot. Gaz.* 103:543–58
111. Hansen, E. 1966. *Ann. Rev. Plant Physiol.* 17:459–80
112. Hansen, E. 1967. *Proc. Am. Soc. Hort. Sci.* 91:863–67
113. Hansen, E., Blanpied, G. D. 1968. *Am. Soc. Hort. Sci.* 93:807–12
114. Harvey, R. B. 1928. *Minn. Agr. Exp. Sta. Bull.* 247:1–36
115. Herrero, F., Hall. W. C. 1960. *Physiol. Plant.* 13:736–50
116. Heslop-Harrison, J., Heslop-Harrison, Y. 1957. *New Phytol.* 56:352–55
117. Holm, R. E., Abeles, F. B. 1967. *Plant Physiol.* 42:1094–1102
118. Holm, R. E., Abeles, F. B. 1967. *Planta* 78:293–304
119. Holm, R. E., O'Brien, T. J., Key, J. L., Cherry, J. H. 1970. *Plant Physiol.* 45:41–45
120. Horton, R. F., Osborne, D. J. 1967. *Nature* 214:1086–88
121. Hulme, A. C. 1958. *Advan. Food Res.* 8:297–413
122. Hulme, A. C., Jones, J. D., Wooltorton, L. S. C. 1963. *Proc. Roy. Soc. Ser. B* 158:514–35
123. Hulme, A. C., Rhodes, M. J. C., Wooltorton, L. S. C. 1971. *Phytochemistry* 10:749–56
124. Hyodo, H., Yang, S. F. 1971. 143:338–39
125. Imaseki, H. 1970. *Plant Physiol.* 46:170–72
126. Imaseki, H., Asahi, T., Uritani, I. 1968. *Phytopath. Soc. Jap.* 1968:189–201
127. Imaseki, H., Pjon, C. J. 1970. *Plant Cell Physiol.* 11:827–29
128. Imaseki, H., Uchiyama, M., Uritani, I. 1968. *Agr. Biol. Chem.* 32:387–89
129. Imaseki, H., Uritani, I., Stahman, M., 1968. *Plant Cell Physiol.* 9:757–68
130. Ivanov, N. N. 1932. *Biochem. Z.* 254:71–87
131. Ivanov, N. N., Prokoshev, S. M., Gabunya, M. K. 1930–31. *Bull. Appl. Bot. Genet. Plant Breed.* 25:262–78
132. Jacobsen, D. W., Wang, C. H. 1968. *Plant Physiol.* 43:1959–66
133. Jansen, E. F. 1963. *J. Biol. Chem.* 238:1552–55
134. Ibid 1964. 239:1664–67
135. Jansen, E. F. 1965. *Plant Biochemistry,* ed. J. Bonner, J. Varner, 641–64. New York: Academic
136. Jansen, E. F. 1969. *Food Sci. Technol.* 1:475–81
137. Jones, J. D., Hulme, A. C., Wooltorton, L. S. 1965. *New Phytol.* 64:158–67
138. Jones, R. L. 1968. *Plant Physiol.* 43:442–44
139. Kaltenbach, D. 1938. *Int. Rev. Agr.* 29:81T–116T
140. Kang, B. G., Newcomb, W., Burg, S. P. 1971. *Plant Physiol.* 47:504–9
141. Ketring, D. L., Young, R. E., Biale, J. B. 1968. *Plant Cell Physiol.* 9:617–31
142. Kidd, F., West, C., Hulme, A. C. 1939. *Gt. Brit. Dep. Sci. Ind. Res., Food Invest. Board Rep.* 1938:119–25
143. Killian, H., Moritz, H. 1931. *Z. Gesamte Exp. Med.* 79:173–83
144. Klein, R. M. 1967. *Gard. J.* 17:126–35
145. Kraynev, S. I. 1937. *Proc. Agr. Inst. Krasnodar* 6:101–94
146. Krishnamoorthy, H. N. 1970. *Plant Cell Physiol.* 11:979–82
147. Ku, H. S., Leopold, A. C. 1970. *Plant Physiol.* 46:842–44
148. Ku, H. S., Leopold, A. C. 1970. *Biochem. Biophys. Res. Commun.* 41:1155–60
149. Ku, H. S., Pratt, H. K. 1968. *Plant Physiol.* 43:999–1001
150. Ku, H. S., Suge, H., Rappaport, L., Pratt, H. K. 1969. *Planta* 90:333–39
151. Ku, H. S., Yang, S. F., Pratt, H. K. 1969. *Phytochemistry* 8:567–75

152. Kursanov, A. L., Kryukova, N. 1938. *Biokhimiia* 3:202–17
153. Laan, P. A. van der 1934. *Trav. Bot. Neer.* 31:691–742
154. LaRue, T. A. G., Gamborg, O. L. 1971. *Plant Physiol.* 48:394–98
155. Leake, C. N., Rapp, H., Tenney, J., Waters, R. M. 1927. *Proc. Soc. Biol. Med.* 25:93–94
156. Leblond, C. 1967. *Fruits* 22:543–55
157. Leighton, P. A. 1961. *Photochemistry of Air Pollution.* New York: Academic
158. Leopold, A. C. 1967. *Symp. Soc. Exp. Biol.* 21:507–16
159. Lewis, L. N., Varner, J. E. 1970. *Plant Physiol.* 46:194–99
160. Lieberman, M., Craft, C. C. 1961. *Nature* 189:243
161. Lieberman, M., Hochstein, P. 1966. *Science* 152:213–14
162. Lieberman, M., Kunishi, A. T. 1967. *Science* 158:938
163. Lieberman, M., Kunishi, A. T. 1968. *Phytopath. Soc. Jap.* 1968:165–79
164. Lieberman, M., Kunishi, A. T. 1971. *Plant Physiol.* 47:576–80
165. Lieberman, M., Kunishi, A. 1971. *HortScience* 6:355–58
166. Lieberman, M., Kunishi, A. T., Mapson, L. W., Wardale, D. A. 1965. *Biochem. J.* 97:449–59
167. Lieberman, M., Kunishi, A., Mapson, L. W., Wardale, D. A. 1966. *Plant Physiol.* 41:376–82
168. Lieberman, M., Mapson, L. W. 1964. *Nature* 204:343–45
169. Lieberman, M., Spurr, R. A. 1955. *Proc. Am. Soc. Hort. Sci.* 65:381–86
170. Looney, N. E., Patterson, M. E. 1967. *Nature* 214:1245–46
171. Lougheed, E. C., Franklin, E. W. 1970. *Can. J. Plant Sci.* 50:586–87
172. Lyon, C. J. 1970. *Plant Physiol.* 45:644–46
173. Lyons, J. M., Pratt, H. K. 1964. *Arch. Biochem. Biophys.* 104:318–24
174. Madeikyte, E., Turkova, N. S. 1965. *Lietuvos TSR Mokslu Akad. Darbai Ser C* 1965:37–45; *Chem. Abstr.* 1966, 64:13303d
175. Mapson, L. W. 1969. *Biol. Rev.* 44:155–87
176. Mapson, L. W. 1970. *Endeavor* 29:29–33
177. Mapson, L. W., Hulme, A. C.

178. Mapson, L. W., March, J. F., Wardale, D. A. 1969. *Biochem. J.* 115:653–61
179. Mapson, L. W., Mead, A. 1968. *Biochem. J.* 108:875–81
180. Mapson, L. W., Robinson, J. 1966. *J. Food Technol.* 1:215–25
181. Mapson, L. W., Self, R., Wardale, D. A. 1969. *Biochem. J.* 111:413–18
182. Mapson, L. W., Wardale, D. A. 1967. *Biochem. J.* 102:574–85
183. Mapson, L. W., Wardale, D. A. 1971. *Phytochemistry* 10:29–40
184. Marei, N., Romani, R. 1971. *Plant Physiol.* 48:806–8
185. Marloth, R. H. 1933. *Farming S. Afr.* 8:17–18
186. Matoo, A. K., Modi, V. V. 1969. *Plant Physiol.* 44:308–10
187. Matoo, A. K., Modi, V. V., Reddy, V. V. R. 1968. *Indian J. Biochem.* 5:111–14
188. Maxie, E. C., Rae, H. L., Eaks, L. L., Sommer, N. F. 1966. *Radiat. Bot.* 6:445–55
189. McCready, R. M., McComb, E. A. 1954. *Food Res.* 19:530–35
190. Mehard, C. W., Lyons, J. M. 1970. *Plant Physiol.* 46:36–39
191. Meheriuk, M., Spencer, M. 1964. *Nature* 204:43–45
192. Meheriuk, M., Spencer, M. 1967. *Phytochemistry* 6:535–43
193. Ibid, 545–49
194. Meigh, D. F. 1962. *Nature* 196:345–47
195. Meigh, D. F., Jones, J. D., Hulme, A. C. 1967. *Phytochemistry* 6:1507–15
196. Michener, H. D. 1938. *Am. J. Bot.* 25:711–20
197. Miller, E. V. 1946. *Bot. Rev.* 12:393–423
198. Miller, E. V. 1947. *Sci. Mon.* 65:335–42
199. Miller, E. V. 1958. *Bot. Rev.* 24:43–59
200. Miller, E. V., Winston, J. R., Fisher, D. F. 1940. *J. Agr. Res.* 60:269–78
201. Moreno Calvo, J. 1961. *Ion* 21:561–67
202. Moreno Calvo, J. 1963. *Rev. Frio* 7:73–78
203. Morgan, P. W. 1967. *Proc. Beltwide Cotton Prod. Res. Conf.* 1967:151–55
1970. *Progr. Phytochem.* 2:343–84

204. Morgan, P. W. 1968. *Tex. Agr. Progr.* 14:4–5
205. Morgan, P. W., Beyer, E., Gausman, H. W. See Ref. 17, 1255–71
206. Morgan, P. W., Gausman, H. W. 1966. *Plant Physiol.* 41:45–52
207. Nakagaki, Y., Hirai, T., Stahmann, M. A. 1970. *Virology* 40:1–9
208. Neljubow, D. 1901. *Beih. Bot. Zentralbl.* 10:128–39
209. Nelson, R. C. 1939. *Food Res.* 4:173–90
210. Nichols, R. 1966. *J. Hort. Sci.* 41:279–90
211. Ibid 1968. 43:335–49
212. Nitsch, J. P. 1965. *Handb. Pflanzenphysiol.* 15:1552–53
213. Nord, F. F., Franke, K. W. 1928. *J. Biol. Chem.* 79:27–51
214. Nord, F. F., Weicherz, J. 1929. *Hoppe-Seylers Z. Physiol. Chem.* 183:191–217; *Chem. Abstr.* 23:5475
215. Olson, A. O., Spencer, M. 1968. *Can. J. Biochem.* 46:277–82
216. Ibid, 283–88
217. Osborne, D. J. 1968. *Soc. Chem. Ind. Monogr.* 1968:236–50
218. Osborne, D. J. See Ref. 17, 815–40
219. Osborne, D. J., Mullins, M. G. 1969. *New Phytol.* 68:977–91
220. Owens, L. D., Lieberman, M., Kunishi, A. 1971. *Plant Physiol.* 48:1–4
221. Palmer, O. H., Halsall, D. M. 1969. *Physiol. Plant.* 22:59–67
222. Phan-Chon-Ton 1960. *C. R. Acad. Sci.* 251:122–24
223. Phan-Chon-Ton 1961. *Advan. Hort. Sci. Appl.* 1:96–111
224. Phan-Chon-Ton 1962. *Rev. Gen. Bot.* 60:505–43
225. Phan-Chon-Ton 1962. *Advan. Hort. Sci. Appl.* 2:238–41
226. Phan-Chon-Ton 1970. *Physiol. Plant.* 23:981–84
227. Porritt, S. W. 1951. *Sci. Agr.* 31:99–112
228. Porutskii, G. V., Luchko, A. S., Matkovskii, K. I. 1962. *Sov. Plant Physiol.* 9:382–84
229. Porutskii, G. V., Matkovskii, K. I. 1963. *Bulg. Akad. Nauk.* 13:147–58
230. Potter, N. A., Griffiths, D. G. 1947. *J. Pomol. Hort. Sci.* 23:171–77
231. Pratt, H. K. 1961. *Int. Bot. Congr. Rec. Advan. Bot.* 9:1160–65
232. Pratt, H. K., Goeschl, J. D. 1969. *Ann. Rev. Plant Physiol.* 20:541–84
233. Priestly, J. H. 1924. *Sci. Progr.* 18:587–96
234. Ram Chandra, G., Spencer, M. 1963. *Nature* 197:366–67
235. Ram Chandra, G., Spencer, M., Meheriuk, M. 1963. *Nature* 199:767–69
236. Ratner, A., Goren, R., Monelise, S. P. 1969. *Plant Physiol.* 44:1717–23
237. Regeimbal, L. O., Harvey, R. B. 1927. *J. Am. Chem. Soc.* 49:1117–18
238. Rhodes, M. J. C., Wooltorton, L. S. C., Galliard, T., Hulme, A. C. 1968. *Phytochemistry* 7:1439–51
239. Rhodes, M. J. C., Wooltorton, L. S. C., Galliard, T., Hulme, A. 1970. *J. Exp. Bot.* 21:40–48
240. Ridge, I., Osborne, D. J. 1970. *J. Exp. Bot.* 21:720–34
241. Ibid, 843–56
242. Ridge, I., Osborne, D. J. 1970. *Nature* 229:205–8
243. Riov, J., Monselise, S. P., Kahan, R. S. 1969. *Plant Physiol.* 44:631–35
244. Rose, D. H., Cook, H. T., Redit, W. H. 1951. *U.S. Dep. Agr. Biblio. Bull.* 13
245. Ross, A. F., Williamson, C. E. 1951. *Phytopathology* 41:431–38
246. Rubinstein, B., Abeles, F. B. 1965. *Bot. Gaz.* 126:255–59
247. Sacher, J. A., Salminen, S. O. 1969. *Plant Physiol.* 44:1371–77
248. Sakai, S., Imaseki, H., Uritani, I. 1970. *Plant Cell Physiol.* 11:737–46
249. Shannon, L. M., Uritani, I., Imaseki, H. 1971. *Plant Physiol.* 47:493–98
250. Shaw, F. H. 1935. *Aust. J. Exp. Biol.* 13:95–102
251. Shcherbakov, A. P. 1939. *Izv. Akad. Nauk SSSR Ser. Biol.* 1939:975–88
252. Shimokawa, K., Kasai, Z. 1966. *Plant Cell Physiol.* 7:1–9
253. Shimokawa, K., Kasai, Z. 1967. *Science* 156:1362–63
254. Shimokawa, K., Kasai, Z. *Plant Cell Physiol.* In preparation
255. Shimokawa, K., Kasai, Z. 1970. *Agr. Biol. Chem.* 34:1633–39
256. Ibid, 1640–45

257. Ibid, 1646–51
258. Shimokawa, K., Yokoyama, K., Kasai, Z. 1969. *Mem. Res. Inst. Food Sci. Kyoto Univ.* 30:1–7
259. Smirnov, A. I., Krainev, S. I. 1940. *Izv. Akad. Nauk SSSR Ser. Biol.* 1940:577–88
260. Smock, R. M. 1944. *Bot. Rev.* 10: 560–98
261. Spencer, M. S. 1956. *Can. J. Biochem. Physiol.* 34:1261–70
262. Ibid 1959. 37:53–59
263. Spencer, M. S. 1959. *Nature* 184: 1231–32
264. Spencer, M. S. See Ref. 135, 793–825
265. Spencer, M. S. 1969. *Fortschr. Chem. Org. Naturst.* 27:32–80
266. Spencer, M. S., Meheriuk, M. 1963. *Nature* 199:1077–78
267. Sprayberry, B. A., Hall, W. C., Miller, C. S. 1965. *Nature* 208: 1322–23
268. Stahl, Q. R. 1969. *Nat. Air Pollut. Contr. Admin. Publ. APTD 69–35*
269. Stahmann, M. A., Clare, B. G., Woodbury, W. 1966. *Plant Physiol.* 41:1505–12
270. Stewart, E. R., Freebairn, H. T. 1969. *Plant Physiol.* 44:955–58
271. Stinson, R. A., Spencer, M. S. 1970. *Can. J. Biochem.* 48: 541–46
272. Stösser, R. 1971. *Z. Pflanzenphysiol.* 64:328–34
273. Takeo, T., Lieberman, M. 1969. *Biochem. Biophys. Acta* 178: 235–47
274. Thomas, M. D. 1951. *Ann. Rev. Plant Physiol.* 2:293–322
275. Thompson, J. E., Spencer, M. S. 1966. *Nature* 210:595–97
276. Thompson, J. E., Spencer, M. S. 1967. *Can. J. Biochem.* 45: 563–71
277. Thornton, N. C. 1940. *Food Ind.* 12(7):48–50, (8)51–52
278. Tolbert, B. M., Lemmon, R. M. 1955. *Radiat. Res.* 3:52–67
279. Turkova, N. S. 1942. *Bull. Acad. Sci. USSR* 6:391–407
280. Turkova, N. S., Vasileva, L. M., Cheremukhina, L. F. 1965. *Sov. Plant Physiol.* 12:721–26
281. Ulrich, R. 1950. *Fruits d'Outre Mer* 5:359–64
282. Ulrich, R. 1952. *La Vie des Fruits.* Paris: Masson et Cie. 370 pp.
283. Ulrich, R. 1958. *Ann. Rev. Plant Physiol.* 9:385–416
284. Valdovinos, J. G., Ernest, L. C., Henry, E. W. 1967. *Plant Physiol.* 42:1803–6
285. Valdovinos, J. G., Jensen, T. E., Sicko, L. M. 1971. *Plant Physiol.* 47:162–63
286. Varner, J. E. 1961. *Ann. Rev. Plant Physiol.* 12:245–64
287. von Guttenberg, H., Beythien, A. 1951. *Planta* 40:36–69
288. von Guttenberg, H., Steinmetz, E. 1947. *Pharmazie* 2:17–21
289. Wang, C. H., Persyn, A., Krackov, J. 1962. *Nature* 195:1306–8
290. Warner, H. L. 1970. PhD thesis. Purdue Univ., Lafayette, Ind.
291. Webster, B. D. 1968. *Plant Physiol.* 43:1512–44
292. Wilkins, H. F. 1965. *Diss. Abstr.* 26:2407–8
293. Wooltorton, L. S. C., Jones, J. D., Hulme, A. C. 1965. *Nature* 207:999–1000
294. Yamaki, T. 1947. *Proc. Jap. Acad.* 23:53–55
295. Yang, S. F. 1967. *Arch. Biochem. Biophys.* 122:481–87
296. Yang, S. F. See Ref. 17, 1217–28
297. Yang, S. F. 1969. *J. Biol. Chem.* 244:4360–65
298. Yang, S. F., Baur, A. H. 1969. *Qual. Plant Mater. Veg.* 19: 201–20
299. Yang, S. F., Ku, H. S., Pratt, H. K. 1966. *Biochem. Biophys. Res. Commun.* 24:739–43
300. Yang, S. F., Ku, H. S., Pratt, H. K. 1967. *J. Biol. Chem.* 242: 5274–80
301. Zimmerman, P. W., Wilcoxon, F. 1935. *Contrib. Boyce Thompson Inst.* 7:209–29

Ann. Rev. Plant Physiol. 1972. 23:293–34

PHYTOCHROME: CHEMICAL AND PHYSICAL   7533
PROPERTIES AND MECHANISM
OF ACTION[1]

WINSLOW R. BRIGGS

Biological Laboratories, Harvard University, Cambridge, Massachusetts

AND

HARBERT V. RICE

New England Aquarium, Boston, Massachusetts

CONTENTS

[1] Supported in part by grants GB-6683 and GB-15572 from the National Science Foundation and a grant from E. I. duPont de Nemours and Co. to the senior author.

293

INTRODUCTION

During the 13 years since the original detection of phytochrome photo-reversibility in etiolated plant tissue and the first isolation of the pigment into aqueous buffer from dark-grown seedlings of *Zea mays* (29), a large amount of biochemical, chemical, and physical information about this chromoprotein has been obtained. Likewise, substantial progress has been made in understanding the mechanism of action of phytochrome. The early hypothesis, that phytochrome played its primary role at some important metabolic crossroads, such as the metabolism of two- and three-carbon fragments (68), has not been supported by more recent evidence. Another hypothesis, that phytochrome functions through gene activation (107), draws its support from extensive physiological, developmental, and enzyme studies mostly with mustard seedlings (*Sinapis alba*). A third recent hypothesis suggests that phytochrome acts in some way to affect membrane permeability (70). The recent experimental evidence in support of this last hypothesis will be considered below.

Because of low levels of phytochrome and difficulties of isolation, progress was relatively slow at first in characterization of the pigment. When Siegelman & Butler (156) reviewed the properties of phytochrome in the *Annual Review of Plant Physiology* in 1965, they could only muster a handful of references, many of them abstracts. Hillman (75) discussed a few studies on partially purified preparations in 1967 but concentrated on spectrophotometric studies of phytochrome in vivo, and the correlation of these measurements (or lack of correlation) with physiological responses.

The enormous growth of the field since 1964 is illustrated by the number of general reviews which have appeared. There were three in 1965 (27, 66, 156), two in 1967 (72, 75), one each in 1968 (55), 1969 (155), and 1970 (161), and another (in addition to the present one) to appear in 1972 (154). A recent comprehensive book by Rollin (136) cites over 450 references. Of further interest will be the proceedings (106) of a NATO Advanced Study Institute on Phytochrome, held in Eretria, Greece, in September 1971, including brief reviews of areas ranging from phytochrome in flowering through chromophore chemistry.

In addition to these general treatments, there have been several reviews devoted to specialized aspects of phytochrome. Mohr (108, 109) published two detailed articles dealing with the high energy reaction (HER) and the gene activation hypothesis. The cogent experiments and reasoning that at least many high energy reactions are phytochrome mediated are given by Hartmann (63) and Wagner & Mohr (182). Light-induced chloroplast movement and localization of phytochrome in the cell are the subjects of three recent reviews (65, 67, 102), dealing mostly with chloroplast movement in the green alga *Mougeotia* and related phenomena. Phytochrome in relation to flowering is considered in numerous articles in the recent book edited by Evans (44). Phytochrome in comparison to other plant bile pigments is treated briefly by Siegelman et al (157). Also of interest is Frankland's detailed treatment of phytochrome biosynthesis and dark reactions (52). Zucker's review of light and enzymes (190), appearing elsewhere in this volume, treats still another aspect of phytochrome physiology.

Considering the extensive burgeoning of the field and the number of recent reviews, the present authors are restricting their coverage to studies aimed at characterization of the chromoprotein in vitro and in vivo, and to recent progress in elucidating the mechanism of phytochrome action. Neither the high energy reaction per se nor chloroplast movement will be considered except where studies impinge directly on pigment characterization or mechanism of action. Physiological events far removed in time from phytochrome phototransformation, e.g. flowering and seed germination, are not discussed. Abstracts and work in progress also are not considered.

PHYTOCHROME PURIFICATION

Since Siegelman & Firer (158) first reported on the partial purification of phytochrome from etiolated oat seedlings, several laboratories have reported purification of the pigment to a high degree of homogeneity. Since the final products obtained differed in molecular weight, amino acid composition, spectral behavior, and other properties, even when the same plant material was used as the protein source, it is worth considering the purification procedures in detail.

Following buffer extraction, filtering and centrifugation for clarification, and a lengthy ultrafiltration step, the Siegelman & Firer procedure (158) involved gel filtration on Sephadex G-50, chromatography on brushite (Ca-$HPO_4 \cdot 2H_2O$), precipitation by 40 percent saturated ammonium sulfate, redissolving in buffer, and Sephadex G-200 gel filtration. These steps preceded a second ammonium sulfate precipitation (50%), redissolving, Sephadex G-50 gel filtration, diethylaminoethyl cellulose (DEAE) chromatography, a third ammonium sulfate precipitation (50% saturation), redissolving in buffer, and a final centrifugation. Recovery was 6%, with about 60-fold purification. The preparation was probably 10–12% pure. The pH throughout the column steps was maintained at 7.8. Siegelman & Hendricks (159) substantially modified the oat phytochrome procedure after the gel filtration step, using gel filtration through sieved 7% agar, brushite chromatography, a second ultrafiltration and centrifugation, a repeat of the agar gel filtration, a second brushite step, and a final ammonium sulfate precipitation (50% saturation), followed by redissolving in buffer. Yields were near 50% and purity probably close to 15%. After the initial harvest and extraction, all operations took place under normal laboratory illumination, and a mild reducing agent, 2-mercaptoethanol, was present throughout.

The first preparation of high purity was that of Mumford & Jenner (112). After extraction and centrifugation, these workers used brushite chromatography, Sephadex G-200 gel filtration, continuous flow electrophoresis, and Biogel P-150 gel filtration. They reported approximately 740-fold purification and a final yield of 3%. The preparations were at least 90% pure, as judged by two different electrophoretic techniques. The pH was maintained at 7.8 except during the continuous flow electrophoresis when it was somewhat higher. Normal laboratory illumination was used following harvesting and extraction. Mercaptoethanol was present at least during the early stages of purification.

Save for some earlier studies by Siegelman and his co-workers on corn and barley phytochrome (e.g. 71) and some work on phytochrome from etiolated peas by Bonner, appearing only in abstracts [see Siegelman & Butler (156) for references], all work published up to this time was with oats. Correll et al (34) chose instead to work with etiolated rye. Their procedure utilized buffer extraction, clarification by filtration and centrifugation, and precipitation with ammonium sulfate (33% saturation). There followed brushite chromatography, ammonium sulfate precipitation, resuspension, and DEAE chromatography. Purification was estimated to be 1200-fold, with a final yield of 25%. Throughout the procedure, the phytochrome was kept as Pr by working only under dim green light. Mercaptoethanol was present throughout. There was no gel filtration step. As will be discussed later, their product was quite different from that of Mumford & Jenner (112).

In the past year and a half, three other laboratories have reported purification of oat phytochrome to apparent homogeneity. The procedure used by Butler's laboratory (e.g. 80) is described in greatest detail by Hopkins (79). Following initial extraction, the crude extract was made $0.015M$ with $CaCl_2$ to precipitate large amounts of pectic compounds and make subsequent operations easier. There followed brushite chromatography, ammonium sulfate fractionation, DEAE chromatography, a second ammonium sulfate precipitation, resuspension, dialysis against phosphate buffer at pH 6.0, binding to a carboxymethyl-Sephadex (CM) column, and step elution at pH 7.8. The last step was gel filtration through Biogel P-150 as a sizing step. Final recovery was between 5 and 10%. Mercaptoethanol was dropped during brushite chromatography, and dim green illumination was used throughout. Judging from the absorption spectrum, the final product was approximately the same as that of Mumford & Jenner (112). The ratio of chromophore absorbance of Pr (A_{665}) to protein absorbance (A_{280}) was one or slightly higher in both cases. Rice and co-workers (132, 135) report the same procedure with minor modifications, and obtained almost exactly the same product from oats. Roux's procedure (137) with oats is similar through DEAE chromatography, though he substitutes hydroxylapatite (HA, $Ca_{10}(PO_4)_6(OH)_2$) for CM as a cation exchanger, and obtains some purification by including streptomycin during one ammonium sulfate precipitation. On some preparations he replaced the DEAE step with a second HA column. His product resembled that of Mumford & Jenner (112) in terms of its $A_{665}:A_{280}$ ratio. Another change involved substitution of $NaHSO_3$ for mercaptoethanol.

Rice and his co-workers (132, 135) undertook the purification of rye phytochrome in an attempt to resolve the differences reported between the rye and oat pigment. They used their oat procedure through DEAE but failed to make the CM step work. Dialysis of the rye phytochrome against pH 6.0 buffer caused a large amount of it to precipitate. They therefore substituted Roux's (137) HA step for the CM step. It was followed by gel filtration on agarose (Biogel A1.5M). Final purification was 3250-fold, with the $A_{665}:A_{280}$ ratio about 0.7. As Correll et al (34, 38) had reported, the final product differed substantially from Mumford & Jenner's oat phytochrome (112). However, it also differed substantially from the rye phytochrome of Correll et al (38).

Kroes and associates (93, 94) have also reported a procedure for purification of oat phytochrome, involving batch absorption of phytochrome on HA, elution, ammonium sulfate fractionation, a second HA batch absorption step, DEAE chromatography, and finally Sephadex G-200 gel filtration. Purification was estimated at 345-fold, and the final product was of the Mumford & Jenner type (112) though not so pure. The Kroes group, however, did thoroughly investigate conditions for binding and elution of phytochrome from the calcium phosphate gel. That brushite shows variable behavior even in the same laboratory is exemplified by our own work. The binding conditions used by Rice et al (135) are not the same as those used by Gardner et al (60). Technical problems of this and other column steps are discussed elsewhere (21).

As mentioned above, workers isolating oat phytochrome usually obtained a similar product, while those working with rye phytochrome usually obtained something quite different. Walker & Bailey obtained phytochrome from oats, however, that differed substantially from anything previously reported (184, 185). One procedure (184) involved alternating gel filtration and DEAE chromatography repeated several times, yielding two photoreversible components with different electrophoretic behavior, absorption spectra, and molecular weight. The second procedure (185) involved gel filtration alternating with brushite chromatography, and only a single DEAE step prior to final electrophoresis. The single product obtained differed from the two they reported previously, but also differed from that of Mumford & Jenner (112). Possible reasons for the many differences will be considered in the following section.

A few studies have attempted to isolate phytochrome from green plants. In 1963, Lane et al (95) reported detecting phytochrome photoreversibility in extracts of a number of light-grown plants. The levels of activity were extremely low, precluding any further purification with the techniques then available. Taylor & Bonner (179) have subsequently isolated phytochrome from a green alga (*Mesotaenium caldariorum*) and a liverwort (*Sphaerocarpos*, species not given). Following breaking of the cells and initial extraction, the preparation was lyophilized. The dried powder was then extracted twice with acetone, once with ether, and then redried. Following resuspension in buffer, gel filtration, ammonium sulfate fractionation, and DEAE chromatography (once as a flow-through step and once with phytochrome binding and elution) yielded pigment sufficiently concentrated and free of chlorophyll to obtain absorption spectra and measure phototransformation kinetics. The absorption maxima for the Pr and Pfr forms were at significantly lower wavelengths than those for phytochrome from higher plants, being 649 nm and 710 nm for *Mesotaenium* and 655 nm and 720 nm for *Sphaerocarpos*, compared to the more familiar 665 nm and 725 nm for higher plant phytochrome. These spectral differences may not be artifacts of the purification procedure, at least for *Mesotaenium*, since Haupt (64) has shown that the action spectrum for phytochrome-controlled chloroplast movement in *Mougeotia*, a close relative of *Mesotaenium*, has an action peak at least for far-red light at considerably shorter wavelengths than that for higher plants (see 69). Giles & von Maltzahn (61) followed the same general procedures as Taylor & Bonner (179) to isolate and partially purify phytochrome from two species of the moss genus *Mnium*. This phytochrome also had absorption maxima at somewhat lower wavelengths than those of higher plant phytochrome.

Phytochrome Characterization

The literature on phytochrome protein chemistry reveals wide disagreement between laboratories on properties such as molecular weight, subunit structure, and amino acid composition, though each report seems sound and internally consistent. In this section we shall examine some of these differences in detail, reviewing recent evidence that the villain is not careless technique, but rather varying degrees of proteolytic degradation occurring during phytochrome purification procedures.

Molecular weight.—The most obvious difference between preparations of phytochrome from different laboratories is molecular weight. Let us consider oat phytochrome first. Siegelman & Firer (158) obtained a sedimentation coefficient ($s_{20,w}$) of 4.5, estimating a molecular weight for their partially purified phytochrome between 90,000 and 150,000. Polydispersity of the preparation precluded a more accurate measurement. Mumford & Jenner (112) estimated the molecular weight of their purified product as 55,000, based on gel exclusion chromatography, a valid technique for globular proteins (see 1, 2). They mention briefly preliminary equilibrium ultracentrifugation experiments yielding a value of 60,000, with evidence for some aggregation. Equilibrium ultracentrifugation values are independent of molecular shape (150), so the reasonable agreement between the two values is evidence that the Mumford & Jenner product is globular. Briggs et al (24) estimated the molecular weight of partially purified oat phytochrome by gel exclusion methods, finding two distinct molecular weight classes, 80,000 and 180,000. Unfortunately they used only two marker proteins, cytochrome c at 12,500 (186) and catalase at 240,000 (5), so the estimates are probably not too accurate.

Roux (137), using gel exclusion chromatography, gave estimated values of 70,000 and 110,000 for small and large molecular weight classes. Hopkins obtained a value of 60,000 by equilibrium ultracentrifugation, S values consistent with this molecular weight by velocity ultracentrifugation, and a value of about 69,000 (79) from sodium dodecyl sulfate (SDS) polyacrylamide electrophoresis (153), again suggesting a globular protein similar to the Mumford & Jenner protein (112). Rice & Briggs (133) similarly obtained values of 62,000 for oat phytochrome both by SDS electrophoresis and gel exclusion chromatography, and Gardner et al (60) report S values consistent with a 60,000 molecular weight globular protein. Kroes (93, 94) reports a value of 55,000 from gel filtration experiments. If one ignores for the moment the larger molecular weight values found by some authors (in addition to the smaller), then the picture for oat phytochrome seems reasonably consistent with numerous authors obtaining values near 60,000. Walker & Bailey, however, obtained drastically different results. In their first paper (184) the two photoreversible proteins gave S values consistent with molecular weights of about 23,000 and 18,000 with evidence for a much larger form of the second (roughly 130,000). In their second paper (185), they reported a value near 26,000 for the single product, suggesting aggregates of between 79,000 and 127,000. [Pea phytochrome purified by the same technique yielded two species, one between 113,000 and 153,000, and the other between 265,000 and 359,000 (185).]

The data from rye phytochrome appear equally contradictory. Correll et al (38) did both velocity and equilibrium ultracentrifugation experiments with their purified rye phytochrome. They obtained both 9 S and 14 S species, though the 14 S species disappeared after storage or handling. Equilibrium measurements of the 9 S protein gave an average value of 180,000. Treatment either with SDS or $8M$ urea yielded a single 42,000 molecular weight species, as did 4-hour urea denaturation followed by carboxymethylation by the procedure of Anfinsen & Haber (6). By contrast, Rice (132) and Rice & Briggs (133) obtained values of 120,000 by SDS gel electrophoresis, using both 5 and 10% gels, and a value of about 375,000 by gel exclusion chromatography. They cite (133) a preliminary equilibrium ultracentrifugation value of about twice the SDS value obtained by Gardner. In addition, Gardner et al (60) report an S value of 9.2, consistent with a molecular weight of about 180,000, for a globular protein.

Resolution for these apparent inconsistencies starts with the work of Pringle (128, 129). In attempting to purify yeast malate dehydrogenase (and studying other proteins), Pringle discovered that proteases, persisting with the enzyme during purification, were systematically degrading the enzyme.

He also found, however, that proteolytically produced fragments retained some enzymatic activity. Furthermore, the level of contaminating protease could be as low as 0.0002% of the protein being studied and still have a detectable effect. Such proteolytic breakdown could frequently be inhibited by using the inhibitor phenylmethanesulfonyl fluoride (PMSF), a reagent known to inhibit proteases such as trypsin which require serine groups for activity (47) [but at least with the protease papain, they are thought to react with sulfhydryl groups required for enzymatic activity (187)]. Significantly, Pringle (128) also showed that neither $8M$ urea nor SDS at concentrations usually used in SDS polyacrylamide electrophoresis would completely inhibit proteolysis.

Pike (119) and Pike & Briggs (121) have reported on the partial purification of a protease from etiolated oat seedlings. The protease is also present in extracts of etiolated rye, but in lower amount. The protease has a pH optimum near 6.5 but substantial activity at pHs as high as 8.0. It is inhibited by the sulfhydryl agent $HgCl_2$ and also strongly inhibited by PMSF. It is activated by low concentrations of the reducing agent mercaptoethanol and also inhibited by a high concentration of inorganic ions.

Gardner et al (60) and Rice and co-workers (132, 133, 135) have explored the role of protease present during phytochrome purification both from etiolated oats and rye. (See also Briggs et al 22.) Using highly purified oat phytochrome of the 60,000 molecular weight class, they made attempts to detect aggregation by varying protein concentration, salt, and pH, using sucrose gradient or velocity ultracentrifugation, or gel exclusion chromatography, as the assay. They could obtain no evidence for aggregation to a higher molecular weight species. Briggs et al (24) had earlier reported two different molecular size classes of oat phytochrome, noting that the larger one disappeared with time. Correll & Edwards (33) confirmed these observations for oat phytochrome and reported also two size classes for rye phytochrome. A nitrogen terminal amino acid determination (143) yielded not one but four end groups, suggesting heterogeneity of the oat preparation (133). Gardner et al (60) showed that the breakdown of large phytochrome was almost quantitatively accompanied by an increase in small, with the breakdown occurring more rapidly in partially purified oat preparations than rye. The breakdown could be prevented in both cases by the inhibitor PMSF. More highly purified rye phytochrome of the large molecular weight kind does not break down to the smaller molecular weight form (60, 132). However, such rye phytochrome could be induced to break down by incubating it with the oat protease of Pike & Briggs (121), or with any of a wide variety of endopeptidases. Exopeptidases such as leucine aminopeptidase or carboxypeptidase A were without effect.

SDS electrophoresis of undegraded material yielded a molecular weight for the polypeptide chain of about 120,000 (133). Overnight incubation at 4°C in the dark with small amounts of oat protease or trypsin (as little as 0.1% trypsin by weight of phytochrome) almost completely converted the

120,000 protein to one about 62,000 (all measured on SDS gels). The rye product was in fact undistinguishable from purified oat phytochrome on the gels (133). It failed to show much further breakdown during a subsequent 24 hours, suggesting that the 62,000 product was for some reason resistant to further proteolytic degradation. What subsequent breakdown did occur produced a chain estimated to be 42,000, plus some smaller products.

Since endogenous proteolysis during purification produces predominantly a 62,000 molecular weight chromopeptide which is quite resistant to further protolysis, it is not surprising that numerous authors have succeeded in purifying this chromopeptide to apparent homogeneity. The reports by Roux (137) and Walker & Bailey (185) of phytochrome species near 120,000 may mean that these authors did obtain some of the undegraded native chain. Rice (132) has shown that a band at 120,000 appears in SDS gels of phytochrome obtained after each of the four column steps in the rye phytochrome purification procedure described by Rice et al (135), suggesting no major degradation during purification. The 9 S phytochrome obtained by Correll et al (38) is probably native material (dimer of the 120,000 polypeptide) and the 14 S form a higher aggregate form. The molecular weight estimates for rye phytochrome from Correll's laboratory are probably in error for two reasons. First, the material was obtained without any sizing step, and may well have been contaminated by other large molecular weight proteins with similar charge properties. Other rye phytochrome preparations with absorbance ratios ($A_{665}:A_{280}$) like those of Correll et al yield multiple bands on SDS gels, while material with a ratio near 0.8 yields only the single 120,000 molecular weight band (133). Second, either proteolysis during incubation in urea or SDS, or contaminants, may account for the 42,000 molecular weight they report for the subunit. Pringle (128) emphasizes that samples to be analyzed in SDS must be boiled and treated with inhibitor *prior* to incubation or during the analysis itself.

How accurate is the 120,000 figure for the native rye phytochrome chain? Weber & Osborn (186) have used the technique of Shapiro et al (153) successfully on a wide variety of proteins, with an accuracy of about 10%. The reliability is based on the binding of a constant amount of SDS per unit of protein (1.4 gm SDS/gm protein) (131). At such binding ratios, migration is determined by the negative charge of the detergent, and is a function of chain size. Recently, Pitt-Rivers & Impiobato (122) have found that glycoproteins may bind at less than the 1.4:1 ratio. The slower migration of such glycoproteins causes a substantial overestimate of the molecular weight. Bretscher (17) recently demonstrated this weakness in the technique for a glycoprotein derived from erythrocyte membranes. The error is more severe for lower acrylamide concentrations than for higher. Bretscher's protein was over 60% sugar, however, while Roux (137) reports his oat phytochrome to contain a maximum of about 3%. Rice (132) has obtained the 120,000 value with both 5 and 10% gels, making it unlikely that phytochrome contains any substantial amount of sugar, and suggesting that the molecular weight estimate is reasonably accurate. However, there are early references suggesting that other plant biliproteins may be glycoproteins (e.g. 54), and native rye phytochrome has not to date been tested for its sugar content.

The smallest molecular weight species of oat phytochrome reported by Walker & Bailey (184, 185) are probably also products of proteolysis. Both Walker & Bailey (185) and Roux (137) caution that DEAE column chromatography may lead to substantial phytochrome denaturation, as it does for some other enzymes (41). Thus any prolonged purification procedure involving DEAE chromatography, lengthy dialyses, etc is prone to proteolytic degradation. The presence of mercaptoethanol, the pH range, and low salt conditions frequently used all favor proteolysis (121). Rice et al (135) mitigated these conditions by omitting mercaptoethanol early in the procedure, and by shortening or omitting dialysis between columns. Differences in phytochrome yield or properties from different seed lots (34, 185) may merely reflect differences in protease levels in the seedlings.

Subunit structure.—If the native size of oat and rye phytochrome polypeptide chains is 120,000, what then is the subunit structure? A reasonable model might be the following. The 9 S material reported by several workers (38, 60) is probably a dimer of the 120,000 molecular weight subunit, while the 14 S protein could be a higher aggregate. If the native phytochrome were not globular, it would yield anomalously high molecular weight estimates by gel filtration and anomalously low ones in velocity sedimentation. Thus the value of about 375,000 reported by Rice & Briggs (133) from gel filtration, and a 9 S form consistent with a molecular weight of about 180,000 reported by Gardner et al (60) and Correll et al (38), are perhaps not out of line. Confirmation of the preliminary report of a molecular weight of about 240,000 from equilibrium ultracentrifugation (133) for undenatured rye phytochrome would support this interpretation. However, aggregation and disaggregation phenomena have not been rigorously excluded in these studies, so the model must be considered as tentative. In any case, the tetramers and hexamers of a basic subunit of 42,000 reported by Correll et al (38) could readily be dimers and trimers of a 120,000 subunit, considering possible sources of error in molecular weight measurement and interpretation of electron micrographs. One must add in conclusion, however, that Correll and his co-workers (34, 38) were clearly the first to obtain undegraded phytochrome. Their choice of rye, with its lower protease level than oats, made this accomplishment possible.

One more question remains: does phytochrome have more than one kind of subunit? Rice & Briggs (133) report obtaining two nitrogen terminal amino acids (glutamate and aspartate) following the Sanger dinitrofluorobenzene method (143), and Rice (132) reports occasionally seeing a double band near 120,000 on 5% SDS acrylamide gels with a low protein load. However, neither report can be considered as conclusive evidence for two different subunits. The two bands observed on SDS gels could be an artifact of loading. Furthermore, Cope et al (32) showed that for *Synechoccus* phycocyanin, the Sanger procedure (143) yielded both glutamate and aspartate as nitrogen-terminal amino acids, while the cyanate procedure of Stark & Smyth (167) yielded only a single one, methionine. Detailed subunit structure of phytochrome must be considered as unresolved pending further study. It clearly differs from that of phycocyanin and other phycobilins, however (9, 133).

Criteria for purity and units of activity.—It should be clear from the above discussion that in the absence of molecular weight information, criteria for purity are difficult to establish. The phytochrome of the Mumford & Jenner type (112), obtained by various workers, has appeared homogeneous by a number of techniques, including perhaps the most sensitive one, SDS polyacrylamide gel electrophoresis (79, 135). Yet analysis for N-terminal amino acids yielded four products (133), suggesting four different kinds of chains. Since this protein is the product of proteolysis (60) by a relatively nonspecific protease (121), the finding of several different end groups is not surprising. All one can say is that the products must all be of approximately the same size and charge.

Many workers (79, 132, 137) have used the ratio $A_{280}:A_{665}$ of Pr as an index of purity, with a ratio near 0.8 being close to purity for the small oat product, and about 1.3 for rye (135). The reciprocal of this ratio is the specific activity of the pigment, units of chromophore absorbance per unit of protein absorbance, as originally used by Siegelman & Firer (158). Specific activity is then a good criterion of purity provided two criteria are met. First, it must be accompanied by a clear demonstration that no proteolytic degradation has occurred, and SDS gels appear to be the method of choice. Second, chromophore absorption both of Pr and Pfr should be normal, that is, with maxima near 665 nm for Pr [or where appropriate for phytochrome from lower plants (179)] and near 725 nm for Pfr, with photobleaching of Pr roughly equal to the absorbance increase of Pfr. Several authors have shown that chromophore absorption is very sensitive to mildly denaturing conditions (30, 36, 71, 119). Extinction coefficients for phytochrome are not useful in the absence of convincing evidence against proteolytic degradation or reasonable demonstration of high purity.

Amino acid analysis.—Several authors have now presented results of amino acid analysis of phytochrome. Both Roux (137) and Rice & Briggs (133) have obtained analyses of 60,000 molecular weight oat phytochrome which are in reasonable agreement with the original one of Mumford & Jenner (112). Walker & Bailey's analyses of their various phytochrome preparations (184, 185), including that of a trypsin-resistant core protein (185), do not agree for the obvious reason that they were probably analyzing only small chromopeptide products of proteolysis, either deliberate or during purification. Their results are interesting, however, for another reason: their small photoreversible phytochrome peptides could be useful in studies on the mechanism of phototransformation.

The two analyses of large molecular weight rye phytochrome (38, 133) do not agree with any of those from oats, nor do they agree with each other. One reason for the diagreement may be that the phytochrome of Correll et al (34) contained significant levels of impurities. [Correll et al's fingerprinting of phytochrome tryptic peptides (38) suffers the same weakness.] The Correll et al claim (38) of a complete absence of cystine, however, was in error. Rice & Briggs (133) found cystine by using the performate oxidation technique of Hirs (78).

Despite the various differences mentioned above, all of the analyses published do show some features in common. Phytochrome whether large or small, from oats or rye, contains high amounts both of acidic and basic amino acids, in addition to at least 11 cystines per 60,000 molecular weight. It is also reasonably high in threonine and serine and is thus potentially a highly reactive protein.

The reported differences between the amino acid compositions of Mumford & Jenner type oat phytochrome (112) and native rye phytochrome (133) may reflect in part protein differences between the two genera. However, they likely also reflect some protein loss by proteolysis during purification of the oat pigment. Siegelman & Firer (158) reported crude oat phytochrome to be insoluble below pH 6.2, while both Hopkins (79) and Rice et al (135) routinely dialyzed the oat pigment to pH 6.0 to obtain binding on CM-Sephadex. Perhaps more convincing (and certainly useful in purification) is the study of Correll & Edwards (33). They showed both for oat and rye phytochrome that the larger molecule was precipitated by lower ammonium sulfate concentrations than the smaller, and that the larger also bound more tightly to DEAE than the smaller. Though these differences could reflect partially differences in protein conformation, they are also consistent with differences in amino acid composition.

One other parameter has been determined from amino acid analyses. Rice & Briggs (133) have calculated the partial specific volume of 62,000 molecular weight oat phytochrome as 0.736 cc/gm, and that for native rye phytochrome as 0.728 cc/gm. Until analyses of proteolytically produced small molecular weight rye phytochrome, or preferably native large molecular weight oat phytochrome (yet to be purified to homogeneity) are done, the question of species differences versus proteolytically produced differences must remain unresolved.

Immunochemical studies.—That small molecular weight oat phytochrome is a powerful antigen has now been demonstrated in three laboratories (80, 127, 134). Hopkins & Butler (80) used antibody techniques in studying protein conformational differences between Pr and Pfr (see below), and Pratt & Coleman (127) used antibody for phytochrome localization studies (see below).

Rice & Briggs (134) used antibody against oat phytochrome for comparing phytochromes from different sources and of different sizes, and have most thoroughly characterized the system. Immunodiffusion on Ouchterlony plates (117) and immunoelectrophoresis (117) indicated that the antibody preparation contained predominantly a single antibody system. The immunoprecipitate showed clear photoreversibility, though the absorption spectra were somewhat altered. The following phytochrome pairs showed lines of identity on Ouchterlony plates: purified oat phytochrome (62,000) and crude rye phytochrome; purified oat phytochrome and crude corn phytochrome; purified oat phytochrome and purified rye phytochrome (native); purified oat phytochrome and trypsinized rye phytochrome (about 60,000); and purified native rye phytochrome against trypsinized rye phytochrome. There were no detectable differences between either Pr and Pfr (purified oat) or glutaraldehyde-fixed Pr and Pfr (purified oat). Hopkins & Butler (80) also failed to detect any differences between Pr and Pfr using double diffusion techniques, turning instead to microcomplement fixation (114) in their study of protein conformational differences between Pr and Pfr. Recognition of glutaraldehyde-fixed phytochrome by the antibody was the basis of the phytochrome localization studies by Pratt & Coleman (127) to be discussed later.

Only when they compared purified oat phytochrome with crude pea phytochrome did Rice & Briggs (134) detect any immunological differences. The Ouchterlony plate shows only partial identity, with interference and a clear spur. Small structural differences between these various phytochromes may be detected by the more sensitive complement fixation procedure used by Hopkins & Butler (80), but such experiments have not been reported.

It is clear that native large molecular weight rye phytochrome shares the same determinants as small molecular weight phytochrome, at least with antibody elicited against the smaller molecule. Large and small phytochrome should be tested with antibody elicited against the larger native molecule which may indeed lose determinants upon proteolysis.

The antigenicity of phytochrome probably does not lie in the chromophore. Rice & Briggs (134) tested the oat antibody system for cross reactivity both with phycocyanin and allophycocyanin, the two phycobiliproteins with chromophores thought to be most similar to that of phytochrome (see below), with negative results. Berns, on the other hand, has reported that antibody elicited against phycocyanin showed cross reactivity with partially purified oat phytochrome (10). Rice & Briggs (134) question his conclusion, however, on the basis that he also obtained cross reactivity with jack bean urease, casting doubt on the specificity of the antibody preparation.

Chromophore chemistry.—A full 9 years before the first spectrophotometric detection of phytochrome (29), workers at the USDA Plant Research Laboratory at Beltsville, Maryland, suggested that the pigment responsible for photomorphogenic responses in plants might be a linear tetrapyrrole, such as the chromophore of phycocyanin or allophycocyanin, two algal accessory photosynthetic pigments (16, 118). Before the chromophore had been cleaved from the protein for even tentative chemical analysis, they were proposing possible molecular mechanisms for the spectral changes in its phototransformation (71).

Following a brief report in 1965 (159), Siegelman and co-workers published details of a relatively low-yield procedure for cleaving the phytochrome chromophore from the protein. The pigment was precipitated with trichloroacetic acid, washed with methanol, and then refluxed for 3 to 4 hours with methanol plus 1% ascorbic acid (160). (Ascorbic acid was apparently not essential.) The extracted chromophore was blue, with absorption maxima in acidic methyl alcohol of 380 nm and 690 nm, compared to 375 nm and 685 nm for phycocyanin and allophycocyanin chromophores [both later shown to be phycocyanobilin (157)]. Absorption maxima in chloroform were 385 nm and 670 nm (159). The chromophore's behavior in three different solvent systems in thin layer chromatography was similar to that of the algal chromophores but not identical. Partial esterification yielded three products in thin layer chromatography, indicating two free acid groups both on the phytochrome and algal chromophores. Treatment with zinc acetate followed by alkaline iodine solution yielded a product with bright red fluorescence for these chromophores, none of which fluoresced prior to the treatment. These workers concluded that the phytochrome chromophore was a bilitriene, probably closely related to the algal pigments but not identical. The isolated chromophore is not photoreversible.

Walker & Bailey (183) used the Siegelman et al procedure (160) to extract chromophore from phytochrome precipitated with trichloroacetic acid either as Pr or Pfr. Unlike Siegelman et al, they reported both products to be fluorescent prior to zinc-iodine treatment. The absorption spectra of the products differed depending upon whether Pr or Pfr had been extracted. Kroes (93) recently questioned Walker & Bailey's claim, noting that he could eliminate the difference by purifying the chromophore fractions on silica gel columns.

Siegelman & Hendricks (159) proposed two possible mechanisms for phototransformation: cis-trans isomerization about a bridge methyne, or hydrogen migration accompanying lactim-lactam tautomerism of the terminal pyrroles (see 72). Siegelman et al (157) proposed a tentative structure for the entire chromophore, based on detailed studies elucidating the structures of the chromophores of the algal phycobilins. (The article summarizes the chemical studies on the algal pigments leading to structure determination, studies which will not be considered here.) Proton migration about ethylidene groups on both terminal pyrroles, leading to an increase from seven to ten double bonds, was suggested to account for the spectral changes observed on phytochrome phototransformation. Simultaneous loss of two assymetric centers would account for circular dichroism changes in the chromoprotein on transformation, to be discussed below. Crespi et al (39) also proposed a model for phytochrome, based on studies of phycocyanobilin, proposing a charge transfer complex mechanism involving both chromophore and protein. Suzuki & Hamanaka (174) likewise propose a model based upon studies of another pigment, in their case rhodopsin. They eliminate lactim-lactam tautomerism on energetic grounds, proposing instead *cis-trans* isomerization.

The only model based on chemical degradation studies is that of Rüdiger & Correll (142). They used Rüdiger's chromic acid oxidation method (140) which yields maleimides (plus other products) from pyrroles or pyrrole pigments without altering β-substituents. [The technique and related literature are discussed elsewhere by Rüdiger (141) and will not be considered further here.] The first phytochrome preparation examined was apparently denatured by alkaline buffer and was greenish yellow like Pfr. It was converted to a blue form like Pr with cold acid, and then back to the greenish yellow form by return to alkaline conditions. Chromic acid oxidation of either of these forms released only rings II and IV from the protein (see Figure 1 for numbering), hot acid being required to release rings I and III. Rings I and III were thus postulated to be linked to the protein. The ring I product was methylethylidene succinamide from the blue form and methylethyl maleimide from the yellowish green. The structures proposed for these two forms are shown in Fig. 1, along with proposed linkages to the protein. Unfortunately, Rüdiger & Correll were unable to obtain a ring I product from native photoreversible phytochrome without prior alkaline denaturation, so they are unable to state whether the different ring I products obtained from the two denatured forms reflect normal changes of phototransformation or not. Note that ring IV is shown as vinyl-substituted rather than ethylidene-substituted, as proposed by Siegelman et al (157).

It is not clear to these reviewers how Rüdiger & Correll (142) could unequivocally distinguish whether ring II or III was bound to the protein. Both have the same β-substituents (methyl and propionic acid), and could form either dialdehydes or imides on chromic acid oxidation, depending upon their conjugation state. In any case, proton migration accompanied by protein complexing could account for the spectral changes observed in the denatured material, and provide a possible model for the mechanism of phototransformation. Simultaneous *cis-trans* isomerization are not precluded, however.

Kroes (92, 93) also reported success in cleaving the phytochrome chromophore using the method of Siegelman et al (160), while reporting failure of another method involving HBr in trifluoroacetic acid. He also verified its diacidic nature by partial esterification and thin layer chromatography, and obtained sufficient chromophore to obtain absorption spectra in chloroform of the phytochrome chromophore dimethyl esters, as compared to the esters of phycocyanobilin and mesobiliverdin. The phytochrome chromophore spectrum is close to but different from that of phycocyanobilin. Its absorption maxima were at 377 nm and 604 nm compared to 368 nm and 595 nm for the algal pigment. The uv absorbancies were almost twice those in the visible. Denaturation of phytochrome with a detergent (SDS) produced a visible spectrum almost indistinguishable from that of the isolated chromophore.

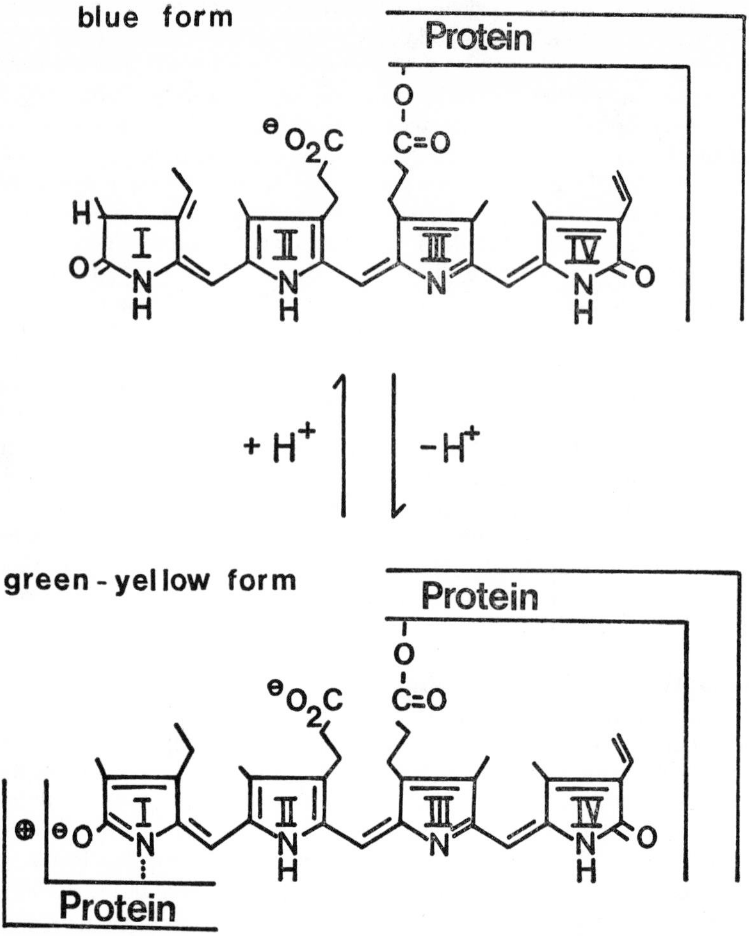

blue form

green-yellow form

FIGURE 1. Proposed chromophore structures, after Rüdiger (141). Yellow-green form obtained from alkaline denatured phytochrome. Blue form obtained by transfering alkaline denatured phytochrome to acid prior to degradation.

[Both Correll et al (36) and Rice (132) have reported a similar visible absorption spectrum for phytochrome in SDS.]

That protein-chromophore interaction is important for the absorption spectrum of native phytochrome is clearly shown by the bleaching effects of a variety of protein denaturants in addition to SDS (30, 36, 93, 133). Thus the simple model propsed by Siegelman et al (157) is not adequate to account for all aspects of the pigment's behavior.

Fry & Mumford (53) have recently reported on the isolation and partial characterization of a nonphotoreversible chromopeptide which they purified after pepsin digestion of oat phytochrome. Its absorption spectrum (in buffer) showed absorbance peaks at 650 nm, 370 nm, and 292 nm, with the 370 nm peak the largest. Sequencing studies starting both at the carbon-terminal and amino-terminal ends of the peptide suggested the following primary structure: leu-arg-ala-pro-his-(ser,cys)-his-leu-gln-tyr. The order of serine and cystine was not resolved. An ester or ether linkage of the chromophore to serine seemed unlikely, since serine was readily released from the chromopeptide by treatment with aminopeptidase M. They suggested instead a possible thioether linkage to cystine, based on the unexpectedly low yield of cystine in their analyses, but concede that direct evidence is still lacking.

Chromophore number.—Most of the literature starting with Siegelman & Firer (158) has assumed a single chromophore per 60,000 molecular weight. However, as Correll et al (36) rightly point out, there is little experimental evidence to support this contention. Their own experiments with what was probably large molecular weight phytochrome from rye present a more complex picture. Dark reversion studies yielded evidence for two populations of Pfr reverting with different first order rate constants to two different populations of Pr. Alteration of the sodium chloride concentration changed the proportions of the two populations present. They proposed four spectrally different chromophoric species, absorbing at 580 nm, 660 nm, 670 nm, and 730 nm. SDS bleached all but the 580 nm band, and abolished photoreversibility, while glutaraldhyde bleached the 730 band, but did not destroy photoreversibility. They also proposed that all four chromophoric species can be bleached either by appropriate irradiation or dark reactions which they suggest as coupled reduction-oxidation between the 580–660 nm and 670–730 nm pairs of chromophores.

While such a detailed hypothesis must remain questionable on the basis of evidence to date, one thing is clear: under the appropriate conditions, chromophores in purified phytochrome from rye may show differing behavior and spectral properties. It is nowhere suggested that these different chromophoric species are chemically distinct. Rather the assumption is that they are in different immediate environments on different parts of the protein. Both phototransformation and further reversion studies to be discussed below support this interpretation. We must caution, however, that all of the evidence is kinetic. Only isolation and characterization of more than one kind of chromopeptide by techniques like those of Fry & Mumford (53) or quantitative recovery of isolated chromophore can provide an unequivocal answer.

PHYTOCHROME PHOTOTRANSFORMATION

A number of studies on phytochrome phototransformation kinetics (and dark reversion) address themselves rather directly to the problem of more than one kind of phytochrome or phytochrome chromophore in intact plants or plant extracts. We must therefore begin with a digression, considering briefly some physiological considerations which seemed to require more than one kind of phytochrome. Almost simultaneous with the discovery by Butler et al (28) and De Lint & Spruit (42) of Pfr destruction in etiolated grass seedlings, phytochrome was first isolated from light-grown plants (95). Lane et al (95) speculate that there must be a form of phytochrome that is in some way protected from the destruction reaction, reasoning that otherwise green plants in the light should contain no spectrophotometrically detectable phytochrome at all.

Subsequently two other paradoxes arose. Briggs & Chon (18) showed that phytochrome-mediated alteration in phototropic sensitivity of corn coleoptiles (31) could be saturated by doses of red light two orders of magnitude too small to transform a measureable amount of phytochrome, yet was fully reversible by far-red light. Hillman (74) found that red light-induced growth of etiolated pea stem sections could be reversed by far-red light long after measurable Pfr had disappeared. These and other paradoxes were considered by Hillman (75). More recently the paradox of corn phototropism has been extended to pea geotropism (103). Bellini & Hillman (8) found no correlation between Pfr level and synthesis of phenylalanine ammonia lyase in mustard or radish seedlings. Finally, Schäfer et al (152) noted that the steady state Pfr level in continuous far-red light changes significantly with age in mustard seedlings, while many interpretations of the high energy reaction assume this value to be constant (63). These paradoxes all suggest that a small fraction of phytochrome in plants somehow differs in its light and dark reactions from the bulk measured by usual spectrophotometry. To date none have been resolved. Against this background we can now consider phototransformation kinetics.

Kinetic analysis of phototransformation.—Butler (25) obtained strict log linearity for phototransformation of partially purified phytochrome (source unspecified), consistent with first order kinetics and a single photoreceptor species. Action spectra for phototransformation in vitro (oat, 26) and in vivo (corn, 123) verified log linearity and first order kinetics. Taylor & Bonner (179) also obtained log linearity in transformation of partially purified *Mesotaenium* phytochrome.

Purves & Briggs (130) obtained phototransformation data from oat, pea, corn, and cauliflower tissue, and from partially purified oat phytochrome, that deviated significantly from log linearity. Save for the cauliflower data, all of the curves could be resolved into two first order components; the cauliflower curves could be resolved into at least three. The authors speculated on the relationship between the fast-transforming component and the physically distinct phytochrome required by the various paradoxes, but concluded that it was present in too high a concentration to be the sought-for "active phytochrome." Unfortunately, it was subsequently found that much if not all of the departure from log linearity could be attributed to a lack of linearity in the response of the spectrophotometer, and the earlier paper was retracted (46). The kinetics as measured with a different spectrophotometer showed strict log linearity both in vivo (pea hooks) and in vitro (oat phytochrome). Pratt & Butler (126) were likewise only able to detect a single first order reaction in the ultraviolet phototransformation of phytochrome in either direction.

The matter was not laid to rest, however. Boisard et al (13) studied phototransformation kinetics in etiolated pumpkin hooks (*Cucurbita pepo*) using three different spectrophotometers, all checked for the linearity of their photometer responses. Their results clearly showed two kinetically distinguishable populations of phytochrome with first order rate constants differing by a factor of about 4. Furthermore, if they first allowed some Pfr destruction to proceed before making the phototransformation measurements, they found that the proportions of the two populations changed significantly.

Additional support for such a model was provided by Schäfer et al (151). Phototransformation measurements revealed two populations of phytochrome, distinguishable by rate constant, in etiolated mustard seedlings (*Sinapis alba*). Furthermore, the relative proportions of the two populations changed with age of the seedlings. Even more interesting, however, were the results of studies of phototransformation kinetics after only partial phototransformation. One might argue that if only half of the phytochrome is transformed, say, by red light, most of that transformed would be that with the greater rate constant. Thus back transformation should show a different proportion of fast to slow components than back transformation of a completely transformed preparation (whether rapidly transformed Pr becomes rapidly or slowly transformed Pfr is irrelevant to the experiment). In fact, such experiments always yielded roughly the same proportions of fast and slow components, regardless of what fraction of phytochrome had been initially transformed! Surprisingly, the rate constants did change for both components. The less the initial phototransformation, the greater the rate constants for both. There is at present no adequate model to account for these results, but rapid achievement of an equilibrium mixture of the two forms following phototransformation seems a minimal requirement. The reversion results of Pike & Briggs (120) to be discussed below require the same rapid achievement of equilibrium between slowly and rapidly reverting forms of relatively pure rye phytochrome. These results have done little to resolve the various paradoxes mentioned above, however.

Intermediates in phototransformation.—Phytochrome intermediates appearing during phototransformation have been studied by several different techniques, including flash photolysis, low temperature (with phytochrome in a buffer-glycerol mixture), and illumination conditions leading to pigment cycling. Except where noted below, all of these studies were done with partially purified oat phytochrome, probably predominantly the 60,000 molecular weight product. This fact should be kept in mind in considering the various studies on intermediates, since there is no assurance that large molecular weight phytochrome will behave in precisely the same way, and no studies of

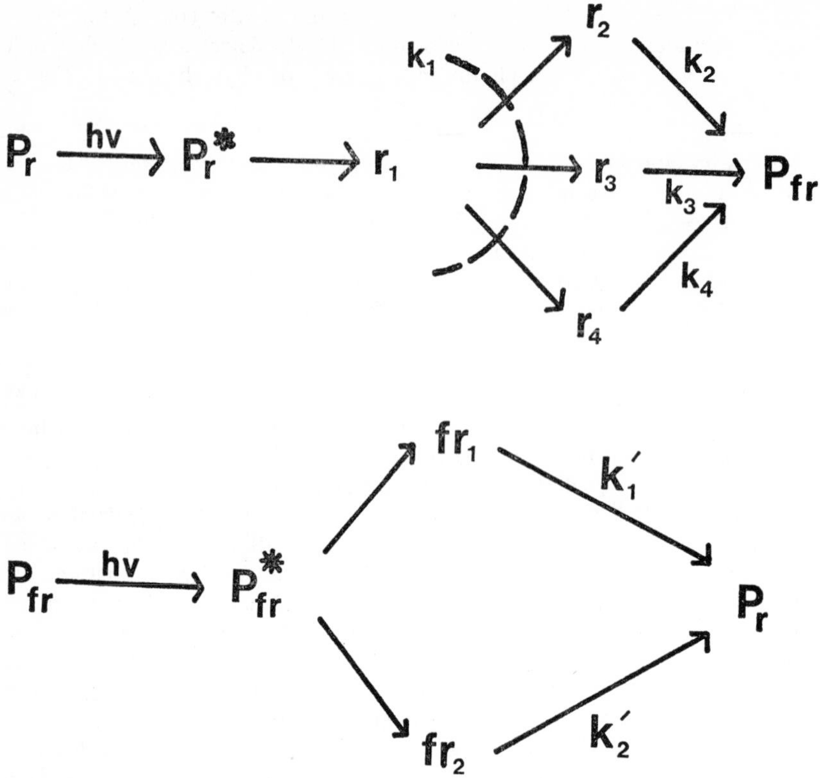

FIGURE 2. Proposed pathways for phytochrome phototransformation, after Linschitz et al (97). Asterisks indicate initial excited states. Other forms are spectrally definable intermediates in the dark reactions leading to the final stable product.

intermediates with authenticated large molecular weight phytochrome have been reported.

The detailed flash photolysis studies of Linschitz et al (97) established several things. First, there are several intermediates both on the Pr to Pfr pathway and the reverse. Second, the two pathways do not involve any common intermediates. Third, both pathways show more than one intermediate decaying in parallel to the final stable product. Linschitz et al established the tentative phototransformation schemes shown in Figure 2. Furthermore, by measuring difference spectra at times ranging from 0.2 msec to several minutes, they were able to determine absorption properties of the intermediates between about 560 nm and 750 nm. On the Pr to Pfr pathway, the initial photoproduct had an absorption maximum at 695 nm. This product then decayed to a lower absorbency form which decayed through three different parallel pathways to Pfr. On the reverse pathway, the initial photoproduct itself decays through two parallel pathways to Pr.

In a further investigation of the Pr to Pfr pathway, Linschitz & Kasche (96) used a second carefully timed flash, far red instead of red, to show that the photobleaching properties of the decay products of all three final intermediates were identical, and identical to those of authentic Pfr, strengthening the concept of parallel pathways.

Where Linschitz and his co-workers had worked with phytochrome in 0.5 M sucrose, Pratt & Butler (125) used 75% glycerol, known to slow down intermediate decay without altering spectral properties at least in the blue and long ultraviolet (20). Varying temperature between $-25°$ and $+5°C$, they confirmed the essential features of the two pathways shown, but obtained evidence for two additional intermediates between r_4 and Pfr, forming a linear sequence of reactions. They also calculated free energies, enthalpies, and entropies of activation for all but one step. All entropies of activation were positive, and none exceeded 25 entropy units. The small values were taken as evidence that a major protein conformational change is probably not involved in phytochrome phototransformation.

Low temperature studies have complicated the picture considerably. Almost simultaneously, Cross et al (40) and Pratt & Butler (124) published results of experiments in which phytochrome in glycerol (up to 75%) was irradiated at low temperatures. Both studies showed that irradiation of Pr below $-150°C$ produced a stable intermediate absorbing at about 695 nm, which decayed to a very low absorbing form as temperature was raised, ultimately forming Pfr. Irradiation of the 695 nm intermediate drove it back to Pr. Cross et al (40) showed in addition a 692 nm intermediate, also photoconvertible to Pr, which had higher absorbency than the 695 nm form, and decayed to it upon warming. In addition, they described a 710 nm intermediate between the 695 nm form and the low-absorbing form (designated Pbl). The consequent modification of their phototransformation scheme now shows only two parallel pathways on the Pr to Pfr pathway.

Pratt & Butler (124) also examined the Pfr to Pr reaction, showing that irradiation of Pfr below $-150°C$ produced an intermediate absorbing maximally at 660 nm. This intermediate decays to a lower extinction form on warming, and eventually Pr is formed. They report that none of these intermediates resemble those on the converse pathway.

Linschitz et al (97) were careful to state that the validity of their scheme depended upon the homogeneity of their phytochrome. Flash photolysis studies and low temperature experiments with large molecular weight native phytochrome are clearly needed.

Briggs & Fork (19) irradiated phytochrome solutions (partially purified oat) with a high intensity mixture of red and far red light, observing accumulation during pigment cycling of intermediates with relatively slow thermal decay constants. Their initial observations were at 543 nm where the intermediates had higher absorbence than the Pr-Pfr mixture left in the dark. Kinetic analysis of the decay patterns of these intermediates were consistent with simultaneous parallel decay to Pfr of two species. With constant light intensity, increasing exposure time yielded a higher proportion of the slower-decaying form. With time constant and intensity decreased, both forms decreased in equal ratio. The Q_{10} values for decay of both intermediates was about 2.0, while that for phototransformation in either direction was close to 1.0.

Though undoubtedly their phytochrome was proteolytically degraded to some extent, Briggs & Fork (20) were able to confirm two parallel pathways for intermediate decay in vivo in oat coleoptile tips. The phototransformation of native phytochrome is thus probably complex, though details remain to be determined. They also obtained difference spectra between the intermediates between 365 nm and 575 nm, comparing them with the dark Pr-Pfr mixture, and showing a strong negative peak at 418 nm, with a positive peak at 380 nm.

Four brief papers by Spruit (162–165) present some results of investigating phytochrome intermediates in vivo in pea plumules and isolated corn phytochrome. Though the data are complicated by possible interference by chlorophyll or protochlorophyll or both (see 76), Spruit obtained clear evidence for an intermediate absorbing at 698 nm, produced by irradiation of the plumules at liquid nitrogen temperature. It was transformed back to Pr by light of appropriate wavelength, and would appear to be the in vivo counterpart of the stable photoreversible intermediates described later by Cross et al (40) and Pratt & Butler (124).

Everett & Briggs (45) directly compared both the intermediate difference spectrum and the transformation difference spectrum for pea phytochrome in vivo and in vitro, between 560 and 380 nm. In both cases, a negative difference peak in vivo was far smaller and shifted to a longer wavelength than that of the in vitro spectra. In this case at least, the spectral properties of phytochrome in vitro do not appear to be precisely the same as those of the pigment in the intact tissue. In vivo measurements of this kind are susceptible to artifact, however, so the results should perhaps be interpreted with caution. Nevertheless, in this connection, oat coleoptile phytochrome has a Pfr absorption maximum in vivo at 735 nm (24), 10 nm to longer wavelengths than any in vitro oat spectra reported.

Chromophore fluorescence and circular dichroism.—In addition to the familiar spectral changes which phytochrome undergoes on phototransformation, two other spectral properties change. These are fluorescence and circular dichroism. Fluorescence has only been examined superficially to date. Hendricks et al (71) observed phytochrome fluorescence both in vivo in an etiolated bean leaf and in vitro in solutions of phytochrome (plant source unspecified, but probably barley or corn). They measured only emission beyond 730 nm and detected an excitation maximum for Pr at about 670 nm. They could detect no fluorescence from Pfr.

In a more detailed study, Correll et al (38) measured the fluorescence of purified rye phytochrome. When fluorescence was measured at 340 nm, the characteristic excitation spectrum of a protein was obtained, with a peak at 290 nm. Excitation of Pr at 370 nm yielded an emission spectrum with a peak at 672 nm, but extending from 600 nm to about 790 nm. When emission was measured at 670 nm, the excitation spectrum showed a maximum at 375, with a small shoulder between 290 nm and 300 nm. The shoulder suggests possible energy transfer between protein and chromophore following protein excitation. Indeed Pratt & Butler (126) measured phytochrome phototransformation by 280 nm light, obtaining a quantum efficiency about one-third that for excitation by 660 nm light for the Pr to Pfr transformation.

Partial transformation of Pr to Pfr merely lowered the relative heights of the excitation and emission peaks of Pr (38). No fluorescence which could be attributed to Pfr was detected. Changes in the protein excitation and emission spectra were not reported. Such spectra might reveal information about protein changes on phototransformation, as might fluorescence studies of intermediates at low temperature.

Both circular dichroism (CD) and optical rotatory dispersion (ORD) have been used to measure phytochrome chromophore and protein asymmetry changes on phototransformation. Kroes (91) found a negative Cotton effect for Pr with a point of inflection near 660 nm in partially purified oat phytochrome. Transformation to Pfr caused the Cotton effect to disappear, suggesting a loss of asymmetry for the 660 nm absorber. Their instrument would not effectively measure ORD above 700 nm. Kroes also measured CD from 300 nm to 800 nm. Pr showed a positive CD band near 380 nm and a negative band near 660 nm. Phototransformation to Pfr reduced the size of the positive and negative bands, and showed a new broad positive band in the far red. Kroes (92, 93) compares the CD spectra of phycocyanin in buffer and in trifluoroacetic acid, and of phycocyanobilin in chloroform with those for phytochrome in buffer, finding positive CD peaks in the visible and negative peaks in the near ultraviolet for the algal chromoprotein and negative peaks in both regions for the isolated chromophore. He suggests that a charge transfer complex between the phytochrome chromophore and the protein could account both for the large absorbance of phytochrome and negative CD effect. Such a complex would be lacking for phycocyanin.

Both Anderson et al (4) and Hopkins & Butler (80) have examined phytochrome CD spectra. The Anderson et al measurements on purified 60,000 molecular weight oat phytochrome only extended up to about 450 nm, but revealed substantial further detail. Pr has a strong positive peak between 380 nm and 400 nm, with a sharp negative peak at about 290 nm. Upon transformation the positive peak disappears and the negative one is decreased substantially. ORD spectra also revealed changes in these spectral regions. The spectra were resolved into a number of Gaussian components, but the authors caution that it is still too early to attempt interpretation of these complex curves.

Hopkins & Butler (80) measured CD spectra from 200 nm to 800 nm, generally confirming Kroes' spectra in the long wavelength visible and far red. Their results differ from those of Anderson et al (4) in the near ultra-violet, however, in that Pfr retains a small broad positive CD band between 350 nm and 400 nm. The difference may simply be that Anderson et al (4) dialyzed their phytochrome against water prior to making the measurements, while the Hopkins & Butler measurements (80) were on phytochrome in phosphate buffer. In our experience, dialysis of phytochrome against distilled water frequently precipitates some of the pigment, suggesting that some denaturation is proceeding. Thus the Hopkins & Butler spectra would seem to be more reliable, though the differences clearly require further investigation. A discussion of CD spectra between 200 nm and 250 nm will be deferred to the next section.

Protein conformational change on phototransformation.—The first suggestion that phytochrome phototransformation might involve protein conformational changes came from the work of Butler, Siegelman & Miller (30). Pfr was far more susceptible to spectral alteration (absorbence loss) than Pr in the presence of protein denaturants such as urea, sulfhydryl reagents such as p-chloromercuribenzoate, and proteolytic enzymes such as trypsin. Since they measured only spectral properties, they could only speak to conformational changes in the chromophore region. Briggs et al (24) subsequently were unable to find any difference between Pr and Pfr in velocity sedimentation in sucrose gradients, electrophoretic mobility, behavior on molecular sieve gels, or binding and elution on brushite. The only difference detected was lability during ammonium sulfate precipitation, with Pfr more sensitive. Both studies used oat phytochrome of fairly low specific activity, and the preparations of Briggs et al (24) clearly contained two molecular weight components of phytochrome.

Better evidence was provided by Roux & Hillman (138), who showed that Pr was more sensitive to glutaraldehyde fixation than was Pfr both in vivo (corn and oat coleoptiles but not pea epicotyl hooks) and in vitro (crude oat phytochrome and partially purified oat and pea phytochrome). The assay was loss of photoreversibility or distortion of the transformation difference spectrum. That the protein itself was directly involved was later demonstrated by Roux (137).[2] Using highly purified 60,000 molecular weight phytochrome, Roux found that out of a total of 27 lysine residues, 13 reacted with glutaraldehyde when Pr (or cycled pigment) was fixed, while only 11 reacted when Pfr was fixed. Similar experiments with trinitrobenzene-sulfonic acid were also consistent with a protein conformational difference between Pr and Pfr, though exact quantitation was not possible. Analysis of tryptic peptides of trinitrobenzenesulfonic acid-fixed phytochrome by mapping also revealed differences between the two forms consistent with the amino acid analyses.

Hopkins & Butler (80) obtained evidence for a small protein conformational change from several different kinds of experiments. First, the detailed

[2] Roux's amino acid studies are now in press in *Biochemistry*, 1972.

ultraviolet difference spectrum for Pr versus Pfr resembled solvent perturbation difference spectra obtained both for proteins and for model mixtures of amino acids (7, 73). (See also Pratt & Butler 126.) Though they failed to find immunological differences between Pr and Pfr using Ouchterlony double diffusion methods (see also Rice & Briggs 134), microcomplement fixation techniques (114) showed that Pfr fixed almost twice as much complement. The circular dichroism spectra also differed, with Pfr showing a minimum at 222 nm, and Pr showing it at 220 nm (with no change in band intensity). Finally, Hopkins (79), in careful differential velocity sedimentation studies, showed Pr to have a larger sedimentation coefficient than Pfr, by about 0.1 S over a temperature range of 5–23°C. Average $s_{20,w}$ values were 5.1 S for Pr and 5.0 S for Pfr.

All of these results are consistent with a relatively small protein conformational change on phototransformation. As mentioned above, Pratt & Butler's (125) thermodynamic studies on decay of phytochrome intermediates were also inconsistent with any large conformational changes, but did not rule out small ones. All published experiments concerning the question of conformational change have involved small molecular weight oat phytochrome (ca. 60,000). Should the same conformational changes characterize native phytochrome of higher molecular weight, they might be more difficult to detect against the higher protein background. On the other hand, the situation might be quite different, and such studies should be made.

On the question of protein conformation in general, one note must be added. Anderson et al (4) saw no changes between Pr and Pfr in their CD spectra below 230 nm. Furthermore, there was only a single negative CD peak just below 220 nm, indicative of a β-helix conformation, rather than the double peak between 205 and 225 nm obtained from an α-helix (62). The Hopkins & Butler (80) CD spectrum was clearly that of an α-helix. Once again, denaturation during dialysis into distilled water could have altered the conformation of the Anderson et al preparation (4). In both studies, relative helicity was calculated. The values presented should be regarded with caution, however, since such calculations may be unreliable when large amounts of random coil are present (62, 149).

Phytochrome Dark Reactions

Restriction of this review to studies aimed at physical and chemical characterization of phytochrome and its possible mechanism of action forces us to be somewhat selective in our treatment of dark reactions. Thus phytochrome dark reversion studies, both in vivo and in vitro, can tell us a fair amount about the actual transformation process. Study of dark destruction of Pfr, on the other hand, has progressed little since Hillman's 1967 review (75), and has not been characterized in vitro at the present writing. It will thus be considered only briefly. Readers are referred to the review of phytochrome biosynthesis and dark reactions by Frankland (52) for the most detailed treatment.

The terminology of phytochrome dark reactions is presently in a highly confused state. Many authors use the word "decay" to refer to dark reversion of Pfr to Pr, while others use it to refer to Pfr destruction, a process involving loss of Pfr absorbance without concomitant appearance of Pr. Still other authors use the term for any unspecified disappearance of Pfr. In the present review, we use the word "reversion" only for thermal conversion of Pfr to Pr, "inverse reversion" for the opposite process (see below), and "destruction" for Pfr absorbance loss without concomitant increase in Pr absorbance. The word "decay" should be abandoned.

Reversion.—Though a host of physiological evidence had appeared supporting the notion that Pfr would revert to Pr in darkness, the earliest in vivo spectrophotometric studies were disappointing. Phytochrome reversion was clearly detected in cauliflower florets (28) but not in etiolated grass seedlings (95, 120, 123). Evidence for reversion in a number of dicotyledons is summarized by Hillman (75) and Frankland (52). Though the grass seedlings seemed to differ from all the dicots examined, Kendrick and co-workers (85–87) reported reversion to be absent in tissues of three dicots of the class Centrospermae. The significance of these taxonomic differences is not clear. What is clear, however, is that in many cases reversion follows the pattern first seen in cauliflower: an initial rapid phase followed by a prolonged slow phase. Even in tissues in which both reversion and destruction occur there may be a lag before the onset of destruction at certain developmental stages (105), as in the case of pea epocotyls, and an initial rapid phase is followed either by a much slower phase or by no reversion at all (the data do not permit an unequivocal discrimination). Thus reversion where it has been found in vivo does not follow simple first order kinetics.

The earliest studies on isolated phytochrome were also puzzling. Hendricks et al (71) reported reversion only by spectrally "altered" barley phytochrome, but not by spectrally normal corn phytochrome. Butler et al (30) later reported an absence of reversion in fresh oat phytochrome. Reversion could eventually be obtained either by aging the preparation or by mildly denaturing it with urea. These phenomena have not at present been further characterized.

The first study of reversion of highly purified oat phytochrome was that of Mumford (111). The pigment, of 60,000 molecular weight, was spectrally normal. Reversion (under nitrogen) followed apparent first order kinetics (though the earliest measurement was at 2 hours after the beginning of the dark period; see below), and showed a half-life of 9 hours at 25°C. The heat of activation for reversion was 24,400 cal.

Taylor (178) investigated reversion in crude and partially purified phytochrome preparations from a number of different sources: etiolated oats, parsnip roots, parsnip leaves, and etiolated peas. He made no effort to exclude oxygen. In all cases, he obtained what the in vivo results suggested—an initial rapid phase followed by a slow phase. In contrast to the observation of Butler et al (30), progressive urea denaturation gradually abolished the rapid phase. In the cases tested, further purification or different extraction techniques failed to alter the reversion behavior of the samples. Taylor states that the wide differences in reversion rates found were a function of the plant material, rather than the conditions of isolation.

As mentioned above, Correll et al (36) also obtained complex reversion kinetics with their purified rye phytochrome preparations (in air), basing their conclusions on multiple chromophoric species of phytochrome on arguments derived from the kinetics. While their detailed scheme for chromophore interactions and transformations remains speculative, departure from log linearity as evidence for more than one chromophoric species is striking.

Pike & Briggs (120) examined the reversion behavior of purified large molecular weight rye phytochrome, and small molecular weight phytochrome obtained from it by mild proteolysis, and also examined reversion by large and small partially purified oat phytochrome. In all four cases, reversion could be resolved into two first order components with rate constants differing by a factor of over 20. The principal finding was that large and small differed only in the relative proportion of the fast component to the slow. The larger molecule had far more of the fast component. Since the amount of fast component in the smaller molecule was low, Mumford (111) might simply have missed the early rapid phase by failing to make measurements less than 2 hours after the beginning of the dark period. On the other hand, Mumford excluded oxygen while Pike & Briggs did not. This difference, however, seems unlikely to account for the different kinetics found since oxidants are without effect on reversion (in air) (119) but reductants of sufficient potential hasten it (see below).

Anderson et al (3) studied the influence of pH on reversion, using phytochrome purified by the Mumford & Jenner procedure (112), and following the reversion techniques of Mumford (111). Between pHs of 6.5 and 8.6, reversion rate was independent of pH (and was reported to be log-linear as before). Below pH 6.5, however, the rate constant increased almost linearly with decreasing pH. They thus propose a new form of phytochrome, PfrH, in equilibrium with $Pfr + H^+$. PfrH then reverts more rapidly to Pr than Pfr. Other evidence for PfrH derives from low temperature studies. Phytochrome in a glycerol-ethylene glycol-buffer mixture (1:1:1) was cooled to $-41.3°C$ as Pfr. Its absorption spectrum showed a peak at 650 nm and enhanced absorption at 380 nm. Upon gradual warming, the normal Pfr spectrum reappeared, and there were three distinct isosbestic points in the transformation, suggesting that only two forms of the pigment were involved. The influence of pH on the equilibrium mixture was determined, and thermodynamic considerations seemed to establish hydrogen ion involvement. The results produced an ingenious temperature compensation model for phytochrome reversion. Though the model is amenable to further testing (and should be tested with large molecular weight phytochrome), its validity in phytochrome physiology must remain open in the absence of information concerning pH at the phytochrome site in the cell. The effects are most dramatic below pH values of 6.0, at which pH large molecular weight phytochrome comes out of solution. Thus Pike & Briggs (120) were unable to confirm the Anderson et al pH effect with purified rye phytochrome. Manabe & Furuya (100) likewise failed to find an effect of pH on reversion of partially purified pea phytochrome.

A significant advance in our understanding of reversion came from the observations of Mumford & Jenner (113) that reducing agents such as dithionite, NADH, and reduced ferredoxin dramatically increased reversion rates of oat phytochrome. Reduced ferredoxin was the most effective, increasing the rate about 800-fold at a concentration of $10^{-6}M$. Both reduced ferredoxin and NADH were acting as true catalysts, since they did not undergo net oxidation during the reaction. Pike & Briggs (120) used dithionite to confirm the phenomenon with purified rye phytochrome. The dithionite was equally effective in promoting reversion both with fast and slow components. It was significantly more effective on proteolytically produced small phytochrome than it was on large. The influence of reducing agents on reversion cannot account for the observation of Klein & Edsall that oxidizing agents mimic far red light and reducing agents red light in expansion of bean leaf discs (89) since the effects are in exactly the wrong direction.

The physical basis for the two component behavior of phytochrome reversion is yet to be determined. Since highly purified phytochrome shows it (120), it cannot be reflection of compartmentalization, or of two distinct kinds of phytochromes. The ratio of the two components remains constant regardless of the percent phototransformation used to establish the initial state (120), and thus two chromophores interacting with each other in some way seems highly unlikely. How these results relate to the phototransformation kinetics mentioned above is as yet unresolved. As a final note of caution, one might note that the only reversion studies on purified phytochrome were done with pigment obtained from plants in which reversion in vivo has never been observed!

"Inverse" reversion.—In 1968, Boisard et al (14) reported an apparent dark transformation of Pr to Pfr in imbibed lettuce (*Lactuca*) seeds. The phenomenon has since been studied in *Lactuca* (12), *Amaranthus* (88), *Cucumis* (166), and other Cucurbitaceae (99). This "inverse" reversion has only been observed in imbibing seeds, not in seedlings or extracted phytochrome. In *Cucurbita* and *Cucumis,* it cannot be detected after 28 hours of imbibition (99). Since dark reversion in the normal Pfr to Pr direction is thermodynamically favored, the various authors are cautious in their interpretation of the phenomenon, though it could account for several physiological aspects of seed germination (52).

One possibility is that inverse reversion could be a hydration phenomenon. Tobin & Briggs (181) found that a low level of phytochrome in dormant *Pinus palustris* embryos could be roughly tripled in less than 2 minutes merely by adding water to the chopped sample. In seeds hydrated normally, added water increased measureable phytochrome in the isolated and chopped embryos until more than 30 hours after the start of hydration. Absorption spectra indicated a large change in the scattering properties of the tissue with increase in hydration. If the Pfr in the dormant embryo was a low absorbing form, and the Pr had more normal absorbancy, then equal hydration of the two forms could appear as inverse reversion in the spectrophotometer, given the magnitude of the scattering changes versus the photoreversibility changes. Careful spectral measurements need to be made to confirm or eliminate this possibility. Unfortunately, the spectral changes in the seeds studied are already near the limit of spectrophotometer sensitivity, so such measurements may have to await still better instrumentation.

Dark destruction.—Shropshire (154) and Frankland (52) discuss phytochrome dark destruction in detail, and Hillman (75) has covered the earlier literature. It is clear that neither the mechanism of destruction in vivo nor the physiological role of the reaction are currently understood. Furuya & Hillman (57) described a low molecular weight substance in extracts of pea tissue that rapidly destroyed Pfr but not Pr (spectral assay). The substance could not be obtained from *Avena,* however, making a general role in dark destruction unlikely. Detailed characterization of this "Pfr killer" has not been published. Since dark destruction studies have shed little light either on physical or chemical properties of phytochrome or its mechanism of action, we make only brief mention of them here.

Boisard et al (13) found that in vivo dark destruction of phytochrome in pumpkin hooks (in the absence of any detectable reversion) was not first order, showing evidence for fast and slow components. Though the data were not precise enough to be resolved into clear log-linear components, they do support the hypothesis of more than one form of Pr, more clearly indicated by the phototransformation kinetics discussed earlier. Studies on *Sinapis* seedlings (101), on the other hand, showed a complex pattern. The evidence suggested formation of a transient low absorbance form of Pfr after saturating red light. The transient form apparently could revert rapidly both to Pfr and Pr. Subsequently, only dark destruction could be observed. The interesting *Sinapis* system is clearly worthy of further investigation.

PHYTOCHROME LOCALIZATION

There have been several studies on phytochrome distribution in etiolated seedlings and changes during seedling growth (e.g. 23, 35, 37, 56, 104). In general, phytochrome levels are highest in meristematic or recently meristematic tissue, both root and shoot, and substantial synthesis occurs during development. We shall not discuss this work in detail, since it relates only indirectly to the focus of this review. In contrast to tissue localization studies, however, knowledge of subcellular localization of phytochrome is essential to our understanding of its mechanism of action. Some progress has been made, though the answers are not yet clear.

Physiological studies.—Physiological studies leading to knowledge of phytochrome localization have to date been limited to two systems: chloroplast movement in certain green algae and polarotropism of germ tubes of certain ferns. Over 10 years ago, Bock & Haupt (11) showed that the chloroplast itself was not the site of the photoreceptor for phytochrome-mediated chloroplast movement. A great deal of subsequent work, much of it using microbeams or polarized light or a combination (66), has established several things. First, the phytochrome mediating the chloroplast movement is located either in the outermost cytoplasm or on the cytoplasmic membrane itself and not free in solution. Second, careful investigations of strong dichroic effects have established that Pr must have its axis of maximum absorption in the plane of the cell surface, while Pfr has its axis perpendicular to the cell surface. Furthermore, Pr chromophores must be located with their axes fixed in parallel, along lines passing helically around the long axis of the cells. This work has been reviewed in detail elsewhere (65, 67, 102).

Etzold (43) came to similar conclusions from his studies on growth of the filamentous germ tubes of spores of *Dryopteris filix-mas*. These filaments orient perpendicular to the electric vector of plane-polarized red light, an effect which could be reduced by unpolarized far-red light, but not by far-red light polarized in the same plane as the initial red. Steiner investigated the system further (170, 171) obtaining both dose-response curves and an action spectrum. The response is complicated by strong polarotropism in blue light. That the blue light response involves a different photoreceptor is strongly suggested by comparable studies (168, 169) on a liverwort, *Sphaerocarpos donnellii*. The blue and ultraviolet action spectrum is the same as that for *Dryopteris,* but there is no evidence for phytochrome involvement. [A later study (172) suggested that red light could alter the response to blue.] Etzold showed that growth reorientation occurred at the extreme tip of the filament (43), and Falk & Steiner (48) confirmed this observation by electron microscopy, but could detect no obvious structural changes such as microtubule reorientation in a very preliminary study. These experiments all suggest localization of the effective phytochrome at the extreme tip of the growing filament, and the dichroic behavior suggests cortical cytoplasm or the plasmalemma as the most probable site. The arguments are the same as for the *Mougeotia* chloroplast system.

Extraction of bound phytochrome.—Rubenstein et al (139) have reported a small fraction of phytochrome in extracts of etiolated oat shoots which is not soluble in aqueous buffer. It sedimented between 1,500 and 40,000 x g, and contained about 4% of the total extractable phytochrome. Though repeated buffer washes failed to solubilize it, it was largely released by treatment with a neutral detergent, Triton X-100. The pelletable phytochrome was more labile during storage at 0°C, irradiation with white light, and detergent treatment than the supernatant fraction. The authors made no attempt to characterize the pelleted phytochrome, but suggested that it may be associated with membranes. Attempts to obtain binding of soluble phytochrome to pellets obtained from plants low in the pigment failed, making nonspecific binding seem unlikely. Clearly further studies on purification of the pellet component containing phytochrome, and on the spectral properties of the bound pigment, are needed.

Microspectrophotometric localization.—Galston (58) has attempted to localize phytochrome in etiolated oat and pea seedlings by using highly sensitive microspectrophotometric techniques. He looked both at living cells in thin sections and cells from tissue which had been frozen, sectioned, and dried. Though baseline drift was a serious hindrance, red and far-red light did appear to bring about some reversible absorbance changes. These changes were only obtained when the microspectrophotometer was focused on nuclei and not on cytoplasm or other organelles. The results presented are difficult to interpret. In living cells, a slight shoulder near 650 nm disappeared on irradiation with red light, and reappeared following far-red treatment. There were, however, no clear spectral changes in the far red. In frozen and dried material, far-red light caused an apparent bleaching near 600 nm, while red light reversed the bleaching. Again, there were no changes in far-red absorbance.

In evaluating these results, one must keep two questions in mind. First, is phytochrome which has been freeze-dried still photoreversible, but with an altered absorption spectrum? The results of Tobin & Briggs (181) suggest that dehydrated phytochrome may not be phototransformable. Second, is the microspectrophotometer used really sensitive enough to detect the minute amount of phytochrome present in a single cell, even if all of it is localized on or in the nucleus? Since the calibration of the microspectrophotometer is arbitrary, one cannot be certain how much absorbance change is needed to account for the percent transmission changes recorded. In view of these questions, one should regard the work with caution. Even if phytochrome is found in the nucleus, its presence elsewhere in the cell is not precluded, as Galston correctly points out.

Immunochemical localization.—Since phytochrome is an effective antigen, and since the antibody elicited is highly specific, as mentioned above, immunochemical techniques for phytochrome localization hold substantial promise. Pratt & Coleman (127) have had some success using the peroxidase-antiperoxidase technique of Sternberger et al (173). Fixed and paraffin-embedded sections of oat shoot were treated sequentially with rabbit antiphytochrome serum, sheep antiserum against rabbit immunoglobulin, and finally a rabbit antiperoxidase-peroxidase complex, which becomes bound to the attached sheep antirabbit immunoglobulin. Actual visualization was achieved by assaying for peroxidase with 3,3'-diaminobenzidine plus dilute H_2O_2. The orange insoluble oxidation product was clearly visible under light microscopy.

At the tissue level, phytochrome was highest just behind the coleoptile tip, decreasing down the coleoptile, and increasing again at the node, extending the pattern previously determined spectrophotometrically (23). A striking observation was the clear dichotomy between cells which did contain phytochrome and those which did not. For example, phytochrome was abundant in parenchyma cells just behind the coleoptile apex, but lacking completely in the apical cells. In lower regions of the coleoptile, it was found only in epidermal cells. In the mesocotyl, it was associated with the procambium only.

At the subcellular level, phytochrome appeared in association both with nuclei and with plastids, in addition to being abundant in the cytoplasm. However, the cytoplasm in the photographs appears somewhat clumped and granular, suggesting that fixation was not optimal. Hence one should regard the subcellular localization results as preliminary. Nevertheless, in view of the other localization studies mentioned above, it would not be surprising to find phytochrome associated with several different membrane systems and free in the cytoplasm as well. Since this technique can be adapted for electron microscopy, one should anticipate substantial progress in this area in the near future.

MECHANISM OF ACTION

In considering the possible mode of action of phytochrome, we concentrate our attention on the most rapid consequences of phytochrome phototransformation. Thus a large literature on phytochrome and protein and nucleic acid synthesis, changes in enzyme level or activity, and morphogenesis is ignored. We concede that complete understanding of the mechanism of action of phytochrome requires consideration of this literature, but we feel that characterization of the earliest events must precede any real understanding of later ones.

Phytochrome and nyctinasty.—In 1966, Fondeville et al (50) clearly demonstrated that the closing of leaflets of *Mimosa pudica*, the sensitive plant, was under phytochrome control. Transfer of plants from high intensity fluorescent light to darkness initiated leaflet closure within 5 minutes, with complete closure within 30 minutes. Far-red treatment immediately before darkness prevented closure, while red treatment after far red allowed it. Photoreversibility was demonstrated through several cycles. A preliminary action spectrum for prevention of closure showed a maximum between 720 and 750 nm, with about 8-fold less action in the blue. Reversal of the far-red response was most effective with 660 nm light. Energies used were in the range for other phytochrome responses. The rapidity of closure in response to Pfr formation was comparable to that described by Haupt (64) for *Mougeotia* chloroplast movement.

Subsequently, Fondeville et al (51) showed that the action maxima for opening of closed leaflets were at 480 and 710 nm. These wavelengths plus the high irradiances needed suggested that opening was under control of the high energy reaction (HER). Thus like the *Mougeotia* chloroplast system and the polarotropism of fern and liverwort germ tubes, the system is complex. However, the regulation of closure of leaflets by phytochrome is clear.

Phytochrome regulation of nyctinastic leaf closure in *Albizzia julibrissin* and several other legumes was then described by Hillman & Koukkari (77). The results with *Mimosa* were completely confirmed. These authors also investigated the role of leaf maturity, pinnule position within a leaf, and system sensitivity as a function of time in the light period. In *Albizzia*, phytochrome control is clearly lost as the light period progresses, with leaflets closing regardless of light treatment after 7 hours of illumination. Photostationary-state studies indicated that about 50% Pfr was sufficient to permit closure [Fondeville et al (50) had calculated about 46%]. Jaffe & Galston (84) likewise confirmed the phenomenon in *Albizzia*, showing in addition an efflux of electrolyte from cut pinna bases upon closure and a lack of sensitivity to actinomycin D. Koukkari & Hillman (90) did various light-shielding experiments which suggested that in *Albizzia*, the pulvinules themselves, at the bases of the closing leaflets, were the site of photoreception, and that the effect was not translocated from an irradiated pulvinule to a nonirradiated one. Further studies on *Samanea saman* (175) confirmed that photocontrol of closure by phytochrome is lost as the light period progressed. A variety of inhibitors of respiration, protein and nucleic acid synthesis, and various growth substances and salts failed to influence photocontrol of closing.

Satter et al (147) have studied the anatomy and fine structure of the so-called motor cells of the *Albizzia* pulvinule, describing parallel fibrils, numerous spherosomes, and a curious multivacoulate condition in contrast to other nearby parenchyma cells. The structural study served as background for later work. Detailed physiological studies (146) revealed several interesting points. First, dark closure was inhibited by NaN_2, dinitrophenol, and anaerobiosis, suggesting involvement of respiration in the closing process. It was also inhibited by salts of monovalent cations, particularly K^+. Electron microprobe analysis showed the motor cells of pulvinules from open leaflets to be very high in K^+. The ion was lost from ventral cells and entered dorsal cells during closure, though quantitative transfer of ion from one site to the other was not indicated. The estimates were based on fixed values for Ca and P, on the assumption that these elements were relatively immobile. Even if the assumption is false to a degree, ionic movements and involvement of K^+ are clearly demonstrated in the motor system. These and other pertinent results are discussed in detail by Galston & Satter (59). These authors (144, 145) have recently extended their microprobe studies to show involvement of potassium flux in light-induced opening, as well as in circadian opening and closing under constant conditions.

Starting with Fondeville et al (50), workers on nyctinasty have used the rapidity of the response and the obvious role turgor changes play in it as evidence against the gene activation hypothesis (107). Hendricks & Borthwick (70) summarized the evidence for membrane changes as being primary, with gene activation relegated to a later role. Unfortunately, most workers have referred only to membrane permeability. The available evidence does not rule out changes in active transport systems or other membrane properties.

Root tip adhesion and electric potentials.—Further evidence of membrane changes as early consequences of phytochrome phototransformation comes from studies on the adhesion of root tips to a negatively charged glass surface. Tanada described the phenomenon both for excised barley root tips (176) and mung bean root tips (177). Root tips were placed in a beaker previously treated with phosphate, and were swirled gently in a solution of carefully defined composition. For barley, it contained ATP, IAA, L-ascorbic acid, $MnCl_2$, KCl, and HCl. Ca^{++} was required in addition by mung bean tips. If the last irradiation was with red light, the root apices adhered to the glass surface; if it was far-red light, they remained detached. The responses could be measured within 30 seconds and could be repeated through up to six red, far-red cycles. If any component of the medium was omitted, the response failed. Tanada suggested (177) the possibility that membrane permeability changes account for the phenomenon, as had workers on nyctinasty.

The phenomenon described by Tanada suggested that the extreme root apex might become electropositive to the base as a consequence of red light treatment. Jaffe (82) therefore attempted a direct measurement of the potential difference between apex and base, measuring both potential change and percent adhesion in parallel for mung bean root tips. During red irradiation, the apex did indeed become more electropositive, with a time course quite similar to the adhesion time course. With far-red light, both effects were reversed. The potential changes were extremely small (just over a millivolt) and were reversible through a R-FR-R cycle, though the second electrical response to red was much reduced. The electrical changes occurred rapidly, with a lag of less than 30 seconds.

Yunghans & Jaffe (189) studied mung bean root tip adhesion in considerable detail. The strongest response was obtained with tips of secondary roots with 6–8 mm tips showing the best response to red light and 3 mm tips the best response to far red. Escape from far-red reversibility occurred within 2 minutes, but spontaneous release of red-treated tips began after about 10 minutes. A variety of anions were found to substitute for phosphate in charging the glass surface. Substitution of GTP, ITP, UTP, or CTP for ATP substantially reduced the response, though CTP was more effective than the other nucleotides.

Are the root tip responses just described a unique property of the excised organs in the required complex medium, or are they representative of more general phenomena in other systems? Newman & Briggs (116) used the flowing drop electrode technique devised by Newman (115) to study electrical potential differences between the upper part of the etiolated oat coleoptile and the base of the plant. Red light produced an increase in positive potential of the coleoptile with a lag of less than 15 seconds, the increase being between 5 and 10 mV. In the absence of any further treatment, the potential declined to the original level within about 12 minutes. Far-red light then induced a clear decrease with a lag of less than 15 seconds, with evidence for recovery. The electric changes could be measured through at least four cycles, though the shapes of the response curves became altered, and the responses damped with successive exposures. Initial far-red light was ineffective, and a second red exposure following saturating red light was likewise ineffective. A far-red response clearly required prior red treatment, and a second red response could only be obtained following an intervening far-red treatment. Largest responses were found 3–6 mm below the coleoptile tip, and a maximal response could be obtained only if the entire coleoptile was irradiated. Clearly the potential changes measured reflected an integration of the responses of many cells, not just those immediately adjacent to the electrode contact drop. Curiously, the shape of the response curve to red light immediately followed by far red approximated the algebraic sum of the curves obtained from separate red and far-red treatments.

Both the root tip adhesion and electrical changes are consistent with cation efflux as a consequence of phytochrome phototransformation, as was so clearly demonstrated for the *Albizzia* system (146). The recent demonstration of potassium movement in connection with stomatal opening and closing (81, 148) strengthens this notion. However, the complexity of the oat coleoptile response precludes any simple model for the present. Intracellular measurements and studies of movements of specific ions are obviously needed.

Phytochrome and acetylcholine.—The presence of acetylcholine (ACh) in plants has been known for a number of years (see 83 for references). Using its high specifity in the clam heart bioassay (49, 98), Jaffe (83) found what appeared to be this neurohumor in mung bean root tips. It co-chromatographed with authentic ACh in two solvent systems, eserine sensitized the clam heart to the extracts, and atropine sulfate and benzoquinonium chloride inhibited the clam heart response. Since there are no other known chemicals besides ACh which show this combination of chromatographic and pharmacological properties, tentative identification as ACh seems reasonable, though chemical characterization is desirable. Red light both caused an efflux of the substance from root tips and increased the endogenous level, while subsequent far-red light reduced the endogenous level to that of dark controls. Authentic ACh could substitute for red light in inhibiting secondary root formation, inducing hydrogen ion efflux, and in causing adherence to a negatively charged glass surface. Both acetylcholine esterase and atropine inhibited red-induced adhesion, and eserine inhibited far red-induced release. Jaffe proposed that rather than being a botanical curiosity, acetylcholine and by implication acetylcholine esterase play central roles in certain phytochrome-mediated phenomena in plants.

We would be less than honest if we did not mention that Jaffe's results have met with considerable (but currently unpublished) scepticism. The burden of proof, however, is on the sceptics: clam hearts are available, and the experiments are relatively simple.

Phytochrome and NAD kinase.—In a brief note in 1969, Tezuka & Yamamoto (180) published some observations on a phytochrome system far removed from membrane phenomena or gene activation. The results were consonant instead with the earlier carbon metabolism hypothesis (68). Red light significantly raised the level of NADP in *Pharbitis* cotyledons, far red partially reversed the effect, and subsequent red partially repromoted it. Pea phytochrome purified according to Siegelman & Firer (158) through the brushite step showed substantial NAD kinase activity, as did crude extract from *Pharbitis*. In both cases, treatment of the solutions with red light significantly raised the NAD kinase activity, and far red reversed the effect. The pea phytochrome effect persisted through a Sephadex G-200 step, with red light more than doubling the enzyme activity. Michaelis constants for the kinase were 1.84×10^{-3} for the dark control, and 0.90×10^{-3} for the red-treated solution.

The authors have extended the observations (188) recently, showing that phytochrome and NAD kinase do not co-chromatograph quite together on brushite columns, though they do on Sephadex G-200. This and other evidence suggested that the phenomenon represented interaction between two discrete proteins, rather than assigning NAD kinase function to phytochrome, though the proteolytic degradation mentioned earlier might confuse the picture. The authors show in addition a number of metabolic alterations as a consequence of supplying exogenous NADP, and propose that phytochrome regulation of NADP level may be central to its mechanism of action.

Like Jaffe's work, the NAD kinase story has met with some scepticism. Since trivial explanations of the results are not obvious, however, further experimentation is clearly needed.

Mechanism of action: conclusions.—It is clear that an early consequence of phytochrome phototransformation may be an alteration of membrane properties, which may quickly lead to extensive biochemical changes. It is premature, however, to conclude that phytochrome must therefore be membrane-bound. There is still a 15-second lag period to be accounted for, and a great deal could happen in the cytoplasm in that time which could change membrane properties.

Borthwick et al (15) summarize all known reactions of Pr and Pfr, including formation of PfrH, and hypothesize reaction of Pfr with some component, X, leading to eventual display. Studies on phytochrome mechanism of action, however, have not yet identified X. Furthermore, as Mohr et al (110) point out, there may be more than one kind of X in the cell. A membrane site and NAD kinase both might be possibilities. It seems reasonable, however, that the nature of the chemical interaction between Pfr and X would be the same, even if the Xs were different.

Epilogue

Siegelman & Butler predicted in 1965 (156) a marked increase in research on phytochrome in the years following their review. Their prediction has been amply justified. Great advances have been made in our understanding of phytochrome protein chemistry, structure, photochemistry, dark reactions, and chromophore chemistry. Though conclusive answers are not yet available, progress in cell localization and mechanism of action has also been substantial. The increasing use of sophisticated physical and chemical techniques plus the large increase in number of physiologists and biochemists studying phytochrome suggest that the years intervening before the next review in this series may be equally productive.

Acknowledgments

The authors wish to thank Dr. David W. Hopkins and Gary Gardner for many useful discussions and for their careful reading of the manuscript.

LITERATURE CITED

1. Ackers, G. K. 1967. *J. Biol. Chem.* 242:3237–38
2. Ackers, G. K. 1970. *Advan. Protein Chem.* 24:343–446
3. Anderson, G. R., Jenner, E. L., Mumford, F. E. 1969. *Biochemistry* 8:1182–87
4. Anderson, G. R., Jenner, E. L., Mumford, F. E. 1970. *Biochim. Biophys. Acta* 221:69–73
5. Andrews, P. 1965. *Biochem. J.* 96: 595–606
6. Anfinsen, C. B., Haber, E. 1961. *J. Biol. Chem.* 236:1361–63
7. Bailey, J. E., Beaven, G. H. Chignell, D. A., Gratzer, W. B. 1968. *Eur. J. Biochem.* 7:5–14
8. Bellini, E., Hillman, W. S. 1971. *Plant Physiol.* 47:668–71
9. Bennett, A., Bogorad, L. 1971. *Biochemistry* 10:3625–34
10. Berns, D. S. 1967. *Plant Physiol.* 42:1569–86
11. Bock, W., Haupt, W. 1961. *Planta* 57:518–30
12. Boisard, J. 1969. *Physiol. Veg.* 7: 119–33
13. Boisard, J., Marmé, D., Schäfer, E. 1971. *Planta* 99:302–10
14. Boisard, J., Spruit, C. J. P., Rollin, P. 1968. *Meded. Landbouwhogesch. Wageningen* 68:1–5
15. Borthwick, H. A., Hendricks, S. B., Schneider, M. J., Taylorson, R. B., Toole, V. K. 1969. *Proc. Nat. Acad. Sci. USA* 64:479–86
16. Borthwick, H. A., Parker, M. W., Hendricks, S. B. 1950. *Am. Soc. Natur.* 84:117–34
17. Bretscher, M. S. 1971. *Nature New Biol.* 231:229–32
18. Briggs, W. R., Chon, H. P. 1966. *Plant Physiol.* 41:1159–66
19. Briggs, W. R., Fork, D. C. 1969. *Plant Physiol.* 44:1081–88
20. Ibid, 1089–94
21. Briggs, W. R., Gardner, G., Hopkins, D. W. See Ref. 106. In press
22. Briggs, W. R., Rice, H. V., Gardner, G., Pike, C. S. 1972. In *Recent Advances in Phytochemistry. Structural and Functional Aspects of Phytochemistry,* ed. V. C. Runeckles, 35–50. New York: Academic
23. Briggs, W. R., Siegelman, H. W. 1965. *Plant Physiol.* 40:934–41
24. Briggs, W. R., Zollinger, W. D., Platz, B. B. 1968. *Plant Physiol.* 43:1239–43
25. Butler, W. L. 1961. In *Progress in Photobiology,* ed. B. C. Christensen, B. Buchmann, 569–71. Amsterdam: Elsevier
26. Butler, W. L., Hendricks, S. B., Siegelman, H. W. 1964. *Photochem. Photobiol.* 3:521–28
27. Butler, W. L., Hendricks, S. B., Siegelman, H. W. 1965. In *Chemistry and Biochemistry of Plant Pigments,* ed. T. W. Goodwin, 197–210. New York: Academic
28. Butler, W. L., Lane, H. C., Siegelman, H. W. 1963. *Plant Physiol.* 38:514–19
29. Butler, W. L., Norris, K. H., Siegelman, H. W., Hendricks, S. B. 1959. *Proc. Nat. Acad. Sci. USA* 45:1703–8
30. Butler, W. L., Siegelman, H. W., Miller, C. O. 1964. *Biochemistry* 3:851–57
31. Chon, H. P., Briggs, W. R. 1966. *Plant Physiol.* 41:1715–24
32. Cope, B. T., Smith, U., Crespi, H. L., Katz, J. J. 1967. *Biochim. Biophys. Acta* 133:446–53
33. Correll, D. L., Edwards, J. L. 1970. *Plant Physiol.* 45:81–85
34. Correll, D. L., Edwards, J. L., Klein, W. H., Shropshire, W. Jr. 1968. *Biochim. Biophys. Acta* 168:36–45
35. Correll, D. L., Edwards, J. L., Medina, V. J. 1968. *Planta* 79: 284–91
36. Correll, D. L., Edwards, J. L., Shropshire, W. Jr. 1968. *Photochem. Photobiol.* 8:465–75
37. Correll, D. L., Shropshire, W. Jr. 1968. *Planta* 79:275–83
38. Correll, D. L., Steers, E. Jr., Towe, K. M., Shropshire, W. Jr. 1968. *Biochim. Biophys. Acta* 168:46–57
39. Crespi, H. L., Smith, U., Katz, J. J. 1968. *Biochemistry* 7:2232–42
40. Cross, D. R., Linschitz, H., Kasche, V., Tenenbaum, J. 1968.

Proc. Nat. Acad. Sci. USA 61: 1095–1101

41. Dalton, H., Morris, J. A., Ward, M. A., Mortenson, L. E. 1971. *Biochemistry* 10:2066–72

42. De Lint, P. J. A. L., Spruit, C. J. P. 1963. *Meded. Landbouwhogesch. Wageningen* 63:1–7

43. Etzold, H. 1965. *Planta* 64:254–80

44. Evans, L. T., Ed. 1969. *The Induction of Flowering.* Cornell Univ. Press, Ithaca, N. Y. 488 pp.

45. Everett, M. S., Briggs, W. R. 1970. *Plant Physiol.* 45:679–83

46. Everett, M. S., Briggs, W. R., Purves, W. K. 1970. *Plant Physiol.* 45:805–6

47. Fahrney, D. E., Gold, A. M. 1963. *J. Am. Chem. Soc.* 85:997–1000

48. Falk, H., Steiner, A. M. 1968. *Naturwissenschaften* 10:500

49. Florey, E. 1967. *Comp. Biochem. Physiol.* 20:365–77

50. Fondeville, J. C., Borthwick, H. A., Hendricks, S. B. 1966. *Planta* 69:357–64

51. Fondeville, J. C., Schneider M. J., Borthwick, H. A., Hendricks, S. B. 1967. *Planta* 75:228–38

52. Frankland, B. See Ref. 106

53. Fry, K. T., Mumford, F. E. 1971. *Biochem. Biophys. Res. Commun.* 45:1466–73

54. Fujiwara, T. 1955. *J. Biochem.* 42:411–17

55. Furuya, M. 1968. In *Progress in Phytochemistry,* ed. L. Reinhold, Y. Liwschitz, 347–405. London: Wiley

56. Furuya, M., Hillman, W. S. 1964. *Planta* 63:31–42

57. Furuya, M., Hillman, W. S. 1966. *Plant Physiol.* 41:1242–44

58. Galston, A. W. 1968. *Proc. Nat. Acad. Sci. USA* 61:454–60

59. Galston, A. W., Satter, R. L. See Ref. 22

60. Gardner, G., Pike, C. S., Rice, H. V., Briggs, W. R. 1971. *Plant Physiol.* 48:686–93

61. Giles, K. L., von Maltzahn, K. E. 1968. *Can. J. Bot.* 46:305–6

62. Greenfield, N., Fasman, G. D. 1969. *Biochemistry* 8:4108–16

63. Hartmann, K. M. 1966. *Photochem. Photobiol.* 5:349–66

64. Haupt, W. 1959. *Planta* 53:484–501

65. Haupt, W. 1970. *Physiol. Veg.* 8:551–63

66. Haupt, W. 1970. *Z. Pflanzenphysiol.* 62:287–98

67. Haupt, W., Schönbohm, E. 1970. In *Photobiology of Microorganisms,* ed. P. Halldal, 283–307. London, New York: Wiley-Interscience

68. Hendricks, S. B. 1964. In *Photophysiology,* ed. A. C. Giese, 1: 305–31. New York: Academic

69. Hendricks, S. B., Borthwick, H. A. See Ref. 27, 405–36

70. Hendricks, S. B., Borthwick, H. A. 1967. *Proc. Nat. Acad. Sci. USA* 58:2125–30

71. Hendricks, S. B., Butler, W. L., Siegelman, H. W. 1962. *J. Phys. Chem.* 66:2550–55

72. Hendricks, S. B., Siegelman, H. W. 1967. *Compr. Biochem.* 27:211–35

73. Herskovits, T. T., Laskowski, M. Jr. 1962. *J. Biol. Chem.* 237: 2481–92

74. Hillman, W. S. 1966. *Plant Physiol.* 41:907–8

75. Hillman, W. S. 1967. *Ann. Rev. Plant Physiol.* 18:301–24

76. Hillman, W. S. 1968. *Biochim. Biophys. Acta* 162:464–66

77. Hillman, W. S., Koukkari, W. L. 1967. *Plant Physiol.* 42:1413–18

78. Hirs, C. W. 1967. *Methods Enzymol.* 11:59–62

79. Hopkins, D. W. 1971. *Protein conformational changes of phytochrome.* PhD thesis. Univ. Calif., San Diego

80. Hopkins, D. W., Butler, W. L. 1970. *Plant Physiol.* 45:567–70

81. Humble, G. D., Hsiao, T. C. 1970. *Plant Physiol.* 46:483–87

82. Jaffe, M. J. 1968. *Science* 162: 1016–17

83. Jaffe, M. J. 1970. *Plant Physiol.* 46:768–77

84. Jaffe, M. J., Galston, A. W. 1967. *Planta* 77:135–41

85. Kendrick, R. E., Frankland, B. 1968. *Planta* 82:317–20

86. Ibid 1969. 86:21–32

87. Kendrick, R. E., Hillman, W. S. 1971. *Am. J. Bot.* 58:424–28

88. Kendrick, R. E., Spruit, C. J. P.,

Frankland, B. 1969. *Planta* 88: 293–302
89. Klein, R. M., Edsall, P. C. 1966. *Plant Physiol.* 41:949–52
90. Koukkari, W. L., Hillman, W. S. 1968. *Plant Physiol.* 43:698–704
91. Kroes, H. H. 1968. *Biochem. Biophys. Res. Commun.* 31:877–83
92. Kroes, H. H. 1970. *Physiol. Veg.* 8:533–49
93. Kroes, H. H. 1970. *Meded. Landbouwhogesch. Wageningen* 70–18:1–112
94. Kroes, H. H., Van Rooijen, A., Geers, J. M., Greuell, E. H. M. 1969. *Biochim. Biophys. Acta* 175:409–13
95. Lane, H. C., Siegelman, H. W., Butler, W. L., Firer, E. M. 1963. *Plant Physiol.* 38:414–16
96. Linschitz, H., Kasche, V. 1967. *Proc. Nat. Acad. Sci. USA* 58:1059–64
97. Linschitz, H., Kasche, V., Butler, W. L., Siegelman, H. W. 1966. *J. Biol. Chem.* 241:3395–3403
98. MacIntosh, F. C., Perry, W. L. M. 1950. *Methods Med. Res.* 3:78–94
99. Malcoste, R., Boisard, J., Spruit, C. J. P., Rollin, P. 1970. *Meded. Landbouwhogesch. Wageningen* 70–16:1–16
100. Manabe, K., Furuya, M. 1971. *Plant Cell Physiol.* 12:95–101
101. Marmé, D., Marchal, B., Schäfer, E. 1971. *Planta* 100:331–36
102. Mayer, F. 1971. In *Structure and Function of Chloroplasts*, ed. M. Gibbs, 35–49. Berlin: Springer-Verlag
103. McArthur, J. A. 1968. *Phytochrome relationships in the dark-grown pea seedling.* PhD thesis. Stanford Univ.
104. McArthur, J. A., Briggs, W. R. 1969. *Planta* 91:146–54
105. McArthur, J. A., Briggs, W. R. 1971. *Plant Physiol.* 48:46–49
106. Mitrakos, K., Shropshire, W. Jr., Eds. 1972. *Phytochrome.* Proc. NATO Advan. Study Inst. Phytochrome, Athens-Eretria, Greece. New York: Academic
107. Mohr, H. 1966. *Photochem. Photobiol.* 5:469–83
108. Mohr, H. 1969. In *An Introduction to Photobiology*, ed. C. P. Swanson, 99–141. Englewood Cliffs, N. J.: Prentice-Hall
109. Mohr, H., 1969. In *Physiology of Plant Growth and Development*, ed. M. B. Wilkins, 509–56. London: McGraw-Hill
110. Mohr, H., Bienger, I., Lange, H. 1971. *Nature* 230:56–58
111. Mumford, F. E. 1966. *Biochemistry* 5:522–24
112. Mumford, F. E., Jenner, E. L. 1966. *Biochemistry* 5:3657–62
113. Ibid 1971. 10:98–101
114. Murphy, T. M., Mills, S. E. 1968. *Arch. Biochem. Biophys.* 127:7–16
115. Newman, I. A. 1963. *Aust. J. Biol. Sci.* 16:629–46
116. Newman, I. A., Briggs, W. R. 1972. *Plant Physiol.* In press
117. Ouchterlony, Ö. 1968. *Handbook of Immunodiffusion and Immunoelectrophoresis*, 139–53. Ann Arbor Sci. Publ.
118. Parker, M. W., Hendricks, S. B., Borthwick, H. A. 1950. *Bot. Gaz.* 111:242–52
119. Pike, C. S. 1971. *Studies on a neutral oat protease and on phytochrome dark reactions.* PhD thesis. Harvard Univ.
120. Pike, C. S., Briggs, W. R. 1972. *Plant Physiol.* In press
121. Ibid.
122. Pitt-Rivers, R., Impiombato, F.S.A. 1968. *Biochem. J.* 103:825–30
123. Pratt, L. H., Briggs, W. R. 1966. *Plant Physiol.* 41:467–74
124. Pratt, L. H., Butler, W. L. 1968. *Photochem. Photobiol.* 8:477–85
125. Ibid 1970. 11:361–69
126. Ibid, 503–9
127. Pratt, L. H., Coleman, R. A. 1971. *Proc. Nat. Acad. Sci. USA* 68:2431–35
128. Pringle, J. R. 1970. *Studies of malate dehydrogenase and of proteases.* PhD thesis. Harvard Univ.
129. Pringle, J. R. 1970. *Biochem. Biophys. Res. Commun.* 39:46–52
130. Purves, W. K., Briggs, W. R. 1968. *Plant Physiol.* 43:1259–63
131. Reynolds, J. A., Tanford, C. 1970. *Proc. Nat. Acad. Sci. USA* 66:1002–7

132. Rice, H. V. 1971. *Purification and partial characterization of oat and rye phytochrome*. PhD thesis. Harvard Univ.
133. Rice, H. V., Briggs, W. R. 1972. *Plant Physiol*. In press
134. Ibid.
135. Rice, H. V., Briggs, W. R., Jackson-White, C. J. 1972. *Plant Physiol*. In press
136. Rollin, P. 1970. *Phytochrome, Photomorphogénès, et Photopériodisme*. Paris: Masson et Cie. 136 pp.
137. Roux, S. J. 1971. *Chemical approaches to the structural properties of phytochrome*. PhD thesis. Yale Univ.
138. Roux, S. J., Hillman, W. S. 1969. *Arch. Biochem. Biophys.* 131: 423–29
139. Rubinstein, B., Drury, K. S., Park, R. B. 1969. *Plant Physiol.* 44:105–9
140. Rüdiger, W. 1969. *Hoppe-Seyler's Z. Physiol. Chem.* 350:1291–1300
141. Rüdiger, W. See Ref. 106
142. Rüdiger, W., Correll, D. L. 1969. *Justus Liebigs Ann. Chem.* 723: 208–12
143. Sanger, F. 1945. *Biochem. J.* 39: 507–15
144. Satter, R. L., Galston, A. W. 1971. *Science* 174:518–20
145. Satter, R. L., Galston, A. W. 1971. *Plant Physiol.* 48:740–46
146. Satter, R. L., Marinoff, P., Galston, A. W. 1970. *Am. J. Bot.* 57:916–26
147. Satter, R. L., Sabnis, D. D., Galston, A. W. 1970. *Am. J. Bot.* 57:374–81
148. Sawhney, B. L., Zelitch, I. 1969. *Plant Physiol.* 44:1350–54
149. Saxena, V. P., Wetlaufer, D. B. 1971. *Proc. Nat. Acad. Sci. USA* 68:969–72
150. Schachman, H. K. 1959. *Ultracentrifugation in Biochemistry*. New York: Academic. 272 pp.
151. Schäfer, E., Marchal, B., Marmé, D. 1971. *Planta* 101:265–76
152. Schäfer, E., Marchal, B., Marmé, D. 1972. *Photochem. Photobiol.* In press
153. Shapiro, A. L., Viñuela, E., Maizel, J. V. Jr. 1967. *Biochem.*
154. Shropshire, W., 1972. In *Photophysiology*, ed. A. C. Giese, 7. New York: Academic. In press
155. Siegelman, H. W., See Ref. 109, 489–506
156. Siegelman, H. W., Butler, W. L. 1965. *Ann. Rev. Plant Physiol.* 16:383–93
157. Siegelman, H. W., Chapman, D. J., Cole, W. J. 1968. In *Porphyrins and Related Compounds*, ed. T. W. Goodwin, 107–20. London: Academic
158. Siegelman, H. W., Firer, E. M. 1964. *Biochemistry* 3:418–23
159. Siegelman, H. W., Hendricks, S. B. 1965. *Fed. Proc.* 24:863–67
160. Siegelman, H. W., Turner, B. C., Hendricks, S. B. 1966. *Plant Physiol.* 41:1289–92
161. Smith, H. 1970. *Nature* 227:665–68
162. Spruit, C. J. P. 1966. *Biochim. Biophys. Acta* 120:454–56
163. Spruit, C. J. P. 1966. In *Currents in Photosynthesis*, ed. J. B. Thomas, J. C. Goedheer, 67–73. Rotterdam: Donker
164. Spruit, C. J. P. 1966. *Biochim. Biophys. Acta* 112:186–88
165. Spruit, C. J. P. 1966. *Meded. Landbouwhogesch. Wageningen* 66–15:1–7
166. Spruit, C. J. P., Mancinelli, A. L. 1969. *Planta* 88:303–10
167. Stark, G. R., Smyth, D. G. 1963. *J. Biol. Chem.* 238:214–26
168. Steiner, A. M. 1969. *Planta* 86: 334–42
169. Ibid, 343–52
170. Steiner, A. M. 1969. *Photochem. Photobiol.* 9:493–506
171. Ibid, 507–13
172. Ibid 1970. 12:169–74
173. Sternberger, L. A., Hardy, P. H. Jr., Cuculis, J. J., Meyer, H. G. 1970. *J. Histochem. Cytochem.* 18:315–33
174. Suzuki, H., Hamanaka, T. 1969. *J. Phys. Soc. Jap.* 26:1462–72
175. Sweet, H. C., Hillman, W. S. 1969. *Physiol. Plant.* 22:776–86
176. Tanada, T. 1968. *Proc. Nat. Acad. Sci. USA* 59:376–80
177. Tanada, T. 1968. *Plant Physiol.* 43:2070–71

178. Taylor, A. O. 1968. *Plant Physiol.* 43:767–74
179. Taylor, A. O., Bonner, B. A. 1967. *Plant Physiol.* 42:762–66
180. Tezuka, T., Yamamoto, Y. 1969. *Bot. Mag. Tokyo* 82:130–33
181. Tobin, E. M., Briggs, W. R. 1969. *Plant Physiol.* 44:148–50
182. Wagner, E., Mohr, H. 1966. *Photochem. Photobiol.* 5:397–406
183. Walker, T. S., Bailey, J. L. 1968. *Biochem. J.* 107:603–5
184. Ibid 1970. 120:607–12
185. Ibid, 613–22
186. Weber, K., Osborn, M. 1969. *J. Biol. Chem.* 244:4406–12
187. Whitaker, J. R., Perez-Villaseñor, J. 1968. *Arch. Biochem. Biophys.* 124:70–78
188. Yamamoto, Y., Tezuka, T. See Ref. 106
189. Yunghans, H., Jaffe, M. J. 1970. *Physiol. Plant.* 23:1004–16
190. Zucker, M. 1972. *Ann. Rev. Plant Physiol.* 23:133–56

Ann. Rev. Plant Physiol. 1972. 23:335-66

ENZYME SYMMETRY AND ENZYME STEREOSPECIFICITY

7534

KENNETH R. HANSON

Department of Biochemistry, Connecticut Agricultural Experiment Station, New Haven, Connecticut

CONTENTS

INTRODUCTION

Biochemists employ stereochemical concepts and terminology in at least four areas: (*a*) in discussing the association of subunits in oligomeric and polymeric proteins; (*b*) in specifying the configurations of intermediates and products of metabolism; (*c*) in analyzing the chemical transformations catalyzed by enzymes; and (*d*) as an aid to the study of metabolic pathways. It is the purpose of this essay to draw attention to aspects of stereochemical theory related to these topics and to illustrate the discussion by examples taken, as far as possible, from plant biochemistry.

The link between the study of plant metabolism and stereochemistry is by no means new. The foundations of stereochemistry were laid in 1853 when Pasteur described his chemical and crystallographic studies of the tartaric acids (178). Of these acids the natural *dextro* form, so named because it rotated the plane of polarized light in a right-handed, i.e. clockwise, sense, had first been isolated by Scheele in 1770 from the tartar incrustation of wine barrels. Pasteur recognized that the optical activity of the natural product implied molecular dissymmetry, and this dissymmetry he perceived as being inherent in the chemistry of the living system from which the acid derived. In 1854 Pasteur observed that *Penicillium glaucum* metabolized the *dextro* but not the *laevo* form of tartaric acid. His subsequent studies of fermentation were viewed initially as a further probing of the fundamental dissymmetry in organic nature.

Chirality, reflective symmetry, and gyrosymmetry.—The type of dissymmetry that fascinated Pasteur is now usually referred to by chemists as chirality (Gk. *cheir*, hand). The word was employed by Lord Kelvin in 1884 and was reintroduced into chemistry at the suggestion of Mislow, who was influenced by Whyte (39). A molecule is *chiral* if it cannot be superposed upon its mirror image (102), i.e. if enantiomers can exist. For a rigid finite molecule, this implies that the structure cannot be superposed on itself as a result of a simple reflection or, more generally, an operation of rotation-reflection (105, 129). If the molecule is flexible, torsions about bonds and changes in bond angles must be allowed in seeking to achieve superposition. A molecule is *achiral* if superposition can be achieved. This terminology frees chemists from the necessity to use dissymmetric in the special technical sense of "lack of reflective symmetry". The human body ideally conceived is achiral as it has reflective symmetry and is dissymmetric in the logical sense that it lacks rotational symmetry. Note that chiral cannot in general be replaced by asymmetric; a propeller or screw lacks reflective symmetry and is thus both dissymmetric and chiral, but it is not asymmetric as it has an axis of rotational symmetry. Because of the "bilateral symmetry" of the brain, human perception and language emphasize reflective symmetry—biochemists need to be on guard against this bias (94, 95).

A rigid molecule, or a flexible molecule set in a single conformation, possesses rotational symmetry if it can be superposed upon itself as a result of a rotation through less than the full circle. If n repetitions of this operation restore the original orientation of the molecule, an n-fold symmetry axis is present. If the whole molecule is superposed in this way with $n < \infty$, groups within the molecule are necessarily superposed. For molecules that are flexible under the conditions of observation, superposition may be achieved by rotation, by conformational changes (torsion about single bonds, changes in bond angles), or by a combination of these. For example, the hydrogens of an $-\ddot{N}H_2$ or, $-CH_2^-$ group (47) may be superposed by a combination of inversion (allowed changes in bond angles) and torsion about the bond linking N or C to the rest of the molecule. It has been proposed recently (8, 58, 96) that groups within a molecule (e.g. $-H$, $-OH$, $-COOH$, $-CHOHCOOH$) that are superposable in any of the above ways should be said to be *homotopic* (Gk. *topos*, place) as they are in equivalent places (environments) within a molecule. (A fuller discussion of topic relationships within molecules is given below.) It has also been pointed out (96) that when superposition is achieved by torsion alone the operation can be regarded as an operation of torsional symmetry. For example, any one of the three hydrogens of the methyl group of ethanol may be superposed upon another by one or more torsions of $2\pi/3$ about the C–C bond; therefore, the methyl group has a three-fold axis of torsional symmetry. Both types of symmetry operation that may be associated with the existence of homotopic groups can be regarded as operations of gyrosymmetry, and if the method of examining a molecule is suitably prescribed, one may say that a molecule cannot contain homotopic groups unless it is *gyrosymmetric.* Likewise one can speak of a group having an axis of torsional symmetry as being gyrosymmetric, e.g. $-CH_3, -NO_2, -CO_2^-, -NH_3^+$, and also $-\overset{..}{N}H_2$ (if its transient planar form is examined for gyrosymmetry).

OLIGOMERIC PROTEINS

The above concepts provide the necessary background for discussing the symmetry of proteins. As the natural amino acids, with the exception of glycine, are chiral molecules, all proteins are chiral. X-ray diffraction studies of a few crystalline proteins have rigorously shown that a protein may be assembled from subunits in such a way that when viewed as a rigid object in, as far as possible, an energetically unique conformation it possesses rotational symmetry, i.e. it possesses two or more homotopic subunits or subunit combinations. These homotopic substructures were named *protomers* by Monod et al (131). (Crystallographers often speak of the "assymmetric subunits" of the protein and say that they are "crystallographically equivalent.") In proteins composed of several different types of subunit the definition of the protomer may be a matter of arbitrary choice. As each protomer is bound to its neighbor in the same manner, there are no extra intraprotomer binding regions available, and the oligomeric protein constitutes a "closed crystal." Once the principles of assembly have been stated it follows from symmetry theory (105) that the possible classes of oligomeric proteins having closed-crystal structures are defined by the rotational point groups (81, 114, 115). These are of five types and in the Schoenflies notation widely used by chemists, but less frequently by crystallographers, their names and symbols are as follows (n=number of protomers):

(*a*) *cyclic* point groups, symbol C_n where $n=2, 3, 4$ etc, only a single n-fold axis of symmetry (Figure 1, line 1);

(*b*) *dihedral,* $D_{n/2}$, hence D_2, D_3, D_4 etc where $n=4, 6, 8$ etc, one $n/2$-fold axis and $n/2$ twofold axes of symmetry (Figure 1, lines 2, 3, 4);

(*c*) *tetrahedral, T,* 12 ($=4\times3$) protomers, 4 three- and 3 twofold axes;

(*d*) *octahedral, O,* 24 ($=6\times4=8\times3$) protomers, 3 four-, 4 three-, and 6 twofold axes;

(*e*) *icosohedral, I,* 60 ($=12\times5=20\times3$) protomers, 6 five-, 10 tri-, and 15 twofold axes.[1]

Given the theory, one must ask whether proteins belonging to all classes exist or whether only the simplest types of point group symmetry are observed. Two recent reviews have concerned themselves with this question (89, 114). The appearance in electron micrographs of proteins with the same symmetry may vary appreciably depending on the shape of the subunits and the distribution of their

---

[1] The symbol C_n is applied to the symmetry operation associated with an n-fold axis of rotational symmetry even when the point-group symmetry of the whole molecule is other than C_n. Crystallographers usually indicate point-group symmetries by the Hermann-Mauguin notation which draws attention to the various C_n axes thus: the C_2 point group as (2), C_3 as (3) etc; D_2 as (222) but D_3 as (32), D_4 as (42) etc; T as (23), O as (432), and I as (532).

CYCLIC

(1)

DIHEDRAL

(2)

(3)

(4)

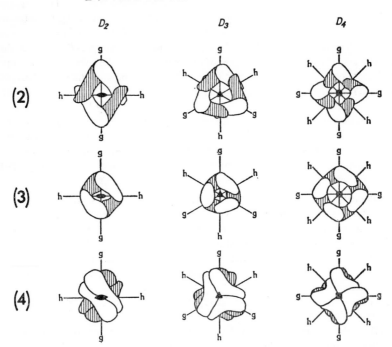

FIGURE 1. Cyclic and dihedral symmetry of proteins (81). The series continue C_5, C_6 etc and D_5, D_6 etc (see text). The symbols ○, △, and □ indicate two-, three-, and four-fold symmetry axes perpendicular to the plane of the paper. The other lines are twofold axes in the plane of the paper with letters g and h added to distinguish different binding modes between protomers. Protomers are shown as white or shaded: shaded protomers are white protomers seen from the opposite side of the plane of the paper, thus rotation through 180° about any twofold axis in the plane of the paper shows the other sides of the protomers but leaves the figure unchanged. Lines 2, 3, and 4 show increasingly compact arrangements of protomers. For the use of toy monkeys to illustrate these principles see (176).

binding areas. Electron micrographs alone are therefore an uncertain guide for deducing quaternary structures (89). The available information concerning the subunit composition of oligomeric proteins (114) reveals that the majority of proteins studied have either 2 or 4 subunits, and X-ray diffraction shows some of these to have respectively C_2 or D_2 point-group symmetries, e.g. hemoglobin (139) has C_2 symmetry with each protomer composed of an α and a β chain—there is only one twofold symmetry axis. Although plant proteins with more than four subunits are known, they have not been studied with sufficient intensity to allow them to be cited as evidence for the validity of the general theory. The better documented cases are of proteins from mammalian, bacterial, and viral sources. Aspartate transcarbamylase from *Escherichia coli* has been shown by X-ray diffraction to have D_3 symmetry with each protomer composed of one catalytic and one regulatory subunit (184); reconstruction of electron micrograph pictures indicates that glutamate dehydrogenase from beef liver also has D_3 symmetry (108, but see 66). Careful electron micrograph studies (177) indicate that *E. coli* glutamine synthetase composed of 12 protomers has D_6 symmetry (not shown in Figure 1). X-ray studies (88) showed that apoferritin has D_5 symmetry. This example is of importance in that the protein is composed of 20 subunits that apparently have identical primary structures. Each protomer, arbitrarily defined, consists of two differently folded subunits (82) and not quasi-equivalent subunits as found in spherical viruses (40).

The remaining three classes of rotational point groups define shells which are likely to be roughly spherical. Computer model building has been used to simulate the appearance in electron micrographs of L-aspartate β-decarboxylase from *Alcaligenes faecalis* (31). It was concluded that this enzyme with 12 subunits has a hollow shell structure with tetrahedral (T) symmetry. The arrangement of the subunits may be visualized as rings of three about each apex of a tetrahedron, each ring having C_3 symmetry. The existence of structures with octahedral (O) and icosahedral (I) symmetry was first recognized in studying the spherical viruses, particularly those from such plant sources as turnip, cucumber, and alfalfa (114). It now appears that such point-group symmetries are encountered in multienzyme complexes (72, 148). The pyruvate dehydrogenase complexes from *E. coli* and mammals have O and I symmetries respectively. In the *E. coli* complex each protomer consists of three enzymes whose separate functions must be closely associated in order to catalyze the overall reaction: additional enzyme activities are built into the mammalian complex. The α-ketoglutarate dehydrogenase complex from *E. coli* and mammalian sources probably has octahedral symmetry. It has been purified from cauliflower mitochondria (144), but its subunit structure has not yet been investigated. The multienzyme complexes of fatty acid, aromatic, and carotenoid metabolism have been studied, but in less detail (50, 72, 148).

Because many of the methods used to investigate the subunit structure of proteins are subject to significant experimental error, symmetry theory provides a necessary check on experimental results. The value of such checks has been demonstrated in the study of the pyruvate dehydrogenase complexes mentioned above. For a long time the stoichiometry of the various constituent enzymes did not suggest simple numerical relationships, and the electron micrographs could be interpreted in more than one way. One must anticipate that symmetry theory will help to make sense in other cases where the number and nature of the subunits is in dispute. For example, stoichiometry studies and Na dodecylsulfate-acrylamide gel studies of spinach ribulose diphosphate carboxylase (157, also 1) suggest that the "native enzyme has 8 heavy catalytic subunits and 8–10 light structural or regulatory subunits." By symmetry theory the simplest structure that could give such results would have equal numbers of the two types of subunit: if there are 8 large there should be 8 small subunits with a protomer consisting of one of each and the structure having D_4 symmetry. The earlier electron micrographs of "fraction I protein" from Chinese cabbage show a protein with a central hole or depression and an apparent four-fold axis of symmetry (90). They were not interpreted as D_4 structures because the stoichiometry data then available suggested a 24 subunit model.

One goal in studying the quaternary structure of proteins is to identify the factors which lead to the evolutionary selection of a particular structure. Such an understanding is likely to come in part from studies of the variations and consistencies of structure in as wide a range of living organisms as possible. For example, there are appreciable differences between the sizes and the subunit structures of a number of enzymes from higher plants and from microorganisms, e.g. ribulose diphosphate carboxylase (1, 5) and citrate synthase (169, 170). The spinach and rabbit muscle fructose diphosphate aldolases resemble each other in quaternary structure (147). Although their molecular weights and amino acid compositions differ significantly, both appear to have fundamentally D_2 structures with minor but random chemical differences between the four subunits. This conservation of quaternary structure is surely not an evolutionary accident.

Two types of argument can be used to explain selection for a particular quaternary structure: (a) those that indicate a particular functional benefit provided by the structure; and (b) those that indicate a preferred pathway of evolution. For (a) Monod et al (131) noted that symmetry allowed the maximum rigidity for the least genetic information and that subunit interactions resulted in the enzyme having regulatory properties (184). For (b) they proposed that C_2 dimers were more likely to form than C_3 trimers or larger cyclic structures as a result of mutational changes. Larger dihedral structures could derive from D_2 structures by mutations leading to successive ring expansions (81, 82).

By combining rotational symmetry with translational symmetry, polymeric proteins with tube structures, e.g. tobacco mosaic virus protein (37), or sheet structures may be defined. This subject is pertinent to discussions of membrane and microtubule function.

ENZYMES AS CHIRAL REAGENTS

It follows from the chiral nature of proteins that enzymes are chiral reagents. They have active sites with specific binding regions and catalytically active groups capable of both motion and precise orientation in three-dimensional space. Certain chiral reagents employed in general organic chemistry show considerable stereoselectivity (32, 124), e.g. isobornyloxy magnesium chloride reducing buty-raldehyde-1-d gives predominantly the salt of (+)-1-butanol-1-d(9, 173). The high degree of stereoselectivity shown by enzymatic reactions is a refinement of that displayed by simpler reagents. It is pertinent to ask, therefore, (*a*) what types of steric discriminations can and cannot be made, in principle, by chiral reagents, and (*b*) how one can describe any differentiations that occur. The answers to these questions require both an analysis of concepts and a review of nomenclature.

Classes of isomerism and topic relationships.—Two types of stereoselectivity may be distinguished (95): (*a*) discrimination between stereoisomers; and (*b*) discrimination between certain constitutionally equivalent groups within a molecule (between the methylene –H atoms of ethanol or the –OH groups of *meso*-tartaric acid) or between the two faces of a planar system (between the two faces of the trigonal carbon of an aldehyde R–CHO where O, R, and H are either seen to be in a clockwise or anticlockwise sequence depending on the face being examined). Discrimination in a chemical reaction occurs if only one member of a pair of atoms, groups, faces, or stereoisomers reacts with the reagent, or if both react, they react with the reagent at different rates.

The classification of isomers advocated in the recent IUPAC paper (102) is that shown in the upper part of Figure 2. Compounds that differ in the bonding connections between their constituent atoms are now termed constitutional isomers rather than structural isomers. Any pair of stereoisomers which are not enantiomers are diastereoisomers; thus D-erythrose and D-threose are diastereoisomers as are fumaric and maleic acids. Formerly the latter *cis-trans* isomers were often excluded from the class of diastereoisomers and termed geometrical isomers.

Constitutionally equivalent groups within molecules may be similarly classified (Figure 2). This analysis of topic relationships proposed by Hirschmann & Hanson (96) extends earlier suggestions by Mislow & Raban (130; see also 8, 58). Homotopic groups have already been discussed. If the environments of two heterotopic groups differ constitutionally, e.g. –OH at C-2 and C-3 of D-glyceraldehyde (**1**, Figure 3), they are constitutionally heterotopic, otherwise they are *stereoheterotopic*. If stereoheterotopic groups can be superposed by operations that include reflection, they are *enantiotopic* (their environments are enantiomeric), otherwise they are *diastereotopic*. Faces of trigonal atoms are handled similarly.

These classifications permit one to generalize about the different effects of achiral and chiral reagents upon isomers, groups, or faces when the solvent is achiral (e.g. water). Achiral reagents (e.g. H$^+$, CN$^-$, OH$^-$, NO$_2^-$, BH$_4^-$, B$_2$H$_6$) will react with equal facility towards homotopic groups or faces, towards identical molecules, towards enantiotopic groups or faces, and towards enantiomers. Chiral reagents (e.g. enzymes) will react with equal facility only towards homotopic groups or faces and towards identical molecules. This, alas, is the only absolute stereochemical generalization based on purely geometrical considerations that can be made about enzyme action. For stereoelectronic considerations (187) see below.

Classes of Isomerism

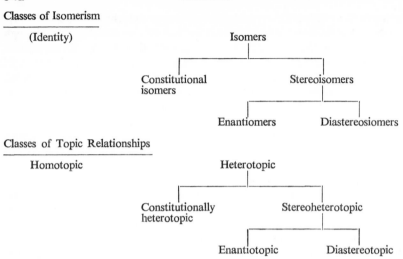

FIGURE 2. Classes of isomerism and classes of topic relationships (96). Substitution in turn of homotopic groups in a molecule by an achiral test group leads to the same compound, substitution of heterotopic groups leads to isomers, substitution of stereoheterotopic groups leads to stereoisomers, etc as indicated by the corresponding positions of the classes in the upper and lower portions of the figure. For example, in Figure 3, $-H_R$ and $-H_S$ of 1 are diastereotopic as are $-H_Z$ and $-H_E$ of 9, whereas $-H_R$ and $-H_S$ of 7 are enantiotopic. Either pair of hydrogens, one at C-2 and one at C-4 of 14, are constitutionally heterotopic. The $-CO_2^-$ groups and the hydrogens of 4 are homotopic. The superposition tests for topic relationships (see text) apply both to groups and to the faces of trigonal atoms. The faces of each carbonyl and olefinic carbon atom in Figure 3 are enantiotopic, whereas those of the olefinic carbons of (2S)-squalene-2,3-oxide (Figure 4) are diastereotopic.

Steric elements and steric descriptors.—Nearly all molecules can be represented by simple models of the ball and stick variety, and such molecular models may be factorized into steric elements or larger steric units. The biochemist normally encounters only a few types, four of which will be mentioned here.

(*a*) Two of these are elements of stereoisomerism (96, 97): the tetrahedral *chiral center* Xghij (where g, h, i. j are distinct ligands of X, there being a tetrahedral distribution of the atoms immediately attached to X and either no enantiomeric pairs of ligands or two such pairs), and the *achiral element* of the olefinic double bond ghX=Yij. The stereoisomerism associated with these elements has been discussed in all textbooks of organic chemistry since the time of van't Hoff (181). The quest for a more systematic understanding of steric concepts, however, has led to changes in nomenclature. The general term "chiral center" or "center of chirality" is preferred to "asymmetric center," although there are only a few instances where a carbon atom at a chiral center could not be said to be an asymmetric carbon atom (39). Also, an attempt has been made to provide comprehensive definitions of the various classes of steric elements (97).

The term "chiral center" has been closely linked with the development of the sequence rule by Cahn, Ingold & Prelog (38, 39,102). The rule assigns priorities to the ligands of a steric element on the basis of their constitutions and, if necessary, configurations. For a chiral center the complexity of the molecule is reduced to a tetrahedral assembly of four ordered points $a > b > c > d$ (differentiated atoms) about the central atom. There are only two possible enantiomeric arrangements of this assembly. Looking towards the face of the tetrahedron remote from d the points $a > b > c$ show a clockwise (right-handed, *rectus*) arrangement if the chirality is R and an anticlockwise (left-handed, *sinister*) arrangement if it is S. The procedure for assigning R/S descriptors has been illustrated with numerous simple examples by Cahn (38) (see Figure 3).

For many purposes the terms *cis* and *trans* will continue to be perfectly satisfactory for specifying the configurations of double bonds; however, the descriptors Z and E (sequence rule priorities *cis* and *trans*) have been introduced to cover difficult cases and for use in *Chemical Abstracts* nomenclature (26, 102). The R/S and Z/E systems are designed to be used in systematic nomenclature. Where biogenetic or chemical relationships are to be stressed, local systems for specifying chirality will be preferred. For example, the configurations of the amino acids (100), sugars (4), and cyclitols (104) are defined by absolute (D/L) and relative (*threo, erythro, myo, allo* etc) descriptors. The R/S and Z/E systems may be used to extend these local systems in complex cases and where it is desired to specify the chirality of centers in amino acids etc that are chiral as a result of isotopic labeling. The use of the R/S nomenclature in conjunction with the *Chemical Abstracts* and square bracket systems for naming labeled compounds has been discussed (60).

(b) Some stereoheterotopic groups can be distinguished with the aid of the R/S and Z/E systems, but many such groups can only be specified by reference to steric elements which are not elements of stereoisomerism but would be so designated if one of the groups to be specified were held to be different from any other such group in the molecule. The most important of these elements of prosteroisomerism (96) is the tetrahedral *prochiral center* Xggij (where none of the ligands g, i, or j is the enantiomer of the other). If one g is formally replaced by h, where h differs constitutionally from g, i, and j, a chiral center Xghij is obtained. If one of two groups associated with superposable ligands of the same prochiral center is given sequence-rule priority over the other, the group may be specified as being *pro-R* or *pro-S* to the prochiral center according to the formal chirality of the center produced (80). Where the groups are single atoms or methyl groups directly attached to the same center (e.g. as in the methylene group of ethanol) it is convenient to use subscripts: $-H_R$ indicates the *pro-R* hydrogen, $-H_S$ the *pro-S* hydrogen (Figure 3). These should never be referred to as the R-hydrogen and the S-hydrogen. In cyclic compounds the *pro-R/pro-S* system is not necessarily the most convenient, and thus steroid chemists frequently distinguish methylene hydrogens as being α or β (101). Stereospecific numbering is used to distinguish the $-CH_2OH$ groups of glycerol (103) and may be used as a supplementary method for differentiating the *pro-3R* and *pro-3S* $-CH_2COOH$ groups of citric acid 11 (96, 169).

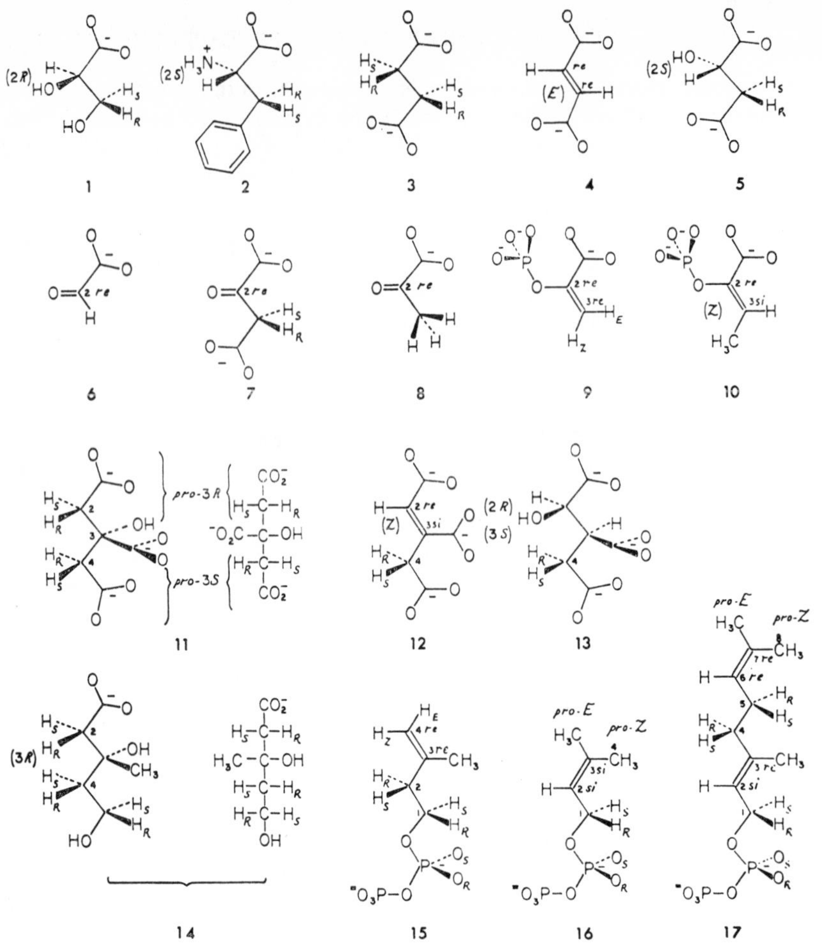

FIGURE 3. Examples of the use of sequence rule dependent steric descriptors for specifying configurations and distinguishing stereoheterotopic groups and faces. Note that the descriptors apply only to the formulae shown and do not, except by accident, reflect biochemical relationships (cf C-4 of **14** with C-2 of **15**). The faces designated *re* or *si* are those presented to the reader. Wedges show bonds above the plane of the paper, i.e. towards the reader; dashed lines show bonds away from the reader. In Fischer projections (**11** and **14**) horizontal bonds are towards the reader and vertical bonds away from the reader (i.e. each molecule has been twisted from the staggered conformation on the left to an eclipsed conformation). The staggered conformations shown are not necessarily the most stable ones. The benzene ring of **2** and carboxylate ions which are not constrained by conjugation or hydrogen bonding, e.g. in **2**, must be turned through a torsional angle of 90° or ±30° to reach an energy minimum. **1**, glyceraldehyde; **2**, L-phenylalanine; **3**, succinate; **4**, fumarate; **5**, L-malate; **6**, glyoxylate; **7**, oxaloacetate; **8**, pyruvate; **9**, phosphoenolpyruvate; **10**, (Z)-phosphoenol-α-ketobutyrate (28, 174); **11**, citrate [note stereospecific numbering (96)]; **12**, *cis*-aconitate; **13**, (2R,3S)-isocitrate (i.e. *threo*-D$_s$-isocitrate); **14**, (3R)-mevalonate; **15**, isopentenyl pyrophosphate; **16**, dimethylallyl pyrophosphate; **17**, geranyl pyrophosphate. The stereoheterotopic methyl groups of **16** and **17** may be distinguished by stereospecific numbering (96).

A second element of prostereoisomerism which is of occasional importance to biochemists is the *proachiral element* (96) of the olefinic double bond ggX=Yhi. The stereoheterotopic ligands attached to X may be distinguished as *pro-Z* and *pro-E* following the same type of reasoning used to assign the *pro-R/pro-S* descriptors (Figure 3, **9, 15–17**).

In many enzymatic reactions hydrogens at prochiral centers are eliminated or displaced. The stereochemistry of such reactions is normally investigated by using substrate "analogs" in which the prochiral center is replaced by a center which has been made chiral by isotopic labeling—either with deuterium (e.g. 99% substitution of protium) or with tritium (tracer amount). Provided that care is taken to ensure that kinetic isotope effects do not lead to erroneous conclusions, the stereochemistry of the enzymatic reaction may be deduced from a knowledge of the chirality of the labeled center. Except in rare instances, labeling of the *pro-R* hydrogen results in *R* chirality, labeling of the *pro-S* hydrogen, *S* chirality. The fate of *pro-Z* and *pro-E* groups may be likewise followed by using *Z* and *E* labeled compounds.

Stereoheterotopic faces of trigonal atoms Xghi may be distinguished as *re* or *si* according to the observed clockwise (*rectus*) or anticlockwise (*sinister*) sequence-rule order of the ligands g, h, i. The descriptor assigned to the adjacent face of Y is used to name the face of X in the system ggX=Yhi (80) (Figure 3).

The discussion of the steric course of enzymatic reactions also requires nomenclature for specifying conformations about single bonds (102). Groups are *synperiplanar* if the torsional angle relating them is 0°, *synclinal* if it is ±30°, *anticlinal* if it is ±60°, and *antiperiplanar* if it is 180°. By an extension of this terminology, addition reactions are said to be *syn* if the attack is on the same face of the double bond and *anti* if attack is on opposite faces (125). Reactions of systems with a definable plane, e.g. butadiene, are said to be *suprafacial* if only one face is involved and *antarafacial* if both faces are involved (187). This usage overlaps with the *syn/anti* terminology, but it can be applied where *syn* and *anti* would be inappropriate.

STEREOSPECIFICITY AND ENZYME CATALYSIS

The evolution of stereospecificity.—The generalization that enzymes are chiral reagents does not explain why they are highly stereospecific. One needs to discern the selective pressures that have caused stereospecific enzymes to evolve from other enzymes and from primative proteins. The following arguments presuppose the existence of discreet metabolic and genetic systems.

(*a*) A great deal of stereospecificity can be attributed to selective binding. Without binding catalysis cannot take place, and without selective binding intermediary metabolism would be impossible. Laboratory selection for mutant enzymes can mimic the evolutionary process.

(*b*) Some specificity must be attributed to chemical inevitability. The success of organic chemists in carrying out "biogenetic type" syntheses of natural products shows that many enzymatic transformations, particularly those involving cyclic molecules, are obliged to be stereospecific. An enzyme may have evolved in such a way that it selects between options each of which is highly stereospecific. Again survival of a particular enzyme depends upon the requirements of intermediary metabolism.

An excellent example of enzymatic selection between chemical alternatives is provided by the formation of sterols from (3S)-squalene-2,3-oxide. The process is so complex and the "Zurich" mechanism (62, 186), with a possible minor variant, is supported by such detailed chemical studies, particularly those by van Tamelen (179, 180), Corey (44), and their associates, that a great deal can be said about the relative contributions of mechanism and enzyme to the outcome. In animals and fungi squalene oxide yields lanosterol as the first steroid intermediate (45, 46, 188), whereas in higher plants cycloartenol is formed and acts as a precursor of most phytosterols (74, 75). An essential role of the cyclase enzymes (180) appears to be to fold the squalene oxide so that the correct π-electron overlap (Figure 4) is achieved for C-2 and the double bonds Δ^6, Δ^{10}, Δ^{14}, and Δ^{18}. In the first stage protonation at the correct position initiates epoxide opening and cyclization thus generating 7 or 8 chiral centers. The product is either a carbonium ion, which may be formulated in alternative ways, or the product of its reaction with a suitably placed group on or bound to the enzyme (44, 45). The further transformation (45, 62) of this intermediate into the nonclassical carbonium ion shown in Figure 4 results in a new orientation of the Me- and R-groups at C-19. It is not yet clear what contribution the enzyme makes to this debated process. In the third stage a sequence of stereospecific methide and hydride shifts, each producing inversion, is terminated by H^+ removal at C-6 or at C-11. The sequence should be influenced mainly by the location of the basic group that abstracts H^+. In the presence of $SnCl_4$ and in benzene solution, C-15 of squalene oxide does not spontaneously fall into position. Rings A and B are formed and Me^- and H^- migrations are terminated by the expulsion of the *pro-8R* hydrogen (162).

(*c*) There remain enzymes where only a requirement for maximum catalytic efficiency can account for the stereospecificity of the reaction. The selection pressure for efficiency must be very great as many enzymes representing a diversity of mechanisms have turnover numbers per active site that approach the theoretical upper limit for acid-base catalysis—about 10^8 sec^{-1} (57, 77). The argument considers the few enzymes in which one enantiotopic face of a trigonal atom is attacked to give a prochiral center with enantiotopic groups, or the reverse of this takes place. As long as there has never been in the course of the enzyme's evolution a free chiral product that could have been a basis for selection, metabolic selection cannot account for the observed stereospecificity and selection for catalytic efficiency must be invoked.

The least debatable cases appear to be the enzymes catalyzing the various 3*si* carboxylations (43, 156) of phosphoenolpyruvate (**9**→**7**) (the list includes phosphoenolpyruvate carboxylase from peanuts) and also the 3*si* protonation of **9** by pyruvate kinase (28, 152, 174) (**9**→**8**). Glycolate oxidase, e.g. from tobacco, yields glyoxylate **6** by removing the *pro-R* hydrogen (107, 151), but the argument is complicated by the fact that the enzyme will also act on L-lactate. Similarly the alcohol dehydrogenases from various sources (21) will remove the *pro-1R* hydro-

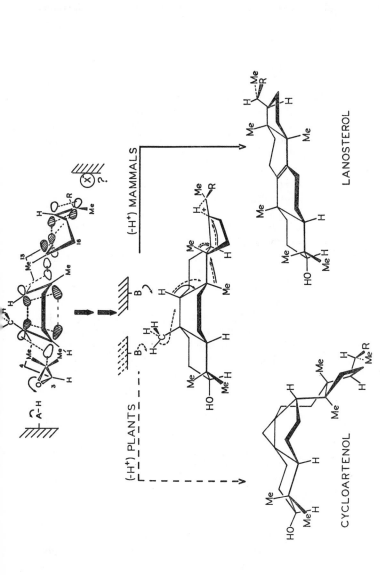

FIGURE 4. Outline of the probable enzymatic mechanisms of cycloartenol and lanosterol biosynthesis (45, 75, 180). The upper formula is (3S)-squalene-2,3-oxide bound in folded form to the enzyme. The overlap between the electron clouds of the double bonds are indicated by drawing their contributing p-orbitals as double balloons. $-A-H$ and $-B$ are acidic and basic catalytic groups on the enzyme with $-A$ and $-BH$ their conjugate basic and acidic forms, $R = -CH_2CH_2CH = C(CH_3)_2$. Cyclization is followed by a reorientation of $-Me$ and $-R$ at C-19 to give the carbonium ion intermediate shown or its immediate X-substituted precursor. Small arrows indicate electron flows; the migrating $-H$ and $-CH_3$ groups accompany the electrons, each chiral center being inverted (hydride and methide shifts). For convenience the biosynthetic fates of the atoms of squalene oxide are indicated in the text by reference to the numbering of the principal chain of this compound rather than to the numbering of the products and intermediates.

gen from both ethanol and from certain chiral substrates. Another possible example is citrate (*si*)-synthase (169), but the enzyme as it now occurs could have evolved from a metabolic system in which (3*S*)-citryl CoA was a metabolic intermediate.

It is impossible to deduce a firm conclusion about the evolution of stereospecificity from conjectures about evolutionary sequences. Fortunately the view that efficiency requires stereospecificity is in keeping with the picture of enzyme catalysis derived from X-ray diffraction studies of proteins, NMR studies of metaloenzymes, etc. Physical organic chemists, in general, agree that the alignments of interacting groups affect rate constants although there is some debate as to the rate decrease associated with departures from ideal transition state geometry (34, 125, 128). Atomic trajectories during a reaction are governed by the necessity for maintaining maximum bonding throughout the process (98) but this condition may be satisfied by least-motion trajectories (93).

Catalytic groups distributed in a specific manner at an active site usually cannot stretch to allow a reaction to show alternative stereospecificity. Occasional exceptions to this are known, e.g. 3-methyl-L-aspartate ammonia-lyase will act on the *erythro* diastereoisomer with 1/100th the V_{max} of the *threo* isomer (15). [The result suggests the elimination proceeds by a stepwise process (33, 85).] In most cases where enzymes show alternative stereospecificity towards different substrates, or attack the same substrate in two different ways, the results are explicable in terms of an *alternative fit* of the substrate to the active site (see below). Rose (154) has suggested that the lack of stereospecificity in the protonation of pyruvate when stereospecifically labeled phosphoenolpyruvate is hydrolyzed by PEP carboxyphosphotransferase implies that enolpyruvate is released from the enzyme and is then protonated nonenzymatically at C-3. Similarly, Kaeppeli & Rétey (109) have shown that the immediate product of urocanase action on urocanate is a 5′ substituted 4′-hydroxyimidazole. This enol undergoes a rapid nonenzymatic tautomerization to the 4′-keto racemate: (5′*RS*)-3-(imidazole-4′-one-5′-yl)-propionate.

(*d*) If a high degree of catalytic efficiency is slow in evolving, the essential steric features of a reaction are likely to be conserved in evolution once they have appeared even though the substrate may be modified or a cosubstrate changed.— This conjecture has been explored by Rose (154). Although the problem is raised by the stereochemical data, firm conclusions will have to be based on sequence and X-ray studies of enzymes.

Active sites and mechanisms.—The relevance of stereochemical information to the study of enzyme mechanisms has been discussed in a number of recent reviews. Many of these are concerned with correlations of configurations and problems of metabolism as well as with enzyme mechanisms. For general and comprehensive discussions see (8, 20, 120, 141, 153). Reviews emphasizing particular groups of enzymes, areas of metabolism, or probable mechanisms of biosynthesis include the following: biosynthesis of steroids (42, 45, 46, 140, 143), biosynthesis of carotenoids, plant triterpenes and phytosteroids (74, 75), biosynthesis of sesquiterpenes (138), biosynthesis of alkaloids (16, 17, 160), enzymes of aromatic metabolism (27, 121, 167), citric acid cycle enzymes (60), enzymes catalyzing elimination reactions (85), pyridoxal phosphate containing enzymes (54). In addition, the *Annual Reports of the Chemical Society* should be consulted, e.g. (171). In this section no attempt is made to review all the available information relating to particular classes of enzymes or to discuss the determination of the configurations of stereospecifically labeled compounds. Instead I have tried to state what types of conclusions about enzyme mechanisms can and cannot be deduced from the results of stereochemical studies. The examples given have been selected to illustrate the arguments; closely related examples have not, in general, been cited. Tables recording the stereochemistry of similar enzymatic reactions have been published (20, 153, 154).

(*a*) The distribution of catalytic groups and substrate molecules at the active site relative to such binding features as counter ions may be inferred. This has already been noted for the complex reaction shown in Figure 4, but it also applies to simpler cases. Fumarate hydratase (fumarase), the ubiquitous citric acid cycle enzyme, catalyzes the addition of D_2O to fumarate **4** to give (3*R*)-L-malate-3-*d* (6, 70). The reaction, as catalyzed by the enzyme from pig heart, is at least 99.99% stereospecific (65). It follows that, viewed as an elimination, the enzyme catalyzes the removal of the *pro*-3*R* hydrogen. If at the active site the conformation of L-malate approximates to the structure of fumarate, the substrates being stretched between two counter ions (Figure 5 A), then the overall stereochemistry of the elimination is *anti*. The basic and acidic catalytic groups on the enzyme must be so positioned as to abstract the *pro*-3*R* hydrogen and protonate the –OH group. This argument does not depend upon knowing whether the reaction proceeds by a concerted or a stepwise mechanism. The same arguments may be used for many other simple elimination-addition reactions. For example, phenylalanine ammonia-lyase in higher plants and certain fungi converts L-phenylalanine **2** to *trans*-cinnamate (as **4** but a benzene ring in place of one $-CO_2^-$) (84, 85, 117, 175). The enzyme catalyzes the first "committed reaction" in the biosynthesis of lignin and many other phenylpropanoid compounds. If the amino acid is assumed to be stretched between a counter ion and a hydrophobic region binding the benzene ring in a conformation approximating to the structure of *trans*-cinnamate, then the elimination is *anti* as the *pro*-3*S* hydrogen is released (87, 99, 183). In this case a prosthetic group on the enzyme combines with the α-amino group prior to the elimination step (83). The reaction may be viewed as a collapse of two planar systems into a single plane, cf A1→A2 (85).

The distribution of binding and catalytic groups may in some instances be difficult to infer. Another citric acid cycle enzyme aconitate hydratase (aconitase) catalyses additions of water to *cis*-aconitate **12** to give either citrate **11** or (2*R*,3*S*)-isocitrate **13**, and in both cases the addition is *re-re*, i.e. apparently *anti* (10, 59, 70, 86). Furthermore, an essentially quantitative transfer of hydrogen, though not of OH, occurs when citrate goes to isocitrate without the intervention of free aconitate (155). To explain these results some form of alternate-fit hypothesis is necessary. Glusker (73) has proposed that the carboxylate ions adjacent to the

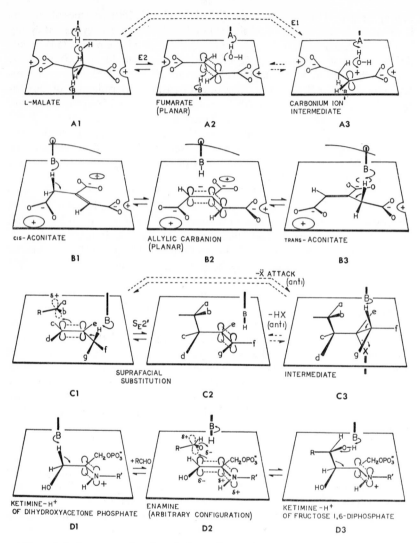

FIGURE 5. Some postulated and alternative enzyme mechanisms. Not all of the *p*-orbitals contributing to the π-orbitals of the unsaturated systems are indicated, e.g. conjugation in **A2** extends to the carboxylate ions. It is assumed that (*a*) eclipsed conformations are avoided; (*b*) in the transition states torsional angles between *p*-orbitals formed by C–C, C–H, or C–O bond breaking and *p*-orbitals of adjacent systems, as in **B1→B2**, approximate to zero; and (*c*) when adjacent *p*-orbitals are generated concurrently, as in **A1→A2**, the torsional angle between them is zero [or 180° (125)].

(A) Fumarate hydratase, i.e. fumarase. **A1⇌A2**, simple collapse of two planes to a single plane; synchronous bond breaking; central E2 transition state. **A1⇌A3⇌A2**, stepwise carbonium ion mechanism (78). If C–H is first cleaved a carbanion intermediate double bond of *cis*-aconitate are both coordinated to Fe^{2+} and that the other –COO^- is attracted to a counter ion on the enzyme. The essential feature of the hypothesis is that the central –COO^- may pass from one position on the octahedral coordination sphere of Fe^{2+} to an adjacent position vacated by H_2O. This takes place, to a first approximation, by turning **12** by about 1/3 of a full circle about the C-1, C-5 axis. The site formerly occupied by –COO^- is then filled by H_2O from the medium. In one conformation the H^+ donating group attacks the 2*re* face and coordinated H_2O the 3*re* face, in the other it attacks the 3*re* face and H_2O from the alternate coordination position attacks the 2*re* face, both additions being *anti*. The sequence from citrate to isocitrate thus involves elimination, a shift in coordination, then addition.

The observation that hydrogen transfer occurs, and occasionally the failure to observe such transfer, provides important stereochemical information about many enzymatic reactions (153). In addition to hydride shifts (Figure 4) and the above proton transfer, the list of examples includes 1,2 shifts (e.g. aldo-keto isomerases: $O=C-CHOH \rightleftharpoons HOHC-C=O$) and 1,3 shifts (e.g. transaminases: $C=N-CH \rightleftharpoons CH-N=C$). The subject could be illustrated by the transaminases (54), but a recent study of aconitate isomerase (112) provides a simpler example that also happens to be of relevance to plant biochemistry. Although *cis*-aconitate is utilized in the citric acid cycle, *trans*-aconitate is the major isomer stored in maize roots (122). It is likely that aconitate isomerase, an enzyme shown to be present in sugar-cane leaves (3), plays an essential role in the mobilization of stored aconitate in higher plants. The enzyme from *Pseudomonas putida* converts the *cis* to the *trans* isomer by an allylic migration of the double bond (Figure 5 **B**) and a small amount of stereospecific hydrogen transfer occurs (^3H labeling). This transfer cannot be an uncatalyzed side reaction as 1,3-migrations, according to the Woodward-Hoffmann orbital symmetry rules (187), are *antarafacial* (H moves from above to below the plane). Such a migration of the *pro*-4*S* hydrogen of *cis*-aconitate (**B1**) leads either to identically labeled *cis*-aconitate or to *trans*-aconitate labeled in the wrong position; likewise migration of the *pro*-4*S* hydrogen of *trans*-aconitate (**B3**) leads to the same *trans* compound or the wrong *cis* com-

---

is formed. (**B**) Aconitate isomerase (112). Proton transfer may, but need not, be assisted by hydrogen bonding to the allylic carbanion (2). (**C**) Alternative explanations for an overall *suprafacial* S_E2' allylic displacement (C1$\rightleftharpoons$C2) (45, 141, 143); a mechanism involving a carbonium ion rearrangement has also been discussed (45). Although a carbonium ion RC^+ab is shown, an S_N2 displacement at this center may occur; the leaving group is not shown. Example: **16**+**15**→**17**+pyrophosphate (see text). Electrophilic displacements of phosphate (instead of H^+) from phosphoenolpyruvate, **9**, by CO_2, aldehydes, or H^+ may be formulated as S_E2' reactions ($C=O$ formed instead of $C=C$), but it is not known whether the process is *suprafacial* or *antarafacial*. (**D**) Steps in the reaction squence for Class I aldolases (133, 165). A lysyl residue on the enzyme (R') forms a Schiffs base with dihydroxyacetone phosphate and this, on protonation, gives **D1**. RCHO=D-glyceraldehyde 3-phosphate. The sequence is completed by deprotonation of **D3** and hydrolysis. Compare **D1**$\rightleftharpoons$**D2** with **B1**$\rightleftharpoons$**B2**.

pound. The fact of transfer rules out an *anti-anti* addition-elimination mechanism involving H^+ and a group $-\ddot{X}$ on the enzyme, and migration is therefore *suprafacial* as shown. One can thus infer the probable distribution on the enzyme of counter ions or sites of coordination, the positioning of the basic transfer group, and the minor out-of-plane motions of the substrate during the reaction process.

The problem of establishing the conformations of flexible substrates and transition-state structures has been diligently studied for glutamate synthetase by Meister and his associates (69, 126). They have employed a variety of inhibitors and modified substrates and expressed their findings with stereopair drawings and photographs. The alternate fit has been described that allows the enzyme to act on both D-and L-glutamate (this dual specificity was first observed with the enzyme from peas). The inhibitor L-methionine-(S)-sulfoximine appears to function as an analog of a tetrahedral transient intermediate in the reaction (69). The importance of mimicking transition-state structures or such transient structures when trying to develop drugs (by implication also agricultural chemicals) has been stressed (185). Rigid ester substrates have been successfully used to investigate the shape of the active site of chymotrypsin (92). A diamond lattice coordinate system has been used for specifying the geometry of the active site of certain dehydrogenases (22, 76, 145, 150).

(*b*) The intermediate formation of a gyrosymmetric group in an enzymatic reaction may be detected or disproved. The principle that enzymes cannot distinguish between homotopic groups, e.g. the hydrogens of $-CH_3$, free $-CH_2^+$, or $-CH_2^-$ groups, has been stated above. DeLeo & Sprinson (51) noted that when phosphoenolpyruvate **19** and D-erythrose-4-phosphate combine in $H_2{}^{18}O$ to form 3-deoxy-D-*arabino*-heptulosonic acid 7-phosphate (a precursor of shikimic acid), the ^{18}O is not found in the inorganic phosphate released (enzyme from *Salmonella typhimurium*). This ruled out an S_E2' allylic displacement (Figure 5 C, legend) of phosphate by the carbonyl carbon (*re* face) of erythrose-4-phosphate as C–O cleavage must occur in the release of phosphate. A "ping-pong" mechanism was therefore postulated in which enolpyruvate is transferred to the enzyme to form enzyme-X-enolpyruvate and $^-OPO_3^=$, followed by an S_E2' displacement of the enzyme X group by attack of the carbonyl carbon on C-3. There was no speculation (19) as to why this should be catalytically advantageous. When the reaction was carried out in tritiated water, tritium was incorporated into the product. If enzyme-enolpyruvate is formed by way of an addition-elimination sequence in which $H_3C-C(OPO_3^=)(X\text{-Enzyme})CO_2^-$ is an intermediate, there is a one-third chance that a proton taken up by the original $=CH_2$ would be released from the $-CH_3$ to give phospho- or enzyme-enolpyruvate. This, however, cannot be the normal intermediate, as Floss and co-workers (67, 135) have shown that when phosphoenolpyruvate stereospecifically labeled in the stereoheterotopic hydrogens is used as a substrate for the enzyme from *Aerobacter aerogenes*, most of the stereospecificity is preserved in the product. The stereochemistry is equivalent to a simple attack on the 3*si* face of **9**. To reconcile these observations it is suggested that a carbanion $^-CH_2-C(OPO_3^=)(X\text{-Enzyme})CO_2^-$ is transiently formed as an intermediate or a transition state and that this is solvated by a suitably placed group on the enzyme so that only a small fraction of the $-CH_2^-$ groups formed undergo inversion and torsion or protonation and deprotonation. It is not known if the enzyme in higher plants conserves stereospecific labeling.

If a free $-CH_2^+$ formed in an enzymatic reaction is associated with an unsaturated system, it is unlikely to be gyrosymmetric. Cornforth (45) and Popják (141) have argued that because stereospecific labeling at C-1 of 16 is preserved when prenyl transferase catalyzes the attack of 16 on 15 to give 17, the reaction is more likely to involve an S_N2 displacement of pyrophosphate than the intermediate formation of $(CH_3)_2C\cdots CH\cdots CH_2^+$. The argument is only valid if one has good reason to suppose that such a carbonium ion intermediate, which has enantiotopic faces, would rapidly undergo rearrangement or torsion about the allylic C–C bonds. This is not to say that a concerted displacement does not occur, but rather to point to the uncertainty in the stereochemical argument.

It seems probable that a free $-C^+(CH_3)_2$ group is formed during the biosynthesis of sesquiterpenes of the picrotoxane group. When mevalonate-2-^{14}C (see 14) was supplied, the terminal carbons of a $-C(CH_3)=CH_2$ side chain of the isolated coriamyrtin were equally labeled (25). The label must have been exclusively in the pro-11E position of the assumed open chain precursor farnesylpyrophosphate (cf 16, 17). Loss of pyrophosphate from C-1 and attack of C-1 upon C-10 of the 10, 11 double bond would generate the postulated carbonium ion. It would have to exist as a true catalytic intermediate and not merely as a formal entity (138) in a highly integrated process. Random labeling would not be expected to occur at the $-CH(CH_3)_2$ level in the postulated biosynthetic sequence as the isopropyl methyls are stereoheterotopic (25). If the interpretation is correct, it bears on the debate (45) concerning the mechanism of sterol formation (Figure 4). The analogous equipartitioning of the ^{14}C label in certain terpenes and indole alkaloids (17, 160) derived from 17 could arise in the same way or through the formation of gyrosymmetric carbanion $-C^-(CHO)_2$ at a late stage of biosynthesis (171).

If a free $-NH_3^+$ group is formed as an intermediate carrier in an enzymatic proton transfer reaction (e.g. Figure 3 B), there is at best only a one-third chance that the H^+ abstracted will be donated again. Even if the migrating proton is hydrogen bonded to the carbanion intermediate, complete transfer would not be probable as "tumbling" from one hydrogen to the next would be expected to occur (2). Therefore, if essentially quantitative transfer occurs, it is unlikely that the catalytic base is $-\ddot{N}H_2$. The problem of correcting for isotope effects in transfer reactions has been discussed (112, 153).

(c) The observed stereochemistry may suggest that a group on the enzyme usually nucleophilic ($-\ddot{X}$), acts first as an attacking and then as a leaving group. This type of argument appears to have been first advanced to explain the retention of configuration observed in many glycoside transfer reactions. In 1954 Koshland (116; see also 19, 119) suggested that enzyme-X-glycoside is first formed by inversion at C-1 with displacement of phosphate, an aglycone, or a carbohydrate residue, and then a second inversion takes place to yield either a new glycoside or the hydrolysis product. Direct evidence to support the hypothesis was not obtained until 1970 when Voet & Abeles (182) demonstrated by two methods that an enzyme-X-β-glucoside could be formed when sucrose phosphorylase acted on sucrose (an α-glucoside). The authors note, however, that the linkage in the isolated compound need not correspond exactly to that in the normal reaction sequence, e.g. the true intermediate could be a C-1 carbonium ion stabilized by an adjacent $-CO_2^-$ group on the enzyme. When the enzyme is chemically modified with periodate or denatured, the ion pair collapses to an ester linkage. The problem of distinguishing between mechanisms involving a stabilized carbonium ion (or a carbanion) intermediate and functionally equivalent mechanisms involving covalent bonding to a group on the enzyme is a recurring one. In the transglycosylation reactions of egg white lysozyme, the side chain carboxyl of Asp 52 could stabilize a carbonium ion intermediate or it could act as a displacing group, as could the C-2 acetamido group of the substrate (41). A study of secondary deuterium isotope effects suggests that a carbonium ion intermediate is formed ($sp^3 \rightarrow sp^2$ hybridization) when certain aryl glucosides are hydrolyzed by lysozyme but not when they are hydrolyzed by the β-glucosidase from almonds (49). In a simple hydrolysis the distinction between a stabilized carbanion mechanism and a single displacement mechanism may be difficult to formulate.

The possibility that an $-\ddot{X}$ group is involved in cycloartenol and lanosterol biosynthesis has already been noted. The advantages of such a mechanism have been discussed by Cornforth (45). Popják & Cornforth (45, 141, 143) have also argued that when prenyl transferase catalyzes the attack by C-1 of **16** on the 4*re* face of **15** to give **17** with loss of the *pro-2R* hydrogen of **15** (also when the *cis* double bond of rubber is formed in an analogous manner with loss of the *pro-2S* hydrogen of **15**) the reaction is probably not a *suprafacial* allylic S_E2' displacement but an *anti* addition of C-1 of **16** and $-\ddot{X}$ followed by an *anti* elimination of HX (Figure 5 C). The immediate objection to this two-step mechanism is that the elimination of HX leads to an isolated double bond, whereas most enzymatic eliminations form double bonds which are conjugated (e.g. to $-CO_2^-$, $-COSCoA$, phenyl). For the mechanism to function $-X-$ must be an unusually good leaving group; perhaps $-S^+MeR$ derived from the side chain of methionine could function in this way. The S_E2' process, or the equivalent process involving a carbonium ion intermediate, requires less motion than the X-group mechanism. It is only objectionable if there exists a stereoelectronic prohibition against *suprafacial* S_E2' displacements. This possibility is discussed below (*e*). As the X-group mechanism involves $sp^2 \rightleftharpoons sp^3$ transitions, a secondary deuterium isotope effect at C-3 would be evidence for such a stepwise process. [In the case of aconitate isomerase (Figure 5 B) an X-group mechanism or its carbonium ion equivalent was ruled out because tritium transfer was observed (see above).]

A closely related problem arises in the formation of chorismate (67, 135). Floss has suggested that the apparent 1,4-*antarafacial* (near axial-axial) elimination of phosphate at C-3 and the *pro-6R* hydrogen from 5-enolpyruvoyl-3-phosphoshikimate [IUPAC numbering (104)] could be realized by an S_N2' *suprafacial* displacement of phosphate by $-\ddot{X}$ attack at C-1 followed by an *anti* elimination of HX from C-6 and C-1. As C-1 is already substituted by $-CO_2^-$ the mechanism cannot be probed by looking for a secondary isotope effect. It might be possible to demonstrate the first step in the two-step reaction if the *pro-6R* hydrogen were replaced by, for example, fluorine. Again the mechanism is only attractive if there is a stereoelectronic prohibition against the apparent overall reaction. Note, however, that a *suprafacial* allylic displacement was discounted above (S_E2') and here a *suprafacial* displacement is involked (S_N2').

Other examples of X-group mechanisms with stereochemical significance could be cited. Now that the possibility of such mechanisms is well recognized, the emphasis needs to be placed on the search for experimental methods to support or disprove the individual proposals.

(*d*) The observed stereochemistry may permit certain mechanisms to be excluded on purely steric grounds. For example, it was suggested (30, 55, 123, 169) that because citrate (*si*)-synthase failed to catalyze an exchange of the hydrogens of acetyl CoA, the enzyme acts in a concerted manner, and that the electrophilic attack by the 2*si* face of the carbonyl carbon of oxaloacetate 7 could be accompanied by a removal of H⁺ by the C-4 carboxylate ion (Figure 6 E). As the carboxylate ion could not reach to the back of the methyl group, such a reaction must occur without inversion. Eggerer (55) suggested that the slow exchange of the hydrogens of acetyl CoA that occurs in the presence of L-malate takes place because the enzyme binds L-malate at the oxaloacetate binding site with its C-4 carboxylate group positioned to catalyze H⁺ removal. The reaction has been shown to proceed with inversion (56, 113, 119a) and the cyclic mechanism, therefore, is excluded. L-Malate is believed to act by stabilizing the conformation of the active site with the proton abstracting group of the enzyme in the correct position (168, 169). The displacement reaction may well not be concerted (see *f* below).

In the rejected mechanism **E** there is nothing to commend the C-4 carboxylate ion as a catalytic group beyond its proximity. If a mechanism involving a concerted cyclic transition state can be written, it is likely to have a catalytic advantage over its noncyclic equivalent. Rétey et al (11, 146, 149) have shown that the enzymatic carboxylation by carboxybiotin of propionyl CoA takes place without inversion. Rose (152) has shown that the analogous carboxylation of pyruvate 9 also occurs without inversion. These results are compatible with the mechanism **F** (without and with Mn²⁺ respectively) in which the carbonyl oxygen of biotin serves to abstract the proton being replaced by carboxylate (134). The possibility of a cyclic transition state was also considered in investigating the elimination of ammonia by phenylalanine ammonia-lyase. It seemed possible that the incompletely defined prosthetic group of the enzyme (83), having combined with the α-amino group of L-phenylalanine 2, might then abstract the β-proton at the same time that the N–C$^\alpha$ bond is broken. Many amides on pyrolysis yield olefins by way of 6-membered transition states, the elimination being highly concerted and *syn* (mechanism **G**) (52). When the enzymatic elimination was found to be *anti* (84, 87, 99, 183) it became unnecessary to try and write structures for the dehydroalanine-containing prosthetic group that would lead to an amide-like *syn* elimination.

(*e*) The observed stereochemistry may permit certain mechanisms to be excluded on stereoelectronic grounds. Orbital-symmetry rules (71, 98, 187) have been remarkably successful in predicting the course of certain types of chemical reactions; typically those involving cyclic transition states. Reference has already been made to these rules in discussing aconitate isomerase. In addition, one may note that if the transformation of chorismate to prephenate is a [3,3]-sigmatropic

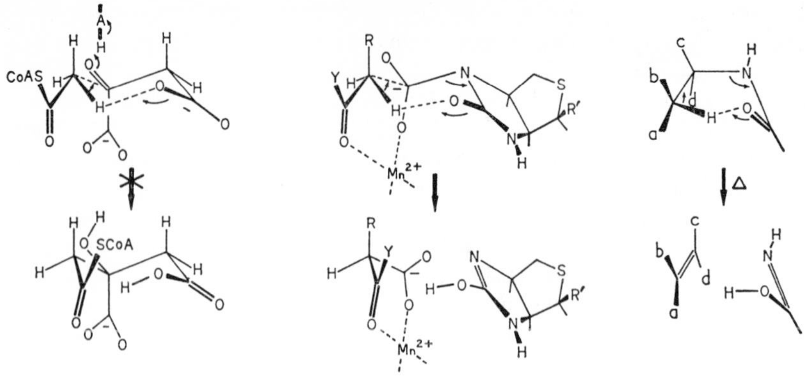

FIGURE 6. Some cyclic processes. (E) postulated (30, 55) but disproved (56, 113) mechanism for the conversion of acetyl CoA and oxaloacetate by citrate (*si*)-synthase to (3*S*)-citryl CoA [a presumed enzyme-bound intermediate in the overall reaction (169)]. (F) Enzymatic carboxylations by carboxybiotin (23, 134). Formation of (2*S*)-methylmalonyl CoA by propionyl CoA carboxylase; Y=SCoA, R=CH₃, no metal (11, 146, 149). Also, formation of oxaloacetate by pyruvate carboxylase; Y=COO⁻, R=H, metal present (127, 152). The biotin tautomer formed readily reverts to the urea form. (G) Pyrolytic *syn* elimination by an amide to yield an olefin and an amide tautomer which readily reverts to the usual form (52). The conformations of the cyclic transition states of F and G are conjectural as a detailed analysis (187) of such processes has not yet been made.

rearrangement, it must be *suprafacial* with respect to both parts and should occur through a chair-shaped transition state (106). If stereospecific labeling of the enolpyruvoyl hydrogens of chorismate is obtainable, and if the distribution of label can be checked, then the stereochemistry implied by the mechanism could be verified by examining the labeled L-phenylalanine-3-*t* (or *d*) produced with L-phenylalanine ammonia-lyase. Unfortunately, although the rules prevent one postulating such symmetry forbidden mechanisms as uncatalyzed 1,2-proton shifts (110), there are not many enzymes whose mechanisms are likely to fall within the area already explored by Woodward & Hoffmann (187). The major impact of the rules on enzymology may therefore be an indirect one, namely, to increase awareness of molecular orbitals and their stereoelectronic properties.

There exists a molecular orbital approach to chemical reactions (highest-occupied, lowest-unoccupied molecular orbital theory) which is closely related to that of Woodward & Hoffmann. It has been applied to such noncyclic reactions as 1,2 and 1,4 concerted eliminations and to allylic displacements, both S_E2′ and S_N2′ (7, 68). One would like to have a theoretically justified rule which says, e.g. that if the stereochemistry is *syn* the mechanism cannot be concerted. The stereochemistry predicted, however, varies according to the nature of the transition state, i.e. according to whether both bonds are equally formed and cleaved in the transition state (a synchronous reaction) or whether one bond is more broken than the other so that a swing in charge distribution occurs. For nonsynchronous eliminations, the system is either carbanion like or carbonium ion like. The representation of the alternative stepwise and concerted elimination processes in terms of a potential energy diagram has been discussed (85, 132). To summarize: synchronous eliminations if 1,2, are preferentially *anti*, and if 1,4,*syn*. Synchronous allylic displacements of either type are *antarafacial*. Nonsynchronous 1,2 eliminations that are carbanion-like are *syn* and carbonium ion-like *anti*, and nonsynchronous allylic displacements of either type are *suprafacial* (e.g. Figure 5, **C1→C2**). If the enzymologist can fit any overall stereochemistry to an appropriate concerted transition state, the theory is of little help when he is trying to determine if the reaction is indeed concerted. The theoretical treatment is in keeping with the earlier results of DePuy and co-workers (53, 125, next paragraph), who studied *syn* and *anti* 1,2 concerted eliminations from phenyl substituted *trans-* and *cis-*2-phenylcyclopentyl tosylates. They concluded, on the basis of Hammett plots and primary kinetic isotope effects, that the E2 transition states were carbanion like (E1cB-like) for *syn* elimination and carbonium ion like (E1-like) for *anti* elimination.

The picture of elimination reactions assembled experimentally is more likely to be of value to the enzymologist, although the model reactions studied by physical organic chemists are not very similar to those catalyzed by enzymes. For reviews see (35, 36, 85, 125, 137). Two matters are worthy of particular comment: the importance of the torsional angle and the importance of the base. For E2 eliminations of HX from $C^{\alpha}X$-$C^{\beta}H$ where X is a good leaving group, where a strong base is used, and where the torsional angle between $C^{\alpha}X$ and $C^{\beta}H$ is controlled by using cyclic compounds, very similar elimination rates have been observed when the angle is 0° and when it is 180° (53, 125). Rates for intermediate angles, of course, are very much slower, e.g. 1000-fold. The observed preference for E2 concerted eliminations from flexible molecules to be *anti* may be a direct consequence of the eclipsing strain required to align the groups for a *syn* elimination. Recent studies (24, 136, 137) of elimination reactions catalyzed by weak (soft) bases, albeit in aprotic solvents, suggest that such bases, B⁻, tend to approach C^{α} first and generate an olefin-like transition state with BH leaving from one face as X⁻ leaves from the other, i.e. this "E2C" process is *anti*. Only if the latter situation universally prevails in enzyme catalysis, and there is no evidence on the matter, could one argue that a *syn* elimination must be a stepwise process. One cannot conclude anything about the details of the process if an *anti* elimination is observed. The stereochemistry of concerted 1,4-eliminations (48) and allylic displacement reactions (172) have been much less intensively studied, and the very existence of S_N2' reactions has been questioned (29). They cannot be less complex than 1,2-eliminations. It seems unwise, therefore, to try to deduce from the overall stereochemistry of enzyme catalyzed eliminations and displacements anything about their concerted or stepwise natures. The X-group mechanisms discussed above should be regarded as valuable conjectures, not as received truth.

All but a few (12, 33, 86, 159, 164) enzymatic eliminations that have been examined are *anti*. Rose (153, 154) has pointed out that this can be explained without invoking stereoelectronic considerations. Even for stepwise eliminations, acidic and basic groups in sufficient proximity to catalyse 1,2 *syn* eliminations would be likely to form a salt bridge and thus be ineffective. This factor would operate in the most primitive enzyme. When D_2O adds to fumarate nonenzymatically, a mixture of *threo* and *erythro* racemates is obtained (61; cf 13, 14). In H_2O the trigonal carbanion ($-CH=CO_2^-$) arising from OH^- attack on the other olefinic carbon would have stereoheterotopic faces which would be protonated at similar rates in free solution. However, they would be distinguished in enzyme catalysis, and by the above argument a primitive enzyme with a carbanion mechanism would be expected to develop *anti* stereochemistry. As the present evidence (78) suggests that fumarate hydratase has a carbonium ion mechanism (Figure 5, $A1 \rightleftharpoons A3 \rightleftharpoons A2$), the enzyme may well have evolved to the present form via a concerted E2 mechanism ($A1 \rightleftharpoons A2$) retaining the same stereochemistry throughout. Stereoelectronic factors, of course, influence the varying details of the transition state geometries.

(f) The observed stereochemistry may indicate nothing of fundamental importance about the electronic details of the reaction. This statement is not the same thing as saying that we cannot use the stereochemical data we have because we lack other relevant information (the dilemma discussed under e). The issue is raised by the large group of electrophilic displacement reactions that involve attack upon enols or metal coordinated enolate ions [e.g. Class II metal containing aldolases (133)], simple enamines [e.g. Class I aldolases as found in mammals and higher plants (133, 147)], or enamine-pyridoxal combinations (54, 165). For example, a number of lines of evidence indicate that Class I fructose-1,6-diphosphate aldolases form a Schiff's base with dihydroxyacetone phosphate (Figure 5 D1). The stereospecific exchange of the *pro-3S* hydrogen (107, 151), which may be formulated as $D1 \rightleftharpoons D2$, can be observed without the cosubstrate being present. It seems likely that in the complete reaction sequence, $D1 \rightleftharpoons D2$ is distinct from $D2 \rightleftharpoons D3$. If this is the case, it is energetically irrelevant to the $D2 \rightleftharpoons D3$ step which of the stereoheterotopic hydrogens at C-3 is eliminated. The *si* electrophilic attack by the carbonyl carbon of D-glyceraldehyde-3-phosphate on the *3si* face of the enamine is necessary to produce the required metabolic product. The overall retention of configuration is consistent with the hypothesis that a single proton-donor proton-acceptor group on the enzymes functions in both steps, but such an economic use of resources is not a fundamental electronic feature of the reaction. It is possible that a positively charged group on the enzyme is positioned close to the *3re* face of the enamine thus increasing δ^- at C-3. This would, of course, be of electronic significance.

The same possibility must be considered in discussing certain electrophilic displacements involving C-α of thio-esters and of metal coordinated carboxylate ions. (The carboxylations formulated as Figure 6B represent a contrary point of view.) Citrate (*si*)-synthase in the presence of L-malate slowly catalyzes the exchange of the hydrogen of acetyl CoA (55, 168). Therefore, it has been suggested that when oxaloacetate, instead of L-malate, is bound, the active site takes on the correct conformation and a thio-ester stabilized carbanion is generated. There are significant primary deuterium and tritium isotope effects for the overall synthesis of citrate (113, 169). Carbanion formation, therefore, may be rate determining for the overall reaction. As carbanions generated in nonenzymatic reactions may exist as chiral ion pairs, or as chiral solvated ions, and these may collapse or react to give overall inversion or retention of configuration (47), the fact that inversion is observed in this case and retention observed in the aldolase case may be of minor electronic significance. There is a considerable amount of evidence that a β-carbanion intermediate is formed in the reaction catalyzed by aspartate ammonia-lyase (64, 85). The carbanion is stabilized by the coordination of the adjacent carboxylate ion to a divalent metal. It is possible, therefore, that in the synthetic action of isocitrate lyase a similar carbanion is derived from succinate 3 with removal of a *pro-S* hydrogen (79, 166). This then reacts with the *si*-face of the carbonyl carbon of glyoxylate 6 to give (2R,3S)-isocitrate 13. Again, the overall inversion of configuration observed may be of minor electronic significance.

Metabolic pathways.—Information about the metabolic sequences leading to the formation of natural products in plant tissues has been obtained by supplying stereospecifically labeled precursors and isolating and degrading the steroids, terpenoids, carotenoids, or alkaloids produced. Much of this work has involved double labeled mevalonate, e.g. (3R,4S)-mevalonate-4-*t*-2-^{14}C [= (3R)-mevalonate (14) + (3R)-mevalonate-2-^{14}C + (3R,4S)-mevalonate-4-*t*]. Deuterium labeling has been used but more usually in conjunction with mammalian tissues. Recently chemical synthesis has made available (3R)- and (3S)-L-phenylalanine-3-*t* (84, 85, 87, 99), (3R)- and (3S)-L-tyrosine-3-*t* (111), and (2R)- and (2S)-2-phenylethylamine-2-*t* (18), and these compounds are being used to probe aspects of alkaloid biosynthesis. In the double labeling experiments, ^{14}C serves to measure the extent of incorporation of the precursor, and the fate of its hydrogen is deduced from the extent of ^{3}H retention, the point of ^{3}H attachment, assuming that it is retained, and the isotopic chirality of the center to which it is attached. Many of the problems that can arise as a result of primary isotope effects are eliminated by carrying out parallel experiments using compounds having at the same methylene center both (R) and (S) isotopic chirality (17, 163). Some inferences that can be made in such studies are obvious, e.g. a compound that is labeled with ^{3}H and ^{14}C cannot be metabolically derived from one which is only labeled with ^{14}C. Metabolic relationships between compounds may be more complex than ^{14}C labeling data alone indicates, e.g. loss of ^{3}H in the conversion of stereospecifically tritiated lanosterol to tritiated cholesterol led to the identification of an unsuspected metabolic intermediate (142). Sometimes ^{3}H:^{14}C labeling data can be used to choose between alternative pathways. For example, although it can be shown that cinnamate-^{14}C as well as L-phenylalanine-^{14}C is incorporated into certain alkaloids, the fact that cinnamate-^{14}C will react with the amino-enzyme intermediate generated when phenylalanine ammonia-lyase acts on L-phenylalanine (83-91) makes it impossible to be certain that the metabolic sequence is from L, phenylalanine→cinnamate→alkaloid. As in this sequence the *pro*-3S hydrogen of L-phenylalanine is eliminated, tritium would not be incorporated from (3S)-L-phenylalanine-3-*t*-3-^{14}C but would be from the (3R) diastereoisomer if cinnamate is an intermediate. On carrying out such experiments with the autumn crocus, the results were consistent with the hypothesis that a part of the colchicine molecule is formed by way of cinnamate (183).

The above examples are a somewhat arbitrary selection from a rapidly growing literature. In view of the discussion of steric concepts earlier in this essay, it seems pertinent to conclude by drawing attention to the various ways in which gyrosymmetric intermediates can be of importance in the study of metabolic pathways.

In the first place, the occurrence of an intermediate with a particular type of gyrosymmetry can lead to a scrambling of labeling. It has been shown (118, 158) using shikimate-^{14}C and also (2R)- and (2S)-shikimate-2-*t*-7-^{14}C that in English walnuts juglone (5-hydroxynaphtho-1,4-quinone) is formed by way of an intermediate which is gyrosymmetric about an axis that makes C-5 and C-8 homotopic.

The formation of a methyl group in a metabolic pathway may be the occasion for the loss of stereospecific labeling. If the methyl group is formed in a side reaction the loss may be only partial, e.g. the =CH$_2$ of **15** is converted to the *pro*-3E methyl group of **16** by isopentenyl pyrophosphate isomerase (161). Both compounds are substrates for prenyl transferase. If **16** is utilized as fast as it is formed, there will be no back isomerization and any differential labeling of the =CH$_2$ hydrogens of **15** will be preserved in the products of prenyl transferase action (geranyl, farnesyl, or geranyl-geranyl pyrophosphates, as appropriate). This appears to be the case in lanosterol biosynthesis, but in sesquiterpene and cyclo-artenol biosynthesis in plants the labeling is randomized (75).

It might be possible to use the information that scrambling of label occurs in one pathway to look for a contribution by an alternative pathway. For example, if (2R,3S)-isocitrate-2-^{14}C were infiltrated into a plant tissue and L-malate-^{14}C isolated and degraded, an excess of ^{14}C at C-2 of L-malate would suggest that the glyoxylate cycle enzymes were present. In the citric acid cycle, free succinate and fumarate occur as intermediates, and there is an equal chance that C-2 of isocitrate will end up as C-2 or C-3 of L-malate, whereas by the glyoxylate cycle route C-2 ends up as C-2 of L-malate.

Lastly, adjacent active sites in a multienzyme complex could form and act upon a sequestered intermediate that in free solution would be gyrosymmetric in such a manner that the gyrosymmetry is not realized. For example, the nonsymmetrical head-to-head reductive condensation of farnesyl pyrophosphate leads to the C$_{30}$ compound squalene which is gyrosymmetric with respect to a rotational axis perpendicular to the C-12, C-13 bond. If one end of squalene is transferred to the enzyme forming (3S)-squalene-2,3-oxide (Figure 4) before the other has peeled off the squalene-forming enzyme, ^{2}H or ^{3}H introduced into one end of squalene during synthesis would appear in one end of the oxide. The position of the label in the oxide may be determined by degrading its later metabolic products. For lanosterol biosynthesis in pig liver homogenates most, possibly all, of the squalene formed passes to free squalene before being oxidized (63). It is possible that in cycloartenol biosynthesis free squalene is not an intermediate; the experiment needs to be repeated with plant tissue extracts.

Recent genetic studies (50) with the mold *Phycomyces blakensleeanus* indicate that a multienzyme complex is responsible for condensing two geranylgeranyl pyrophosphate molecules to give the C_{40} compound phytoene and transforming this to β-carotene by way of five sequestered intermediates. The first four steps involve *anti* eliminations (75) of H_2 (represented as d→D) and the last two steps involve cyclyzations (c→C) at the two ends of the chain; thus if the phytoene chain is represented as cdd.ddc, the probable sequence is →cdD.ddc→cdD.Ddc →cDD.Ddc→cDD.DDc→CDD.DDc→CDD.DDC. Four of these compounds in free solution are gyrosymmetric with respect to an axis perpendicular to the C-16, C-17 bond. The genetic evidence shows that although enzymes coded by the same gene catalyze both cyclizations, only one is so placed that it will accept one end of the lycopene molecule cDD.DDc, i.e. the two enzymes are heterotopic in the complex. This suggests that the C_{40} compounds "walk" from active site to active site along the complex. To explore the details of the walk it would be necessary to supply earlier intermediates lacking rotational symmetry labeled at one end (cdD.ddc* and cDD.Ddc*) and degrade the γ-carotene (e.g. CDD.DDc*) formed by the mutant complex. Of course, it may be difficult to get the complex to accept a free intermediate as a substrate. The genetic evidence would allow the whole complex to have, for example, C_n symmetry provided that the enzymes associated with one chemical sequence cannot utilize substrate from an independent sequence. All this remains to be investigated; nonetheless, it is difficult to imagine a more fugal combination of enzyme specificity and enzyme symmetry.

Concluding Remarks

The last 10 years have seen a remarkable development in our knowledge of the overall stereochemistry of the more familiar biochemical transformations. The technical vocabulary for expressing this information aids the initiated in recognizing problems that would formerly have been overlooked but tends to mystify the outsider. I have therefore tried to present an overview of the stereochemical concepts and nomenclature necessary for understanding the current literature. As more is learned about plant enzymes and the biosynthesis of natural products formed in plants, increasing reference to stereochemical matters will appear in the context of plant physiology.

A second purpose of the essay has been to enumerate some of the limitations as well as the successes of the stereochemical approach to the study of enzyme mechanisms. Stereochemical correlations and X-ray diffraction studies of enzymes are a necessary complement to the kinetic analysis of enzyme action. A general understanding of the relationship between catalysis and the detailed conformations of active sites and substrate molecules will be of increasing importance to those concerned with the chemical control of plant growth and plant pathogens. In addition, I have discussed briefly the evolution of enzyme quaternary structures and enzyme mechanisms. Information relevant to both topics should be obtainable from the comparative study of plant enzymes.

ACKNOWLEDGMENTS

I am indebted to Drs. A. R. Battersby, I. A. Rose, and P. A. Srere for making manuscripts of their recent reviews (17, 154, 169) available prior to publication, and to Mrs. Ruth DiLeone for preparation of the figures. The writing of this review has been supported in part by Grant GB 29021X from the National Science Foundation.

LITERATURE CITED

1. Akazawa, T. 1970. *Progr. Phytochem.* 2:107–41
2. Almy, J., Cram, D. J. 1969. *J. Am. Chem. Soc.* 91:4459–68
3. Altekar, W. W., Bhattacharyya, P. K., Rangachari, P. N., Maskati, F. S., Rao, M. R. R. 1965. *Indian J. Biochem.* 2:132–33
4. American and British Chemical Societies: Rules of Carbohydrate Nomenclature 1963. *J. Org. Chem.* 28:281–91. Updated 1972. See Ref. 103a
5. Anderson, L., Price, G. B., Fuller, R. C. 1968. *Science* 161:482–84
6. Anet, F. A. L. 1960. *J. Am. Chem. Soc.* 82:994–95
7. Anh, N. T. 1968. *Chem. Commun.* 1089–90
8. Arigoni, D., Eliel, E. L. 1969. *Top. Stereochem.* 4:127–243
9. Ibid, 161, 166
10. Ibid, 193–95
11. Arigoni, D., Lynen, F., Rétey, J. 1966. *Helv. Chim. Acta* 49:311–15
12. Avigad, G., Englard, S. 1969. *Fed. Pro.* 28:345
13. Bada, J. L., Miller, S. L. 1969. *J. Am. Chem. Soc.* 91:3948–49
14. Ibid 1970. 92:2774–82
15. Barker, H. A., Smyth, R. D., Wilson, R. M., Weissbach, H. 1959. *J. Biol. Chem.* 234:320–28
16. Battersby, A. R. 1970. *Natural Substances Formed Biologically From Mevalonic Acid. Biochem. Soc. Symp.* 29:157–68
17. Battersby, A. R. 1972. *Accounts Chem. Res.* 5:148—54
18. Battersby, A. R., Kelsey, J. E., Staunton, J. 1971. *Chem. Commun.* 183–84
19. Bell, R. M., Koshland, D. E. Jr. 1971. *Science* 172:1253–56
20. Bentley, R. 1969, 1970. *Molecular Asymmetry in Biochemistry.* New York: Academic. 566 pp. and 322 pp.
21. Ibid. 2:22
22. Ibid. 2:22–50
23. Ibid. 2:435
24. Biale, G., Parker, A. J., Smith, S. G., Stevens, I. D. R., Winstein, S. 1970 *J. Am. Chem. Soc.* 92:115–22
25. Biollaz, M., Arigoni, D. 1969. *Chem. Commun.* 633–34
26. Blackwood, J. E., Gladys, C. L., Loening, K. L., Petrarca, A. E., Rush, J. E. 1968. *J. Am. Chem. Soc.* 90:509–10
27. Bohm, B. 1965. *Chem. Rev.* 65:435–66
28. Bondinell, W. E., Sprinson, D. B. 1970. *Biochem. Biophys. Res. Commun.* 40:1464–67
29. Bordwell, F. G. 1970. *Accounts Chem. Res.* 3:281–90
30. Bové, J., Martin, R. O., Ingraham, L. L., Stumpf, P. K. 1959. *J. Biol. Chem.* 234:999–1003
31. Bowers, W. F., Czubaroff, V. B., Haschemeyer, R. H. 1970. *Biochemistry* 9:2620–25
32. Boyd, D. R., McKervey, M. A.1968. *Quart. Rev. Chem. Soc.* 22:95–122
33. Bright, H. J., Lundin, R. E., Ingraham, L. L. 1964. *Biochemistry* 3:1224–30
34. Bruice, T. C., Brown, A., Harris, D. O. 1971. *Proc. Nat. Acad. Sci. USA* 68:658–61
35. Bunnett, J. F. 1962. *Angew. Chem. Int. Ed.* 1:225–80
36. Bunnett, J. F. 1969. *Surv. Progr. Chem.* 5:53–93
37. Butler, P. J. G. 1971. *Nature* 233:25–27
38. Cahn, R. S. 1964. *J. Chem. Educ.* 4:116–25
39. Cahn, R. S., Ingold, C. K., Prelog, V. 1966. *Angew. Chem. Int. Ed.* 5:385–415
40. Caspar, D. L. D., Klug, A. 1962. *Cold Spring Harbor Symp. Quant. Biol.* 27:1–24
41. Chipman, D. M., Sharon, N. 1969. *Science* 165:454–65
42. Clayton, R. B. 1965. *Quart. Rev. Chem. Soc.* 19:168–201
43. Cohn, M., Pearson, J. E., O'Connell, E. L., Rose, I. A. 1970. *J. Am. Chem. Soc.* 92:4095–98
44. Corey, E. J., Lin, K., Yamamoto, H. 1969. *J. Am. Chem. Soc.* 91:2132–34
45. Cornforth, J. W. 1968. *Angew. Chem. Int. Ed.* 7:903–11
46. Cornforth, J. W. 1969. *Quart. Rev. Chem. Soc.* 23:125–40
47. Cram, D. J. 1965. *Fundamentals of Carbanion Chemistry.* New York: Academic. 289 pp.
48. Cristol, S. J. 1971. *Accounts Chem. Res.* 4:393–400
49. Dahlquist, F. W., Rand-Meir, T., Raftery, M. A. 1969. *Biochemistry* 8:4214–21
50. De La Guardia, M. D., Aragón,

G. M. G., Murillo, F. J., Cerdá-Olmedo, E. 1971. *Proc. Nat. Acad. Sci. USA* 69:2012–15

51. DeLeo, A. B., Sprinson, D. B. 1968. *Biochem. Biophys. Res. Commun.* 32:873–77

52. DePuy, C. H., King, R. W. 1960. *Chem. Rev.* 60:431–57

53. DePuy, C. H., Morris, G. F., Smith, J. S., Smat, R. J. 1965. *J. Am. Chem. Soc.* 87:2421–28

54. Dunathan, H. C. 1971. *Advan. Enzymol.* 35:79–134

55. Eggerer, H. 1965. *Biochem. Z.* 343:111–38

56. Eggerer, H. et al 1970. *Nature* 266:517–19. See also Ref. 119a

57. Eigen, M., Hammes, G. G. 1963. *Advan. Enzymol.* 25:1–38

58. Eliel, E. L. 1971. *J. Chem. Educ.* 48:163–67

59. Englard, S. 1960. *J. Biol. Chem.* 235:1510

60. Englard, S., Hanson, K. R. 1969. *Methods Enzymol.* 13:567–601

61. Ericson, L. E., Alberty, R. A. 1959. *J. Phys. Chem.* 63:705–9

62. Eschenmoser, A., Ruzicka, L., Jeger, O., Arigoni, D. 1955. *Helv. Chim. Acta* 38:1890–1904

63. Etemadi, A. H., Popják, G., Cornforth, J. W. 1969. *Biochem. J.* 111:445–51

64. Fields, G. A., Bright, H. J. 1970. *Biochemistry* 9:3801–9

65. Fisher, H. F., Frieden, C., McKinley McKee, J. S., Alberty, R. A. 1955. *J. Am. Chem. Soc.* 77:4436

66. Fiskin, A. M., van Bruggen, E. F. J., Fisher, H. F. 1971. *Biochemistry* 10:2396–2408

67. Floss, H. G., Onderka, D. K., Carroll, M. 1972. *J. Biol. Chem.* 247:736-44

68. Fukui, K. 1971. *Accounts Chem. Res.* 4:57–64

69. Gass, J. D., Meister, A. 1970. *Biochemistry* 9:1380–90

70. Gawron, O., Glaid, A. J. III, Fondy, T. P. 1961. *J. Am. Chem. Soc.* 83:3634–40

71. Gill, G. B. 1968. *Quart. Rev. Chem. Soc.* 22:338–89

72. Ginsburg, A., Stadtman, E. R. 1970. *Ann. Rev. Biochem.* 39:429–67

73. Glusker, J. P. 1968. *J. Mol. Biol.* 38:149–62

74. Goad, L. J. 1970. *Natural Substances Formed Biologically From Mevalonic Acid. Biochem. Soc. Symp.* 29:45–77

75. Goodwin, T. W. 1971. *Biochem. J.* 123:293–329

76. Graves, J. M. H., Clark, A., Ringold, H. J. 1965. *Biochemistry* 4:2655–71

77. Hammes, G. G. 1968. *Accounts Chem. Res.* 1:321–29

78. Hansen, J. N., Dinovo, E. C., Boyer, P. D. 1969. *J. Biol. Chem.* 244:6270–79

79. Hanson, K. R. 1965. *Fed. Proc.* 24:229

80. Hanson, K. R. 1966. *J. Am. Chem. Soc.* 88:2731–42

81. Hanson, K. R. 1966. *J. Mol. Biol.* 22:405–9

82. Ibid 1968. 38:133–36

83. Hanson, K. R., Havir, E. A. 1970. *Arch. Biochem. Biophys.* 141:1–17

84. Hanson, K. R, Havir, E. A. 1972. *Recent Advan. Phytochem.* 4:45–85

85. Hanson, K. R., Havir, E. A. 1972. *The Enzymes*, 3rd ed. 7:75-166

86. Hanson, K. R., Rose, I. A. 1963. *Proc. Nat. Acad. Sci. USA* 50:981–88

87. Hanson, K. R., Wightman, R. H., Staunton, J., Battersby, A. R. 1971. *Chem. Commun.*:185–86

88. Harrison, P. M. 1963. *J. Mol. Biol.* 6:404–22

89. Haschemeyer, R. H. 1970. *Advan. Enzymol.* 33:71–118

90. Haselkorn, R., Fernández-Morán, H., Kieras, F. L., van Bruggen, E. F. J. 1965. *Science* 150:1598–1601

91. Havir, E. A., Hanson, K. R. 1968. *Biochemistry* 7:1904–14

92. Hayashi, Y., Lawson, W. B. 1969. *J. Biol. Chem.* 244:4158–67

93. Heine, J. 1966. *J. Am. Chem. Soc.* 88:5525–28

94. Hirschmann, H. 1960. *J. Biol. Chem.* 235:2762–67

95. Hirschmann, H. 1964. *Compr. Biochem.* 12:236–60

96. Hirschmann, H., Hanson, K. R. 1971. *Eur. J. Biochem.* 22:301–9

97. Hirschmann, H., Hanson, K. R. 1971. *J. Org. Chem.* 36:3293–3306

98. Hoffmann, R., Woodward, R. B. 1970. *Science* 167:825–31

99. Ife, R., Haslam, E. 1971. *J. Chem. Soc. C* 2818–21

100. IUPAC Definitive Rules for the Nomenclature of Natural Amino Acids and Related Substances. 1960. *J. Am. Chem. Soc.* 82:5575–77. Addendum 1963. *J. Org. Chem.* 28:291–82

101. IUPAC/IUB Revised Tentative Rules for the Nomenclature of

Steroids. 1969. *J. Org. Chem.* 34: 1517–32

102. IUPAC Tentative Rules for the Nomenclature of Organic Chemistry: Section E, Fundamental Stereochemistry. 1970. *J. Org. Chem.* 35:2849–67

103. IUPAC/IUB CBN: The Nomenclature of Lipids. 1967. *J. Biol. Chem.* 242:4845–49

103a. IUPAC/IUB Tentative Rules for Carbohydrate Nomenclature. 1972. *J. Biol. Chem.* 247:613–35

104. IUPAC/IUB Tentative Rules for Cyclitol Nomenclature. 1968. *J. Biol. Chem.* 243:5809–19

105. Jaffe, H. H., Orchin, M. 1965. *Symmetry in Chemistry.* New York: Wiley. 191 pp.

106. Jefferson, A., Scheinmann, F. 1968. *Quart. Rev. Chem. Soc.* 22:391–421

107. Johnson, C. K., Gabe, E. J., Taylor, M. R., Rose, I. A. 1965. *J. Am. Chem. Soc.* 87:1802–4

108. Josephs, R. 1971. *J. Mol. Biol.* 55:147–53

109. Kaeppeli, F., Rétey, J. 1972. *Eur. J. Biochem.* 23:198–202

110. Kemp, D. S. 1971. *J. Org. Chem.* 36:202–4

111. Kirby, G. W., Michael, J. 1971. *Chem. Commun.* 187–88

112. Klinman, J. P., Rose, I. A. 1971. *Biochemistry* 10:2259–67

113. Ibid, 2267–72

114. Klotz, I. M., Langerman, N. R., Darnall, D. W. 1970. *Ann. Rev. Biochem.* 39:25–62

115. Klug, A. 1967. *Formation Fate Cell Organeles* 6:1–18

116. Koshland, D. E. 1954. *Mechanism of Enzyme Action,* ed. W. D. McElroy, B. Glass, 608–41. Baltimore: Johns Hopkins. 819 pp.

117. Koukol, J., Conn, E. E. 1961. *J. Biol. Chem.* 236:2692–98

118. Leduc, M. M., Dansette, P. M., Azerad, R. G. 1970. *Eur. J. Biochem.* 15:428–35

119. Lemieux, R. U. 1954. *Advan. Carbohyd. Chem.* 9:1–57

119a. Lenz, H. et al 1971. *Eur. J. Biochem.* 24:207–15

120. Levy, H. R., Talalay, P., Vennesland, B. 1962. *Progr. Stereochem.* 3:299–349

121. Lingens, F. 1968. *Angew. Chem. Int. Ed.* 7:350–60

122. MacLennan, D. H., Beevers, H. 1964. *Phytochemistry* 3:109–13

123. Marcus A., Vennesland, B. 1958. *J.*

124. Mathieu, J., Weill-Raynal, J. 1968. *Bull. Soc. Chim. Fr.* 1211–44

125. McLennan, D. J. 1967. *Quart. Rev. Chem. Soc.* 21:490–506

126. Meister, A. 1968. *Advan. Enzymol.* 31:183–218

127. Mildvan, A. S., Scrutton, M. C. 1967. *Biochemistry* 6:2978–94

128. Milstien, S., Cohen, L. A. 1970. *Proc. Nat. Acad. Sci. USA* 67:1143–47

129. Mislow, K. 1966. *Introduction to Stereochemistry.* New York: Benjamin. 193 pp.

130. Mislow, K., Raban, M. 1967. *Top. Stereochem.* 1:1–38

131. Monod, J., Wyman, J., Changeux, J. P. 1965. *J. Mol. Biol.* 12:88–118

132. More O'Ferrall, R. A. 1970. *J. Chem. Soc. B* 274–77

133. Morse, D. E., Horecker, B. L. 1968. *Advan. Enzymol.* 31:125–81

134. Moss, J., Lane, M. D. 1971. *Advan. Enzymol.* 35:321–442

135. Onderka, D. K., Floss, H. G. 1969. *J. Am. Chem. Soc.* 91:5894–96

136. Parker, A. J. 1969. *Chem. Rev.* 69:1–32

137. Parker, A. J. 1971. *Chem. Tech.* 1:297–303

138. Parker, W., Roberts, J. S., Ramage, R. 1967. *Quart. Rev. Chem. Soc.* 21:331–63

139. Perutz, M. F. 1969. *Eur. J. Biochem.* 8:455–66

140. Popják, G. 1970. *Natural Substances Formed Biologically from Mevalonic Acid. Biochem. Soc. Symp.* 29:17–33

141. Popják, G. 1970. *The Enzymes,* 3rd ed. 2:115–215

142. Ibid, 204

143. Popják, G., Cornforth, J. W. 1966. *Biochem. J.* 101:553–68

144. Poulsen, L. L., Wedding, R. T. 1970. *J. Biol. Chem.* 245:5709–17

145. Prelog, V. 1964. *Chem. Natur. Prod.* 3:119–30

146. Prescott, D. J., Rabinowitz, J. L. 1968. *J. Biol. Chem.* 243:1551

147. Rapoport, G., Davis, L., Horecker, B. L. 1969. *Arch. Biochem. Biophys.* 132:286–93

148. Reed, L. J., Cox, D. J. 1970. *The Enzymes,* 3rd ed. 1:213–40

149. Rétey, J., Lynen, F. 1965. *Biochem. Z.* 342:256–71

150. Robertson, J. S., Hussain, M. 1969. *Biochem. J.* 113:57–65

151. Rose, I. A. 1958. *J. Am. Chem. Soc.* 80:5835–36

152. Rose, I. A. 1970. *J. Biol. Chem.* 245:

6052–56

153. Rose, I. A. 1970. *The Enzymes*, 3rd ed. 2:281–320
154. Rose, I. A. 1972. *Crit. Rev. Biochem.* 1:33–57
155. Rose, I. A., O'Connell, E. L. 1967. *J. Biol. Chem.* 242:1870–79
156. Rose, I. A. et al 1969. *J. Biol. Chem.* 244:6130–33
157. Rutner, A. C. 1970. *Biochem. Biophys. Res. Commun.* 39:923–29
158. Scharf, K. H., Zenk, M. H., Onderka, D. K., Carroll, M., Floss, H. G. 1971. *Chem. Commun.* 576–77
159. Ibid, 765–66
160. Scott, A. I. 1970. *Accounts Chem. Res.* 3:151–57
161. Shah, D. H., Cleland, W. W., Porter, J. W. 1965. *J. Biol. Chem.* 240:1946–56
162. Sharpless, K. B., van Tamelen, E. E. 1969. *J. Am. Chem. Soc.* 91:1848–49
163. Simon, H., Palm, D. 1966. *Angew. Chem. Int. Ed.* 5:920–33
164. Smith, B. W., Turner, M. J., Haslam, E. 1970. *Chem. Commun.* 842–43
165. Snell, E. E., DiMari, J. 1970. *The Enzymes*, 3rd ed. 2:335–70
166. Sprecher, M., Berger, R., Sprinson, D. B. 1964. *J. Biol. Chem.* 239:4268–71
167. Sprinson, D. B. 1960. *Advan. Carbohyd. Chem.* 15:235–70
168. Srere, P. A. 1967. *Biochem. Biophys. Res. Commun.* 26:609–14
169. Srere, P. A. 1972. *Current Topics in Cellular Regulation*, 5. In press
170. Srere, P. A., Pavelka, S., Das, N. 1971. *Biochem. Biophys. Res. Commun.* 44:717–23
171. Staunton, J. 1970. *Ann. Rep. Chem.*

Soc. London 67:535–56
172. Stork, G., White, W. N. 1956. *J. Am. Chem. Soc.* 78:4609–19
173. Streitwieser, A., Jr., Wolfe, J. R., Jr., Schaefer, W. D. 1959. *Tetrahedron* 6:338–44
174. Stubbe, J. A., Kenyon, G. L. 1971. *Biochemistry* 10:2669–77
175. Towers, G. H. N., Subba Rao, P. V. 1972. *Recent Advan. Phytochem.* 4:1–44
176. Valentine, R. C., Chignell, D. A. 1968. *Nature* 218:950–53
177. Valentine, R. C., Shapiro, B. M., Stadtman, E. R. 1968. *Biochemistry* 7:2143–52
178. Vallery-Radot, R. 1923. *The Life of Pasteur*, Transl. R. L. Devonshire, 24, 70–73. New York: Doubleday. 484 pp.
179. van Tamelen, E. E. 1968. *Accounts Chem. Res.* 1:111–20
180. van Tamelen, E. E., Freed, J. H. 1970. *J. Am. Chem. Soc.* 92:7206–7
181. van't Hoff, J. H. 1891. *Chemistry in Space*, ed. J. E. Marsh. Oxford: Clarendon. 128 pp.
182. Voet, J. G., Abeles, R. H. 1970. *J. Biol. Chem.* 245:1020–31
183. Wightman, R. H., Staunton, J., Battersby, A. R., Hanson, K. R. 1972. In preparation
184. Wiley, D. C., Lipscomb, W. N. 1968. *Nature* 218:1119–21
185. Wolfenden, R. 1972. *Accounts Chem. Res.* 5:10–18
186. Woodward, R. B., Bloch, K. 1953. *J. Am. Chem. Soc.* 75:2023–24
187. Woodward, R. B., Hoffmann, R. 1969. *Angew. Chem. Int. Ed.* 8:781–853
188. Yamamoto, S., Bloch, K. 1970. *Natural Substances Formed Biologically From Mevalonic Acid.* *Biochem. Soc. Symp.* 29:35-43

Ann. Rev. Plant Physiol. 1972. 23:367–88

SALT TRANSPORT BY PLANTS IN RELATION TO SALINITY[1]

7535

D. W. RAINS

Department of Agronomy and Range Science
University of California, Davis

CONTENTS

INTRODUCTION

The title of this review necessitates, if not a rigid definition, at least a limited description of the term salinity.

Salinity is described in many ways, depending on the level at which an individual relates to this particular environmental condition. There is chloride salinity, sulfate salinity, oceanic salinity, salinity as defined by the agriculturist, and the quantitative chemical definition as it relates to plant productivity (110).

People concerned with productivity would agree that salinity is an unwanted problem. Historically we are inclined to believe that encroachment of salinity was a certain indication of the future decline and demise of an agricultural society (122). The decline in productivity was related to the inability of plants to withstand excessive salt in their environment. However, salinity is not incompatible with plant life. The question is one of degree, not whether the salt present is or is not excessive. If plant life were incompatible with high levels of salt, the ocean would indeed be a sterile body, as would many terres-

---

[1] Abbreviations used: ADP (adenosine diphosphate); ATP (adenosine triphosphate); ATPase (adenosine triphosphatase); DNP (2,4 dinitrophenol); EDTA (ethylene diaminetetraacetic acid).

trial areas with high salt burdens. Although salinity is inimical to the growth of some plants, at least one half or more of the total plant productivity of the biosphere is found in the ocean, in the form of phytoplankton and other microscopic and macroscopic plant life (128). In terrestrial areas of the earth, plants are found living in saline deserts, a situation where water is a limiting factor, as well as moist salt marshes, an environment with unlimited amounts of water. The imposition of salinity on high forms of plant life is not unequivocally coupled with a restricted supply of water. The regulation of salt as well as of water is necessary in a saline environment.

This review addresses the question of how plants acquire the necessary nutrients from a highly saline environment, that is, a situation in which the required nutrient (e.g. K) is present in small amounts relative to a chemically similar ion (Na) not generally required for physiological reactions. Also discussed are the mechanisms involved in regulating and distributing a required ion in a high-salt system. The discussion is restricted primarily to terrestrial plants since these organisms are of immediate economic and esthetic concern. Coupled with this is the interesting aspect of the survival of plant life in a terrestrial system, a highly variable environment when compared with the relatively stable marine environment. One can only speculate how evolutionary pressures brought about by a highly fluctuating environment have led to flexibility in maintenance of a stable ionic environment within the plant cell, a condition necessary for normal physiological processes.

It is beyond the scope of this review to discuss long-distance translocation in plants. A complete review is available in this volume (64). For a more comprehensive treatment of ion transport, not restricted to saline environments, the reader is referred to an article by Anderson (1), also presented in this volume. Lüttge (69) reviewed the function of salt glands in plants and discussed this aspect of salt relations of cells in a very recent review article.

Ion Transport in a Saline Environment

If we assume that plants are not completely at the mercy of the surrounding media but, within limits, are able to regulate their intercellular ionic composition, then it seems logical and necessary to invoke means by which plants maintain a suitable ionic environment for a myriad of physiological and biochemical processes that proceed within the cell. To paraphrase Steward (120), plants do not put their vital functions in free contact with the vicissitudes of the environment. A suitable internal environment must be maintained in the presence of high concentrations of potentially harmful salts and under conditions of adverse water relations as expressed in a saline environment (8, 121).

The concept of dual mechanisms of ion transport in plants has been especially useful in understanding the way in which plants acquire necessary mineral nutrients from a high salt substrate and has provided a working model for investigations of this process. Some of the ideas expressed in the following section on the characteristics of dual mechanism of ion transport were recently reviewed by Epstein (24). Laties (63) also contributed an extensive review of the relationship of this conceptual and working model of dual mechanisms to absorption and long-distance translocation. Leggett (66) incorporated the dual-mechanism concept in a recent review of ion transport by plants.

This section considers application of the dual mechanisms to ion transport in a high-salt environment.

Dual mechanisms of ion transport in relation to salinity.—The concept of dual mechanisms of ion transport implies the operation of two distinct entities for the movement of ions across the plant cell membranes into the cell. The definition of two mechanisms was derived from data collected on the kinetic response of ion absorption to substrate concentrations. The concept, presented initially by Epstein & Hagen (26), was developed further by Epstein et al (29). Similar observations have included numerous plants and ionic species (16, 17, 21, 50, 59, 71, 89, 96, 101, 103, 104, 123).

The rate of absorption was characterized by two distinct kinetic responses over a wide range of concentrations, one mechanism operating below approximately 1 mM and a second mechanism operating within a concentration range of 1 to 50 mM (29). The properties of these kinetically separable mechanisms are very important in regulation of the intercellular ionic milieu of cells exposed to a high-salt environment. The differences and similarities in these two mechanisms are described below, with special emphasis placed on the interaction between two cations—K and Na. When applicable, however, other cations and anions are incorporated into the discussion of these mechanisms.

The mechanism operating in the low concentration range has been designated as mechanism 1 (29). Numerous studies (29, 103, 104, 126) have demonstrated a very important characteristic of mechanism 1. It shows a higher specificity and a higher affinity for K than the mechanism operating at high concentrations, mechanism 2. A high affinity for K means that mechanism 1 mediates the transport of this ion across the cell membrane from a substrate medium of low K concentrations and is selective for K in the presence of high concentrations of chemically similar ions. For example, an average soil solution contains 0.2 mM K (109), a concentration at which mechanism 1 operates at maximal rates. Even in many saline soils, the concentrations of K are within the range restricting K uptake to transport by mechanism 1.

This high affinity for ions by mechanism 1 has been shown for barley roots (29, 104, 105, 126), corn roots (71, 123), corn leaves (117), storage tissues (89, 96), stem tissue (101), halophytes (7, 103), and legumes (50). Specificity is manifested in the observation that little interaction exists between a chemically similar ion (e.g. Na) and the transport of K in this low concentration range (16, 19, 29, 104). These characteristics depend to a large extent on the presence of Ca (22, 62, 65). A more detailed description of the influence of Ca is given later herein.

A high specificity for K in the concentration range of mechanism 1 is not ubiquitous. In the absence of Ca and K, Na is transported by barley roots at substantial rates from solutions of low concentrations of Na (104). In freshly sliced bean stem tissue, Na is absorbed selectively, whereas K is not transported from solutions containing concentrations of K and Na within the range of mechanism 1 (<1 mM) (101).

Investigations of aging of storage tissue have also shown alterations in patterns of ion transport processes (89, 96). A selective Na-transport system developed as red beet root tissue was washed for a period of days (96). The tissue selectively absorbed Na over the entire range of concentrations encompassing mechanisms 1 and 2 after beet root discs were washed for 8 days in distilled water.

The development of Na absorption capacity by sugar beet roots took 20 to 25 days after germination (16). The Na-transport capacity was not selective, however, but appeared to be mediated by the low-affinity high-concentration mechanism 2.

The examples discussed above are exceptions to the numerous observations of high-affinity mechanisms for transport of K from substrate of low concentrations. It would seem important to our understanding of specific-ion transport to characterize these phenomena further.

Mechanism 1 does not appear to be influenced by concomitant anions when K is transported, but accompanying anions can alter the absorption of K over the higher concentration range (29). Contrary to this, Na absorption is depressed over the entire concentration range (0.005 $-$50 mM) when SO$_4$ replaces Cl as the accompanying anion (104).

Kinetic evidence suggests then that mechanism 1 when transporting K has a relatively high affinity and specificity for K, it is not influenced by anions, and it delivers ions into the cytoplasm (34, 126, 127).

There are some important differences between mechanism 2 and mechanism 1. In mechanism 2, transport of ions proceeds at concentrations much higher than the range shown for mechanism 1. Experimentally and kinetically this has been shown to represent a lower affinity for the ion being absorbed (29, 104). Specificity for a particular ion is reduced, and in contrast to mechanism 1, Rains & Epstein (105) observed a higher affinity for Na than for K in the high range of concentrations. This has been demonstrated also by Bange (3) and Epstein et al (26, 29).

Another significant difference is the influence of concomitant anions. When compared with Cl, SO$_4$ severely depresses K absorption, with the rate only slightly greater than that found over the concentration range of mechanism 1 (29). At low concentrations (mechanism 1 range) these two anions have little influence on K uptake. The differences observed in Cl and SO$_4$ salinity (8, 40, 41, 121) might be partially explained by the differential effect that these two anions have on the uptake of an essential ion such as K. This speculation should be considered, along with the observation that SO_4 depresses Na absorption over the entire concentration range encompassing mechanisms 1 and 2 (104).

Mechanism 2 has another characteristic quite different from that of mechanism 1. When the concentration of substrate ion is increased over the high range, instead of the smooth curve observed for mechanism 1 there are a series of smaller hyperbolic curves superimposed on the curve describing the rate of uptake over the entire high range of concentrations. With only a few points scattered over the higher concentration range the function was a smooth curve with no obvious heterogeneity (29). A more careful evaluation of the response range demonstrated this rather unique property of heterogeneity of mechanism 2 (21, 26, 28, 71, 105).

There are many characteristics of selective ion transport that can be related to an outwardly directed flow of ions. The imposition of an outpump for certain ions has been utilized in interpreting some aspects of selective transport of ions in green algae (15, 75).

In higher plants bidirectional pumps are envisioned at both the plasmalemma and the tonoplast. Pitman & Saddler (94) located inwardly directed K and Cl pumps and an outwardly directed Na pump at the plasmalemma. Selectivity between Na and K is maintained by these two pumps operating in opposite directions. Potassium can build up while Na is constantly being depleted inside the cell. These ideas were extended by Pitman et al (92) and others (45, 90) to include the tonoplast membrane. Increased selectivity for K, as salts are accumulated, results from the transport of K inward and Na outward at the plasmalemma while at the tonoplast K and Na are both transported into the vacuole. The cytoplasmic phase is very effectively depleted of any Na as this ion is continually transported out of this compartment, either to the external media or to the vacuole. This model is very similar to the one proposed by Jennings (57).

The relevance of these models to salt transport in a saline system is that this is another mechanism whereby a selective, intercellular ionic environment is maintained. Instead of exclusion of a particular ion from the interior of a cell, the ion is pumped out after diffusing into the cell. As long as the pump can keep pace no Na will accumulate. This model, however, seems to be very inefficient in conserving energy. Energy would have to be expended to maintain low levels of Na concentration inside the cell, and it would seem much more advantageous for a plant to exclude a potentially toxic ion (e.g. Na) by transporting a biologically important ion such as K, preferentially. Energy is required only for the transport of one ion instead of two ions.

Application of the dual mechanisms of ion transport in saline media.— With the elaboration and definition of the characteristics of ion transport over a wide range of concentrations—including salt concentrations corresponding to saline situations—a comprehensible hypothesis began to evolve on the salt relations of plants in a saline environment.

How do these observations relate to a plant exposed to a saline environment? One interpretation is that the plant can obtain an essential ion such as K even in the presence of large amounts of Na, a chemically similar ion. When salt concentrations are increased, however, there is a mechanism available to absorb ions in excess of the rate limited by the maximal absorption due to the operation of mechanism 1. This so-called "luxury absorption" would allow osmotic adjustment so that the potential gradient of water transfer does not result in outflow and dehydration of a cell exposed to saline conditions. Bernstein (5, 6) was able to demonstrate osmotic adjustment by plants exposed to increasing levels of salt. When compared with other ions in the nutrient substrate, K was found to have a fairly specific part in this adjustment. It would seem, as pointed out by Epstein (25), that the term "luxury absorption" may be an inaccurate description of this phenomenon since luxury implies a frivolous and unessential function.

This aspect of salt regulation and maintenance of a favorable water balance was suggested by Black, working with *Atriplex vesicaria* (7). This species was exposed to increasing levels of NaCl in nutrient solution. Although K levels were depressed, the K content was not reduced to a level below which K was considered to be limiting. When Na was not added or present in low amounts, K was accumulated at levels considered to be luxurious. Osmond observed this characteristic in *Atriplex* when divalent cations were used in experiments (87). Greenway (35, 36) also demonstrated similar relationships with barley and halophytes.

Absorption characteristics of the mangrove *Avicennia marina* were studied by Rains & Epstein (103). This halophytic plant is found in tidal flats and estuaries constantly exposed to sea water. The Na/K ratio was altered by the plant from a ratio of 40/1, the composition of the solution surrounding the plant, to approximately 7/1 inside the plant. Leaf tissue was utilized as an experimental tissue, and it was found that K absorption was described by the dual mechanism of transport and the uptake of K was highly specific. A Na concentration of 0.5 M had virtually no effect on K transport when K and Ca were present at 0.01 M. The plant could accumulate the required K and still maintain osmotic water for physiological processes by absorbing Na and other ions from the media.

Exposure of plants to elevated levels of salt in surrounding media results in two potentially inimical situations. 1. An unfavorable water balance, with the potential gradient for water transfer into the cells considerably reduced. Nonselective ion absorption by mechanism 2 from high-salt concentrations mediates a buildup of osmotica within the cell from mixed ionic media. 2. With increasing salt content the ratio between required and unrequired ions is reduced. For maintaining a favorable ionic environment within the cell, plants must possess a means by which the necessary ions are accumulated selectively from an environment containing many unessential ionic species.

The dual mechanism of ion transport encompasses both of these characteristics and is of significant advantage for plants exposed to saline conditions.

Ion transport by high- and low-salt status roots.—Much of the information on ion transport by higher plants has been obtained utilizing low-salt status roots, i.e. roots from plants grown on water or dilute $CaSO_4$ solutions. These roots are then exposed to media ranging from low to high salt concentrations. There has been some question as to whether this information can be extrapolated to a system in which a plant has continually been exposed to high salt.

Pitman (91) suggested a change in selective ion transport as roots increased their salt status. The salt status was increased by exposing the roots to 2.5 mM K and 7.5 mM Na for a period up to 30 hours. Within the first 15 hours the tissue switched from preferentially absorbing Na to preferentially absorbing K. The alteration resulted from a decrease in Na influx relative to K. With time the influx of K decreased until the rate of absorption by both ions approached zero, but only after the K content of the tissue exceeded the Na content. The tissue seems to have a greater capacity for K ions than for Na ions; however, it takes longer to reach this capacity because of the reduced rate of K transport. A slower rate of K transport could be a reflection of the lower concentration of K relative to Na—2.5 mM vs 7.5 mM, respectively.

A change in specificity for K vis-à-vis Na need not be invoked to explain this phenomenon. A more rigorous approach would be to maintain Na and K at equal concentrations during the absorption period, and any change in rate or specificity would not be complicated by different concentrations of these two ions.

El-Sheikh et al (16) studied the absorption of K, Rb, and Na by beet seedlings grown in nutrient solutions. The plants were cultured up to 48 days and could be considered to be high-salt status plants. Experiments on the selective uptake of K, Rb, and Na by intact seedlings for long periods of time were comparable with experiments on excised roots over short periods. Potassium and Rb were mutually competitive at all concentrations except when the Rb/K ratio was excessive. This was considered to be a toxicity due to Rb since it is not entirely a substitute for K in physiological reactions (23). For the first 20 days of growth Na was virtually excluded from uptake into the plant when exposed to solutions containing both K and Na. Sodium was absorbed after the first 20 days only when K was depleted from the solution to a point of being virtually eliminated. Another treatment was the addition of Na and K (8 + 8 mM) at concentrations within the range of mechanism 2. Sodium and K transport were expected to be mutually competitive (29, 104, 105). This was not observed until days 20–25. A Na-transport system became operative with time, contrary to what other workers found for high salt barley roots (91, 92).

The previous discussion does suggest caution when extrapolating information collected from excised roots over short periods to an intact plant growing in a saline system. Recent information (16), however, lends support to the use of short-term quantitative studies as a procedure for developing information applicable to the transport of ions in a saline system.

Bioelectropotentials and ion transport in saline media.—Measurement of potential differences between external salt concentrations and intercellular ion concentrations across a plant cell membrane has proven to be a valuable technique. One can determine whether ions are transported into the cell against a concentration gradient or are passively distributed according to diffusion potentials set up by membrane permeability and ionic mobility (12, 45, 46, 74, 90, 93, 114, 118).

Dainty reviewed the subject and outlined some criteria for measuring and interpreting results obtained with this technique (12). A potential difference between the outside and inside of a cell can be maintained by relative distribution of anions and cations which is governed by their permeability and mobility across the plant membranes. Suitable equations are utilized to calculate theoretical electropotentials at given concentrations. If the measured electropotentials are very much different from the expected values, then an active transport either into or out of the cell has to be invoked to allow for the observed discrepancy.

A major difficulty in using this technique is the question of placing the electrode. When microelectrodes are inserted into a tissue it is sometimes questionable as to where the tip of the electrode is actually located (e.g. cell wall, cytoplasm, vacuole, etc). In giant green algal cells this does not seem to be a limitation and this procedure has proven very informative. Spanswick & Williams (118) utilized giant cells and concluded that K and Cl are actively transported into the cell and Na was transported, actively, out of the cell.

Studies with higher plant tissues have been less conclusive. Chloride has been found to be actively transported into the cell and Na actively moved out of the cell (45, 90). The driving force for K transfer across a plant cell membrane is less clear. Dainty has suggested passive movement (12), Higinbotham and co-workers initially were undecided (45, 90), and Pitman et al (93) recently have suggested K absorption against an electrochemical gradient.

Pitman et al (93) compared electropotential measurements on low-salt plants with high-salt plants. They also investigated the effect of aging (time after excision of roots) on cell electropotentials. Aging in $CaSO_4$ resulted in an increase in electronegativity of the cell potential. They suggested two possibilities for the observed increase. Plasmodesmata might seal with time reducing possible leakage from a cut surface and thereby increasing the potential difference. Also hormones that normally diffuse from the apex would no longer be available to the excised root and a reduction in hormone levels might alter ionic distributions in the cell increasing potential differences.

Alterations in cell electropotentials were related to selective ion transport

(93). At high external K concentrations, within the range of mechanism 2, K was found to be accumulated against an electrochemical gradient and was not at diffusive equilibrium. This has also been shown by Gerson & Poole (34) to apply to Cl transport by mung bean roots.

Electropotential measurements have been essential in determining whether an ion is transported uphill against a concentration gradient, thus requiring the expenditure of metabolic energy. If not actively transported then ion movement is dictated by passive distribution of ions according to diffusion potentials. In saline environments where the concentration gradient between the inside and outside of a cell is less than that found in media of lower concentrations, the use of electropotential techniques will prove to be of value. We need to ascertain whether ionic regulation is under metabolic control. This information is useful in guiding our thinking on the possible mechanisms responsible for regulating ionic content of cells in high salt systems and whether or not physiological processes need be invoked in describing these mechanisms.

The next three sections will be concerned with the dual mechanisms of ion transport in relation to organic acids and ion balance, salt respiration, and Ca and other ions found to influence salt regulation. Finally, a section will be presented on the physiological and ecological significance of salt relations of plants in a saline environment.

Organic Acids and Salt Regulation

The absorption of ions from a salt solution is commonly associated with alterations in solution pH. This alteration has been related to unequal rates of absorption of cations and anions (42, 48, 51). It has been shown that pH increases when anions are absorbed in excess of cations. Conversely, pH decreases when cation absorption exceeds anion uptake (42, 51).

Reduced growth of cereals resulting from unbalanced nutrient supplies has been correlated with a decrease in the level of organic acids in the plants (14). Also the level of organic acids is generally found to be reduced in plants adversely affected by saline environments (5).

The organic acid level in plants is influenced by rate of ion absorption as well as the species of ion being absorbed (55). The amounts and species of ions in a saline environment are extremely varied, and it is logical to assume that the organic acids would vary considerably in a system where SO_4 or Cl might be a predominant anion or HCO_3 is present in excessive amounts. Halophytes and other plants associated with dry climates can have elevated levels of organic acids (88, 130), and it would seem pertinent to discuss the interaction of salt transport with organic acids in a saline environment.

The stoichiometry between the excessive cation uptake and a decrease in solution pH has prompted suggestions of a causal relation between these two processes (42, 54, 88). The demonstration of a dual mechanism for ion transport generated a new round of research on the interaction of organic acid regulation and ion transport. Emphasis was centered on the high concentration range because of the influence of anions on cation transport (29, 104). Torii & Laties (124) assumed that cation transport by mechanism 2 was responsible for delivery of ions across the tonoplast into the vacuole. Since cation transport by mechanism 2 is influenced by the accompanying anion (29) they reasoned that this unbalanced ion uptake would alter organic acid synthesis. By using $^{14}CO_2$ incorporation into organic acid and varying the salt concentration and species of ions absorbed they concluded delivery of ions into the vacuole by mechanism 2 effectively removed organic acids from the cytoplasm by sequestering the salt of the acid in the vacuole. This results in more acids being synthesized (124).

Hiatt & Hendricks (44) determined the fixation of $^{14}CO_2$ as a function of salt concentration. An increase in fixation was observed with the addition of K_2SO_4 from 0.01 to 1.0 mM. Between 1.0 and 10 mM no further increase in fixation occurred, and at concentrations exceeding 10 mM fixation of CO_2 was decreased. Fixation of CO_2 was positively correlated with K transport via mechanism 1. At concentrations within the range of mechanism 2, fixation was independent of increased concentrations. Since K transport by mechanism 2 is greatly reduced when SO_4 is the accompanying anion (29) it would seem reasonable to assume in this instance that organic acid balance is associated with ion transport mediated by mechanism 1 and that the contribution by mechanism 2 is minor.

The first assumption of Torii & Laties (123, 124), the placing of mechanism 2 at the tonoplast, has been a matter of controversy. Laties and his coworkers (63, 70–73, 89,123) contributed considerable work defending this viewpoint. Welch & Epstein (126, 127) proposed that both mechanisms reside in the plasmalemma. Kannan (59) recently showed dual mechanisms in *Chlorella pyrenoidosa,* an organism with no vacuoles. Gerson & Poole (34) used electropotential measurements on mung bean roots to demonstrate accumulation of Cl against a concentration gradient at low and high Cl concentrations in the external media. At high concentrations, within the range of the operation of mechanism 2, absorption was found to be mediated by an active process in vacuolated root segments as well as in unvacuolated root tips. They placed mechanism 1 and 2 in the plasmalemma of this tissue.

The second assumption, that $^{14}CO_2$ incorporation into the organic acid fraction represented organic acid synthesis, was shown to be untenable. Hiatt (42) demonstrated a high percentage of CO_2 exchange, resulting in an increase in $^{14}CO_2$ incorporation into organic acids even though the actual amounts of organic acid decreased. He investigated CO_2 incorporation into organic acids by determining the level of organic acids in plants exposed to several concentrations of salt. He suggested that the synthesis of acids responded only to unbalanced ion uptake, not concentrations, and therefore was associated with mechanism 1. That conclusion is also reached in recent papers by Osmond & Laties (89) and Jacoby & Laties (54). These workers (54) proposed, however, that the organic acid synthesis is controlled by the level of HCO_3 in the cytoplasm, which regulates PEP carboxylase activity.

Stoichiometry is the result of accumulation of malate in the vacuole mediated by salt transport to the vacuole via mechanism 2. Transport by mechanism 1 is not responsible for stoichiometry between ion transport and organic acid synthesis.

Regulation of organic acid synthesis by HCO_3 (54), which in turn influences ion transport, could be an important aspect of salt regulation in a saline environment, an environment with a demonstrable high level of HCO_3.

Hiatt (43) has suggested an alternative to carrier mechanisms for the accumulation of ions over the range of concentrations encompassing mechanism 1 and 2. At concentrations less than 1 mM he envisions an important role of organic acids and amino acids in the acquisition of ions. The organic and amino acids in the cell act as fixed, nondiffusible charges which bind inorganic ions thereby mediating the accumulation of ions until equilibration is achieved with the external solution. Plant membranes are considered to be selectivity barriers excluding some ions (e.g. Na) and allowing passage of other ions (e.g. K).

At high concentrations (> 1 mM) ions enter as neutral salts mediated by a diffusion potential produced by the Donnan phenomenon. Cations are attracted to an excess of fixed negative ions (organic acids) where they are retained creating an excess of cations to be accumulated relative to anions. Diffusion into Donnan space is restricted to neutral salts, and any one ion of the ion pair which has difficulty in transversing the membrane could conceivably retard the movement of the other ion. Since SO_4 uptake is slow at high concentrations of K_2SO_4, K uptake would be reduced if the Donnan phenomenon governed transport. A reduced transport has been observed at high K_2SO_4 concentrations (29, 72, 104).

Donnan-mediated transport does not allow for selectivity between two similar ions such as K and Na, a specificity which has been demonstrated at high concentrations (29, 104, 105), and this might suggest a more complicated system at high external salt concentrations than diffusion according to the Donnan phenomenon.

The preceeding discussion reemphasizes a suggestion made previously, that at concentrations of ions found in saline environments alterations in organic acids levels could have considerable influence on salt transport, thereby proving to be an important aspect of salt regulation in a saline system.

High levels of organic acids have been reported in a number of plants associated with dry climates and high salt. Williams (130), investigating the toxicity of the halophyte *Halogeton glomeratus,* found a strong correlation between organic acid levels, ionic species of salts, and concentrations of salts in the nutrient media. Sodium chloride was the principal salt absorbed, and the level of oxalic acid present was correspondingly high. Potassium could partially substitute for Na, though the oxalate content was reduced when the substitution was made. The causal relation is not clear, but under conditions of high salt uptake a cation-anion balance must be maintained and could alter considerably the response of a plant to saline conditions. The observation by Meeuse & Campbell (84) that Cl inhibited oxalic acid oxidase could be an important part of the maintenance of intercellular ionic balance in situations of high-Cl substrates. Accumulation of oxalic acid would attend the reduced oxidase activity, and anions available for cationic balance would be increased. Osmotic relations of the cells might be improved by this physiological response.

Osmond (87) studied the oxalate content of *Atriplex* leaves and related this to the balancing of cations absorbed by this plant. Since reduced growth in a saline medium has been correlated with a low organic acid level (5, 6), he suggested that organic acid synthesis is related to the metabolism of nutrient anions. Removal of NO_3 or PO_4 by incorporation into metabolites promotes organic synthesis, which is necessary to balance the remaining excess cations. The intolerance of some plants to salinity could be attributable to a competition of Cl with NO_3 or PO_4 in the absorption process, resulting in a reduced potential for organic acid synthesis. The inability to balance cations could conceivably reduce the survival potential of these plants when exposed to high salinity. It might prove illuminating to compare the absorption of nutrient anions (NO_3 and PO_4) in salt-tolerant and salt-intolerant plants and study the interaction of Cl with this uptake (9).

Balancing of osmotica within the cell is a necessary function of plants exposed to high salt. The inherent coupling between cellular compartmentation of ions (11, 77, 78), organic acid synthesis (54), and regulation of turgor by cellular compartments (30) are aspects of ionic regulation in a saline environment that will undoubtedly lead to a greater understanding of salt transport by plants exposed to high salt levels.

SALT TRANSPORT AND RESPIRATION

The term "salt-respiration" is used to describe the observed increase in respiration as salt in the medium is increased (68, 112). Lundegardh (68) linked the enhancement of respiration to an increase in anions in the substrate, a relationship characterized by the term "anion respiration." In recent years, considerable controversy has developed over the meaning of salt respiration and the source and form of energy utilized in transport of ions.

One school of thought, originally developed by Lundegardh (68) and more recently promoted and expanded by Mitchell (86) and Robertson (111, 112), follows the concept that the transfer of electrons along the electron chain supplies the potential for accumulation of anions against a concentration gradient, while cations follow in a passive manner to maintain electrical neutrality. Charge separation by means of a "proton pump" supplies energy for movement of ions or formation of ATP, and these processes are competitive. If ions are transported, ATP synthesis is decreased, and vice versa. A comprehensive review by Schwartz (113) supports the chemiosmotic theory in chloroplast and mitochondrial studies.

A second view holds that ATP or high-energy intermediates, generated by respiratory and photosynthetic processes, are utilized by plants to accumulate ions (70, 99, 116).

The views expressed by both groups have merit and are argued with considerable enthusiasm by each school. Demonstration of the dual mechanism of ion transport has suggested a new approach to this problem, and information is now being developed on the link between respiratory processes and ion transport over the range of concentrations mediated by mechanisms 1 and 2.

Respiration is generally enhanced at concentrations of salt well above the concentration range of mechanism 1 (112). It seemed obvious that the respiratory response must be the result of transport by mechanism 2, transport taking place at high salt concentrations.

Polya & Atkinson studied the absorption of K, Na, and Cl by aged beet discs as a function of time and antimetabolites (95). Concentration was maintained at 0.5 mM, which effectively limits uptake to the high-affinity mechanism 1 (89). By relating ATP levels with ion fluxes and studying the effect of various antimetabolites they concluded that transport by mechanism 1 was dependent upon electron transport and not on utilization of ATP or high-energy intermediates. It would have been of interest to compare the effect of inhibitors on ion transport by both mechanisms.

Rains & Epstein (104) compared DNP, an uncoupler of oxidative phosphorylation, with anaerobic conditions over both ranges of concentration. DNP had more of an effect than anaerobiosis over both ranges of concentration. By manipulating the ratios of K, Ca, and Na, the contribution by mechanism 1 to Na absorption can be reduced virtually to zero (102). When this condition was achieved and the two mechanisms compared, Na uptake via mechanism 2 was depressed more by anaerobiosis than was Na transport mediated by mechanism 1. Transport of Na by both mechanisms was inhibited to the same extent by DNP. It was concluded that energy utilized in transport (ATP) was not significantly different when the two mechanisms were compared, neither in source nor the amount required for absorption of ions.

Lüttge & Laties (73) compared the relative sensitivities of the two mechanisms to inhibitors and concluded that absorption by mechanism 1 is more sensitive to inhibitors than is absorption by mechanism 2. The two sets of conflicting data may be partially explained by the differences in plant species and tissue—respectively, barley roots versus corn seedlings—and the utilization of different ions in the investigations. However, it is possible that the energy required to accumulate ions from high concentrations is less since the accumulation ratios (concentration of ions inside/concentration of ions outside) are lower at the higher concentrations found in the range of mechanism 2. This might be manifested in the observed lower sensitivities to antimetabolites.

Controversy continues over the manner and form in which energy is transmitted to the site of ion accumulation. Inhibitor studies have given support to both the ATP hypothesis and the electron-transport pump. The problem, simply stated, is: how do you separate net electron flux from the production of ATP or an equivalent high-energy intermediate? A photosynthetic process, cyclic photophosphorylation, is responsible for the generation of ATP without net electron flux. MacRobbie (76), investigating ion transport in algae, concluded that light-mediated K transport was driven by ATP whereas Cl was taken up in response to electron transport. Rains (99) suggested that ATP was responsible for supplying the necessary energy for light-enhanced uptake of K by corn leaf slices. The potassium concentrations used in those experiments did not exceed 1 mM, and it was concluded that the same site (mechanism 1) was responsible for K transport in both light and dark treatments.

The implication of ATP in the transport of ions, along with the demonstration of ATPases in plant tissue (2, 31, 37, 61), has prompted considerable research on the possible link between these enzymes and ion transport.

A monovalent ion-stimulated ATPase has been characterized in oat roots (31). Enzyme activity was positively correlated with K transport. As the K concentration was increased K-stimulated ATPase activity increased. For every 0.7 to 0.8 μmoles of K transported, 1 μmole of ATP was hydrolyzed by K-stimulated ATPase.

Kylin & Gee (61) demonstrated a Na, K-stimulated ATPase in the mangrove *Avicennia nitida*, Jacq. The ATPase activity of this halophyte was dependent upon the ratio of Na/K and the total concentration of the two cations. Enzyme activity was reduced when K or Na were present individually in the incubation media. When the cations were added in combination the activity was enhanced. This is similar to animal systems.

Investigations conducted so far on plant ATPases have not conclusively linked ATP hydrolysis to ion transport; however, strong correlative evidence presented by Fisher & Hodges (31) supports a possible connection between ATPase activity and transport of K. If such a connection exists, biochemical procedures can be developed to define and characterize ion transport in a saline environment and greatly increase our understanding of the process governing salt regulation.

Jennings (57) proposed a "unified theory" for halophytes, succulence and Na. Previous investigations had shown a connection between absorption of Na and increased succulence. Also succulence is related to aridity and light. He suggested that the one unifying factor is ATP since all of these processes influence the level of ATP in plant tissue. Light increases ATP through photophosphorylation. Aridity increases Na concentration in the environment, which reverses a Na out-pump, resulting in synthesis of ATP through the activity of a membrane-bound Na ATPase. It was proposed that when Na is pumped out, ATP is required. When Na is forced in (high Na in the external medium) ATP is produced. Sodium ions are to provide a homeostatic mechanism reducing the detrimental effect of high concentrations of toxic ions.

The information on salt respiration in plants indicates uncoupling of ATP, and it is possible that high salt content in the medium results in an increase in utilization of ATP to maintain an ionic balance favorable for water uptake by the cells (5, 6). This is in direct conflict with Jennings' suppositions, since an increase in Na uptake should promote ATP synthesis. Jennings does suggest the presence of a Na pump at the tonoplast which consumes ATP as Na is transported to the vacuole. This system would utilize ATP while maintaining turgor in the cell by vacuolar regulation of salt osmotica. If ATP levels increase in response to increased Na in the media, one would expect a reduction in respiration as a result of reduced ADP levels. In most instances, however, respiration is increased. Livne & Levin (67) have results demonstrating a reduced ATP/ADP ratio when pea seedlings were salinized by NaCl. An increase in ADP could conceivably enhance respiration by removing the rate-limiting factor of low ADP concentrations, although they concluded that the change in ratio was a response to alteration in various metabolic pathways.

Reduction of the ATP/ADP ratio (41) as well as an alteration in the ratios between various adenine nucleotides (40) have been observed as salinity is increased. The content of ATP decreased and ADP increased when pea seedlings were exposed to elevated levels of salt (41). Phosphatase activity was enhanced by NaCl while the phosphorylating activity of mitochondrial fractions decreased. A combination of these two biochemical processes was suggested as the possible cause of the decrease in ATP/ADP ratios.

The relationship between salt-stimulated respiration and the dual mechanisms of salt transport was investigated by Lüttge et al (70). They suggested that salt-stimulated respiration was an energy-source energy-sink relationship; that is, energy required for ion uptake influences the rate of respiration. With this in mind they studied uptake and respiration by aged carrot slices over a wide range of concentrations. Respiration was stimulated much more at high concentrations (mechanism 2) than at low concentrations (mechanism 1). When respiration was maximally uncoupled there was no change in respiration as a function of increasing salt concentrations. It was concluded that salt transport is not linked directly to electron fluxes of the respiratory chain but is linked indirectly through the consumption of ATP or high-energy intermediates during salt transport. In their experiments, respiration increased as concentration was increased over the entire range of mechanisms 1 and 2. The magnitude of the increase was not as great at lower ionic concentrations, but respiration paralleled uptake over the entire range as shown by their data. The data presented do not justify their conclusion that salt respiration is in response to ion transport across the tonoplast mediated by mechanism 2. That conclusion was based on earlier suppositions on the location of the two mechanisms (63, 71, 123). The location of those mechanisms has not been established with any certainty (34, 59, 126, 127), and more evidence is certainly needed before the spatial location of salt transport in relation to salt respiration can be assigned. It seems logical to assign the site of respiratory response to the cytoplasm since it is agreed that mitochondria are found in this fraction of the cell. Changes in the salt content of the vacuole would not be expected to have a direct effect on mitochondria, since that cellular fraction is devoid of those organelles.

The question of why respiration increases as a function of increasing salt in the environment has not been satisfactorily answered. Unbalanced metabolism resulting from changes in ATP/ADP ratios (41, 67) is a possibility, as well as specific effects on enzymes controlling metabolic pathways. Such effects have been demonstrated by Porath & Poljakoff-Mayber (97, 98).

One is inclined to correlate salt respiration more closely with energy relationships, mainly because of the coupling between respiration and energy conservation. Expenditure of energy for ion transport has been found to be related to the specific ion being transported (32, 38, 96, 106). Specificity of ion absorption was found to alter in aging bean stems. Initially the tissue absorbed Na. With aging, K absorption capacity was increased and Na capacity was reduced to nil (101). These changes were attended by fluctuations in rates of respiration as well as variations in the coupling of energy-conserving processes with respiration. In red beet slices, aging resulted in the development of a Na-selective system (96) instead of the K-selective mechanism shown for the bean stem. This difference may be partially accounted for by the differences in functions of the two tissues utilized in the studies. Bean stems are involved in excluding Na from the tops of beans, a known Na-sensitive species (52, 53), and tissue surrounding the conducting tissue was proposed as the site of specific Na accumulation (100). Beets, on the other hand, do not exclude Na from their shoots, and distribution of Na within the plant appears to be handled in a manner very different from that in beans.

An important aspect of the results described above is that a change in ion-uptake specificity is accompanied by a change in respiratory processes and attendant biochemical responses. Additional information on the biochemical aspects of selective ion transport is essential to understand the interaction of a plant with its environment. That is particularly true in a saline environment, where specific control of internal ion environment is put to the ultimate test, and it is mandatory for a plant's survival to have some control in a high-salt situation.

One response of plants to saline conditions is to expend energy to maintain a favorable water balance as well as a relatively stable ionic environment. There is a limitation; Bernstein (5) has suggested that when a plant is no longer able to adjust its internal salt content in response to increasing salt levels in the external media, physiological drought occurs, terminating in death. This situation arises when the plant becomes energy deficient and can no longer supply energy necessary for accumulating salt. The plant has literally worked itself to death.

CALCIUM AND ION TRANSPORT IN SALINE MEDIA

The influence that Ca has on ion transport has been of considerable interest in recent years, and the literature is replete with reports on this interaction (4, 20, 22, 47, 56, 58, 62, 65, 80, 82, 85, 106–108, 125).

In what are now considered to be classical experiments, Viets (125) observed an increase in ion uptake when Ca or other polyvalent cations were added to experimental solutions. However, absorption of ions is not always accelerated by the presence of Ca. The effect of Ca on absorption of monovalent ions depends on the ionic species, the concentration of Ca and of the ion being affected, and on the plant species. Calcium has also been shown to influence the interaction between various ions.

Selective ion absorption by plants is altered by Ca. Epstein (22) and Jacobson et al (56) found that K was absorbed specifically from a solution containing a mixture of K and Na if Ca was included in the solution. In the absence of Ca, Na and K were absorbed in a nonselective manner.

Absorption of K is usually accelerated by Ca, though this depends on the concentration of K (82, 105) and the physiological state of the tissue (106). In the low concentration range of K (<1.0 mM), Ca was shown to accelerate K absorption (106). When K concentrations exceeded 1 mM, Ca depressed Na and K uptake (105). Conversely, Elzam & Hodges (20) found that Ca inhibited K uptake even at low K concentrations when corn was used as an experimental plant. This illustrates the differences among plant species. The data discussed previously were collected from barley roots (22, 56, 82, 105, 107).

Of particular interest in this discussion is the interaction of Ca with Na and the influence that Ca has on selective transport of ions. Calcium has also been assigned a role in maintaining the integrity of plant cell membranes, the physiological barrier to free diffusion of potentially toxic ions prevalent in a saline environment. Evidence that Ca plays such a role has been derived from two experimental approaches.

Direct evidence has been developed from studying the effect that Ca has on the physical characteristics of plant cell membranes. Marinos (79), in microscopic examination of Ca-deficient barley roots, observed disarranged, damaged membranes. Marschner & Gunther (81) limited the supply of Ca to roots and correlated a reduced capacity for ion absorption with changes in the structure of cell membranes.

Indirect evidence relating Ca with membrane integrity originated from measurements of ion fluxes in the presence and absence of Ca. Alterations in membrane structure are partially responsible for changes in membrane permeability. Van Steveninck (119) found considerable efflux of ions from beet root tissue when 75% of the total Ca was removed by EDTA. Of the cations tested, calcium was the most effective in reducing efflux, and small additions of Ca reversed the EDTA-induced efflux of K. Removal of Ca from soybean roots has also been found to be detrimental to ion uptake (33). Wildes & Neales (129) concluded that Ca was required to prevent loss of K during aging of beet tissue, and the capacity to absorb B was increased when Ca was included in the aging solutions. Findings were similar with bean stem tissue. The K-absorbing capacity developed during aging was enhanced when Ca was included in the aging medium (106). Evidence presented by Jackman & van Steveninck (49) indicates that marked alterations in endoplasmic reticulum membrane fractions attend aging processes in storage tissue. This, coupled with the requirement of Ca during certain phases of aging, suggests a close connection between membrane integrity and Ca.

The relation between Ca and ion transport is resolved to an even finer degree when plants are exposed to an environment high in salt. Increasing the level of salts in the medium surrounding the roots of plants results in greater demands on the salt-regulating processes of plants. The ratio between required ions and unessential ions is reduced, with the unessential ions predominating in many saline systems. Selective ion transport can become paramount to survival. The presence of potentially toxic ions will increase the possibility of membrane damage. The role of Ca therefore becomes even more important as the system becomes increasingly saline.

Elzam & Epstein (18) investigated the effect of increasing NaCl on the growth and salt content of two species of *Agropyron*. One species (*A. elongatum*) was salt tolerant, and the other (*A. intermedium*) was salt sensitive. *A. elongatum* grew reasonably well in the presence of 50 mM NaCl, while the growth of *A. intermedium* was reduced substantially in the presence of 5 mM NaCl. Analyses of salt content indicated a strong correlation between growth and Ca levels in the roots. It was concluded that Ca is extremely important in saline environments and influences the response of these plants to salinity.

Kelley (60) found that saline irrigation water could be utilized to better advantage if the water contained relatively more Ca.

La Haye & Epstein (62) demonstrated a definite role of Ca in increasing the tolerance of beans to high salt. Beans, a salt-sensitive species, were grown in nutrient solutions containing various levels of NaCl and $CaSO_4$. In the absence of Ca, beans exposed to 50 mM NaCl were severely damaged. When Ca was increased to 1 mM the plants were damaged only slightly, and at 3 mM Ca growth was not affected noticeably. Sodium content of the roots was reduced by 30% as Ca increased in the external solution, and very little Na was found in the leaves even though 50 mM NaCl was present in the solution.

Those results, coupled with many others reported in this section, have given Ca a central role in regulating ion transfer into plant cells exposed to a saline environment. The effect of Ca is not only on transport processes at cell surfaces (20, 22, 56, 82, 107); Ca also influences physiological reactions (106, 129), membrane structure (79, 80, 129), and translocations and redistribution of ions in plants exposed to salinity (10, 62).

The role of Ca in ion transport systems has not been clearly defined. The standard suggestion that Ca influences the integrity of membranes is still an excuse for not understanding the mechanistic aspects of the role of Ca. The interaction of Ca and ion transport in systems with high salt content reinforces the centrality of Ca in this process. A greater understanding of this interaction will undoubtedly increase our knowledge of the survival ability of plants in high-salt media.

Conclusions

As population increases, the need for food production rises concomitantly. Utilization of marginal land (e.g. salt-affected regions) requires a buildup of fundamental information on the management of plants growing in such saline areas.

Another rather recent pressure brought to bear on salt-plant relationships is the need to recycle onto agricultural lands waste water from sewage treatment plants as well as drainage water from irrigation of semiarid and arid regions. Water from such sources will have a greater load of salts than does water normally applied to agricultural soils.

The stores of fundamental knowledge presently available to deal with the problem of salt transport in a saline environment are not adequate. We are just beginning to comprehend how plants transport ions. This must now be extended to include the many facets of an environment which will influence this process. The partitioning of metabolic energy of a plant into various functions could be greatly influenced by environmental parameters. In the presence of salt, plants reflect a change in partitioning of energy as observed by the rise in respiration (salt-respiration). Is this efficient energy coupled to anabolic processes and salt-regulating functions? Are there some species of plants that are more efficient at coupling these responses to environmental changes?

Succulence is often related to high salt and water deficiency. This is correlated with the survival of plants exposed to a substrate with restricted water availability. It now seems reasonable to investigate how the avoidance of water loss by increased succulence is related to regulation of salt content of cells. Strogonov (121) has discussed the effect of SO_4 or Cl salinity on succulence. Plants exposed to Cl salinity have a much greater degree of succulence than plants exposed to SO_4 salinity.

Interrelations between salinity, organic acid metabolism, and Ca are all part of the quest to understand ion transport in a saline environment, and this quest has proven fruitful. Our approach must be broadened to include the geneticist. Understanding alone will not solve the problem. We must convert this knowledge into a usable product: an economically important plant growing in a saline soil. A combination of plant physiologists, plant nutritionists, and geneticists, working together, might supply the impetus to carry out a program on adapting plants to saline environments. Such a plea was made by Singh in a recent review (115), and breeding for salt tolerance does have precedence (13). There is much evidence for genetic control of ion transport by plants (27), and ecotypic adaptation to salinity by *Typha* (83) and by *Festuca* and *Agrostis* species (39) has been demonstrated. If a plant cannot regulate cellular ionic contents, its survival potential is reduced.

It may be too optimistic to expect a solution in the near future to the problem of how to use the millions of acres now considered to be too saline for optimal plant growth, but it is certainly not beyond possibility. Scientists are an inquisitive lot, and the quest for knowledge not only should lead to an understanding of how plants survive in saline environments but should forward application of this knowledge to the many related problems of recycling our diminishing resources.

ACKNOWLEDGMENTS

The assistance of Hector Schoo in preparation of this manuscript is greatly appreciated.

LITERATURE CITED

1. Anderson, W. P. 1972. *Ann. Rev. Plant Physiol.* 23:51–72
2. Atkinson, M. R., Polya, G. M. 1967. *Aust. J. Biol. Sci.* 20: 1069–86
3. Bange, G. G. J. 1959. *Plant Soil* 11:17–29
4. Ibid 1968. 28:177–81
5. Bernstein, L. 1961. *Am. J. Bot.* 48:909–18
6. Ibid 1963. 50:360–70
7. Black, R. F. 1960. *Aust. J. Biol. Sci.* 13:249–66
8. Boyko, H. 1966. *Salinity and Aridity; New Approaches to Old Problems*, 408. The Hague: Junk
9. Carter, O. G., Lathwell, D. J. 1967. *Agron. J.* 59:250–53
10. Collins, J. C., Lindstead, P. J. 1969. *Planta* 84:353–57
11. Cram, W. J. 1968. *Biochim. Biophys. Acta* 163:339–53
12. Dainty, J. 1962. *Ann. Rev. Plant Physiol.* 13:379–402
13. Dewey, D. R. 1962. *Crop Sci.* 2: 403–7
14. DeWit, C. T., Dykshoorn, W., Noggle, J. C. 1963. *Versl. Landbouwk. Onderzoek.* 68.15. 68 pp.
15. Dodd, W. A., Pitman, M. G., West, K. R. 1966. *Aust. J. Biol. Sci.* 19:341–54
16. El-Sheikh, A. M., Broyer, T. C., Ulrich, A. 1971. *Plant Physiol.* 47:709–12
17. Elzam, O. E., Epstein, E. 1965. *Plant Physiol.* 40:620–24
18. Elzam, O. E., Epstein, E. 1969. *Agrochimica* 13:187–95
19. Ibid, 196–206
20. Elzam, O. E., Hodges, T. K. 1967. *Plant Physiol.* 42:1483–88
21. Elzam, O. E., Rains, D. W., Epstein, E. 1964. *Biochem. Biophys. Res. Commun.* 15:273–76
22. Epstein, E. 1961. *Plant Physiol.* 36:437–44
23. Epstein, E. 1965. *Mineral Metabolism. Plant Biochemistry,* ed. J. Bonner, J. E. Varner, 438–68. New York: Academic. 1054 pp.
24. Epstein, E. 1966. *Nature* 212: 1324–27
25. Epstein, E. 1969. *Mineral Metabolism of Halophytes. Ecological Aspects of Mineral Nutrition of Plants,* ed. I. H. Rorison, 345–55. Oxford: Blackwell. 484 pp.
26. Epstein, E., Hagen, C. E. 1952. *Plant Physiol.* 27:457–74
27. Epstein, E., Jefferies, R. L. 1964. *Ann. Rev. Plant Physiol.* 15: 169–84
28. Epstein, E., Rains, D. W. 1965. *Proc. Nat. Acad. Sci. USA* 53: 1320–24
29. Epstein, E., Rains, D. W., Elzam, O. E. 1963. *Proc. Nat. Acad. Sci. USA* 49:684–92
30. Fineran, B. A. 1971. *Protoplasma* 72:1–18
31. Fisher, J., Hodges, T. K. 1969. *Plant Physiol.* 44:385–95
32. Floyd, R. A., Rains, D. W. 1971. *Plant Physiol.* 47:663–67
33. Foote, B. D., Hanson, J. B. 1964. *Plant Physiol.* 39:450–60
34. Gerson, D. F., Poole, R. J. 1971. *Plant Physiol. Suppl.* (Abstr.) 47:26
35. Greenway, H. 1963. *Aust. J. Biol. Sci.* 16:616–28
36. Greenway, H., Gunn, A., Thomas, D. A. 1966. *Aust. J. Biol. Sci.* 19:741–56
37. Gruener, N., Neumann, J. 1966. *Physiol. Plant.* 19:678–82
38. Handley, R., Vidal, R. O., Over-

street, R. 1960. *Plant Physiol.* 35:907–12
39. Hannon, N., Bradshaw, A. D. 1968. *Nature* 220:1342–43
40. Hasson-Porath, E., Poljakoff-Mayber, A. 1970. *J. Exp. Bot.* 21:300–3
41. Hasson-Porath, E., Poljakoff-Mayber, A. 1971. *Plant Physiol.* 47:109–13
42. Hiatt, A. J. 1967. *Plant Physiol.* 42:294–98
43. Ibid 1968. 43:893–901
44. Hiatt, A. J., Hendricks, S. B. 1967. *Z. Pflanzenphysiol.* 56:220–32
45. Higinbotham, N., Etherton, B., Foster, R. J. 1967. *Plant Physiol.* 42:37–46
46. Higinbotham, N., Graves, J. S., Davies, R. F. 1968. *Plant Physiol. Suppl.* (Abstr.) 43:51
47. Hooymans, J. J. M. 1964. *Acta Bot. Neer.* 13:507–40
48. Hurd, R. G. 1958. *J. Exp. Bot.* 9:159–73
49. Jackman, M. E., van Steveninck, R. F. M. 1967. *Aust. J. Biol. Sci.* 20:1063–68
50. Jackman, R. H. 1965. *N. Z. J. Agr. Res.* 8:763–77
51. Jackson, P. C., Adams, H. R. 1963. *J. Gen. Physiol.* 46:369–86
52. Jacoby, B. 1964. *Plant Physiol.* 39:445–49
53. Jacoby, B. 1965. *Physiol. Plant.* 18:730–39
54. Jacoby, B., Laties, G. G. 1971. *Plant Physiol.* 47:525–31
55. Jacobson, L. 1955. *Plant Physiol.* 30:364–69
56. Jacobson, L., Hannapel, R. J., Moore, D. P., Schaedle, M. 1961. *Plant Physiol.* 36:58–61
57. Jennings, D. H. 1968. *New Phytol.* 67:899–911
58. Johansen, C., Edwards, D. G., Loneragan, J. F. 1968. *Plant Physiol.* 43:1722–26
59. Kannan, S. 1971. *Science* 173:927–29
60. Kelley, W. P. 1963. *Soil Sci.* 95:385–91
61. Kylin, A., Gee, R. 1970. *Plant Physiol.* 45:169–72
62. La Haye, P. A., Epstein, E. 1969. *Science* 166:395–96
63. Laties, G. G. 1969. *Ann. Rev. Plant Physiol.* 20:89–115
64. Lauchli, A. 1972. *Ann. Rev. Plant Physiol.* 23:197–218

65. Lauchli, A., Epstein, E. 1970. *Plant Physiol.* 45:639–41
66. Leggett, J. E. 1968. *Ann. Rev. Plant Physiol.* 19:333–46
67. Livne, A., Levin, N. 1967. *Plant Physiol.* 42:407–14
68. Lundegårdh, H. 1955. *Ann. Rev. Plant Physiol.* 6:1–24
69. Lüttge, U. 1971. *Ann. Rev. Plant Physiol.* 22:23–44
70. Lüttge, U., Cram, W. J., Laties, G. G. 1971. *Z. Pflanzenphysiol.* 64:418–26
71. Lüttge, U., Laties, G. G. 1966. *Plant Physiol.* 41:1531–39
72. Lüttge, U., Laties, G. G. 1967. *Planta* 74:173–87
73. Lüttge, U., Laties, G. G. 1967. *Plant Physiol.* 42:181–85
74. Macklon, A. E. S., Higinbotham, N. 1968. *Plant Physiol.* 43:888–92
75. MacRobbie, E. A. C. 1962. *J. Gen. Physiol.* 45:861–78
76. MacRobbie, E. A. C. 1965. *Biochim. Biophys. Acta* 94:64–73
77. MacRobbie, E. A. C. 1970. *J. Exp. Bot.* 21:335–44
78. MacRobbie, E. A. C. 1971. *Ann. Rev. Plant Physiol.* 22:75–96
79. Marinos, N. G. 1962. *Am. J. Bot.* 49:834–41
80. Marschner, H. von 1964. *Z. Pflanzenernaehr. Dueng. Bodenk.* 107:19–32
81. Marschner, H. von, Gunther, I. 1964. *Z. Pflanzenernaehr. Dueng. Bodenk.* 107:118–36
82. Marschner, H. von, Ossenberg-Neuhaus, H. 1970. *Z. Pflanzenernaehr. Bodenk.* 127:217–28
83. McMillan, C. 1959. *Am. J. Bot.* 46:521–26
84. Meeuse, B. J. D., Campbell, J. M. 1959. *Plant Physiol.* 34:583–86
85. Mengel, K. 1967. *Z. Pflanzenphysiol.* 57:223–34
86. Mitchell, P. 1966. *Biol. Rev.* 41:445–502
87. Osmond, C. B. 1966. *Aust. J. Biol. Sci.* 19:37–48
88. Ibid 1967. 20:575–87
89. Osmond, C. B., Laties, G. G. 1968. *Plant. Physiol.* 43:747–55
90. Pierce, W. S., Higinbotham, N. 1970. *Plant Physiol.* 46:666–73
91. Pitman, M. G. 1967. *Nature* 216:1343–44
92. Pitman, M. G., Courtice, A. C., Lee, B. 1968. *Aust. J. Biol. Sci.* 21:871–81

93. Pitman, M. G., Mertz, S. M. Jr., Graves, J. S., Pierce, W. S., Higinbotham, N. 1971. *Plant Physiol.* 47:76–80
94. Pitman, M. G., Saddler, H. D. W. 1967. *Proc. Nat. Acad. Sci. USA* 57:44–49
95. Polya, G. M., Atkinson, M. R. 1969. *Aust. J. Biol. Sci.* 22: 573–84
96. Poole, R. J. 1971. *Plant Physiol.* 47:735–39
97. Porath, E., Poljakoff-Mayber, A. 1964. *Isr. J. Bot.* 13:115–21
98. Porath, E., Poljakoff-Mayber, A. 1968. *Plant Cell Physiol.* 9: 195–202
99. Rains, D. W. 1968. *Plant Physiol.* 43:394–400
100. Rains, D. W. 1969. *Experientia* 25:215–16
101. Rains, D. W. 1969. *Plant Physiol.* 44:547–54
102. Rains, D. W., Epstein, E. 1965. *Science* 148:1611
103. Rains, D. W., Epstein, E. 1967. *Aust. J. Biol. Sci.* 20:847–57
104. Rains, D. W., Epstein, E. 1967. *Plant Physiol.* 42:314–18
105. Ibid, 319–23
106. Rains, D. W., Floyd, R. A. 1970. *Plant Physiol.* 46:93–98
107. Rains, D. W., Schmid, W. E., Epstein, E. 1964. *Plant Physiol.* 39:274–78
108. Ramana, K. V. R., Rao, G. R. 1971. *Z. Pflanzenphysiol.* 64:91–92
109. Reisenauer, H. M. 1966. *Environmental Biology,* ed. P. L. Altman, D. S. Dittmer, 507. Bethesda, Md.: Fed. Am. Soc. Exp. Biol.
110. Richards, L. A., Ed. 1954. *Diagnosis and Improvement of Saline and Alkali Soils.* U.S. Dep. Agr. Handbook No. 60
111. Robertson, R. N. 1967. *Endeavour* 26:134–39
112. Robertson, R. N. 1968. *Electrons, Protons, Phosphorylation, and Active Transport.* Cambridge Univ. Press
113. Schwartz, M. 1971. *Ann. Rev. Plant Physiol.* 22:469–84
114. Scott, B. I. H., Gulline, H., Pallaghy, C. K. 1968. *Aust. J. Biol. Sci.* 21:185–200
115. Singh, L. B. 1970. *Econ. Bot.* 24: 439–42
116. Slater, E. C. 1967. *Eur. J. Biochem.* 1:317–26
117. Smith, R. C., Epstein, E. 1964. *Plant Physiol.* 39:992–96
118. Spanswick, R. M., Williams, E. J. 1964. *J. Exp. Bot.* 15:193–200
119. van Steveninck, R. F. M. 1965. *Physiol. Plant.* 18:54–69
120. Steward, F. C. 1971. *Ann. Rev. Plant Physiol.* 22:1–22
121. Strogonov, B. P. 1962. *Physiological Basis of Salt Tolerance of Plants.* USSR Acad. Sci., Inst. Plant Physiol.
122. Thorne, D. W., Peterson, H. B. 1949. *Irrigated Soils, Their Fertility and Management.* Philadelphia: Blakeston
123. Torii, K., Laties, G. G. 1966. *Plant Physiol.* 41:863–70
124. Torii, K., Laties, G. G. 1966. *Plant Cell Physiol.* 7:395–403
125. Viets, F. G. Jr. 1944. *Plant Physiol.* 19:466–80
126. Welch, R. M., Epstein, E. 1968: *Proc. Nat. Acad. Sci. USA* 61: 447–53
127. Welch, R. M., Epstein, E. 1969. *Plant Physiol.* 44:301–4
128. Weyl, P. K. 1970. *Oceanography. An Introduction to the Marine Environment.* New York: Wiley
129. Wildes, R. A., Neales, T. F. 1971. *Aust. J. Biol. Sci.* 24:397–402
130. Williams, M. C. 1960. *Plant Physiol.* 35:500–5

Ann. Rev. Plant Physiol. 1972. 23:389–418

MYCOPLASMAS AS PLANT PATHOGENS: 7536
PERSPECTIVES AND PRINCIPLES[1]

R. O. HAMPTON

Research Plant Pathologist, Plant Science Research Division, Agricultural Research Service, U.S. Department of Agriculture, Corvallis, Oregon

CONTENTS

INTRODUCTION

Since the suggestion by Doi et al (36) and Ishiie et al (77) that mycoplasmas might be the causal agents of plant diseases, many papers have been published with similar evidence that tetracycline-susceptible, mycoplasma-like organisms are associated with diverse plant diseases. Many of these papers have already been reviewed by Whitcomb & Davis (147) and by Maramorosch et al (104). In the former, comprehensive information on the plant viruses and "yellows agents" transmitted persistently by arthropod vectors was drawn together, generalized, and effectively summarized. In the latter review, a brief resumé of general mycoplasma information was presented as background for an evaluation of each plant mycoplasma publication then available. The purpose of the present review is twofold: first to consider the profitable information concerned with zoopathogenic mycoplasmas, and secondly to apply selected principles and concepts from this literature to problems presently confronting phytopathogenic mycoplasma research. Accordingly, neither comprehensive coverage of all mycoplasma information, nor

---

[1] Cooperative Investigations of the Plant Science Research Division, Agricultural Research Service, U.S. Department of Agriculture, Corvallis, Oregon, and the Oregon State University, Corvallis.

critical analyses of specific data or interpretations concerning plant mycoplasmas, will be undertaken in this review.

There is an obvious risk of improperly applying principles concerning
mycoplasmas from animals or humans to plant mycoplasmas, thus stimulating misdirected efforts. There are two factors, however, which encourage the
judicious application of these principles. First, despite specific distinctions
among the known mycoplasmas, this group of organisms is remarkably homogeneous in terms of their basic ultrastructure, colony morphology, cultural
requirements, and nucleic acid composition. If the yellows agents and organisms investigated by Hampton et al (69) and Cousin et al (28) are indeed
mycoplasmas, it seems expedient and profitable to assume that their properties approximate those of known mycoplasmas. Secondly, mycoplasma information is the only base of information which relates consistently to the ultrastructural and antibiotic response characteristics of the yellows agents. The
likelihood that these agents are mycoplasmas is substantial, and they are considered as such for the purposes of this review.

Nearly all reports of plant mycoplasmas to date have concerned diseases
of the "yellows type" whose causal agents are essentially confined to phloem
tissue and are leafhopper transmitted. When a few diseases of this group were
suspected of having a mycoplasma etiology, it is understandable that all such
diseases would collectively become suspect. The veritable explosion of reports
following those of Doi et al and Ishiie et al was due largely to this quest-
invoking suspicion. At present no one has reversed the trend of thought concerning yellows disease etiology which they initiated, i.e. no one has suggested that any yellows-type disease is caused by a virus or by an interaction
between a virus and a mycoplasma.

Interestingly, recognition that mycoplasmas were causes of animal and
human diseases was in each instance an outgrowth of successfully cultivating
the organism from diseased hosts. The development of plant mycoplasma information has been an exception, and in fact thoroughly successful cultural
methods have not yet been reported for these agents.

HISTORICAL PERSPECTIVES TO THE ESTABLISHMENT OF
MYCOPLASMAS AS PATHOGENS

It is the reviewer's purpose to relate salient mycoplasma research achievements which apply to current problems of research with plant mycoplasmas.
In a very real sense this history is an accounting of the impediments, innovations, and achievements in culturing mycoplasmas on cell-free media. The
collective factors impeding this accomplishment for plant mycoplasmas have
indeed been profound. And yet a great resource of remarkably relevant information awaits application to this important task. Perhaps the most brilliant
achievements within this history were the very first: the work with bovine
pleuropneumonia begun at Pasteur Institute more than 75 years ago. It

seemed appropriate, therefore, in this review to examine other works and present problems in relation to those premiere achievements.

The development of historical outlines (Tables 1, 2, 3) involve personal tastes and are also subject to personal perspectives. It was not my intention to omit any important research or to misplace the emphasis on any achievement. For outline comprehensiveness, I relied upon contributions by Hayflick (73), Edward & Freundt (50), Sharp (133, 134), Adler (1), Fabricant (54), Yoder & Hofstad (151), Chanock et al (24), Graystone et al (68), Whitcomb & Davis (147), and Maramorosch et al (104).

For the purpose of this review, three examples of information development have been chosen. The specific scope of information outlined—i.e. avian mycoplasmas, *Mycoplasma pneumoniae*, and aster yellows disease— was chosen to illustrate diverse principles as well as to facilitate information unity and manageability. Numerous other diseases could have been selected and would have been profitable, in view of the fact that comparable development obtained for the infectious mycoplasmas of sheep, swine, and laboratory animals. In each example presented, much of the information was developed during the last 30 years, i.e. within the bounds of relatively modern scientific dogma and technology. It is hoped, therefore, that those research achievements will be viewed as eminently relevant to today's research, rather than as merely historically interesting.

THE GENUS *Mycoplasma*

Borrel et al (17) proposed the binomial *Asterococcus mycoides* for the filterable, culturable agent of bovine pleuropneumonia in 1910. By 1940 this term was commonly used by virtue of its priority and was still retained in 1954 by Edward (46). During these years, however, the terms PPLO (pleuropneumonia-like organism) and L organisms (L = Lister Institute) came into general usage and were applied to all organisms morphologically resembling *A. mycoides*, including L-phase variants (L forms) of bacteria. Complex and confusing interpretations of relationships among pleuropneumonia organisms, bacteria, and bacterial L-phase variants created widespread misunderstandings among researchers concerned with these organisms during the 1940s and 1950s.

In 1955 (43) the Editorial Board of the *International Bulletin on Bacterial Nomenclature and Taxonomy* pointed out that the term *Asterococcus* had been proposed as a genus for fresh water algae by Scherffel (129) in 1908, thus making its use illegitimate for organisms of the pleuropneumonia group. Alternative terms, *Mycoplasma* or *Borrelomyces*, were suggested to replace *Asterococcus*. The term *Mycoplasma*, proposed as the genus for the bovine pleuropneumonia agent by Nowak (115) in 1929, clearly had priority and was advocated by Freundt (59) and Edward (47). *Mycoplasma mycoides* var. *mycoides*, the agent of bovine pleuropneumonia, was the type

TABLE 1. Developmental History of Information on Avian Mycoplasmas

Landmark	Developmental steps	Workers	Year
Chronic respiratory disease of poultry recognized as a unique disease.	(a) Agent transmitted to healthy poultry by sinus exudate.	Graham-Smith (65) Tyzzer (144)	1907 1926
	(b) Agent transmitted by natural aerosols from infected fowls; infectious agent retained in recovered fowls for 46 days; not retained in contaminated cages.	Beach & Schalm (12)	1936
Agent of "slow onset" coryza consisted of filterable, gram-negative, coccobacilliform bodies; birds recovered from "rapid onset" coryza susceptible to this agent, thus distinguishing separate diseases and agents.		Nelson (111)	1935
Cultivation of the coccobacilliform bodies.	(a) Agent cultured in chick embryo, in embryo tissue culture, and in blood with blood culture.	Nelson (112, 113)	1936–39
	(b) Mycoplasma incidentally encountered in egg; pure-cultured on agar, characterized; infectious to chickens, cotton rats.	Van Herick & Eaton (145)	1945
	(c) Formulation of mycoplasma-selective cultural medium.	Edward (44)	1947
	(d) Cultivation of agent from Nelson on mycoplasma-selective medium; identification of agent as mycoplasma (PPLO).	Smith et al (139)	1948
Cultural purification of infectious mycoplasma from sinus exudate; mycoplasma etiology of respiratory disease of turkeys and chickens established.		Markham &	1952

TABLE 1—(Continued)

Landmark	Developmental steps	Workers	Year
Diverse mycoplasma types from infected poultry separated, characterized.	(a) Type diversity established.	Wong (106)	
		Adler & Yamamoto (4)	1957
		Adler et al (5)	1957
	(b) Types separated, characterized.	Adler et al (3)	1958
		Yamamoto & Adler (148, 149)	1958
	(c) One avian mycoplasma species named.	Freundt (60)	1957
	(d) Three species named replacing that of Freundt: *M. gallisepticum* is agent of chronic respiratory disease.	Edward & Kanarek (51)	1960
	(e) Two other pathogens named: *M. synoviae, M. meleagridis.*	Olson & Kerr (118)	1964
		Yamamoto et al (150)	1965
	(f) Avian mycoplasma types progressively characterized.	Kleckner (79)	1960
		Yoder & Hofstad (151)	1964
		Fabricant (54)	1969
		Barber & Fabricant (9)	1971
	(g) Seven species from avian sources reviewed.	Fabricant (54)	1969

species. The genus *Mycoplasma* was implemented for classification in *Bergey's Manual of Determinative Bacteriology* by Freundt (60) in 1957, and had come into written and popular use near the time of the Second International Conference on the Biology of the Mycoplasma (48, 72) in 1967.

Edward & Freundt (49) and Freundt (60) listed 15 species of *Mycoplasma* and have recently amended the list (50) to include 20 additional species. Hayflick (73) listed 13 species known to product diseases in animals and man. Sharp (133) listed 12 species with established pathogenicity and 10 from animal sources with irregular or doubtful pathogenicity. Hayflick (73) and Edward & Freundt (49) eloquently and authoritatively discussed the fundamental biology and classification critieria for this group of organisms. Concepts concerning mycoplasma structure and function were effectively reviewed by Razin (121).

BOVINE PLEUROPNEUMONIA

The ingenious work with bovine pleuropneumonia by French bacteriologists at the Pasteur Institute in the 1890s is unparallelled in the history of microbiology. Nocard & Roux (114) by 1898 had found that this disease was caused by a filterable organism, now designated as *Mycoplasma mycoides* var. *mycoides*. They had defined its ultramicroscopic morphology by light microscopy and had succeeded in culturing this agent on cell-free medium.

As will be obvious in this review, comparable information development for other mycoplasma-induced diseases has required the cumulative and protracted research efforts of numerous individuals, usually operating in a framework of advancing technology. Shortly after this initial work, Dujardin-Beaumetz (38) cultured the agent on solid medium and described the tiny, delicate colonies which it produced, and Bordet (16) described its cellular pleomorphism. The more recent work with *M. mycoides* var. *mycoides* has thoroughly confirmed their concepts and results and only slightly refined their techniques.

Current mycoplasma information may erroneously convey the impression that the work of these early researchers was readily accepted and adopted. If this had been true, research with diseases eventually found to be mycoplasma-induced might have been accelerated by many years. Emmy Klieneberger-Nobel (82) did implement their findings and was responsible for isolating and characterizing numerous mycoplasma types in the 1930s, three of which are now recognized as distinct species (80). Fortunately, Nelson (111, 112; Table 1) in the 1930s also was objective in his viewpoint concerning fowl coryza etiology. However, some veterinary researchers in America, the professional group primarily concerned with bovine pleuropneumonia, considered this disease to be virus-induced. This opinion, fortunately, did not limit the unequaled effectiveness with which the disease was eradicated from this country by 1892 (56). Foster (56) in 1934 wrote: "Contagious pleuropneumonia is caused by a filterable virus. Pöls and Nolen in 1886, Nocard and Roux in 1896, and Arloing in 1897 isolated certain organisms in connection with the diesase, none of which has been accepted as an etiological agent." This particular skepticism apparently evolved from an acceptance of the bacteriological filtration results of Nocard and Roux, but rejection of the cultural work reported. It was Foster's assumption that organisms cultured were incidental to the disease.

Disease researchers and microbiologists who placed confidence in the cultural and etiological accomplishments at Pasteur Institute regarded the causal agent of bovine pleuropneumonia as a virus (46) because it had been demonstrated to be filterable. But it was thought of as a unique virus since it could be cultured. One cannot help contemplating the possibly productive influence of the culturable virus concept in plant pathology during the twentieth century. Relatively great benefit could have been realized by concerted efforts to "cultivate" all plant viruses by the methods elaborated by the imaginative bacteriologists at Pasteur Institute.

AVIAN MYCOPLASMAS

Information development.—The agents causing poultry sinusitis and chronic respiratory disease were known to be contagious probably well before 1900, and they were later found to be readily transmissible among birds when they were in normal contact within the same enclosures. This ease of transmission then not only caused attention to be focused on such diseases in early times (35), but also enabled early investigators to infect birds experimentally and to test factors enhancing and reducing natural spread of the diseases (144). More precise determination of conditions under which natural spread occurred and communicability from contaminated cages were later reported (12).

Two unrelated microorganisms were encountered in the sinus exudate from diseased chickens, each eventually found to be pathogenic. Schalm & Beach (128) observed these two organisms: a bacterium designated as *Haemophilus gallinarum*, and small gram-negative spheres which they concluded were unable alone to induce infectious coryza. The important work of Nelson (111) established that these organisms each caused a type of infectious coryza, and that these were distinguishable by incubation period and disease duration. He reported that rapid-onset, short-duration coryza was caused by *H. gallinarum,* and that slow-onset, long-duration coryza was caused by filterable, gram-negative, coccobacilliary bodies. The latter were cultured on mycoplasma-selective medium in 1948 by Smith et al (139), were characterized in 1958 by Yamamoto & Adler (148, 149), and in 1960 were named *Mycoplasma gallisepticum* by Edward & Kanarek (51). Nelson reproduced slow-onset coryza with filtrate produced by passing diluted sinus exudate through a Berkfeld V bacteriological filter, but he could culture no organism from this infectious filtrate in vitro. Later (112, 113) he implemented the useful egg-inoculation techniques reported for fowl pox virus and established bacteria-free cultures of the infectious agent in embryonated eggs, in chick embryo tissue cultures, and in horse blood subtending blood-agar slants. Based on morphological and cultural characteristics, he concluded that the coccobacilliary bodies he had implicated as the coryza agent were related to bacteria rather than to viruses, and noted their similarities to the agent of bovine pleuropneumonia and to the organism which had been found in sewage (*M. laidlawii*) by Laidlaw & Elford (90).

Nelson (112) mentioned that exudate from certain infected chickens contained virutally pure coccobacilliary bodies (mycoplasma), and he further purified this inoculum by Berkfeld filtration. Such inoculum contained high concentrations of this organism. For a number of reasons, including his virtually unlimited access to excellent quality inoculum and his use of a filter for purifying this agent, Nelson had accomplished more with avian mycoplasmas by 1935 than has been possible with plant mycoplasmas to date. Other factors contributing significantly to his success were: (a) his careful use of the light microscope to examine sinus exudate and monitor filtered exudate, thus visually establishing that the coccobacilliary bodies were filterable; (b) his separation of two microorganisms, determination of pathogenicity for each, and differentiation of the diseases induced by each; (c) his awareness of the definitive French work with the agent of bovine pleuropneumonia and recognition that the coccobacilliary bodies were probably related to that agent; (d) his persistent attempts and final success in culturing the coryza agent in artificial media. He might conceivably have emphasized the role of the bacterium, which could be cultured and have been satisfied upon finding that this organism caused coryza. This, in fact, had essentially been the conclusion of Schalm & Beach (128). Nelson instead accepted the challenge of the coccobacilliary bodies which could not be cultured, and persisted on the premise that Nocard & Roux (114) had cultured the pleuropneumonia agent. In brief, his work represented unusual discernment, carefulness, awareness of relevant published work, and thoroughness.

The fate of Nelson's results within the scientific community, however, was similar to that of the French work with bovine pleuropneumonia. It received far less attention than it deserved and was virtually unkown to veterinary researchers for many years. Almost 15 years had passed before work of comparable quality was published (106), establishing mycoplasmas (PPLO) as the cause of coryza (chronic respiratory disease) in chickens. Perhaps it was too incredible that any filterable agent could also be cultured.

A decade after the work of Nelson, Edward (44) formulated an excellent mycoplasma-selective medium containing fresh yeast extract and antibacterial agents, thallium acetate and penicillin. Freshly prepared yeast extract has since been one of three basic components in most mycoplasma media and has been crucial for successfully culturing and maintaining numerous strains and species of zoopathogenic mycoplasmas, including M. pneumoniae (21, 92). Apparently unaware of this new medium, Smith et al (139) were successful in culturing the fowl coryza agent obtained from Nelson on agar containing peptic digest and ascitic fluid. They designated it as a mycoplasma (PPLO) on the bases of staining characteristics, cell morphology, and colony morphology.

Quite incidentally and virtually unnoticed, van Herick & Eaton (145) were the first workers to demonstrate that an avian mycoplasma was transmitted in chicken eggs. They were using embryonated eggs for cultivating the agent of primary atypical pneumonia (a human pathogen then assumed to be a virus; see Table 2), and they suspected an egg-borne contaminant because of occasional sudden increases in numbers of embryo deaths. They also observed mycoplasma-like organisms associated with embryo tissues and found that inocula from such eggs were virulent to cotton rat and nonvirulent to hamster (the agent of primary atypical pneumonia was virulent to hamster). They determined that the organism from eggs fulfilled the critieria for mycoplasma (PPLO) specified by Sabin (127), cultured it on mycoplasma-selective medium, and described its morphology in broth and on agar medium. They tested sera from the laying hens producing eggs for their laboratory and found that more than half contained antibodies against the agent they had cultured from the eggs, establishing that the agent was egg transmitted.

More recently all three avian mycoplasma species known to be pathogenic have been implicated in turkey and/or chicken egg transmission. *Mycoplasma gallisepticum* was determined to be egg transmitted in turkeys in 1957 by Hofstad (76), who recovered the agent on agar medium. This species was later found to be egg transmitted in chickens by Heishman et al (74). A mycoplasma found to be egg transmitted in turkeys in 1958 by Adler et al (3) was later named *M. meleagridis* by Yamamoto et al (150). Carnaghan (19) in 1961 determined that the chicken synovitis agent was egg transmitted, but he did not culture *M. synoviae* from infected tissues on cell-free media.

Van Herick & Eaton (145) were probably the first workers to report the induction of fatal pulmonary lesions in cotton rats with sterile Seitz filtrates from broth cultures of mycoplasmas (cultures of the egg-borne agent). This phenomenon has been reported since but is not understood. It is reviewed as an ancillary factor by Ward & Cole (146), and should be carefully considered in studies in which plant mycoplasmas are tested for pathogenicity in laboratory animals. Van Herick & Eaton may also have been the first workers to adapt a mycoplasma (the egg-borne agent) to a medium containing only rabbit serum, so that mycoplasma antigen preparations would not elicit antibodies against foreign serum proteins.

An ironic aspect of the excellent work of van Herick & Eaton was their closeness to successfully culturing the agent of primary atypical pneumonia. They had in fact cultured at least two mycoplasma species, *M. pneumoniae* and probably *M. gallisepticum,* in the same embryonated eggs. Their suspicions concerning an egg-borne contaminant had stimulated them to examine the embryos for microorganisms, and therefore they had observed *M. gallisepticum* which grows abundantly throughout infected embryos. Had they similarly examined embryos infected by *M. pneumoniae,* they most likely would not have observed this agent, since it is primarily localized in the bronchial epithelium of chick embryos (97). Except for the deficiency of fresh yeast extract in their mycoplasma-selective medium, however, they might easily have cultured both mycoplasmas on solid medium. The thoroughness of their infectivity studies suggests that they would have differentiated these agents pathologically, had they cultured them both and separated them on the basis of colony morphology.

These workers carefully considered the relationship between the agent they had cultured and that of primary atypical pneumonia, and correctly con-

TABLE 2. DEVELOPMENTAL HISTORY OF INFORMATION ON MYCOPLASMA
PNEUMONIAE

Landmark	Developmental steps	Workers	Year
Disease defined as being distinct from other superficially similar diseases.	(a) Unique disease, unknown etiology.	Reinmann (124) Dingle & Finland (34)	1938 1942
	(b) Designated as primary atypical pneumonia (PAP).	Official statement (116)	1942
	(c) Many patients with PAP developed cold agglutinins.	Peterson et al (119) Turner et al (143)	1943 1943
	(d) Cause assumed to be viral.	general	1942–62
	(e) Many unreported attempts to isolate agent.		1942–62
Propagation of filterable agent in chick embryo, agent labile at room temperature and transmission to laboratory animals.		Eaton et al (41, 42)	1944–45
Transmission of agent to human volunteers by filtered inoculum from patients with cold-agglutinin-positive PAP.		Team (26)	1946
Agent found to be sensitive to aureomycin, chloromycetin and streptomycin.		Eaton et al (39, 40)	1950–57
	Development of crucial techniques and definitive information concerning agent. (a) Immunofluorescent localization of PAP agent in chick embryo; effective antibody titration.	Liu (97) Liu et al (98)	1957 1959
	(b) Suppression of PAP agent by sodium aurothiomalate; association of coccobacilliform bodies with localized chick embryo infection.	Marmion & Goodburn (107)	1961

TABLE 2—(Continued)

Landmark	Developmental steps	Workers	Year
	(c) Agent characterized in tissue culture.	Clyde (25)	1961
Agent causing cold-agglutinin-positive PAP defined.	(a) Cultivated on cell-free agar medium; identified as a mycoplasma (Hayflick).	Chanock et al (21)	1962
		Hayflick (71)	1965
	Identified as causal agent of the disease.	Chanock et al (22)	1962
	(b) Human respiratory illness induced by artificially cultured agent.	Couch et al (27)	1964

cluded that they were different. Their bases for this distinction were that the mycoplasma from eggs (a) grew on cell-free media and had the cellular morphology of mycoplasmas; (b) had not been recovered from previous egg-passages of the agent inducing primary atypical pneumonia; (c) caused egg mortality abnormal for the agent of primary atypical pneumonia; and (d) did not react serologically with human sera (from patients recovering from primary atypical pneumonia) which neutralized the agent of primary atypical pneumonia. Cultivation of the agent associated with primary atypical penumonia (M. pneumoniae) was eventually achieved by Chanock, Hayflick & Barile (21) 17 years later.

A culture of the coccobacilliary organism was obtained from Nelson by Smith et al (139), who cultured it on agar medium containing ascitic fluid and peptic digest, recognized it as a mycoplasma (PPLO), and described its cellular and colony morphology. Despite this work and that of Nelson, however, the agent inducing chronic respiratory disease of chickens and infectious sinusitis of turkeys was generally assumed to be viral (1) until the work of Markham & Wong (106) was published in 1952. These workers obtained isolates from diseased turkeys and chickens and cultured them on the mycoplasma-selective medium of Edward. Each isolate was serially passed at least 13 times to a final dilution of the original inoculum of 10^{26} to eliminate the presence of passive viral contaminants in the culture. Isolates thus processed were inoculated into embryonated eggs and were then recovered from infected embryos and injected into sinuses of turkeys. Three of four inoculated turkeys developed sinusitis, and the agent was recovered from the sinus exudate from these infected birds. Koch's postulates with M. gallisepticum were thus practically satisfied, except that their cultures had not been single-colony purified on agar media, and may have contained other avian mycoplasma species.

Turkeys in this study were also inoculated with the similarly processed agent of chicken chronic respiratory disease after which one of the three inoculated birds developed sinusitis. The agents from infected chicken air sacks and from turkey sinuses appeared to these workers to be morphologically similar and, according to Adler & Yamamoto (4) and Yamamoto & Adler (148, 149), both represented *M. gallisepticum,* a name later designated by Edward & Kanarek (51). Edward's type culture of this species is designated as PG31. The American reference type of this species, S6, was derived from turkey brain by Adler in 1957.

Kleckner (79) assembled 15 avian mycoplasma isolates, briefly characterized these culturally and grouped them into eight serotypes, based on serological cross reactions, designated alphabetically A through H. Fabricant (52) confirmed the distinctness of six of these and retained their identity, A through F. Yoder & Hofstad (151) assembled 98 isolates, including several tested by Kleckner, agreed closely with the eight serotypes of Kleckner, and expanded the system to include 12 serotypes, A through L. Dierks et al (32) expanded this grouping further by adding serotypes M through S. Fabricant (54) reviewed information on these serotypes, and listed species equivalents where appropriate for each sterotype A through S. Barber & Fabricant (9) most recently have suggested that the 18 serotypes A through R, excluding *M. synoviae* (serotype S), fit into 10 serologic groups. Unpublished biochemical tests confirmed the relatedness of isolates within these 10 groups and distinctness of isolates between groups.

Concepts of mycoplasma relationships.—Accompanying a renewed interest in mycoplasma research during the 1940s and 1950s, a diversity of backgrounds and perspectives came to be represented among those working with these organisms, and a diversity of concepts concerning the nature of these organisms was published. Some workers were inclined to group together organisms known to be L-phase variants of bacteria with those of the pleuropneumonia group, under the common name "L organisms." This was based on the assumption that both were the products of bacteria. There was limited incentive for eliminating bacteria from cultures of mycoplasmas among those who considered these entities as equivalents, i.e. if both were intimately related to bacteria. Such mixed cultures were both the products of confusion among workers and the cause of confusion. The unrelatedness of mycoplasmas to bacteria was stressed repeatedly by Freundt (59, 61), Edward (45, 46), Klieneberger-Nobel (80), and Edward & Freundt (49). The concepts expressed by Edward & Freundt during those years of confusion appear steadfast and remarkably accurate.

Gentry (64) reported that half of the 316 avian isolates supplied him by other laboratories contained bacteria, and he concluded that such isolates were probably L-phase variants rather than PPLO. He hypothesized that the pathogenic strains were true mycoplasmas (PPLO) and that the nonpathogenic strains were L-phase variants which had reverted to bacteria. Kelton et al (78), with various cultural manipulations, derived bacteria from all of 28 diverse mycoplasma isolates tested and presented evidence of transition forms between mycoplasmalike cells and bacterial cells. McKay & Truscott (108) believed that the bacteria, *Hemophilus gallinarum* and *Mycoplasma gallinarum* were spontaneously or inducibly interconvertible, and yet the nonbacterial phase was referred to as a PPLO instead of an L-phase variant.

Information was sufficiently confused at the time of the first International Symposium on Pleuropneumonia-like Organisms in 1960 to elicit the following remarks from Dr. Julius Fabricant (53):

> A careful examination of the literature suggests that much of the previous PPLO work could have been done with mixed cultures, and that no reliable techniques exist for the production of pure PPLO cultures or for ascertaining their purity. These fundamental weaknesses stand in the way of present and future PPLO research. It is futile to discuss the interrelationships of the various PPLO types until techniques are devised for producing and identifying pure cultures of the organisms. It is equally futile to argue about the relationship of PPLO and L forms or to talk about the reversion of PPLO or L forms to bacteria for the same reason.
>
> A further serious problem in avian PPLO studies is associated with the inability of the present culture media to support adequately the growth of many pathogenic strains of PPLO.
>
> The reversion to bacterial forms of avian PPLO cultures has several possible explanations: (1) bacteria are present in the original culture; (2) bacteria arise from contamination during the reversion procedure; (3) the so-called PPLO cultures are made up of a mixture of PPLO and L forms; (4) some of the organisms termed avian PPLO are true PPLO and some are L forms; (5) all so-called avian PPLO are L forms; or (6) several of the previous possibilities are true.

It is now generally accepted that mycoplasmas are unrelated to L-phase variants of bacteria, and that their similarities are primarily coincidental with their common lack of a rigid cell wall (50). The purpose of recounting this conceptual evolution is singular: that those in plant mycoplasma research might avoid similar confusion. I believe there are at least five safeguards against such confusion, once mycoplasma-like organisms can be consistently extracted from diseased plants and cultured in vitro.

First, cultural conditions, media, and procedures must be developed by which mycoplasmas can be isolated from other plant pathogens, tested for phytopathogenicity, and maintained for comparative studies and for reference. This may or may not include cultural purification by single-colony cloning, or investigation of possible bacterial L-phase variants, depending upon the experience of the investigator (see Safeguard No. 5).

Secondly, to the extent possible, media used for culturing plant mycoplasmas should contain no bacterial inhibitors, e.g. penicillin or thallium acetate. This reduces the likelihood that bacteria normally encountered in plant sap will be converted to L-phase variants. Dienes (33) has worked with L-phase variants which became adapted to agar media and were maintained in his laboratory for years without reverting to the bacterial form. Plant mycoplasma literature can ill afford the confusion produced by reports of such isolates. No L-phase variant has yet been reported to be pathogenic (50). On the other hand, initial or even prolonged failure to demonstrate clear-cut pathogenicity of mycoplasmas, which eventually turn out to be phytopathogenic, should not be unexpected. An additional disadvantage to the use of thallium acetate as indicated by Shepard (135) is that all known T strains of mycoplasmas (T: tiny colonies) are inhibited by this chemical. Thus, should organisms with similar sensitivity be encountered as plant pathogens, they would not be isolated on media containing thallium acetate.

Thirdly, all possible precautions should be taken to preclude mycoplasma contaminants from nonplant sources during attempts to culture plant mycoplasmas. The refinement of microbiological techniques and precautionary asepsis common to medical microbiology laboratories will not be inherent in most plant virus laboratories where the pioneer research with plant mycoplasmas has begun. Mycoplasma-selective media should be considered analogous to marine-plankton catch nets. These media are designed specifically to "catch" mycoplasmas. Because mycoplasmas are known to be commonly associated with the human body (73) and in the aerosols from human oral and nasal passages, the potential for contamination using selective media is substantial. Unless great care is exercised, therefore, the likelihood of culturing mycoplasma contaminants may exceed that of culturing unkown mycoplasmas from plants.

Fourthly, reference mycoplasma isolates should be used as controls for visual comparisons under all test conditions, and for assaying growth-supporting potential of cultural media. Obviously, all laboratory procedures must be designed to preclude contamination by these reference isolates.

Fifthly and probably most importantly, organisms derived from plants which appear to be mycoplasmas and are culturally stable should promptly be sent to a well-recognized mycoplasma laboratory for confirmation and serological typing. Much of the initial confusion concerning plant isolates could be minimized or even eliminated by our reliance upon highly experienced, renowned mycoplasmologists for this work.

Mycoplasma pneumoniae

While the development of avian mycoplasma information is relevant because of its complexities, ensuing controversies, and the eventual application of sound microbiological principles, information concerning primary atypical pneumonia (Table 2) is singularly more parallel to that of aster yellows (Table 3). In each case there have been elaborately developed studies on the nature of the disease, therapy, and natural transmission, preceding an understanding of the microbiological basis for the diseases. Response of each agent to certain antibiotics has been an impetus for the search for a nonviral etiol-

TABLE 3. Developmental History of Information on Aster Yellows Disease

Landmark	Developmental steps	Workers	Year
Disease defined as being distinct from other diseases which were superficially similar.	(a) Agent leafhopper transmissible (*Macrosteles fascifrons*); long incubation period of agent in leafhopper vector; agent not mechanically transmissible; uniquely wide host range of agent.	Kunkel (83, 84)	1924–26
	(b) Thermolability of agent in the leafhopper and in host plants.	Kunkel (85–87)	1937–41
	(c) Apparent thermoattenuation of agent.	Kunkel (86)	1937
	(d) Mechanical inoculation (injection) of vector with agent.	Black (13)	1940
	(e) Evidence of agent's multiplication in leafhopper.	Black (14) Maramorosch (101)	1941 1952
	(f) Particles larger than known plant viruses.	Black (15)	1943
	(g) Relationships (cross-protection) among strains of the agent, in host plants and in the vector, elaborated.	Kunkel (88, 89) Freitag (57, 58)	1955–57 1964–67
	(h) Dynamics of vector-agent relationships defined.	Maramorosch (104) Sinha & Chiykowski (136)	1950 to present 1967
	(i) Attempts to purify agent by virological methods.	Lee & Chiykowski (91) Steere (140)	1963 1967
Mycoplasma-like ultrastructure of agent in diseased plants.		Doi et al (36)	1967
Mycoplasma-like response of agent in situ to tetracycline antibiotics.		Ishiie et al (77)	1967
Isolation of agent from infected plants; cultural purification on cell-free medium; establish disease with purified agent.		(not yet accomplished)	

ogy. In the case of primary atypical pneumonia, however, *Mycoplasma pneumoniae* was implicated etiologically coincident with successful cultivation of the agent on cell-free cultural medium. The establishment of aster yellows etiology still awaits this cultural achievement.

Information development on primary atypical pneumonia began in the United States during the 1930s, when a new type of pneumonia was recorded with increasing incidence in unpublished medical reports. Neither pathogenic bacteria nor influenza viruses could be isolated from patients affected by this disease. The first formal report of the disease was that of Reinmann (124) in 1938. The illness was officially named primary atypical pneumonia (PAP) by the Director of the Commission on Pneumonia, United States Army, in 1943 (116). As an extension of work describing the disease by Dingle & Finland (34), Peterson et al (119) determined that many patients with PAP developed cold agglutinin antibodies which could be detected reliably by agglutination serology. They proposed the use of this test for diagnosing this disease until the etiology could be determined. The association of cold agglutinin and PAP was also reported in that same year by Turner et al (143). The disease then became known by the specific term, cold-agglutinin-positive PAP.

During the period from 1942 to 1962 there were many unreported attempts to culture the agent responsible for PAP (24). Lack of success in culturing this agent on artificial media was sufficient basis for the general assumption that it was probably viral. The work of Eaton et al (41, 42) established cultures of the agent by bacteriologically filtering diluted sputum from patients with PAP and injecting the filtrate into chick embryos. They determined pathogenicity of the agent in sputum; and following serial passage in chick embryos, by nasal inoculation of hamsters and rats, they determined that the agent survived only a few hours at room temperature in sputum from infected patients.

A team of medical researchers working with human volunteers at Johns Hopkins Hospital in 1946 established transmission of this filterable agent from PAP patients to groups of human volunteers (26). Cold agglutinins were used in this study to monitor the infection process in volunteers who had received filtered inoculum. Inocula were applied as mist to nasal and oral passages. Fifty-two of the 84 volunteers developing PAP also produced cold agglutinin antibodies, whereas only 1 of 27 persons with pneumonia associated with influenza virus A developed these antibodies. None of 25 patients with pneumonia associated with other viral or bacterial pathogens developed cold agglutinins. These studies clearly established that the filterable agent could produce respiratory disease in man and confirmed its definitive property of inducing production of cold agglutinin antibodies.

The first doubts concerning the viral etiology of PAP arose in 1950 when Eaton demonstrated that aureomycin and chloromycetin inhibited growth of the agent (39). Streptomycin was later found also to be inhibitory by Eaton

& Liu (40). However, attempts by these workers to cultivate PPLO (myco-plasmas) from egg material containing strains of the PAP agent were unsuc-cessful. Renewed efforts to cultivate the PAP agent in artificial media were nevertheless stimulated in other laboratories. The exact requirements for cul-turing this agent were first determined by Chanock, Hayflick & Barile (21), who in the same study established by immunofluorescence the homology between colonies of their agent cultured in vitro and PAP antibodies (98). This cultural accomplishment was verified by others (22) in the same year, 1962.

Liu (97) had previously adapted immunofluorescence to the identifica-tion of specific PAP antigen in infected chick embryo bronchial epithelium. Liu et al (98) then delineated three types of antibodies in the sera of patients with PAP and developed a sensitive immunofluorescence method for titrating PAP antibodies in these sera. Also, Marmion & Goodburn (107) had found that lung lesions induced in hamsters inoculated with the PAP agent were suppressed either by sodium aurothiomalate or by sera from PAP-affected patients. They further had found coccobacilliform bodies closely associated with the epithelial areas in which fluorescein-tagged antibodies accumulated further suggesting that the PAP agent was a mycoplasma (PPLO). Clyde (25) had characterized the agent in tissue culture and had suggested it might be a mycoplasma (PPLO).

Following the definitive work of Chanock et al (21), the agent was named *Mycoplasma pneumoniae* in 1963 (23), and human respiratory illness was produced by the agar-propagated mycoplasma the following year (27).

On the basis of the work by Chanock et al (21) and by Couch et al (27), it is believed that *M. pneumoniae* is the agent of PAP. It is thus the only mycoplasma species established to be a human pathogen. It therefore seems appropriate in this review to examine the pathological evidence (27) in rela-tion to analagous evidence that shall be acceptable for establishing a myco-plasma etiology of plant disease.

Within the framework of Koch's postulates, purification of an agent to be tested is undertaken to determine solo-pathogenicity of that agent, i.e. capa-bility of inducing "the disease" unaccompanied by another infectious agent. Mycoplasma isolates are routinely purified by serially transferring single colo-nies to cell-free media, thus simultaneously selecting toward type homogen-iety and diluting virus particles which might passively accompany the myco-plasma in vitro. If pathogenicity is lost or modified before purification is achieved, however, the factor(s) responsible for this change would be inde-terminate, i.e. due to elimination of an accompanying agent perhaps synergis-tic in disease induction, or to inherent adaptive (metabolic) change in the mycoplasma. In defining specific etiology, therefore, it would seem unsafe to discount the importance of pathogenicity loss coincident with purification of the agent. Further, our present concern that virus particles could passively ac-company mycoplasma isolates in early serial passages should soon be extended to viruses which multiply in mycoplasmas (94) and to the possible role of mycoplasmas as vectors (8).

One of two isolates evaluated by Couch et al (27) was nonpathogenic after only four serial passages on agar medium. Although most volunteers inoculated with the other, stable isolate (FH) produced PAP antibodies, none developed the PAP snydrome. Fifteen of 33 volunteers developed no illness, 12 of 33 developed brief, mild cold-like symptoms without fever, 1 of 33 developed upper respiratory illness accompanied by mild fever, and none developed pneumonia. This imperical evidence, particularly serological results, certainly implicated *M. pneumoniae* as a factor in PAP. It appears, however, that solo-pathogenicity of *M. pneumoniae* in the PAP syndrome may not have been established.

In the inherent absence of serological evidence in plant disease development, establishment of solo-pathogenicity for a plant mycoplasma would probably be contingent upon reproducing "the disease" with the purified agent. Failure to do so would presumably elicit the search for other agents participating in disease production. In other words, the symptomological demands upon fulfillment of Koch's postulates for plant mycoplasmas may well be more stringent than upon former evidence (27) implicating *M. pneumoniae* as a human pathogen. These demands would seemingly be based on different types of questions: First, "Is a mycoplasma contributory to primary atypical pneumonia?" The answer was conspicuously yes, and appropriate antibiotic therapy measures were applied to PAP patients. Secondly, once pure cultures of plant mycoplasmas are obtained, the more immediate question would seem to be, "Is a mycoplasma capable of inducing plant disease in the absence of viral or other agents?" The answer here may well depend upon (*a*) whether or not truly pathogenic mycoplasma biotypes exist; (*b*) success in cultivating and purifying one or more of these; and (*c*) the proneness of such types to lose pathogenicity during serial passage by inherent adaptive change.

Hayflick (personal communication 1970) stated that much effort had been expended in attempts to find strains of *M. pneumoniae* which maintained virulence during cultivation on cell-free medium. Adler & Yamamoto (4) had encountered strains of *M. gallisepticum* which were virulent and could be cultured in embryonated eggs, but which produced no growth on cell-free media. However, they were able to find strains such as S6 which not only grew well on agar medium, but which retained virulence during hundreds of serial passages. Plant mycoplasma researchers should, therefore, be prepared to encounter pathotypes representing any position within this behavioral range.

In relating the significance of strain occurrence to etiological studies of plant mycoplasmas, it would seem important to investigate biotype diversity simultaneous with purification. In the early stages of isolation, numerous single-colony isolates should be established, before diversity is lost through serial selection of colonies. Strains which retain virulence during serial passage should then be sought, and reduction in virulence should be compensated for experimentally by (a) searching for hypersusceptible host plants; (b) developing methods for predisposing plants to infection; (c) innovating methods of inoculating plants with purified mycoplasmas, including use of vectors, injection, root infiltration, mechanical abrasion; (d) inoculation of largest feasible numbers of host plants; and (e) intense scrutiny of inoculated plants for detection of mild expressions of disease. When suspect plants are observed unusual effort should be exercised in attempts to transmit the agent from such plants. Simultaneously, attempts should be made to reisolate the agent in cell-free medium from such plants and where satisfactory antiserum to the agent is available demonstrate the presence of the agent in inoculated plants by immunofluorescence (31, 120). Another serological approach for confirming presence of the agent in infected plant tissues would be electron microscopic localization by immuno-autoradiography (67, 130).

ASTER YELLOWS DISEASE

The reviews of Whitcomb & Davis (147) and Maramorosch et al (104) have dealt comprehensively with aster yellows research. I will examine in detail only selected papers included in Table 3.

Due primarily to the creative efforts of one man, Dr. L. O. Kunkel, an abundance of excellent and enduringly accurate information on aster yellows had been developed by 1941 (Table 3). By that time much had been learned about the basic nature of the disease, the nonmechanical transmissibility of the agent, the leafhopper vector and host range of the causal agent, the thermolability of the agent in vector and in infected plant tissue, and about thermotherapy of aster yellows infected plants.

Evidence by Fukushi (62, 63) that the rice dwarf virus multiplied in its leafhopper vector *Nephotettix cincticeps* (Uhler) instituted a protracted controversy on the nature of the relationship between viruses and their vectors. Kunkel (85) felt that rendering leafhoppers (*M. fascifrons*) noninoculative by short-duration heat treatment, followed by their regained ability to transmit the agent, suggested multiplication. Storey (141) and Bawden (10) intuitively opposed the concept of plant viruses multiplying in insect vectors and proposed alternative interpretations. Their explanations were based on the concept of a passive relationship between agent and vector. Black (14), in retrospect, clearly demonstrated multiplication of the aster yellows agent in its leafhopper vector. However, Bawden (11) persisted that those results may only have reflected a changed state of the virus, or an increased accessibility to the virus with time, following acquisition by the leafhopper. The question was successfully settled by Maramorosch (101), who implemented the infectivity assay techniques of Black (13, 14) and serially passed diluted extracts from infective to healthy leafhoppers in a sequence of 10 passages from the original inoculum. Dilution of infective juice at each passage had been 10^{-4}, thus producing a dilution of 10^{-40} of the original inoculum volume. The measured concentration of the agent in the first and ninth passage was the same, thus satisfactorily establishing that the agent multiplied in the leafhopper. An analogous method was used by Merril & Ten Broeck (109) to demonstrate that equine encephalomyelitis virus multiplied in its mosquito vector, and had been suggested by Storey (141) as the type of evidence necessary to demonstrate plant virus multiplication in vectors.

In 1955 Kunkel (88) had determined that a strain of aster yellows described by Severin (132), herein called the C strain, differed from the standard strain in that it was transmissible to *Zinnia elegans* and celery. Using these distinct strains and inoculations by the aster leaf hopper *Macrosteles fascifrons* (Stål), Kunkel (88) inoculated *Vinca rosea* first with C strain and subsequently with the standard strain, and reciprocally, with the standard strain first and subsequently with the C strain. Symptoms induced after these sequence inoculations were always those of the first strain introduced. He considered this behavior as representative of cross-protection, and on this basis suggested that the two agents were closely related. Kunkel (89) later showed that leafhoppers which had fed for 2 weeks on either strain and subsequently fed for 2 weeks on the other, transmitted only the first strain acquired, thus suggesting cross-protection between these strains in the leafhopper vector.

Freitag (57, 58) described three other strains of aster yellows which occurred in California, generalized herein as A, B, and C. He examined cross-protection and interference relationships among these three strains in host plant species (57) and later in the leafhopper vector (58). In *Plantago major* L. and *Nicotiana rustica* L. var. *humilis* he found a peculiar variety of strain relationships ranging from normal cross-protection, as described by Kunkel (88), to partial cross-protection (including instances of interference), to no cross-protection. Two strain combinations in *N. rustica* var. *humilis*, consisting of A/B and A/C inoculation sequences, resulted in disappearance of disease symptoms and recovery of the plants. This phenomenon was interpreted as being due to mutual suppression between strains, and is probably unprecedented in literature pertaining to strain relationships among plant pathogens.

Interestingly, the interpretations and significance of aster yellows strain relationships should not be affected by the likelihood that the causal agent is a mycoplasma. Whatever factors are eventually understood to limit multiplication of the agent in plant and leafhopper hosts (availability of specific host substrates, spatial limitations at sites where the agent multiplies, etc), past determinations of cross-protection and interference among strains remain valid, essentially irrespective of the precise etiology of aster yellows.

Studies by Maramorosch and co-workers on aster yellows agent-vector dynamics began before 1952 (101). In a study of factors influencing incubation period of this agent in leafhoppers (*M. fascifrons*) and in China aster

(102), he stated, "The mutagenic action of heat on the the virus, demonstrated by Kunkel (86), and reasons listed in previous papers (100, 101) support the view that aster-yellows virus is a parasitic organism, highly specialized in its requirements." Later (103) he improvised a procedure by which the agent completed the characteristic incubation period in leafhopper tissue suspended in tissue culture broth. In this study, he fed leafhopper nymphs on infected aster plants for 2 days, anesthetized and surface sterilzed them, cut them into small pieces, rinsed the pieces in tissue-culture broth, and finally suspended bits of tissue in single drops on cover glasses. These were sealed and incubated at 25°C for 10 days. Preparations from incubated tissues were injected into healthy leafhoppers, which were in turn caged on aster test plants in groups of three. Eleven of 80 such groups tested transmitted the aster yellows agent, whereas equal numbers injected with nonincubated tissue extracts from leafhoppers treated the same way did not transmit the agent. Incidentally, the tissue culture medium used in this study contained streptomycin, penicillin, and mycostatin. Streptomycin would presumably but not necessarily have inhibited the growth of mycoplasma in this medium (131).

In discussing results he stated, "The lack of virus recovery from fluids which surrounded the tissues was understandable and predictable. Viruses are not known to multiply in cell-free media, and aster yellows virus loses all infectivity within 2 hours at room temperature" (Black 15). His specific interests and manner of inquiry seem to have taken him close to a view of the nonviral nature of the aster yellows agent. Examining his preparations by light microscopy, Maramorosch (103) commented,

The experiments reported above provide evidence for the first time that under proper conditions a plant virus can complete its incubation in *in vitro* preparations of insect vector tissues. Microscopic examination revealed that most of the cells suspended in hanging drops were alive after 10 days, in so far as could be judged from the normal appearance of their nuclei.

Littau & Maramorosch (95, 96) also extensively examined tissues of infective leafhoppers by light microscopy, and pointed out certain fat body abnormalities in some of the inoculative males. Fat body tissue of inoculative leafhoppers was later found to be apparently void of the aster yellows agent (136). Hirumi & Maramorosch (75) estimated the distribution of the aster yellows agent within the bodies of inoculative leafhoppers. In this study saliva glands, the gut, Malpighian tubules, ovaries, and testes were dissected from leafhoppers previously fed on aster yellows-infected plants. Homogenates from these organs were injected into healthy leafhoppers, the incubation period was fulfilled, and the leafhoppers were tested for infectivity. Salivary glands contained only trace amounts of the agent 5 days after leafhoppers were placed on infected plants, but contained substantial amounts 19 days after placement on these plants.

The work of Sinha & Chiykowski (136) determined that assayable quantities of the aster yellows agent (C strain) were present in the hemolymph, alimentary canal, salivary glands, and ovaries, but not in testes, Malpighian tubules, fat body tissue, mycetomes, or the brain. These workers provided elegant evidence of the chronological increase in concentration of the agent in hemolymph, alimentary canal, and salivary glands. Presence of the agent in each dissected body part, pooled from 30 synchronously fed leafhoppers, was determined by sequential assays during the incubation and inoculative periods. Their data suggested that the agent first multiplied in the alimentary canal, and that hemocytes could be a principal site for multiplication of the agent. Completion of the incubation period was coincident with a sharp rise in concentration of the agent in the salivary glands. It is also true for this contribution that the elucidated nature of the infectious agent does not detract from the excellence or significance of the work.

Both Lee & Chiykowki (91) and Steere (140) recovered fractions containing infectious aster yellows agent when homogenates of inoculative leafhoppers were processed, respectively, by differential centrifugation and gel filtration. Although the clarified supernate from an initial low-speed centrigu-ation was most infectious of the fractions assayed (91), the final fraction which had undergone two cycles of centrifugation also contained detectable quantities of the agent. Five buffer systems evaluated barely influenced the yield of infectious agent. If mycoplasma were indeed being processed by centrifugation (91), the organism should have been sedimented during clarification (9500 g, 30 min). The authors withdrew supernate from the bottom of centrifuge tubes to avoid the floating fat layer (leafhopper body fat), and thus may have swept the agent from the sediment during withdrawal. Steere (140) recovered infectious fractions from gel column filtration under prescribed regimens of buffer system and pH. Apparently samples in neither case were processed for examination by electron microscopy.

By comparison with information development for other mycoplasma-induced diseases, the conspicuous deficiency in the history of aster yellows research is microbiological investigation. There were at least attempts by Swezy & Severin (142), Droscky (37), and Hartzell (70) to visualize the agents of plant disease transmitted by leafhoppers, by light microscopy. Essentially, Droscky lacked only the technology of tissue preparation and electron microscopy to have observed the aster yellows agent in the salivary glands of *M. fascifrons*. However, observation has not always led to recognition. Although structures like those illustrated by Doi et al (36) had been observed by electron microscopy by others before 1967 (104), their nature and significance had not been perceived.

The critical deficiency in the development of aster yellows research was the absence of contributions equivalent to those of Nelson (111-113) or Smith et al (139), or Eaton et al (39, 41, 42), which respectively directed subsequent research toward microbiology. The "mycoplasmalike bodies" now so abundantly visualized in the phloem of aster yellows-infected plants have, consequently, not been cultivated in cell-free media. The case was effectively stated by Razin (121): "Mycoplasma-like bodies have been detected in electron micrographs of plants suffering from corn stunt, aster yellows, and related diseases, and in leafhoppers known to transmit them (66, 105). The electron micrographs leave little doubt that these were really mycoplasmas, as corroborated by the successful treatment of the diseased plants and vectors with tetracyclines and chloramphenicol, and the ineffectiveness of penicillin (30), but the establishment of their etiology must await the isolation, cultivation, identification, and plant inoculation of these agents."

Our efforts in attempting to accomplish these same objectives (69) have yielded limited success. Our cultural methods and media are not yet sufficiently refined to permit a thorough accounting of microorganisms associated with diseased plants, or to derive stable, highly pathogenic mycoplasma isolates. It is apparent to all who have undertaken these tasks that long-term, persistent efforts will be essential for determining the role of mycoplasmas in plant diseases. Fortunately, there are several research groups attempting to cultivate phytopathogenic mycoplasmas, and the necessary accomplishments could be forthcoming within the year of 1972. Indeed, if it has been thoroughly established that the agent cultured and mechanically transmitted by Lin et al (93) was a mycoplasma, the objectives listed by Razin would have been fulfilled by these workers in 1970.

ISOLATION AND CULTIVATION OF MYCOPLASMAS

Historically, nutritional innovations have consistently been key factors in successfully culturing mycoplasma species on cell-free media. The mycoplasma-selective media in use today are products of these innovations. General aspects of mycoplasma nutrition are presented by Hayflick (73), Edward & Freundt (50) and Sharp (134). Specific nutritional requirements of several mycoplasma types and species are reviewed by Rodwell (126) and Smith (138). In the absence of published methodology I will briefly discuss conceptual and mechanical points which, based on our work, appear applicable for attempts to cultivate plant mycoplasmas.

Prerequisite to establishing mycoplasmas as etiological agents is the process of isolating the agent, cultivating it on cell-free medium, and culturally purifying the agent. Success in this objective is dependent upon the concentration and viability of mycoplasma in the initial inoculum, the presence of inhibitory substances or other microorganisms in the inoculum, the provision of suitable cell-free substrates and an adequate cultural environment, and the implementation of adequate cultural procedures. This process probably can be best achieved by (a) inoculating mycoplasma-selective broth medium with available source material (fluid extract from infected plants, inoculative leafhopper vectors, etc); (b) incubating parallel cultures at 25°C and 37°C for 7 to 16 hours; (c) transferring the culture to fresh broth and streaking onto selective agar medium, each 24 hours thereafter; (d) incubating streaked plates at 25°C and 37°C in sealed candle jars (a candle is burned after the lid is sealed to increase CO_2 content in container) and observing streaked plates for mycoplasma colonies every 2 to 3 days; (e) removing individual colonies from plates by capillary micropipettes, and placing these in 0.2 ml of selective broth; (f) incubating these single-colony cultures, and in turn streaking the culture onto agar medium; (g) repeating steps e and f from 5 to 10 times.

Cultures from single mycoplasma colonies could be initially established in 0.2 ml to 0.3 ml of broth medium, checked for freedom from bacteria by light or electron microscopy within 48 hours, streaked onto selective agar medium for colony examination, progressively incorporated into larger broth quantities (with routine checks for freedom from nonmycoplasma contamination), and prepared for phytopathogenicity tests within 72 hours of isolation of single colonies. Subsequent pathogenicity tests following cultural purification would be necessary, however, to determine the possibility that one or more viruses had been passively carried in the culture, and thus had been reintroduced into plants with the mycoplasma. In such a case, if part or all the symptoms developing in inoculated plants were virus induced, the investigator would have erroneously associated the cultured mycoplasma with symptoms induced by virus(es). The dilution of original inoculum generally accepted as sufficient to eliminate infectivity by any known virus is approximately 6.0×10^{23} (Avogadro's number).

As was pointed out by Lemcke (92) mycoplasma-selective media consist of three basic components: nutrient base, yeast extract, and serum or ascitic fluid. Modifications of media have generally been within four general categories: (a) variations in the preparation of yeast extract (92); (b) variations in the quantity (5% to 20%) or source of serum (horse, bovine, swine, rabbit, fowl); (c) omission or modification of bacterial inhibitors; and (d) addition of specific nutritional supplements (126) or biochemical test indicators (6).

The ingredients of seven distinct mycoplasma-selective media were elaborated by Al-Aubaidi & Fabricant (7), and other media, including specific supplements and procedures for biochemical characterization, were presented by Al-Aubaidi (6). Media used at times in our laboratory in attempts to culture plant mycoplasmas are those presented by the following authors: Edward (44), Yoder & Hofstad (151), Adler & Berg (2), Fabricant (54), Al-Aubaidi & Fabricant (7), Al-Aubaidi (6), Chalquest & Fabricant (20), Hayflick (71), Hers (Lemcke 92), and Shepard (135).

Supplements to cultural media for the enchancement of mycoplasma growth have included DNA, urea (essential for T strains of mycoplasma), various tris, phosphate, bicarbonate and citrate buffers, erythrocyte suspensions, various soluble carbohydrates (6), various salts, proteins and protein digests, amino acids, nucleic acid precursors, vitamins, coenzymes, lipids, lipid precursors (126), staphylococcus filtrate (81), and others. Several laboratories are presently engaged in massive evaluations of specific extracts from animal tissues and bacterial cells (110), infusions (117), and steroids (123), as mycoplasma growth factors. Most of these ingredients, or their equivalents, occur naturally in yeast extract or animal serum, or are provided in the nutrient base. Addition of such to standard media have had essentially two purposes: to elicit growth in artificial media of fastidious mycoplasmas during isolation attempts (6, 7), and to determine optimal levels of required substrates for cultivated mycoplasmas (126). More recently (18, 29), cultural modifications have been found to prolong the viability of artificially cultivated mycoplasmas.

Obviously, for unproven plant mycoplasmas at least trace amounts of growth must be stimulated before responses to supplementary nutrients can be measured. Preliminary responses to specific supplements can subsequently lead to increased success and precision in the isolation and culturing processes. Following this stage of achievement, which would be a cultural endpoint for many pathological studies, determination of optimal conditions and substrates would be possible. Rodwell (125) and others (99, 122, 137, 138) have attempted to establish substrate requirements for selected mycoplasma types sufficiently to ultimately permit their cultivation in completely defined media. The thorough cultural characterization of the first phytopathogenic mycoplasma should facilitate the cultivation of other such organisms.

The term cultural characterization as applied to mycoplasmas encompasses substrate utilization, formation of certain metabolic byproducts, reduction of dyes, temperature optima, hemolysis of erythrocytes, and other aspects of cultural behavior. Al-Aubaidi (6) described tests for 17 such characteristics. The confusion produced by attempts to characterize mycoplasma species by nonstandardized procedures, and the importance of test standardization, was recognized by Fabricant & Freundt (55) and by Fabricant (54). For the sake of an orderly development of plant mycoplasma information, therefore, it seems essential that newly isolated plant mycoplasmas be characterized in reputable mycoplasma laboratories experienced in these procedures. Pure cultures of the mycoplasma could in turn be released from that laboratory to other researchers by those submitting the organism for characterization.

Besides nutritional innovations, another factor which has historically contributed to success in culturing mycoplasmas has been readily available, high quality inoculum, i.e. inoculum containing high concentrations of viable mycoplasmas, in the absence of growth inhibitors. This inoculum consisted of fowl sinus exudate in the work of Nelson (111, 112) and of human sputum in the work of Eaton et al (41, 42). The importance of obtaining plant mycoplasma inoculum which contains high concentration of the microorganism, and relative freedom from incidental organisms, should therefore be recognized. The work of Sinha & Chiykowski (136) suggests that the hemolymph of inoculative *M. fascifrons* may constitute the best available inoculum for the aster yellows agent. The quantity of available inoculum in this case should not be a limiting factor, since it would be possible to establish the agent in microcultures. Thus, 10 μliter of hemolymph could be placed into

0.25 ml of mycoplasma-selective broth and incubated aseptically in 1.0 ml sealed vials. In such a system, mycoplasma growth could be monitored by streaking microcultures onto selective agar medium at predetermined incubation periods between 7 and 100 hours.

Simultaneously with streaking, 10 μliter to 15 μliter quantities of culture could be injected into healthy leafhoppers for infectivity assay of the cultured mycoplasma. Leafhopper tolerance to mycoplasma-selective media would be an obvious prerequisite in such experiments. Hemolymph from healthy leafhoppers could also be substituted for the serum component of the medium if necessary. Hemolymph from a single leafhopper could conceivably provide the serum component for a 0.1 ml or 0.2 ml microculture of mycoplasma. Phloem exudate from yellows-infected plants presumably would also contain high concentrations of mycoplasma and should be feasibly obtainable.

Nonpathogenic bacterial species will be encountered in almost all extracts from plants and insects during attempts to isolate plant mycoplasmas. These may or may not interfere with growth of mycoplasmas in broth or on agar medium. If their growth is profuse, their presence will certainly reduce the investigator's ability to detect mycoplasmas which could be present. Filtration of inoculum through bacteriological filters, as was done by Nocard & Roux (114), appears to be the best solution to this problem. Filters with pore sizes of 1.5 μ to 0.5 μ should be tested for this purpose to provide sufficient separation from bacteria but also to prevent loss of the mycoplasma. Filtration of leafhopper hemolymph would necessitate use of microvolume filters or at least those accommodating 1.0 ml volume of hemolymph. Fortunately, inoculum containing aster yellows agent can be directly assayed for infectivity during attempts to establish experimental procedures.

The likelihood that plant mycoplasmas would require or even tolerate animal serum could be appropriately questioned. Another question, therefore, is also appropriate: what is the rationale for attempting to culture plant mycoplasmas in media containing animal sera? First, the use of media prescribed for animal mycoplasmas is not based on an assumption that these media provide optimal substrates for plant mycoplasmas. Indeed, it would be surprising if plant mycoplasmas required the proteins and sterols provided by animal sera. Instead, presently available mycoplasma-selective media should be considered a starting point in the absence of a more logical choice. The precise nature of plant mycoplasma requirements can be learned, unfortunately, only after at least limited success in cultivating them has been achieved.

By using media prescribed for animal mycoplasmas, plant mycoplasma researchers may face two serious risks: (*a*) the possibility of acquiring contaminations of mycoplasmas from nonplant sources; and (*b*) inadvertant selection for mycoplasma types from plants which are nonrepresentative of the natural population. The first risk can be minimized by stringent, effective precautions exercised during all cultural procedures. An understanding of the second risk must await a thorough evaluation of mycoplasma biotypes associated with plant disease.

LITERATURE CITED

1. Adler, H. E. 1970. *The Role of Mycoplasmas and L Forms of Bacteria in Disease,* ed. J. T. Sharp, 240–61. Springfield: Thomas. 338 pp.
2. Adler, H. E., Berg, J. 1960. *Avian Dis.* 4:3–12
3. Adler, H. E., Fabricant, J., Yamamoto, R., Berg, J. 1958. *Am. J. Vet. Res.* 19:440–47
4. Adler, H. E., Yamamoto, R. 1957. *Am. J. Vet. Res.* 18:655–56
5. Adler, H. E., Yamamoto, R., Berg, J. 1957. *Avian Dis.* 1:19–27
6. Al-Aubaidi, J. M. 1970. *Bovine mycoplasma: Purification, Characterization, Classification and Pathogenicity.* PhD thesis. Cornell Univ., Ithaca, NY
7. Al-Aubaidi, J. M., Fabricant, J. 1968. *Cornell Vet.* 58:558–71
8. Atanasoff, D. 1969. *The Role of Mycoplasma in Pathology: A Hypothesis. Biol. Zentralbl.* 88(5):571–74
9. Barber, T. L., Fabricant, J. 1971. *Avian Dis.* 15:125–38
10. Bawden, F. C. 1939. *Plant Viruses and Virus Diseases.* Chronica Botanica. 1st ed. 272 pp.
11. Ibid 1950. 3rd ed. 335 pp.
12. Beach, J. R., Schalm, O. W. 1936. *Poultry Sci.* 15:466–72
13. Black, L. M. 1940. *Phytopathology* 30:2 (Abstr.)
14. Ibid 1941. 31:120–35
15. Ibid 1943. 33:2 (Abstr.)
16. Bordet, J. 1910. *Ann. Inst. Pasteur Paris* 24:161–67
17. Borrel, A., Dujardin-Beaumetz, E., Jeantet, J. 1910. *Ann. Inst. Pasteur Paris* 24:168–79
18. Buttery, S. H. 1967. *Bull. Epizoot Dis. Afr.* 12:433–36
19. Carnaghan, R. B. A. 1961. *J. Comp. Pathol.* 71:279–85
20. Chalquest, R. R., Fabricant, J. 1960. *Avian Dis.* 4:515
21. Chanock, R. M., Hayflick, L., Barile, M. F. 1962. *Proc. Nat. Acad. Sci. USA* 48:41–49
22. Chanock, R. M. et al 1962. *Proc. Soc. Exp. Biol. Med.* 110:543–47
23. Chanock, R. M. et al 1963. *Science* 140:662
24. Chanock, R. M., Steinberg, P., Purcell, R. H. See Ref. 1, 110–36
25. Clyde, W. A. Jr. 1961. *Proc. Soc. Exp. Biol. Med.* 107:715–18
26. Commission on Acute Respiratory Disease, Fort Bragg, N. C. 1946. (a) *Bull. Johns Hopkins Hosp.* 79:97–108. (b) Ibid, 109–24. (c) Ibid, 125–52. (d) Ibid, 153–67
27. Couch, R. D., Cate, T. R., Chanock, R. M. 1964. *J. Am. Med. Assoc.* 187:442–47
28. Cousin, M.-T., Darpoux, H., Faivre-Amiot, A., Staron, T. 1970. *C. R. Acad. Sci. Paris* 271:1182–84
29. DaMassa, A. J., Adler, H. E. 1969. *Appl. Microbiol.* 17:310–16
30. Davis, R. E., Whitcomb, R. F., Steere, R. L. 1968. *Science* 161:793–95
31. Del Giudice, R. A., Robillard, N. F., Carski, T. R. 1967. *J. Bacteriol.* 93:1205–9
32. Dierks, R. E., Newman, J. A., Pomeroy, B. S. 1967. *Ann. NY Acad. Sci.* 143:170–89
33. Dienes, L. See Ref. 1, 285–312
34. Dingle, J. H., Finland, M. 1942. *New Engl. J. Med.* 227:378–85
35. Dodd, S. 1905. *J. Comp. Pathol.* 18:239–45
36. Doi, Y., Teranaka, M., Asuyama, H. 1967. *Ann. Phytopathol. Soc. Jap.* 33:259–66
37. Droscky, I. D. 1929. *Phytopathology* 19:1009–15
38. Dujardin-Beaumetz, E. 1900. *Le Microbe de la peripneumonie et sa culture.* These de Paris, Octava Doin
39. Eaton, M. D. 1950. *Proc. Soc. Exp. Biol. Med.* 73:24–26
40. Eaton, M. D., Liu, C. 1957. *J. Bacteriol.* 74:784–87
41. Eaton, M. D., Meiklejohn, G., van Herick, W. 1944. *J. Exp. Med.* 79:649–68
42. Eaton, M. D., Meiklejohn, G., van Herick, W., Corey, M. 1945. *J. Exp. Med.* 82:317–28
43. Editorial Board. 1955. *Int. Bull. Bact. Nomencl. Taxon.* 5:13–20
44. Edward, D. G. F. 1947. *J. Gen. Microbiol.* 1:238–43
45. Ibid 1953. 8:256–62
46. Ibid 1954. 10:27–64
47. Edward, D. G. F. 1955. *Int. Bull.*

Bact. Nomencl. Taxon. 5:85–93

48. Edward, D. G. F. 1967. *Ann. NY Acad. Sci.* 143:7–8
49. Edward, D. G. F., Freundt, E. A. 1956. *J. Gen. Microbiol.* 14:197–207
50. Edward, D. G. F., Freundt, E. A. 1969. *The Mycoplasmatales and the L-Phase of Bacteria,* ed. L. Hayflick, 147–200. New York: Appleton-Century-Crofts. 731 pp.
51. Edward, D. G. F., Kanarek, A. D., 1960. *Ann. NY Acad. Sci.* 79:696–702
52. Fabricant, J. 1960. *Avian Dis.* 4:505–14
53. Fabricant, J. 1960. *Ann. NY Acad. Sci.* 79:393–96
54. Fabricant, J. See Ref. 50, 621–41
55. Fabricant, J., Freundt, E. A. 1967. *Ann. NY Acad. Sci.* 143:50–58
56. Foster, J. P. 1934. *J. Am. Vet. Med. Assoc.* 84:918–22
57. Freitag, J. H. 1964. *Virology* 24:401–13
58. Freitag, J. H. 1967. *Phytopathology* 57:1016–24
59. Freundt, E. A. 1955. *Int. Bull. Bact. Nomencl. Taxon.* 5:67–68
60. Freundt, E. A. 1957. *Bergey's Manual of Determinative Bacteriology,* ed. R. S. Breed et al, 914–26. 7th ed. 1094 pp.
61. Freundt, E. A. 1960. *Ann. NY Acad. Sci.* 79:312–25
62. Fukushi, T. 1935. *Proc. Imp. Acad. Tokyo* 11:301–3
63. Ibid 1939. 15:142–45
64. Gentry, R. F. 1960. *Ann. NY Acad. Sci.* 79:403–9
65. Graham-Smith, G. S. 1970. *J. Agr. Sci.* 2:227–43
66. Granados, R. R., Maramorosch, K., Shikata, E. 1968. *Proc. Nat. Acad. Sci. USA* 60:841–44
67. Granboulan, N. 1967. *Methods in Virology,* ed. K. Maramorosch, H. Koprowski, 618–36. New York: Academic
68. Graystone, T. J., Hjordis, M. F., Kenny, G. E. See Ref. 50, 651–82
69. Hampton, R. O., Stevens, J. O., Allen, T. C. 1969. *Plant Dis. Rep.* 53:499–503

70. Hartzell, A. 1937. *Contrib. Boyce Thompson Inst.* 8:375–88
71. Hayflick, L. 1965. *Tex. Rep. Biol. Med. Suppl. 1* 23:285–303
72. Hayflick, L. 1967. *Ann. NY Acad. Sci.* 143:5–6
73. Hayflick, L. See Ref. 50, 15–47
74. Heishman, J. O., Olson, N. O., Cunningham, C. J. 1965. *Avian Dis.* 10:189–93
75. Hirumi, H., Maramorosch, K. 1963. *Contrib. Boyce Thompson Inst.* 22:141–52
76. Hofstad, M. S. 1957. *Avian Dis.* 1:165–70
77. Ishiie, T., Doi, Y., Yora, K., Asuyama, H. 1967. *Ann. Phytopathol. Soc. Jap.* 33:267–75
78. Kelton, W. H., Gentry, R. F., Ludwig, E. H. 1960. *Ann. NY Acad. Sci.* 79:410–21
79. Kleckner, A. L. 1960. *Am. J. Vet. Res.* 21:274–80
80. Klieneberger-Nobel, E. 1954. *Biol. Rev.* 29:154–77
81. Klieneberger-Nobel, E. 1959. *Brit. Med. J.* 1:19–23
82. Klieneberger-Nobel, E. See Ref. 50, Forward, xi–xiii
83. Kunkel, L. O. 1924. *Phytopathology* 14:54 (Abstr.)
84. Kunkel, L. O. 1926. *Am. J. Bot.* 13:646–705
85. Ibid 1937. 24:316–27
86. Kunkel, L. O. 1937. *J. Bacteriol.* 34:132 (Abstr.)
87. Kunkel, L. O. 1941. *Am. J. Bot.* 28:761–69
88. Kunkel, L. O. 1955. *Advan. Virus Res.* 3:251–73
89. Kunkel, L. O. 1957. *Science* 126:1233
90. Laidlaw, P. P., Elford, W. J. 1936. *Proc. Roy. Soc. London Ser. B* 120:292–303
91. Lee, P. E., Chiykowski, L. N. 1963. *Virology* 21:667–69
92. Lemcke, R. 1965. *Lab. Pract.* 14:712–15
93. Lin, S., Lee, C., Chiu, R. 1970. *Phytopathology* 60:795–97
94. Liss, A., Maniloff, J. 1971. *Science* 173:725–27
95. Littau, V. C., Maramorosch, K. 1956. *Virology* 2:128–30
96. Ibid 1960. 10:482–500
97. Liu, C. 1957. *J. Exp. Med.* 106:455–67

98. Liu, C., Eaton, M. D., Heyl, J. T. 1959. *J. Exp. Med.* 109:545–56
99. Lund, P. G., Shorb, M. S. 1966. *Proc. Soc. Exp. Biol. Med.* 121: 1070–75
100. Maramorosch, K. 1952. *Nature* 169:194–95
101. Maramorosch, K. 1952. *Phytopathology* 42:59–64
102. Maramorosch, K. 1953. *Am. J. Bot.* 40:797–809
103. Maramorosch, K. 1956. *Virology* 2:369–76
104. Maramorosch, K., Granados, R. R., Hirumi, H. 1970. *Advan. Virus Res.* 16:135–93
105. Maramorosch, K., Shikata, E., Granados, R. R. 1968. *Trans. NY Acad. Sci. Ser. 2* 30:841–55
106. Markham, F. S., Wong, S. C. 1952. *Poultry Sci.* 31:902–4
107. Marmion, B. P., Goodburn, G. M. 1961. *Nature* 189:247–48
108. McKay, K. A., Truscott, R. B. 1960. *Ann. NY Acad. Sci.* 79: 465–80
109. Merril, M. H., Ten Broeck, C. 1935. *J. Exp. Med.* 62:687–95
110. Nakamura, M. 1968. *Jap. J. Bacteriol.* 23:772–82
111. Nelson, J. B. 1935. *Science* 82: 43–44
112. Nelson, J. B. 1936. *J. Exp. Med.* 64:749–58
113. Ibid 1939. 69:199–209
114. Nocard, Roux, Borrel, Salimbeni, Dujardin-Beaumetz. 1898. *Ann. Inst. Pasteur Paris* 12:240–62
115. Nowak, J. 1929. *Ann. Inst. Pasteur Paris* 43:1330–52
116. Official statement. 1942. *War Med.* 2:330–33
117. Ogata, M., Ohta, T. 1968. *Jap. J. Vet. Sci.* 29:259–71
118. Olson, N. O., Kerr, K. M. 1964. *Avian Dis.* 8:209–14
119. Peterson, O. L., Ham, T. H., Finland, M. 1943. *Science* 97:167
120. Purcell, R., Chanock, R., Taylor-Robinson, D. See Ref. 50, 221–64
121. Razin, S. 1969. *Ann. Rev. Microbiol.* 23:317–56
122. Razin, S., Cohen, A. 1963. *J. Gen. Microbiol.* 30:141–54
123. Redman, U. 1968. *Ann. Inst. Pasteur Paris* 114:313–15
124. Reinmann, H. 1938. *J. Am. Med. Assoc.* 111:2377–84
125. Rodwell, A. W. 1969. *J. Gen. Microbiol.* 58:39–47
126. Rodwell, A. W. See Ref. 50, 413–50
127. Sabin, A. B. 1941. *Bacteriol. Rev.* 5:1–66
128. Schalm, O. W., Beach, J. R. 1936. *Poultry Sci.* 15:473–82
129. Scherffel, A. 1908. *Ber. Deut. Bot. Ges.* 26A:762–71
130. Schlegel, D. E., Delisle, D. E. 1971. *Virology* 45:747–54
131. Schnitzer, R. J. 1963. *Drug Resistance in Chemotherapy*, ed R. J. Schnitzer, F. Hawking, 1:1–21. New York: Academic. 3 vols. 2269 pp.
132. Severin, H. H. P. 1945. *Hilgardia* 17:21–59
133. Sharp, J. T. See Ref. 1, 3–9
134. Ibid, 10–28
135. Shepard, M. C. See Ref. 50, 49–66
136. Sinha, R. C., Chiykowski, L. N. 1967. *Virology* 33:702–8
137. Smith, P. F. 1955. *Proc. Soc. Exp. Biol. Med.* 88:628–31
138. Smith, P. F. 1964. *Bacteriol. Rev.* 28:97–125
139. Smith, W. E., Hillier, J., Mudd, S. 1948. *J. Bacteriol.* 56:589–601
140. Steere, R. L. 1967. *Phytopathology* 57:832–33 (Abstr.)
141. Storey, H. H. 1939. *Bot. Rev.* 5: 240–72
142. Swezy, O., Severin, H. H. P. 1930. *Phytopathology* 20:169–78
143. Turner, J. C. et al 1943. *Lancet* 1: 765–69
144. Tyzzer, E. E. 1926. *Cornell Vet.* 16:221–43
145. Van Herick, W., Eaton, M. D. 1945. *J. Bacteriol.* 50:47–55
146. Ward, J. R., Cole, B. C. See Ref. 1, 212–39
147. Whitcomb, R. F., Davis, R. E. 1970. *Ann. Rev. Entomol.* 15: 405–64
148. Yamamoto, R., Adler, H. E. 1958. *J. Infect. Dis.* 102:143–52
149. Ibid, 243–50
150. Yamamoto, R., Bigland, C. H., Ortmayer, H. B. 1965. *J. Bacteriol.* 90:47–49
151. Yoder, H. W. Jr., Hofstad, M. S. 1964. *Avian Dis.* 8:481–512

ADDENDUM

The following important publications were received by the author too late to be included in the present review. They are cited here for their technical advancements and significant progress toward establishment of mycoplasmas as specific plant pathogens. Particularly the results of Giannotti and Vago constitute the most satisfactory compliance with Koch's postulates for a plant mycoplasma presented to date. Reproduction of precisely the original disease by the cultured agent was especially impressive in their work. It appears, however, that neither their procedures nor those of Lima et al. would have precluded viruses carried as passive contaminants with their cultured agents. Solo-pathogenicity presumably could have been established only after the respective agents had been diluted during successive cultural passages at least 6.23×10^{23} to eliminate such contaminants. See discussions in the present review under the section headed *Mycoplasma pneumoniae* and *Isolation and Cultivation of Mycoplasmas*.

Giannotti, J., Vago, C. 1971. *Physiol. Veg.* 9(4):541–53

Giannotti, J., Vago, C., Leclant, F., Marchoux, G., Czarnecky, D. 1972. *C. R. Acad. Sci. Paris* 274:394–97

Giannotti, J., Vago, C., Marchoux, G., Devauchelle, G., Czarnecky, D. 1972.

C. R. Acad. Sci. Paris 274:330–33

Lima, A., Pereira, G., Oliveira, B. 1971. *Arq. Inst. Biol. S. Paulo* 38(4):191–200

Saglio, P., Lafleche, D., Bonissol, C., Bove, J. 1971. *Physiol. Veg.* 9(4):569–82

Ann. Rev. Plant Physiol. 1972. 23:419–36

ELECTRON TRANSPORT IN PLANT RESPIRATION 7537

Hiroshi Ikuma

Department of Botany, University of Michigan
Ann Arbor, Michigan

CONTENTS

INTRODUCTION

Electron transfer in plant mitochondria resembles in many respects that in animal mitochondria: (*a*) room temperature spectra show the presence of *a*-, *b*-, and *c*-type cytochromes, flavoproteins, and pyridine nucleotide (16–19, 24, 25, 40, 41, 46, 56, 60, 69, 87, 114); (*b*) both types of mitochondria contain salt extractable *c*-type cytochromes (60); (*c*) the most likely ubiquinone in both higher plant and mammalian mitochondria is ubiquinone-10 (12, 39); (*d*) the morphology of plant mitochondria in electron photomicrographs is very similar to that of animal mitochondria (10, 18, 30, 31, 37, 48, 59, 79, 85, 86); (*e*) an oxygen electrode trace indicates the presence of five respiratory states (29) as defined for animal mitochondria (22, 47, 50–53, 59, 75, 76, 83, 87, 95, 111, 114, 118, 123); (*f*) the apparent Kms for succinate, inorganic phosphate, and ADP are similar to those of several animal mitochondria (51); (*g*) many respiratory inhibitors and uncouplers function in the respiration of plant mitochondria in a manner similar to that of animal mitochondria (52, 53, 123); and (*h*) there is a certain similarity of activity in response to temperature between plant and mammalian mitochondria (21, 71,

419

90). In addition, plant mitochondria have been shown to contain DNA and RNA as in the case of animal mitochondria (7).

In closer examinations, however, plant mitochondria have been shown to exhibit many respiratory activities different from those of animal mitochondria: (a) low temperature (at 77°K) spectra show three b-type cytochrome peaks, as compared with one cytochrome b in animal mitochondria, and two c-type cytochrome peaks which are different from cytochromes c and c_1 of animal mitochondria (16–21, 25, 60, 102); (b) the flavoprotein composition in plant mitochondria is apparently different from that in animal mitochondria (35, 103, 109); (c) plant mitochondria can be considered more "leaky" than animal mitochondria in view of the presence of 25–30 Å pits in the outer membrane (18, 85, 86), the ability to readily oxidize externally applied NADH (51, 76, 114, 118), apparent low concentrations of endogenous substrates in isolated plant mitochondria (52), and a relatively rapid aging upon storage at 0°C (50); (d) respiration rates of plant mitochondria are in general greater than that by animal mitochondria (51, 60); (e) the respiratory pattern of plant mitochondria as revealed in the respiratory control ratios and ADP:O ratios shows a less coupled appearance than that of animal mitochondria (50, 51); (f) malate oxidation proceeds in plant mitochondria rapidly even in the absence of pyruvate or glutamate (51); (g) plant mitochondria contain ATP specific succinyl-Co A synthetase (84) in contrast to GTP specificity of the enzyme in animal mitochondria (44); (h) the presence of the "cyanide-resistant" pathway has been detected in plant mitochondria (11, 13–15, 28, 40–43, 46, 50, 51); and (i) rotenone, which is by far the most potent inhibitor of electron transport in animal mitochondria, is a weak inhibitor of plant mitochondrial respiration (53). In addition, plant mitochondria contain a linear DNA of about 10 μ (78) in contrast to animal mitochondria which carry circular DNA of 5 μ in circumference (81). These comparisons reveal differences between higher plant and mammalian mitochondria in the nature of the electron transport system, in the activity of the Krebs cycle, and in the morphology of the mitochondrial membranes.

Recently, Packer et al (82) questioned whether the observed differences were experimental artifacts mainly due to problems associated with the preparation of higher plant mitochondria. Obviously, the presence of tough cell walls and large acidic vacuoles, together with the usual very low yield of mitochondria, complicate the mitochondrial preparatory procedure: isolation requires large quantities of plant tissue and isolation medium, as well as an extended preparation time. The absence of an established isolation procedure, available in the case of animal mitochondria (cf 96), may present a further problem in comparing the properties of plant mitochondria prepared in various laboratories with those of animal mitochondria. New isolation procedures of plant mitochondria are thus still appearing in the literature (e.g. 24, 47, 50, 57, 59, 75, 76, 83, 87, 89, 91). The preparatory problems may strengthen the view presented by Packer et al (82). However, detailed examinations of various parameters associated with the preparatory procedures continue to reveal essentially the same respiratory characteristics of plant mitochondria as observed earlier (cf 50). While the basic mechanisms between the two types of mitochondria are similar, and while some differences may be ascribed to preparatory procedures and to rapid mitochondrial aging, it is difficult to deny all differences stated above. In fact, recent careful investigations on electron carriers in plant mitochondria clearly verify the earlier notion that the electron transfer components are largely different from those in mammalian mitochondria (see below).

In this article the ensuing discussions will be limited to the electron transfer system in plant mitochondria and to an intriguing problem on the development of cyanide resistance in storage tissues. Both these topics have recently been carefully investigated. Further comparisons of differences between the two types of mitochondria will be treated minimally. Readers with interest in the mechanisms of oxidative phosphorylation in animal mitochondria are referred to recent discussions (e.g. 77, 100, 101, 107, 108, 115).

COMPONENTS OF THE RESPIRATORY CHAIN

The b- and c-type cytochromes in plant mitochondria have been customarily classified in terms of the positions of α-absorption maxima in a difference spectrum taken at 77°K (16–19, 21, 25). This situation is necessitated by the presence of five peaks in a 15 nm region in low temperature spectra which are not clearly distinguishable in room temperature difference spectra. Bonner (25) presented an argument in his last review in this series that the five absorption maxima should belong to three different b-type and two different c-type cytochromes (see also 60). Recent analyses in Storey's laboratory of the oxidation kinetics and of the oxidation-reduction potentials (see below) provide further strong support for this conclusion. Three b- and two c-type cytochromes are identified in this review as b_{562}, b_{557}, b_{553}, c_{549}, and c_{547}, where the subscripts stand for the positions of α-absorption maxima in nm (± 1 nm) in low temperature difference spectra.

Concentrations of each of the a-, b-, and c-type cytochromes in mitochondria from five different higher plant tissues range between 0.1 and 0.4 nmoles per mg mitochondrial protein, and the relative content is roughly 0.7 for a- and b-type cytochromes with reference to unity for c-type cytochromes (25, 60). This apparent uniformity among cytochrome types appears misleading in view of larger differences in the concentrations among individual components in each cytochrome type (cf 60, 102). The total content of flavoproteins is somewhat higher than the total content of cytochromes in mitochondria (60), but the concentration of each of the substrate-reducible flavoproteins appears quite comparable with those of some cytochromes (cf 103). Clearly, the stoichiometry of individual components of the plant respiratory chain is not yet firmly established.

a-*type Cytochromes.*—Subtle differences exist between plant and animal cytochrome oxidases (16, 19, 20, 25, 60). Inhibitor sensitivity, on the other hand, is generally very similar (14, 53, 104, 123). In room temperature difference spectra of plant mitochondria, an α-absorption maximum appears at 601–603 nm and a Soret peak at 442–445 nm, while in low temperature difference spectra the α-peak is located at 597 nm and double Soret peaks at 438 and 445 nm (14, 20, 25, 60). Under certain conditions the α-peak can be observed to split and give 589 and 597 nm peaks in a low temperature spectrum (20). Appearance of double peaks in the α- and Soret regions is clearly different from the spectrum of animal mitochondria (cf 20, 60). From studies of oxidation kinetics, Storey (104) estimated that a component corresponding to a_3 contributes less than 10% of cytochrome oxidase absorption at 602 nm and 40% at 445 nm, while the contribution of another component a is over 90% at 602 nm, 60% at 445 nm, and 100% at 438 nm. Separation of a and a_3 can also be achieved in the measurements of the oxidation-reduction potentials: in mung bean mitochondria the midpoint potential of a at pH 7.2 is +190 mV; of a_3, +380 mV (33). This difference of nearly 200 mV between midpoint potentials of a and a_3 strongly suggests that the third site of energy conservation lies between these two cytochromes (33), as proposed earlier by Bonner & Plesnicar (20) on the basis of crossover analysis. This possibility agrees well with that obtained with mammalian mitochondria (116, 117). Thus recent spectrophotometric, potentiometric, and crossover analyses further confirm the presence of two components, a and a_3, in plant mitochondria (cf 25).

b-*type Cytochromes.*—Cytochrome b_{557} is readily and completely reduced by succinate or ascorbate plus TMPD at anaerobiosis in the presence or absence of uncoupler (60, 102), and the reduced form is readily oxidized by a pulse of oxygen in sequence to c-type cytochromes (102). This cytochrome was tentatively identified as cytochrome b_7 by Bendall & Hill (15) in view of the position of its α-absorption maximum in a difference spectrum (25, 60) and of unchanged rates of oxidation of reduced b_{557} regardless of the presence of cyanide (103). This identity was questioned by Storey (103) because of differences in response to treatment with antimycin A: the oxidation of b_{557} is strongly inhibited by antimycin A (103), while b_7 remains oxidized in the presence of the inhibitor (13).

Cytochrome b_{553} remains oxidized in the aerobic steady-state in the presence of succinate and azide (60), but it does not appear extensively oxidized in the aerobic steady-state in the absence of azide nor completely reduced at anaerobiosis by ascorbate plus TMPD (N, N, N′, N′-tetramethylphenylene diamine) (102). It is, however, completely reduced at anaerobiosis with succinate in the absence of respiratory inhibitors. In a kinetic study of oxidation of this cytochrome with an oxygen pulse, it was shown that this cytochrome behaved more like ubiquinone and flavoproteins than like b- or c-type cytochromes (102). The oxidation-reduction potentials of b_{557} and b_{553} were measured at pH 7.2: b_{557} showed a potential of +42 mV, while b_{553} gave + 75 mV. From the potentials alone, it appears that the order of electron transfer should be from b_{557} to b_{553}, but a slow oxidation rate of the former may make this ordering questionable (see next section).

The oxidation-reduction behavior of b_{562} is not straightforward, and the position of this cytochrome in the respiratory chain is at present unclear. It remains oxidized in coupled mitochondria at anaerobiosis in the presence of azide (60) and also in uncoupled mitochondria at anaerobiosis (102). It becomes reduced by succinate in coupled mitochondria upon anaerobiosis (102). The reduction of this cytochrome appears to be closely associated with the energy conservation capacity of plant mitochondria (60, 102). The oxidation-reduction potential of -77 mV measured for b_{562} (33) supports the notion that this cytochrome is different from the other two b-type cytochromes.

c-type Cytochromes.—Cytochrome c_{549} is oxidized very slightly faster than c_{547}, suggesting that it is oxidized by cytochromes $(a+a_3)$ and reduced by c_{547} (102). However, the order of oxidation reverses when the flow of electrons to c-type cytochromes is minimized by antimycin A (102). Furthermore, c_{547} has been repeatedly shown to be reduced together with cytochromes $(a+a_3)$ in the presence of cyanide or azide but in the absence of added substrate (60, 103). These observations strongly suggest that c_{547} is more closely associated with cytochrome oxidase than c_{549}. The measurement of oxidation-reduction potentials does not resolve the two opposing possibilities in the ordering of electron flow between the two c-type cytochromes: both cytochromes have a midpoint potential at pH 7.2 of $+235$ mV (33). A difference between c_{547} and c_{549} can, however, be found in salt extractability: c_{547} is readily extracted from plant mitochondria by salt solutions in a manner similar to cytochrome c of animal mitochondria, while c_{549} remains membrane-bound under these conditions as does cytochrome c_1 of animal mitochondria (60).

Flavoproteins.—Rotenone, one of the most potent inhibitors of electron transport in animal mitochondria, inhibits plant mitochondrial respiration only partially and ineffectively, while amytal inhibition of respiration is found almost equally effective for both plant and animal mitochondria (53, 118). These observations strongly suggest a difference in the composition of flavoprotein species at the NADH dehydrogenase level between plant and animal mitochondria. Studies with malonate do not reveal a difference in the nature of inhibition between the two types of mitochondria (53). The content of flavoproteins in highly "cyanide-resistant" mitochondria is much higher than that in cyanide-sensitive mitochondria (60).

Simultaneous measurements of fluorescence and absorption changes between the oxidized and reduced states have revealed several flavoprotein species distinguishable in the respiratory chain of mammalian mitochondria (cf 27, 38). Similar studies with mung bean and skunk cabbage mitochondria have indicated at least four distinguishable species which appear different from those of animal mitochondria (35, 105). They are designated by Storey (105) as FP_{ha}, FP_{hf}, FP_M, and FP_{1f} whose ratios of fluorescence to absorption changes are 0, 1.4, 0, and 3.8, respectively. The first two species, FP_{ha} and FP_{hf}, are reduced by succinate and exogenously applied NADH in fully uncoupled mitochondria; FP_M in these mitochondria is reduced by malate, but not by succinate at anaerobiosis; and FP_{1f} is reduced by succinate when the reversed electron transport from succinate to pyridine nucleotide is allowed to proceed. The site of malonate or TTFA (1-thienoyl-3, 3, 3-trifluoroacetone) inhibition seems to lie between succinate and flavoproteins (35, 105), while the site of amytal inhibition may come after FP_M (105). In addition to these four flavoproteins reducible by substrates, there is in mung bean mitochondria a highly fluorescent flavin-containing component rapidly reducible by dithionite (105).

Anaerobic measurements of the oxidation-reduction potentials of the flavoproteins of skunk cabbage mitochondria have recently revealed five components with midpoint potentials at pH 7.2 of + 170 mV, + 110 mV. + 20 mV, −70 mV, and −155 mV (109). In these studies the most negative component, perhaps FP_{1f}, is highly fluorescent with a ratio of fluorescence to absorption change of 12, but all other species can only be identified by absorption changes. The presence of fluorescent species FP_{hf} is now questioned and is considered to be the same as FP_{1f}. A flavoprotein with a potential of + 110 mV is present in skunk cabbage mitochondria at a high concentration and is tentatively identified as FP_{ha}. Flavoprotein FP_M may be a species with a potential of −70 mV or + 20 mV. These results are different from earlier studies cited above. Obviously the identification of flavoprotein species in plant mitochondria requires further investigation. Nevertheless, these studies implicate the presence of several flavoprotein species in plant mitochondria which are different from those in animal mitochondria.

Organization of the Respiratory Chain

Cyanide-resistant pathway.—The term "cyanide-resistant respiration" has been used to describe cellular respiration which is insensitive to inhibition by terminal inhibitors (e.g. cyanide, azide, CO) and by inhibitors which act between *b*- and *c*-type cytochromes (e.g. antimycin A, HOQNO) (see 11, 40, 41, 46). Mitochondria isolated from resistant tissues contain both the cyanide-sensitive electron transfer system which is coupled to phosphorylation and the cyanide-insensitive pathway which is not phosphorylative (14, 42, 111, 119). Varying degrees of cyanide insensitivity have been found in isolated mitochondria; for example, skunk cabbage mitochondria are highly cyanide resistant (13–15, 55, 98, 99, 124), mung bean mitochondria are partly (ca. 25%) cyanide-resistant (14, 53), and white potato mitochondria are highly cyanide sensitive (14).

Both spectrophotometric and kinetic (regenerated flow) analyses of cytochromes in skunk cabbage mitochondria strongly suggest that *a*-, *b*-, and *c*-

type cytochromes contribute only slightly to cyanide-resistant respiration (14, 110). Flavoproteins, on the other hand, are likely involved in the alternate pathway insensitive to cyanide and antimycin A (14, 60, 110). The content of flavoproteins in skunk cabbage mitochondria is in fact much higher than that in cyanide-sensitive mitochondria (35, 60). Flavoproteins, however, do not appear to function as the terminal oxidase of the alternate pathway, because (a) hydrogen peroxide has not been detected (14, 88); (b) the oxidation rate of flavoproteins is too slow in the presence of cyanide or antimycin A to account for the rapid rate of inhibitor-resistant respiration (35, 105, 110); and (c) the apparent Km for O_2 is possibly lower than that of flavoprotein oxidase (14, 19, 54). The alternate pathway is specifically inhibited by iron-complexing agents, such as hydroxamic acids, potassium thiocyanate, 8-hydroxylamine, and α,α'-dipyridyl (14, 97). These inhibitors have little effect on electron transfer through the cyanide-sensitive pathway or phosphorylative activities of the mitochondria. Furthermore, difference spectra in the presence and absence of these inhibitors do not show any characteristic absorption peaks (14, 102).

These studies point out that the oxidase in the cyanide-resistant pathway is not the flavoproteins themselves, but a different protein which mediates electron transfer between reduced flavoproteins and oxygen (14, 97). Since submitochondrial particles prepared with an osmotic shock method (121) also show cyanide-resistant respiration (14, 97, 110), the oxidase is likely to be closely associated with the mitochondrial membranes. A high affinity second oxidase, which was kinetically detected in both skunk cabbage and bean mitochondria (19, 54), may function as the oxidase in the alternate pathway. Since the EPR signals at $g = 1.94$ and 2 increases in the presence of a hydroxamic acid, this oxidase may perhaps be a nonheme iron protein (97). These considerations allow us to conclude that the cyanide-resistant pathway branches out at the flavoprotein level and contains an oxidase which mediates electron transfer between the reduced flavoprotein and oxygen (14, 35). In view of the finding that the oxidation rates of the highly fluorescent flavoprotein, or FP_{1f}, are not affected by antimycin A or cyanide treatment but are strongly inhibited by mClam (m-chlorobenzhydroxamic acid) (35, 109), the flavoprotein which provides electrons to the alternate oxidase is likely this highly fluorescent species (35).

Several hypotheses which had been proposed earlier to explain cyanide-resistant respiration were recently evaluated carefully by Bendall & Bonner (14). The above new hypothesis explains many observations associated with cyanide-resistant respiration in mitochondria. These recent developments, in particular the finding of specific inhibitors of the cyanide insensitive oxidase, will no doubt stimulate research towards unraveling of the molecular mechanisms and physiological significance of cyanide-resistant respiration.

Electron transfer chain.—Bonner pointed out in 1968 that very little work had been done to elucidate the time sequence of cytochrome action in

TABLE 1. Half-times of Oxidation of Electron Transfer Components in Intact Mitochondria Isolated from Mung Bean Hypocotyls and Skunk Cabbage Spadices.

Oxidation kinetics were measured on energy-depleted anaerobic mitochondria with O_2 pulses in the presence or absence of antimycin A (1 μg AA/mg protein), cyanide (0.2–0.3 mM) or m-chlorobenzhydroxamic acid (0.8 mM, mClam).

Electron Transfer Component	Half-time of oxidation in msec						
	Mung bean mitochondria			Skunk cabbage mitochondria			
	Control	+AA	+KCN	Control	+AA	+KCN	+mClam
FP_{ha}	300[e]			120[f]	550[f]	120[f]	220[f]
FP_{hf} (FP_{lf}?)	300[e]			160[f]	150[f]	160[f]	2000[f]
UQ	500[c]		1000[c]	400[a]	2000[a]	2000[a]	
b_{553}	400[c]			400[a]	3500[a]	1000[a]	
	500[b]		>500[b]	180[f]			180[f]
b_{557}	5.8[c]		5.8[c]	15–18[a]	3500[a]	200[a]	
				<23[f]			<23[f]
c_{547}	2.6[b]	2.6[b]		3.2[a]	3.4[a]	3.8[a]	
	3.1[c]		2.5[c]				
c_{549}	1.9[b]	3.0[b]					
	2.5[c]		2.5[c]				
a	2.0[d]		2.0[d]				
a_3	0.9[d]		1.3[d]				

[a] Measured at 18°C (Storey & Bahr 110).
[b] Measured at 18°C (Storey 102).
[c] Measured at 18°C (Storey 103).
[d] Measured at 24°C (Storey 104).
[e] Measured at 24°C (Storey 105).
[f] Measured at 24–25°C (Erecinska & Storey 35).

plant mitochondria (25). The time sequence of oxidation of the electron carriers is important, as illustrated by the work on mammalian mitochondria, in order to characterize their positions in the sequence of the respiratory chain as well as to provide insight into the molecular mechanisms of electron transfer (25, 26). During the last few years, Storey and his colleagues in the Johnson Research Foundation(33, 35, 102–106, 109–111) have focused on this problem using an improved rapid mixing regenerated flow apparatus developed by Chance (26). Their studies were performed with both mung bean and skunk cabbage mitochondria which were depleted of energy conservation activity with an uncoupler and brought to an anaerobic state with a limited flux of reducing equivalents from succinate. The kinetics of oxidation of an electron carrier was followed spectrophotometrically in response to a pulse of oxygen. Table 1 summarizes the results of their work in terms of oxidation half-times in milliseconds in the presence and absence of respiratory inhibitors. Since b_{562} is mostly oxidized under the experimental conditions, the oxidation rate of this cytochrome has not been measured extensively.

Although not complete, the data in this table show close similarities in oxidation half-times between cyanide-resistant skunk cabbage mitochondria and partially resistant mung bean mitochondria. Furthermore, the oxidation rates of a- and c-type cytochromes and flavoproteins are similar to those in mammalian mitochondria (cf 25). Electron transport in the respiratory chain of plant mitochondria can be grouped into two time domains, as in the case of rat liver mitochondria (cf 25)—a chain of a short time domain consisting of cytochromes a_3, a, c_{547}, and b_{557}, and a chain of a much longer time domain composed of cytochrome b_{553}, ubiquinone, and flavoproteins.

The components in the short time domain react with oxygen rapidly with relatively small time differences, as seen particularly in mung bean mitochondria (Table 1). The order of oxidation of a_3 is faster than a both in the presence and in the absence of cyanide. Reduced a_3 becomes fully oxidized whether uncomplexed or complexed with cyanide, but the dissociation constant for the reduced a_3-cyanide complex is 15 times higher (30 μM) than that for the oxidized a_3-cyanide complex (2 μM) (104). A small difference in oxidation half-times exists between c_{547} and c_{549}, but the order of oxidation from c_{549} to c_{547} in the absence of inhibitor is reversed in the presence of antimycin A. Since the antimycin A concentration used was high enough to completely inhibit electron transfer between b- and c-type cytochromes (53, 102), the inhibitor treatment can be considered to effectively isolate a-c-type cytochromes from other members of the respiratory chain. If this is the case, c_{547} must be considered more closely associated with a and a_3. This conclusion agrees well with the observed reduction of c_{547} with a and a_3 in the presence of cyanide or azide but in the absence of added substrate (60, 103). The oxidation of b_{557} proceeds slower than that of c- or a-type cytochromes (Table 1). In addition, the midpoint potential of oxidation-reduction of this cytochrome is less positive than that of c- and a-type cytochromes (33). These considerations allow us to conclude that the oxidation sequence in the short time domain is probably $O_2 \rightarrow a_3 \rightarrow a \rightarrow c_{547} \rightarrow c_{549} \rightarrow b_{557}$.

The sequencing of the long time domain components is difficult in view of the presence of the "cyanide-resistant" pathway (see above). Furthermore, the data with mung bean mitochondria are incomplete, and FP_{hf} (highly fluorescent flavoprotein) may be FP_{1f} (cf 109). Nevertheless, the results, particularly with skunk cabbage mitochondria (Table 1), can be summarized as follows: (a) in the absence of inhibitor, the oxidation half-times of both b_{553} and ubiquinone are slightly larger than those of flavoproteins; (b) FP_{ha} oxidation is little affected by mClam and cyanide but is partially inhibited by antimycin A, whereas the FP_{hf} oxidation is strongly inhibited by mClam, but unaffected by antimycin A and cyanide; and (c) the oxidation rates of ubiquinone and b_{553} are strongly inhibited by antimycin A and cyanide, but b_{553} oxidation is unaffected by mClam. It appears that b_{553} and ubiquinone behave as a unit, while the oxidation kinetics of FP_{ha} is separated from that of FP_{hf} by antimycin A and mClam treatment. In the presence of antimycin A and mClam, the pattern of FP_{ha} oxidation is similar to b_{553}, but cyanide treatment makes it follow the FP_{hf} pattern. These considerations tend to indicate that ubiquinone is closely associated with b_{553}, FP_{hf} with the cyanide-resistant pathway, and FP_{ha} with the antimycin A sensitive pathway. In his studies on the kinetics of reduction of electron carriers by exogenous NADH, Storey (106) concluded that ubiquinone may function as a storage pool for reducing equivalents entering the respiratory chain on the substrate side of coupling site 2. This further suggests that ubiquinone is branched out of the main electron transport chain at b_{553}.

These discussions of studies by the Johnson Research Foundation group lead us to a possible sequence of the electron transfer chain in plant mitochondria with succinate as substrate. Such a sequence is shown in Figure 1 where the sites of action of respiratory inhibitors are also included. It should be pointed out that this figure is somewhat different from that presented by Erecinska & Storey (35): the differences are seen in the positions of b_{557} and ubiquinone. In their scheme b_{557} is placed outside of the short time domain and considered to carry out electron transfer from FP_{ha} and FP_{hf} to the short time domain components bypassing b_{553}. The presence of the cyanide-resistant pathway from FP_{hf} does not seem to make this bypass necessary, since b_{553} and b_{557} behave as a unit in the presence of inhibitors (Table 1). Ubiquinone is associated with FP_{ha} in their scheme, while it is branched out in Figure 1 from b_{553} on the basis of discussions given above.

Figure 1 explains a majority of data obtained so far. However, a striking discontinuity in the oxidation time sequence between the two time domain components cannot be explained clearly, though it may suggest a possible structural break point. Remarkable uniformity in oxidation half-times within each domain may indicate physical closeness in the arrangement of the components in the mitochondrial membrane. Obviously salt extractability of c_{547} is not reflected in the oxidation time sequence (cf 25). It is of interest to note that cyanide treatment slows the oxidation rates of b-type cytochromes and ubiquinone, but not others, in a manner similar to antimycin A treatment. This may indicate a second site of cyanide inhibition between b- and c-type cytochromes. Although this figure emphasizes that FP_{ha} is associated with the cytochrome chain and FP_{hf} with the cyanide-resistant pathway, a question can be raised in view of both earlier and recent observations: is FP_{hf} indeed associated with the cyanide-resistant pathway? As presented earlier, the flavoprotein content in skunk cabbage mitochondria is markedly higher than that in mitochondria from cyanide-sensitive tissues (35, 60). In a recent paper Storey (109) tentatively identified FP_{hf} as FP_{1f} whose concentration was far lower than

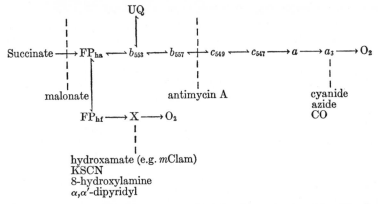

FIGURE 1. Possible electron transfer chain in plant mitochondria with sites of action of respiratory inhibitors. UQ = ubiquinone; X = oxidase of cyanide-resistant pathway.

FP_{ha}. In addition, the oxidation-reduction potential of FP_{1f} is 265 mV more negative than that of FP_{ha} (109). Since the oxidation rates of the two flavoproteins are practically identical, the low concentration and highly negative oxidation-reduction potential of FP_{hf} or FP_{1f} appear to cast doubt as to its presence in the cyanide-resistant pathway. This doubt seems even more plausible by the observations that antimycin A inhibits completely NADH oxidation, but incompletely inhibits succinate oxidation of mung bean submitochondrial particles (121). It is possible, therefore, that the second oxidase, X, may be associated with a flavoprotein but not FP_{hf} or FP_{1f}. In addition to these, a few other questions still remain unanswered in the scheme: (a) how does b_{562} interact with the electron transfer chain; (b) where are at least two other flavoproteins to be placed in the scheme; and (c) how do the reducing equivalents from NAD-linked substrate enter the respiratory chain? Clearly the electron transport chain in plant mitochondria still cannot be presented in a complete form. It is hoped that Figure 1 and the above discussions will help further in clarification of the electron transfer chain in plant mitochondria.

Sites of phosphorylation.—Tightly coupled mitochondria isolated from a variety of plant tissues show respiratory control and ADP:O ratios similar to, but with somewhat lower values than, mammalian mitochondria (50, 51, 76, 114, 118). The number of sites of phosphorylation is two with succinate and exogenous NADH, three with malate, pyruvate, and isocitrate, and four with α-ketoglutarate. With mung bean mitochondria, the phosphorylation efficiency is 75% of the value expected from mammalian work; this lowered efficiency is ascribed to the presence of cyanide-resistant respiration (50, 51).

It is generally accepted in analogy with the case with mammalian mitochondria that phosphorylation occurs at three sites. Spectrophotometric determinations of ferricyanide reduction (36) give ADP:2e ratios of about 2 with malate as substrate and about 0.7 with succinate, suggesting that two sites of phosphorylation exist between NADH and c-type cytochromes (Ikuma, unpublished work). The oxidation of ascorbate plus TMPD (N,N,N',N'-tetramethylphenylene diamine) in the presence of antimycin A results in an ADP:O ratio of about 0.7, indicating that one site of phosphorylation is located between c-type cytochromes and oxygen (Ikuma, unpublished work). Dutton & Storey (33) attempted to identify the sites of phosphorylation with measurements of oxidation-reduction potentials, but were unable to locate the exact sites. Their work, however, suggests that coupling site II is located between b- and c-type cytochromes and coupling site III between cytochromes a and a_3.

Spectrophotometric analyses of electron transfer carriers in the presence or absence of ADP or uncoupler indicated crossover points between highly fluorescent flavoprotein (FP_{1f}?) and highly absorbing flavoprotein (FP_{ha}?), between b- and c-type cytochromes (20, 111), and between cytochromes a and a_3 (20). The three sites thus suggested in plant mitochondria agree well with those in mammalian mitochondria (29).

As referred to earlier, exogenously applied NADH gives rise to an ADP:O ratio of 2 less as compared to the ratio of 3 for endogenous NADH oxidation, suggesting that its oxidation pathway is different from that of endogenous NADH. Inhibition of respiration by rotenone and amytal also shows differences between the oxidation of exogenous and endogenous NADH (118).

These studies on phosphorylation have not shown unique features of plant mitochondria, and the basic mechanisms of phosphorylation are generally considered to be the same for both plant and mammalian mitochondria (cf 82, 120). On the other hand, certain differences between the two types of mitochondria are seen in energy-linked functions of mitochondria, such as ion translocation (45) and reversed electron transport (49). Further characterization of the differences should await future work. Exact sites of phosphorylation, in particular sites I and II, still remain unresolved.

DEVELOPMENT OF CYANIDE-RESISTANT RESPIRATION IN STORAGE TISSUES

Another curious observation is the development upon aerobic incubation of cyanide resistance in slices of storage tissues such as potato tubers, sweet potato roots, carrot roots, and beet roots (see 11, 40, 41, 62). In fact, an addition of cyanide to aerobically aged slices sometimes stimulates respiration above control values (cf 58). Mitochondria isolated directly from storage tissues are highly sensitive to cyanide and antimycin A (14, 42, 43), while the mitochondria from aged slices show resistance to cyanide inhibition in much the same manner as do the slices themselves (42, 80). Similar to skunk cabbage mitochondria, the inhibitor-resistant mitochondria from aged slices contain both the phosphorylative, cyanide-sensitive pathway and the nonphosphorylative, cylinder-resistent alternate pathway (42, 43).

The development of cyanide-resistant respiration requires energy, probably derived from oxidative phosphorylation, as evidenced by the lack of development when slices are treated with anaerobiosis, DNP (2,4-dinitrophenol), and cyanide (42). Furthermore, the development is accompanied by an increase in a number of metabolic activities, such as glucose and phosphate uptake (4, 42, 58, 63, 70, 92), the activity of the pentose phosphate pathway (4), and the activity of glycolysis-Krebs cycle (1–3, 63, 64, 92).

Recently the increased respiration of slices has been shown to be related to an increase in the number of mitochondrial particles without an accompanying increase in cell number (5, 6, 67, 80). The newly formed mitochondria, apparently produced from fission of pre-existing mitochondria (68), are heavier than the pre-existing ones and more resistant to cyanide (80). Their aerobic biogenesis is expectedly accompanied by the synthesis of new RNA species and proteins (9, 32, 34, 66, 113).

What physiological significance does the development of cyanide-resistant respiration have for the welfare of the cell? Although information is still too scanty to provide an explicit answer to this question, certain aspects of the physiological importance can be summarized from the literature.

Using potato slices whose malonate-resistant, cyanide-sensitive respiration becomes predominantly malonate-sensitive, cyanide-resistant with aging, Laties & Hoelle (65) concluded that malonate-resistant respiration reflects insensitivity to the inhibition, while cyanide resistance results from a compensatory respiratory activity (cf 63). Rapid esterification of inorganic phosphate in aged pototo slices (42, 62, 70) supports the notion that the increased respiration may be largely phosphorylative. Furthermore, treatment with ADP and ATP have been shown to stimulate the respiration of aged carrot disks (2). The cyanide-resistant pathway thus appears inoperative in vivo, while the presence of cyanide or antimycin A allows the insensitive pathway to express itself.

The earlier data of Hackett et al (42) indicated that mitochondria from aged slices contain more NADH dehydrogenase, but not succinate dehydrogenase, than those from fresh tissue (cf 94). In addition, the work on skunk cabbage mitochondria strongly suggests a possible presence of a second non-heme iron oxidase (14, 97). Higher concentrations of NADH dehydrogenase and the second oxidase may likely contribute to the heavier weight of newly formed mitochondria in aged slices. Increased content of cyanide-resistant pathway components would increase the metabolic flux through the catabolic glycolysis-Krebs cycle pathway. The increased metabolic flux would provide necessary intermediates more readily to the associated pathways. Since adenylates can act as allosteric effectors for a number of enzymes (cf 8, 93), the presence of the cyanide-resistant pathway in newly formed mitochondria may perhaps play an important role in controlling cellular adenylate concentrations (cf 1–3, 41, 58) without altering increased metabolic flux. That is, if the requirement for ATP is high in the cell, ADP and/or AMP may act to inhibit the alternate pathway in order to allow oxidative phosphorylation to proceed maximally. Under the conditions where the cellular ATP concentration becomes high, however, the mitochondrial electron transfer may proceed largely through the resistant path and the activity of the regular cyanide-sensitive path would proceed at the state 4 (reduced) rate. In this connection it should be added that the cyanide-resistant pathway appears to be present in tissues that are rapidly growing or which liberate large quantities of heat (14). This apparent correlation between the type of tissues and the degree of cyanide resistance would further support the above possibility.

Oxygen plays an important role in the development of cyanide-resistant respiration in slices of storage tissues (42, 61, 112). In fact, peeling of a potato and subsequent incubation in air induce about a threefold respiratory increase at room temperature, whereas the respiratory rate of unpeeled potato does not change during the incubation period (61). The oxygen concentration in the gas phase around the center of a sizable potato tuber is 50–70% that of air at room temperature (23). The initial cutting or damaging of cells must thus trigger the development of cyanide-resistant respiration and oxygen aids the developmental process (61). In view of an observation that potato slices which have been placed under anaerobic conditions immediately after slicing develop cyanide-resistant respiration even more rapidly upon transfer to air than in freshly cut slices (42), the triggering substance may be postulated to be produced by fermentative pathways. Laties (61, 62) considers that a volatile substance such as acetaldehyde, but not CO_2, may act to trigger the development.

Thus the development of cyanide-resistant respiration appears to be stimulated first by the phosphorylative activity of existing mitochondria which acts to increase the activity of glycolysis and Krebs cycle. The pentose phosphate pathway appears much later in the developmental process (1–3, 41, 58, 70, 92). It is, however, not clear whether the increase in the glycolysis-Krebs cycle activity precedes or succeeds the development of cyanide-resistant respiration. Nevertheless, once triggered by initial wounding, the resistant respiration appears to be stimulated in an autocatalytic manner. The increased respiration is correlated with an increased number of heavy mitochondria which contain the cyanide-resistant pathway. It is possible that the cyanide-resistant pathway is needed in these mitochondria to increase metabolic flux of the cell and that its activity is controlled by adenylates.

Epilogue

The basic structure and function of higher plant mitochondria are very similar to those of animal mitochondria. Subtle but distinct differences have, however, been noted. Amongst them the composition and the ordering of the respiratory chain have been carefully analyzed recently. The plant respiratory chain is not only different from the chain of animal mitochondria in the composition of cytochromes and flavoproteins, but is also complicated by the presence of the cyanide-resistant pathway. A recent discovery of specific inhibitors of the cyanide-resistant oxidase will no doubt help further in unraveling the function of the plant respiratory chain and also stimulate further research of the development of cyanide-resistant respiration in storage tissue slices.

Subtle differences are also noted between the energy conservation activities of plant mitochondria and those of animal mitochondria, but they are less well characterized than the electron transfer system. An increasing use of submitochondrial particles will help further characterize the energy conservation activities (cf 12, 87, 121, 122).

Although not reviewed here, another aspect which is receiving increasing attention is the activity of the Krebs cycle, in particular with respect to malate oxidation (47, 72–74, 125). Plant mitochondria appear to contain NAD-specific malic enzyme which is absent in animal mitochondria (73) and malate oxidation seems to be under adenylate control (125). Furthermore, plant mitochondria do not contain detectable amounts of NADP (49), but their NAD content is comparable to that in animal mitochondria (49, 60). Together with ATP-specific succinyl CoA synthetase (84), these studies may lead to another difference between plant and animal mitochondria.

LITERATURE CITED

1. Adams, P. B. 1970. *Plant Physiol.* 45:495–99
2. Ibid, 500–3
3. Adams, P. B., Rowan, K. S. 1970. *Plant Physiol.* 45:490–94
4. Ap Rees, T., Beevers, H. 1960. *Plant Physiol.* 35:839–47
5. Asahi, T., Honda, Y., Uritani, I. 1966. *Arch. Biochem. Biophys.* 113:498–99
6. Asahi, T., Majima, R. 1969. *Plant Cell Physiol.* 10:317–23
7. Ashwell, M., Work, T. S. 1970. *Ann. Rev. Biochem.* 39:251–90
8. Atkinson, D. E. 1966. *Ann. Rev. Biochem.* 35:85–124
9. Bacon, J. S. D., MacDonald, I. R., Knight, A. H. 1965. *Biochem. J.* 94:175–82
10. Baker, J. E., Elfvin, L.-G., Biale, J. B., Honda, S. I. 1968. *Plant Physiol.* 43:2001–22
11. Beevers, H. 1961. *Respiratory Metabolism in Plants.* Evanston, Ill.: Harper Row. 232 pp.
12. Beyer, R. E., Peters, G. A., Ikuma, H. 1968. *Plant Physiol.* 43:1395–1400
13. Bendall, D. S. 1958. *Biochem. J.* 70:381–90
14. Bendall, D. S., Bonner, W. D. Jr. 1971. *Plant Physiol.* 47:236–45
15. Bendall, D. S., Hill, R. 1956. *New Phytol.* 55:206–12
16. Bonner, W. D. Jr. 1961. *Haematin Enzymes,* ed. J. E. Falk, R. Lemberg, R. K. Morton, 479–97. London: Pergamon
17. Bonner, W. D. Jr. 1964. *Sixth Int. Congr. Biochem.* Abstr. 4:291
18. Bonner, W. D. Jr. 1965. *Plant Biochemistry,* ed. J. Bonner, J. E. Varner, 89–123. New York: Academic
19. Bonner, W. D. Jr. 1965. *Fed. Proc.* 24: Abstr. 928
20. Bonner, W. D. Jr., Plesnicar, M. 1967. *Nature* 214:616–17
21. Bonner, W. D. Jr., Slater, E. C. 1970. *Biochim. Biophys. Acta* 223:349–53
22. Bonner, W. D. Jr., Voss, D. O. 1961. *Nature* 191:682–84
23. Burton, W. G. 1950. *New Phytol.* 49:121–34
24. Carmeli, C., Biale, J. B. 1970. *Plant Cell Physiol.* 11:65–81
25. Chance, B., Bonner, W. D. Jr., Storey, B. T. 1968. *Ann. Rev. Plant Physiol.* 19:295–320
26. Chance, B., DeVault, D., Legallais, V., Mela, L., Yonetani, T. 1967. *Novel Symposium 5. Fast Reactions and Primary Processes in Chemical Kinetics,* ed. S. Claesson, 437–68. New York: Interscience
27. Chance, B., et al 1967. *Proc. Nat. Acad. Sci. USA* 57:1498–1505
28. Chance, B., Hackett, D. P. 1959. *Plant Physiol.* 34:33–49
29. Chance, B., Williams, G. R. 1956. *Advan. Enzymol.* 17:65–134
30. Chrispeels, M. J., Simon, E. W. 1964. *J. Roy. Microsc. Soc.* 83:271–76
31. Chrispeels, M. J., Vatter, A. E., Hanson, J. B. 1966. *J. Roy. Microsc. Soc.* 85:29–44
32. Crick, R. E., Hackett, D. P. 1963. *Proc. Nat. Acad. Sci. USA* 50:243–50
33. Dutton, P. L., Storey, B. T. 1971. *Plant Physiol.* 47:282–88
34. Edelman, J., Hall, M. A. 1965. *Biochem. J.* 95:403–10
35. Erecinska, M., Storey, B. T. 1970. *Plant Physiol.* 46:618–25
36. Estabrook, R. W. 1961. *J. Biol. Chem.* 236:3051–57
37. Frey-Wyssling, A., Mühlethaler, K. 1965. *Ultrastructural Plant Cytology.* Amsterdam: Elsevier
38. Garland, P. B., Chance, B., Ernster, L., Lee, C. P., Wong, D. 1967. *Proc. Nat. Acad. Sci. USA* 58:1696–1702
39. Green, D. E., Brierley, G. P. 1965. *Biochemistry of Quinones,* ed. R. A. Morton, 405–31. New York: Academic
40. Hackett, D. P. 1959. *Ann. Rev. Plant Physiol.* 10:113–46
41. Hackett, D. P. 1963. *Control Mechanisms in Respiration and Fermentation,* ed. B. Wright, 105–27. New York: Ronald
42. Hackett, D. P., Haas, D. W., Griffiths, S. K., Niederpruem, D. J. 1960. *Plant Physiol.* 35:8–19
43. Hackett, D. P., Rice, B., Schmid, C. 1960. *J. Biol. Chem.* 235:2140–44
44. Hager, L. P. 1962. *The Enzymes* 6:387–99

45. Hanson, J. B., Hodges, T. K. 1967. *Curr. Top. Bioenerg.* 2: 65–98
46. Hartree, E. F. 1957. *Advan. Enzymol.* 18:1–64
47. Hobson, G. E. 1970. *Phytochemistry* 9:2257–63
48. Honda, S. I., Hongladarom, T., Laties, G. G. 1966. *J. Exp. Bot.* 17:460–72
49. Ikuma, H. 1967. *Science* 158:529
50. Ikuma, H. 1970. *Plant Physiol.* 45:773–81
51. Ikuma, H., Bonner, W. D. Jr. 1967. *Plant Physiol.* 42:67–75
52. Ibid, 1400–6
53. Ibid, 1535–44
54. Ikuma, H., Schindler, F. J., Bonner, W. D. Jr. 1964. *Plant Physiol.* 39:suppl. 1x
55. James, W. O., Elliott, D. C. 1955. *Nature* 175:89
56. Keilin, D. 1925. *Proc. Roy. Soc. London* B98:312–39
57. Killion, D. D., Grooms, S., Frans, R. E. 1968. *Plant Physiol.* 43: 1996–2000
58. Kolattukudy, P. E., Reed, D. J. 1966. *Plant Physiol.* 41:661–69
59. Ku, H. S., Pratt, H. K., Spurr, A. R., Harris, W. M. 1968. *Plant Physiol.* 43:883–87
60. Lance, C., Bonner, W. D. Jr. 1968. *Plant Physiol.* 43:756–66
61. Laties, G. G. 1962. *Plant Physiol.* 37:679–90
62. Laties, G. G. See Ref. 41, 129–55
63. Laties, G. G. 1964. *Plant Physiol.* 39:391–97
64. Ibid, 654–63
65. Laties, G. G., Hoelle, C. 1965. *Plant Physiol.* 40:757–64
66. Leaver, C. J., Key, J. L. 1967. *Proc. Nat. Acad. Sci. USA* 57: 1338–44
67. Lee, S. G., Chasson, R. M. 1966. *Physiol. Plant.* 19:194–98
68. Ibid, 199–206
69. Lieberman, M., Baker, J. E. 1965. *Ann. Rev. Plant Physiol.* 16: 343–82
70. Loughman, B. C. 1960. *Plant Physiol.* 35:418–24
71. Lyons, J. M., Raison, J. K. 1970. *Plant Physiol.* 45:386–89
72. Macrae, A. R. 1971. *Phytochemistry* 10:1453–58
73. Ibid, 2343–47
74. Macrae, A. R., Moorhouse, R. 1970. *Eur. J. Biochem.* 16:96–102
75. Malhotra, S. S., Spencer, M. 1971. *J. Exp. Bot.* 22:70–77
76. Matlib, M. A., Kirkwood, R. C., Smith, J. E. 1971. *J. Exp. Bot.* 22:291–303
77. Moore, C. L. 1971. *Curr. Top. Bioenerg.* 4:191–236
78. Mukulska, E., Odintsova, M. S., Turischeva, M. S. 1970. *J. Ultrastruct. Res.* 32:258–67
79. Nadakavukaren, M. J. 1964. *J. Cell Biol.* 23:193–95
80. Nakano, M., Asahi, T. 1970. *Plant Cell Physiol.* 11:499–502
81. Nass, M. M. K. 1969. *Science* 165:25–35
82. Packer, L., Murakami, S., Mehard, C. W. 1970. *Ann. Rev. Plant Physiol.* 21:271–304
83. Palmer, J. M. 1967. *Nature* 216: 1208
84. Palmer, J. M., Wedding, R. T. 1966. *Biochim. Biophys. Acta* 113:167–74
85. Parsons, D. F., Bonner, W. D. Jr., Verboon, V. G. 1965. *Can. J. Bot.* 43:647–55
86. Parsons, D. F., Williams, G. R., Thompson, W., Wilson, D., Chance, B. 1967. *Mitochondrial Structure and Compartmentation*, ed. E. Quagliariello, S. Papa, E. C. Slater, J. M. Tager, 29–70. Bari: Adriatica
87. Passam, H. C., Palmer, J. M. 1971. *J. Exp. Bot.* 22:304–13
88. Plesnicar, M., Bonner, W. D. Jr., Storey, B. T. 1967. *Plant Physiol.* 42:366–70
89. Raison, J. K., Lyons, J. M. 1970. *Plant Physiol.* 45:382–85
90. Raison, J. K., Lyons, J. M., Mehlhorn, R. J., Keith, A. K. 1971. *J. Biol. Chem.* 246:4036–40
91. Romani, R. J., Yu, I. K., Fisher, L. K. 1969. *Plant Physiol.* 44: 311–12
92. Romberger, J. A., Norton, G. 1961. *Plant Physiol.* 36:20–29
93. Rowan, K. S. 1966. *Int. Rev. Cytol.* 19:301–91
94. Sakano, K., Asahi, T., Uritani, I. 1968. *Plant Cell Physiol.* 9: 49–60
95. Sarkissian, I. V., Srivastava, H. K. 1968. *Plant Physiol.* 43:1406–10
96. Schneider, W. C. 1948. *J. Biol. Chem.* 176:259–66
97. Schonbaum, G. R., Bonner, W. D. Jr., Storey, B. T., Bahr, J. T. 1971. *Plant Physiol.* 47:124–28
98. Simon, E. W. 1957. *J. Exp. Bot.* 8:20–35

99. Ibid 1959. 10:125–33
100. Skulachev, V. P. 1971. *Curr. Top. Bioenerg.* 4:127–90
101. Slater, E. C. 1971. *Quart. Rev. Biophys.* 4:35–71
102. Storey, B. T. 1969. *Plant Physiol.* 44:413–21
103. Ibid 1970. 45:447–54
104. Ibid, 455–60
105. Ibid 1970. 46:13–20
106. Ibid, 625–30
107. Storey, B. T. 1970. *J. Theor. Biol.* 28:233–59
108. Ibid 1971. 31:533–52
109. Storey, B. T. 1971. *Plant Physiol.* 48:493–97
110. Storey, B. T., Bahr, J. T. 1969. *Plant Physiol.* 44:115–25
111. Ibid, 126–34
112. Thimann, K. V., Yocum, C. S., Hackett, D. P. 1954. *Arch. Biochem. Biophys.* 53:239–57
113. Vaughan, D., MacDonald, I. R. 1967. *Plant Physiol.* 42:456–58
114. Wakiyama, S., Ogura, Y. 1970. *Plant Cell Physiol.* 11:835–48
115. Williams, R. J. P. 1969. *Curr. Top. Bioenerg.* 3:79–156
116. Wilson, D. F., Chance, B. 1967. *Biochim. Biophys. Acta* 131:421–30
117. Wilson, D. F., Dutton, P. L. 1970. *Arch. Biochem. Biophys.* 136:583–84
118. Wilson, R. H., Hanson, J. B. 1969. *Plant Physiol.* 44:1335–41
119. Wilson, S. B. 1970. *Biochem. J.* 116. 20 pp.
120. Wilson, S. B., Bonner, W. D. Jr. 1970. *Plant Physiol.* 46:21–24
121. Ibid, 25–30
122. Ibid, 31–35
123. Wiskich, J. T., Bonner, W. D. Jr. 1963. *Plant Physiol.* 38:594–604
124. Yocum, C. S., Hackett, D. P. 1957. *Plant Physiol.* 32:186–91
125. Zimmerman, E. J., Ikuma, H. 1970. *Plant Physiol.* 46:suppl. 37

Ann. Rev. Plant Physiol. 1972. 23:437–64

PHYSIOLOGICAL ASPECTS OF GRAIN YIELD 7538

Shouichi Yoshida[1]

The International Rice Research Institute
Los Baños, Laguna, The Philippines

A high grain yield of any crop can be achieved only when a proper combination of variety, environment, and agronomic practices is obtained. Understanding the physiological processes involved in grain production, such as vegetative growth, formation of storage organs, and grain filling, helps determine the best combination of the above three factors, and also suggests what

[1] I wish to express my sincere thanks to Dr. Akira Tanaka, Dr. Yoshio Murata, Mr. Steven A. Breth, and Dr. L. T. Evans for going through the manuscript and making valuable suggestions.

improvements can be made to achieve a further increase in grain yield under a given condition.

Most physiological processes may be studied best in single plants in a controlled environment. Crop production, however, usually occurs in a community in which the plants differ in many ways from single plants, and under a variable environment. Crop species also differ from each other in their morphological and physiological characters, so they differ in their response to the environment. For these reasons, analysis of cause-and-effect relationship in crop grain yield is extremely complex.

In this review, I use rice to illustrate most points because of my familiarity with this crop. At the same time, physiological similarities and dissimilarities of rice and other grain crops are examined. In the past, several excellent discussions were attempted on physiological aspects of grain yield (52, 66, 146, 162, 227, 241, 255, 275), and the reader is advised to refer to these articles.

DRY MATTER PRODUCTION AND GRAIN YIELD
Leaf Area and Dry Matter Production

In his article on photosynthesis and the theory of obtaining high crop yields, Nichiporovich (171) introduced the terms "biological yield" and "economic yield." The biological yield (Ybiol.) refers to total dry matter, and the economic yield (Yecon.) refers to the economically useful part of biological yield. These two yields can be related by one parameter (Kecon.), which was originally called "the coefficient of effectiveness of formation of the economic part of the total yield," and is now more widely known as "harvest index" (65):

$$\text{Yecon.} = \text{Kecon.} \times \text{Ybiol.}$$

This simple equation tells us that the economic yield, grain yield for instance, can be increased either by increasing total dry matter production or by increasing harvest index.

In his early studies on the physiological causes of variation in crop yield Watson (273–275) reached the conclusion that variation in leaf area and leaf area duration was the main cause of differences in yield; variation in net assimilation rate was of minor importance. In other words, the area of leaf surface that intercepts solar radiation is the most important factor, and the photosynthetic efficiency of leaf per unit area is of secondary importance.

As a result, the importance of leaf area index (LAI) as a determinant of dry matter production, and hence yield, has been widely accepted, and LAI has been extensively used in subsequent studies on analysis of dry matter production.

In recent years, however, researchers have closely examined photosynthetic rate of single leaves, and have shown that leaf photosynthetic rate differs greatly among different species (105, 106, 163). If so, it is odd that great variation in leaf photosynthetic rate should be of minor importance in dry matter production by a crop community. Indeed, Loomis & Williams (135) have shown that leaf photosynthetic rate is a powerful determinant of crop growth rate in their mathematical model for canopy production. And when Buttery (43) compared growth of corn and soybean by growth analysis technique, he demonstrated that difference in crop growth rate between corn and soybean was caused by difference in net assimilation rate, which is likely related to difference in leaf photosynthetic rate in this comparison.

Light interception by a canopy of leaves is strongly influenced by the leaves' size and shape, angle, and azimuthal orientation, vertical separation and horizontal arrangement, and by absorption by nonleaf structure (135, 136, 154, 156). The optimum geometry in terms of light distribution for maximum crop photosynthesis varies with climatic conditions such as sun angle and the proportion of direct and diffuse light. The optimum crop geometry may also be considered from other considerations such as crop ventilation, CO_2 profiles, and the microclimate of the sink organs.

Of the factors that affect light interception by a canopy of leaves, leaf angle has attracted special attention in terms of total photosynthesis. Monsi & Saeki (154) introduced leaf angle as an important determinant into their mathematical model. Later and more refined models for photosynthesis in plant communities (63, 72, 135, 155) also have established leaf angle as an important parameter in the dry matter production of a community. These models demonstrate that erect leaves are the most efficient arrangement for maximum photosynthesis when LAI is large.

When sun angle is high and LAI is large, an erect-leaved canopy has a larger sunlit leaf surface than a droopy-leaved canopy, but it receives lower light intensity per unit leaf surface according to the cosine law. Since photosynthetic efficiency is high at low light intensity as seen from light-photosynthesis curve of a single leaf, and since the major portion of daily photosynthesis is attained when sun angle is high, it follows that an erect-leaved canopy gives a higher rate of daily photosynthesis than a droopy-leaved one. But the erect-leaved arrangement can be beneficial only when LAI is large. Several people, by different reasoning, have reached the conclusion that plants with erect upper leaves grading to droopy ones at low canopy levels appear to be the most desirable (115, 149, 189). Thus, in terms of total dry matter production by a crop community, LAI, leaf photosynthetic rate, and leaf angle appear to be the major determinants of crop growth rate. Of these three parameters LAI is the most variable and it can be widely changed by manipulating plant density and application of fertilizers. Indeed, a major objective of agronomic practice is to attain a sufficiently large LAI for maximum crop production.

Increasing LAI raises dry matter production, but this relationship does not hold indefinitely because of increased mutual shading of the leaves so the mean photosynthetic rate per unit leaf area decreases. Investigators disagree

whether an optimum LAI value exists at which crop growth rate reaches
its maximum and beyond which crop growth rate is decreased, or whether
this maximum is approached asymptotically with increasing LAI.

Several reports support the existence of a clear optimum LAI (34, 59,
102, 123, 160, 172, 220, 226, 227, 276, 278), and others do not confirm
these conclusions and indicate the presence of asymptotic plateau in the relationship of photosynthesis or crop growth rate and LAI (41, 43, 130, 137,
173, 186, 207, 216, 266, 277, 281, 285–288, 296).

The reason for this discrepancy is not yet clear. Obviously, gross photosynthesis of a canopy increases curvelinearly with increasing LAI because as
LAI increases, lower leaves are more shaded so the mean photosynthetic rate
of all leaves is decreased. As a result, net photosynthetic production is determined by the nature of respiration. In early mathematical models for canopy
production, respiration was assumed to increase linearly with increasing LAI,
and therefore the existence of optimum LAI was expected (60, 125, 154,
205, 258).

This assumption appears to have been accepted as such or with slight
modification by many investigators (155, 160, 220, 227, 293). Linear increase in canopy respiration with increasing LAI would mean that light intensity does not affect respiration. Ample evidence, however, indicates that low
light intensity or shading reduces the respiration of leaves considerably (137,
160, 198). Therefore, mean respiration rate of leaves should decrease as
the degree of mutual shading increases with increasing LAI. Moreover, the
existence of photorespiration increases the gradient in respiration rates down
the canopy still more (130), and therefore contributes to reducing the occurrence of an optimum LAI under most conditions.

Direct measurement of respiration rate of rice, wheat, alfalfa, subterranean clover, white clover, and cotton canopies has, in fact, shown that the respiration increases not linearly but asymptotically with increasing LAI (130,
137, 152, 296). Since dry matter production is the balance between photosynthesis and respiration, it appears that crop growth rate increases asymptotically with increasing LAI, or at least there would not be any pronounced optimum LAI in dry matter production by a crop community. If the leaf angle of
a canopy decreases at high LAI values, however, gross photosynthesis must
decrease. As a result, an optimum LAI value may exist. Tall indica rice varieties tend to have droopier leaves at higher nitrogen levels and hence they may
have an optimum LAI level (295).

If we define critical LAI for convenience as the LAI value beyond which
crop growth rate does not increase or increases only very slowly, species with
erect leaves should have much higher critical LAI or optimum LAI values
than species with flat leaves. Indeed, direct measurement has shown a critical
LAI (or optimum LAI) of about 3.2 for soybean (207), 5 for corn (287,
288), 6 to 8.8 for wheat (130, 216, 266, 281), and 4 to 7 for rice (102, 160,
226, 266, 296).

It should be noted that the surface area of the leaf sheath and exposed
stem is usually added to the surface area of the leaf blades to obtain LAI in

barley, wheat, corn, and other crops, while it is customary to measure only the surface area of the leaf blade alone for LAI in rice. The proportion of the surface area of the leaf blade to LAI varies with plant species and age. For instance, this proportion in wheat is about 0.5 at flowering and less than 0.1 at maturity (5). Thus when LAI values of rice and wheat are directly compared, LAI values of rice tend to be underestimated.

DRY MATTER PRODUCTION AND GRAIN YIELD

Total dry matter production is the integral of crop growth rate over the entire growth period, and it is related to grain yield by the harvest index. Although it is possible to show experimentally to some extent how harvest index can be varied by restricting formation of storage organs (262), it is usually difficult to drastically change harvest index of a given variety under most conditions. Harvest index of rice tends to be lower as total dry matter production increases (165, 208).

It has been shown in rice that an allometric relationship holds between grain yield and total dry matter production up to about 10 t/ha of rough rice (132). That is, the grain yield of rice increases more slowly than total dry matter does.

Thus, in general, increased total dry matter production results in increased grain yield for a given variety.

TOLERANCE TO HIGH PLANT DENSITIES

Higher LAI values can be achieved by increasing plant density and nutrient supply. Crops differ in their response to increasing plant density, however. Rice appears to be highly tolerant to high plant densities (290). Wheat is less tolerant (195), and corn is the least tolerant (84, 286). The grain yield of rice increased with increasing plant density up to about 182 to 242 plants/m^2 beyond which it leveled to 909 plants/m^2. The rice plant appears to be capable of producing at least one panicle per plant even at very high densities. In corn, the crop growth rate was shown to increase with increasing LAI up to values as high as 18 (287, 288). The grain yield, however, was positively correlated with crop growth rate only to the optimum population density (4.8 plants/m^2). Beyond that the grain yield was negatively correlated with population density (286) because the percentage of barren stalks increases with increasing population density (134, 203, 286). As a result, total dry matter production increases asymptotically with increasing plant population and LAI, but grain yield reaches a maximum at a finite population level (84). The incidence of barren stalks is closely correlated with the sugar content of stalks at silking time (286).

YIELD CAPACITY

DEVELOPMENT OF YIELD CAPACITY

Storage organs of cereal crops form after a period of vegetative growth and before panicle emergence. Following Murata's expression (162), yield capacity or potential yield of cereal crops can be formulated as:

Yield capacity = (number of panicle per m² of land) × (number of spikelet per panicle) × (number of grain per spikelet) × (potential size of grains)

= (number of grains per m² of land) × (potential size of grains)

The number of panicles per square meter can be varied by varying plant density and tillering performance. Beyond a certain density, negative correlations exist between the components of yield capacity, thus the yield tends to become the constant under a given set of conditions (290). Such correlations appear to be developmental rather than genetical (1).

Most rice varieties have only one floret per spikelet whereas modern wheat varieties may set four or more florets per spikelet.

The potential size of rice grain is physically restricted by the size of hull. Accordingly, grain weight is a quite stable varietal character with a variation coefficient of less than 5 percent among different years (145, 146). On the other hand, yearly variation of grain weight of barley is as large as 50 percent (241), and the variation of wheat grain weight as affected by temperature is as large as 30 percent (19). However, the relative magnitude of varietal difference still exists under varying conditions.

The development of the panicle has been well studied in rice (145, 146), barley (36), wheat (37), and oat (38). Among these Matsushima's work on rice is most extensive and informative. According to Matsushima, differentiation of panicle neck-node starts about 32 days before flowering. The differentiation of spikelets proceeds for the period from about 23 to 15 days before flowering, during which time the maximum number of spikelets is determined. Degeneration of formed spikelets, however, occurs afterwards; the reduction division stage, which occurs about 11 to 13 days before flowering, is the most sensitive to degeneration. Lastly, the size of hull is determined by 1 week before flowering.

These developmental stages are valid only for a single panicle. Since a rice crop is composed of hills, individual plants, and tillers, the developmental stages of a crop as a whole are more variable. Differentiation of spikelets, for instance, takes about 8 to 9 days for a single panicle but about 2 weeks for a whole crop. Understanding such variation is highly important when the effect on grain yield of stress at different stages of growth is to be studied in the field.

Degeneration of spikelets can be as high as 40 to 50 percent under certain circumstances (146) but under most conditions it is slight (263). The degeneration of spikelets is greatly affected by nitrogen status at the differentiation stage of spikelets, and degeneration is affected by solar radiation at the reduction division stages (145, 146).

Thus the number of spikelets or grains per unit land area of a rice crop is positively correlated with the amount of nitrogen absorbed by the end of spikelet initiation stage (263) or by flowering (165, 209, 263). On the other hand, the number of degenerated spikelets per unit land area is negatively correlated with dry matter production per differentiated spikelet during the period from spikelet differentiation to flowering (263). The degeneration of the differentiated spikelet may continue until about 5 days before flowering (145).

Leaf area growth is closely correlated with spikelet formation and grain yield. A close correlation exists between LAI at flowering and number of spikelets per unit land area. This is because the amount of nitrogen absorbed by flowering is nearly proportional to LAI at flowering (162). When grain filling proceeds normally (so grain yield is largely determined by the number of grains per unit land area), grain yield of rice is closely correlated with LAI at flowering (296).

A close correlation is also found between grain yield and LAI in corn at silking (75, 178, 230). The existence of a close correlation between grain yield and LAI flowering (or silking) implies that LAI at flowering is closely correlated with formation of yield capacity or with production of carbohydrates for grain filling, or with both.

DURATION OF PANICLE GROWTH

The length of growth period affects the growth of the panicle. Possibly the number of spikelets per ear in cereal crops can be increased by increasing the length of growth period for the panicle (143). Rawson (199) showed that number of spikelets per ear of 12 wheat varieties was closely correlated with the length of period from double ridge formation to terminal spikelet formation. The length of this period, however, was also correlated with length of vegetative growth period or with number of days from sowing to heading. Thus wheat varieties of similar growth duration appear to have a similar length of period from spikelet initiation to ear emergence (244).

In rice, although a great variation is found in growth duration, relatively small difference exists in the period from panicle initiation to heading under normal crop conditions (4, 13, 114, 145, 147, 228, 257). The variation in growth duration is largely due to differences in vegetative growth period. There is, however, a positive correlation between growth duration and the length of period from panicle initiation to heading. Thus an early maturing rice crop has a relatively short period for panicle growth (4). The shortened duration for panicle growth is often accompanied by decreased grain yield (4, 182). The obvious question then is whether the period of panicle growth can be extended independent of whole growth duration.

In general, the growth of panicle is the product of growth rate and growth duration, so extention of growth duration is one way to increase panicle size. In grain crops, however, panicles and leaves grow at the same time. Therefore, the distribution of assimilates between panicles and leaves will also determine the size of panicle. Not surprisingly, most high yielding rice varieties have a small flag leaf. This is probably because the flag leaf competes with the developing panicle for assimilates. Similarly, in wheat the ear size may be negatively correlated with flag leaf area (199).

GRAIN FILLING

CONTRIBUTION OF STORED CARBOHYDRATE IN VEGETATIVE PARTS

Carbohydrates such as sugars, starch, and other polysaccharides reach a maximum concentration in the plant's vegetative part around heading time after which it starts to decrease (9, 113, 168, 248). The stored carbohydrate could be translocated into the grain, thus contributing to grain carbohydrate, or it could be consumed as a substrate for respiration. Therefore, the loss of carbohydrate from the vegetative parts during the grain filling gives only the maximum estimate of the contribution of the stored carbohydrate to the grain. The reported estimates are 0 to 40 percent for rice, depending on the rate of nitrogen application and growth duration (168, 211, 222, 248, 263, 294), 20 percent for barley (11), 5 to 10 percent to less than 50 percent for wheat (14, 25, 271), and 12 to 14 percent for corn (70, 224). Direct evidence that stored carbohydrate is translocated into the grain has been obtained for rice and wheat by labeling the stored carbohydrate with ^{14}C (55, 167, 180, 215). Cock & Yoshida (55) showed that under normal field conditions 68 percent of the stored carbohydrate was translocated into the grain, 20 percent was respired during the ripening period, and 12 percent stayed in the vegetative parts. The amount of the carbohydrate translocated was equal to about 21 percent of the grain carbohydrate, or equivalent to about 2 tons of grain per hectare. When photosynthesis during the ripening period is restricted by shading or defoliation, the stored carbohydrate appears to be able to support the grain growth of rice and corn at almost a normal rate for some time (70, 168). Perhaps the stored carbohydrate can serve as a buffer to support normal grain growth despite the fluctuations of weather.

The possible contribution of the stored carbohydrate to the grain is large at low nitrogen levels (168, 218, 294) or when light intensity after heading is low (211).

Since large amounts of nitrogen must be applied to achieve high yields, and increased application of nitrogen tends to reduce the stored carbohydrates (9, 89, 168, 218, 219, 294), the relative contribution of the stored carbohydrates to the grain becomes less significant when high yields are produced by heavy application of nitrogen. Thus it appears safe to state that the grain carbohydrates of high yield crops are mostly derived from photosynthesis after heading.

CONTRIBUTION OF DIFFERENT PLANT PARTS TO THE GRAIN DURING RIPENING

Possible contribution of photosynthesis of different plant parts to the grain are based on (a) potential photosynthetic activity, (b) longevity of the tissue during the ripening period, and (c) light environment in a crop canopy. The photosynthetic rates of the plant parts of different crops are not the same. In rice and corn, compared with the leaf blades, net photosynthesis of ear and leaf sheath is very low; sometimes it is negative (145, 222, 223, 231,

295). On the other hand, net photosynthesis of ear, leaf sheath, and stem is relatively high in barley and wheat (29–32, 82, 141, 190, 235, 240). Thorne (240) showed that net photosynthesis of barley ear is about the same as that of the flag leaf while net photosynthesis of the ear is much less than that of the flag leaf in wheat. Thus the relative importance of the ear and flag leaf in grain filling differs from crop to crop.

Use of $^{14}CO_2$ has helped the identification of the direct source of grain carbohydrate. In rice, not only the flag leaf (12, 90, 223) but the third leaf (223) from the top export assimilates to the ear. Lower leaves send their assimilates mostly to the roots (223). In wheat and barley the assimilates by the ear stay mostly in the grain (46, 140). Among leaves the flag leaf appears to be the major source of grain carbohydrate (196, 215). The second and third leaf also export their assimilates to the grain, but to a lesser extent (109, 141, 201, 268). In corn, assimilates by the leaves above the ear are translocated efficiently into the kernel, but the translocation of the assimilates by the leaves below the ear sharply decreases, the lower the leaf position (74, 185). In other words, not only top leaves but the middle leaves above the ear in corn contribute much to the grain filling (6, 74, 110, 185). In many grain crops the upper leaves send their assimilates mostly to the grains and stem, the lower leaves send them mostly to the roots and tillers, and leaves in an intermediate position may send assimilates in either or both directions (74, 185, 201, 223, 269). Such differential functioning between the upper and lower leaves appears to be only relative and is affected by both internal and external conditions (131, 267, 268). When the lower leaves were shaded, and hence their photosynthesis was restricted, the upper leaves increased their supply to the roots (131). The lower leaves of the main culm send their assimilates to the ear of the tiller but not to the ear of the main culm (201). Thus movement of the assimilates appears to be regulated by the proximity and size of sink (131, 201, 269).

The longevity of the green tissue of different plant parts during the ripening period also must be considered. Rice leaves remain green almost until maturity, while the ear becomes yellow at relatively early stages of ripening (222). On the other hand, in wheat yellowing occurs in order—in leaves, stem, and ear (18). Allison (5) demonstrated that senescence of leaf blades occurs quicker in wheat than in corn, thus in corn the leaf blade area is about 80 percent of total green surface area at both anthesis and maturity, but in wheat it is only 50 percent at anthesis and less than 10 percent at maturity. Thus the relative importance of different plant parts to grain filling can be different not only from one crop to another but at different stages of ripening within the same crop species. Birecka et al (29–32) demonstrated that the relative contribution of ear, stem, and sheath increases in barley and wheat as ripening proceeds whereas that of leaf blades decreases. At later stages of ripening, however, photosynthesis may not make much of a net contribution to grain production (17, 145, 222). The reason relates to the general S-shape growth curve.

The light environment of different plant parts in a crop canopy is extremely important for determining the real photosynthetic activity of a given part. Obviously ears of barley and wheat are fully exposed to sunlight, so they have a chance to exhibit their maximum photosynthetic potential. Ears of improved rice varieties, however, tend to bend and are positioned below the flag leaf, so they are heavily shaded by the leaf canopy. Therefore, along with their low potential photosynthetic activity, ears of these varieties are unable to make a significant contribution to grain filling.

In a canopy with flat leaves the whole area of the top leaf is more exposed to sunlight than lower leaves. But in a canopy with very erect leaves, such as that of improved rice varieties, the tips of the lower leaves may receive more sunlight than the basal part of flag leaf, thus contributing to the total photosynthesis of the crop canopy.

To estimate overall contribution of different plant parts to grain production, various techniques have been devised: ear and leaf shading, leaf removal, kernel competition, and the short or long term measurement of CO_2 exchange rate. The combination of the first two techniques has been most extensively used because they are the easiest to use.

A vast amount of data are reported for the contribution of ear, leaves, and sheath plus stem to grain carbohydrate in wheat (15, 16, 33, 39, 44, 46, 82, 133, 139, 193, 194, 197, 210, 216, 236), in barley (10, 32, 45, 85, 190, 237, 238, 240, 279, 280), in rice (78, 222), in oats (120), in corn (6, 20, 57, 110, 188, 224), and in sorghum (92, 213, 214).

In rice, complete defoliation at flowering decreased the ripening percentage to 36 percent of the control (145), and the grain weight to 55 percent in one example (222) and 81 to 88 percent in another (181). In wheat, the same treatment along with ear shading produced 75 percent of the grain yield of the intact plant (193). Such estimation, however, is subject to large variation due to such sources as the amount of the stored carbohydrate, timing of defoliation treatment, and panicle size. Therefore, a relatively high yield for defoliated plant does not necessarily mean that the leaves contribute little to grain filling. On the other hand, a low yield indicates that photosynthesis by leaves contributes much to the grain yield.

Direct measurement of CO_2 exchange indicates that photosynthetic activity of the second and third rice leaves is higher than that of the flag leaf at early stages of ripening. The removal of the second to fourth leaves decreased the grain yield much more than removing the flag leaf (222). This, along with ^{14}C translocation studies (223), suggests that the top three leaves are important for grain filling in rice. In one measurement the leaf area of the top three leaves of an improved indica variety comprised about 74 percent of the total leaf area at flowering when the LAI value was 5.5 (296). Other evidence also indicates that all the leaves or leaf surface is not necessary for grain filling. The grain yield of an improved indica variety increases with increasing LAI until LAI becomes about 6, beyond which the grain yield levels off with further increase in LAI unless the crop lodges (296). In other words, increasing LAI values above a certain point are neither beneficial nor detrimental to the grain yield. The detrimental effects of large LAI values on grain yield may come from other directions such as lodging, diseases, and pests.

In wheat and barley the photosynthetic activity of the ear and the flag leaf appears to meet the carbohydrate requirement for grain filling (82, 240). Welbank et al (282) showed that grain yield of wheat was closely correlated with duration of green surface above the flag leaf node. A comparison of the differential functioning of the ear, the flag leaf, and lower leaves between rice and wheat or barley suggests that erect leaf arrangement is more important for rice than for wheat or barley during the grain filling period.

A remarkable aspect of studies on the contribution of different plant parts to the grain is the great variation among the reported values. For instance, the estimated contribution of ear photosynthesis to the grain ranges from 8 to 23 percent for rice (78, 222), 10 to 49 percent for wheat (39, 133), and 26 to 76 percent for barley (85, 280). Such great variation could be attributed to differences in techniques employed, varietal differences, and differences in growing conditions.

Clearly no method can determine true rates of photosynthesis of different plant parts in a canopy. The limitations of various techniques have been discussed elsewhere (133, 194, 241). Thus Thorne (241) concluded that measurement of CO_2 exchange would give the most reliable estimate of the contribution of different parts to the grain carbohydrates. Among the objections discussed so far, changes in light environment by defoliation and operation of compensation mechanism by defoliation or shading appear to be important. In a cotton canopy in which the leaves are disposed horizontally, it was shown that removal of lower leaves had no effect on the photosynthesis of the upper leaves (137). However, in crop canopies with erect leaves such as rice and wheat, removing the lower leaves affects the light environment of the higher parts as well as increasing the light intensity available to the leaf sheath or stem.

Compensation is another important factor. If a part of green tissue is removed or shaded, the photosynthetic rate of the remaining green tissue increases (30, 129, 131, 272). The existence of such compensation makes it difficult to estimate the contribution of different parts in a crop canopy precisely under natural conditions.

DURATION OF GRAIN FILLING PERIOD

There are some correlations between longer duration of grain filling period and larger grain yield in rice and corn (6, 61, 229, 255). But whether the extended duration of grain filling really caused larger grain yields in these examples is not clear. In rice, grain size is physically limited, and hence yield capacity is largely determined by number of grains per unit land area. Therefore, the extended duration of grain filling period can be meaningful only when the number of grains per unit land area does not limit grain yield. On the other hand, in crops such as wheat and corn, grain size is loosely restricted, so extending duration of grain filling period or maintaining higher photosynthetic activity during this time might increase grain yield.

PLANT CHARACTERS IN RELATION TO YIELDING ABILITY
MORPHOLOGICAL CHARACTERS

Since Tsunoda's pioneering work (251–255) in which he compared high and low yielding rice varieties, considerable attention has been paid to the relationship between morphological characters and yielding ability. The close association between certain morphological characters of rice varieties and yielding ability in response to nitrogen application led to the "plant type concept" as a guide for breeding high yielding varieties (22, 23, 26, 118, 119, 227, 228, 255). Table 1 summarizes certain desirable characters for high yielding rice varieties. Recently, Donald (66) discussed the breeding of crop ideotypes in which he described the morphological requirement for wheat ideotype. Although these concepts are concerned with morphological characters of rice or wheat, judgment of desirable characters is based on physiological considerations. The plant type concept has proved extremely effective for breeding high yielding indica rice varieties at the International Rice Research Institute in the past decade. Few attempts have been made, however, to establish direct cause-and-effect relationships between these characters and nitrogen response or yielding ability.

Plant height.—No factor is more important in determining the nitrogen responsiveness of a rice plant than the length and stiffness of its culm. Tall, weak-strawed varieties lodge early and severely at high nitrogen; and lodging decreases the rice yield (50). Among the plant characters associated with lodging, plant height is the predominant factor affecting lodging resistance (53). Lodging reduces the cross-sectional area of vascular bundles which in turn disturbs the movement of photosynthetic assimilates and absorbed nutrients via roots. In addition, lodging disturbs leaf display which results in increased shading, and eventually increases the percentage of unfilled grains (108).

The introduction of semidwarf genes into rice and wheat varieties has spectacularly increased the yielding ability of these crops largely because of increased resistance to lodging (21). The close association between plant height and other plant characters such as leaf erectness and grain-to-straw ratio must not be overlooked, however (102, 227). In relation to photosynthesis-respiration balance, shorter culm may minimize respiration loss by the culm, thereby improving net gains (227). On the other hand, tall stature would be more advantageous than short stature for light penetration (160). Clearly, extremely short stature would be disadvantageous because leaves are very closely spaced on a short culm, resulting in serious shading within the plant. That may explain the lack of sorghum cultivars with all four known dwarf genes (66). Thus an optimum plant height for a given plant species must exist; the short stature presently preferred for rice and wheat varieties is related to lodging resistance, and is not necessarily the optimum height for these crops.

TABLE 1. MORPHOLOGICAL CHARACTERS ASSOCIATED WITH HIGH
YIELDING POTENTIAL OF RICE VARIETIES

Plant part	Desirable characters	Effects on photosynthesis and grain production	References
Leaf	Thick	Associated with more erect habit. Higher photosynthetic rate per unit leaf area.	23, 102, 118, 160, 225, 227, 252
	Short and small	Associated with more erect habit. Even distribution of leaves in a canopy.	23, 118, 124, 149, 225, 227, 252
	Erect	Increase sunlit leaf surface area, thereby permitting more even distribution of incident light.	23, 101, 102, 116, 118, 124, 149, 222, 225, 227, 232, 233, 252
Culm	Short and stiff	Prevents lodging.	22, 53, 102, 116, 118, 228, 232, 254
Tiller	Upright (compact)	Permits greater penetration of incident light into canopy.	227, 252
	High tillering	Adapted to a wide range of spacings; capable of compensating for missing hills; permits faster leaf area development (transplanted rice).	22, 295
Panicle	Low sterility or high ripening percentage at high nitrogen rates	Permits use of larger amounts of nitrogen.	23, 119
	High grain-to-straw ratio (high harvest index)	Associated with high yields.	23, 50, 98, 99, 228

Leaf characters.—Among several leaf characters associated with high yielding ability, erect leaf habit seems the most important. Leaf angle has been closely correlated with nitrogen response in rice, barley, and wheat (91, 102, 116, 124, 221, 225, 227, 232, 234, 252).

Direct evidence of effect of erect leaves in increasing photosynthesis and hence yields have been reported for rice (149, 233). T. Tanaka et al (233) demonstrated by mechanical manipulation that a horizontally leaved canopy showed a low photosynthetic rate and a plateau type response by LAI to photosynthesis while an erect-leaved canopy showed a high photosynthetic rate and increased its photosynthesis with increasing LAI. The higher photosynthetic activity of an erect-leaved canopy produced a higher grain yield. Pendleton et al (189) also showed that the corn canopies with leaves positioned upright by mechanical manipulation gave higher yields than the untreated canopy. The effect of upright leaves, however, may have been caused by greater illumination of leaves adjacent to the developing ears rather than by increased crop growth rate.

Leaf angle has been used successfully as a selection criterion for breeding high yielding rice varieties at the International Rice Research Institute. All the varieties released from IRRI have erect leaves. In barley, wheat, and oats,

Tanner et al (234) have shown the extreme usefulness of leaf angle and leaf width for selection of high yielding varieties. Out of 300 varieties, 50 varieties were high yielding, and 48 out of 50 high yielding varieties could be picked up by leaf angle and leaf width.

In rice, leaf length is much more variable than leaf width, and leaf length is closely associated with leaf angle. The longer the leaves, the more droopy the leaves. As a result, short and small leaves are associated with erect leaves (225, 227, 252). Theoretically, short and small leaves can be more evenly distributed than long and large leaves in a canopy. More even distribution of leaves should increase use of incident light by a canopy. By mechanical manipulation of shoot number and leaf size, Matsushima et al (149) showed that the photosynthesis of a canopy which has larger number of shoots but smaller size of leaves was greater than that of a canopy which has fewer shoots but larger leaves and the same LAI.

The size of an individual shoot is a function of plant density. Increasing plant density reduces the size of individual shoots. Thus the above experiment not only supports the idea that short and small leaves are desirable, but suggests that broadcast direct seeded rice at high plant densities would have a higher potential canopy photosynthesis than transplanted rice.

Leaf thickness has been often mentioned as an important morphological character. Leaf thickness can be measured directly under the microscope, but it is conveniently expressed as specific leaf area or specific leaf weight. In wheat, leaf thickness as measured by micrometer is well correlated with specific leaf area (88). The association between thick leaves and high yielding potential of rice varieties is inconsistent (100–102, 127, 252). Some high yielding varieties have thick leaves, and others have thin ones. Perhaps leaf thickness itself is not an important leaf character. Nevertheless, thick leaves seem to be desirable. Leaf thickness is positively correlated with leaf photosynthetic rate (100, 112, 127, 160, 187). In Hayashi's experiment (101), however, the tested high yielding varieties had higher LAI values and more erect but thinner leaves. With the same amount of dry weight, leaf area development is inversely related to leaf thickness. Therefore, it appears that when leaf area development is likely to be limiting to growth as in transplanting rice cultivation, varieties that have thin leaves but large LAI tend to be high yielding.

Tillering habit.—Tillering habit has two aspects: spatial arrangement of tillers and tillering capacity. Tsunoda (252) described a "gathering type" and "dispersing type" of leaf arrangement in rice, the gathering type being considered desirable for high yield. This description involves both tiller angle and leaf angle. A. Tanaka et al (227) also compared two isogenic lines that differed in tiller angle. The "open tillered" line yielded better at a low nitrogen level and at a wide spacing than the "upright tillered" line. At a high nitrogen level and at a close spacing, the "upright tillered" line performed better than the "open tillered" one.

For the same reason as for leaf angle, upright tillered plants can be accommodated in larger numbers and with less mutual shading in the same land area.

Whether tillering capacity is advantageous or not needs a careful examination. In rice, medium tillering capacity has been considered desirable for a high yielding variety (26). This was because low yields of rice varieties were believed to be caused by faster growth rate and excessively large LAI beyond an optimum LAI, which in turn are closely related to high tillering capacity (221, 228, 229, 255). This conclusion was based on comparison of unimproved indica varieties and improved japonica varieties. With an improved indica variety, no optimum LAI exists with respect to dry matter production although a critical LAI does. A LAI as large as 12 is not detrimental to grain yield unless the crop lodges (296). At wide spacings it is quite obvious that high tillering varieties yield more than low tillering varieties. Thus high tillering varieties can be high yielders at close spacings as well as at wide spacings.

Donald (66) believes that a single culm is a desirable characteristic for a wheat ideotype. Duncan (69) also stated that tillering hills are the worst arrangement in plant geometry according to his model for corn. At high plant densities, however, even multiculm rice plants would not produce any tillers or they would produce only a limited number of tillers, thus approaching a single culm plant population. It is at high plant densities when high yield is usually achieved. On the other hand, a single culm plant cannot compensate for missing plants which may be caused by poor germination, pests, diseases, and other stresses; a multiculm plant can. Thus tillering capacity appears to be a desirable plant character for rice and wheat.

Panicle.—Low floret sterility at high nitrogen rate is considered one of the important selection criteria for nitrogen response varieties of rice (119). It was reported that some indica rice varieties had high sterility when grown with high nitrogen supply in culture solution (175). Recently, however, rice breeders have made selections from plants grown at high nitrogen levels, so that recent improved indica rice varieties do not become sterile at high nitrogen rates. Even so, a varietal difference in ripening percentage in response to increased application of nitrogen exists (23).

The ratio of grain weight to straw (grain-to-straw ratio) or total dry weight (harvest index) is another important criterion for selecting high yielding varieties. These are basically the same measures of the relative weight of grain to total dry matter.

There is great variation in harvest index or grain-to-straw ratio among varieties of rice (23, 50, 65, 98, 227, 228, 261), wheat (54, 64, 122, 281), barley (237, 239), corn (7), sorghum (93), and peas (65, 265). Although the grain-to-straw ratio of a rice variety varies with the rate of nitrogen applied, spacing, and season (65, 98, 228, 261), the relative magnitude of varietal difference remains unchanged, and hence it is considered as a varietal character. An increase in the yielding potential of a variety is usually associated with increased grain-to-straw ratio or harvest index.

Grain-to-straw ratio is closely correlated to the nitrogen responsiveness of rice varieties. The grain-to-straw ratio of five highly nitrogen-responsive varieties averaged 1.13 while that of six poorly nitrogen-responsive varieties was 0.56. In other words, of the total dry matter produced, the nitrogen-responsive varieties put twice as much into grain production as the poorly responsive varieties (50).

The physiological cause for variation in harvest index of a variety and of different varieties is not well understood. When the number of spikelets of rice was reducd artificially by shading before flowering, a large amount of carbohydrate produced after flowering was accumulated in the culm because of a limited number of grains. As a result, the harvest index of the crop was decreased (145, 262). In one variety of sorghum, a large portion of dry matter produced after flowering accumulated in the stem because of a small number of grains, and thus the variety had a low harvest index (93). These examples indicate that the number of spikelets is a major determinant of yield capacity and a major cause for variation in harvest index.

Plant height and grain-to-straw ratio are closely related in rice (227) and wheat (122). Thorne (241) states that a large ratio of grain to total dry weight is characteristic of the new high yielding varieties, but it may be the result of conscious selection for short, stiff straw rather than for large yields of grain with the minimum production of total dry matter.

In wheat and barley, ear photosynthesis of the awned varieties appears to be greater than that of the awnless ones (8, 29, 45, 82). The awn contains chlorophyll and stomates, and hence is capable of assimilating carbon dioxide.

Spatial arrangement of the ear relative to the leaves appears to be of some importance. Since, in wheat and barley, ear photosynthesis is quite high, the ear should be exposed to light. On the other hand, photosynthetic activity of panicles of rice and tassels of corn is very low or even negative. If the tassels shade the leaves, photosynthesis of the leaves may be reduced (73). In fact, Hunter et al (111) increased grain yield by removing tassels. The panicles of most old rice varieties are above the canopy of leaves. The panicles of new improved rice varieties, however, usually are below the canopy of leaves. Considering the shading effect of panicles and their low photosynthetic activity, the spatial arrangement of panicles in improved rice varieties appears to be desirable.

Changes in morphological characters of high yielding varieties.—The most important morphological character that has contributed to breeding high yielding race and wheat varieties in recent years is a short, stiff culm, giving lodging resistance. This is true in rice in Japan (208), wheat in Japan, USA, Mexico (40, 144, 191, 259, 260), and tropical rice (49).

A. Tanaka et al (232) studied changes in the morphological characters of rice varieties in Hokkaido, Japan, that became commercially available in the last 50 years. They found that better varieties have been selected for shorter plant height, higher tillering capacity, and more erect leaves. A similar trend was also observed for rice varieties in southern Japan (116). The selection must have been based on "selection for yield" (27, 66). The outcome of this selection, however, is in good agreement with present knowledge of the physiological aspects of high yields in transplanted rice.

PHOTOSYNTHETIC RATE OF LEAF

The physiological and biochemical aspects of differences in leaf photosynthetic rate among different species have been studied extensively in recent years (76, 77, 95–97, 106, 121, 163). Among grain crops, corn and sorghum have higher photosynthetic rate than rice, wheat, soybean, and peas.

Since photosynthesis by a single leaf is the basis for dry matter production, and hence economic yield, it appears reasonable to look for varietal differences as a basis for raising crop yields. Varietal differences in photosynthetic rate of leaves exist in rice (51, 159, 179), wheat (128), corn (71, 103), soybean (58, 67, 68, 176, 177), and pea (117). The magnitude of these varietal differences range from about 50 percent in rice to as much as 200 percent in corn.

Theoretically, the dry matter production of a single plant must correlate with the product of leaf area and photosynthetic rate. Duncan & Hesketh (71) compared growth rate of 22 races of corn grown as single plants and found that dry matter production was more dependent on leaf area development than on leaf photosynthetic rate. Khan & Tsunoda (128) obtained similar results with six wheat varieties. A high yielding semidwarf wheat variety, Mexi-Pak, has a high leaf photosynthetic rate but shows a low relative growth rate because of low leaf area ratio.

In physiological studies of the evolution of wheat, it has been shown that modern cultivated varieties have been selected for larger leaf area and larger grain size. The photosynthetic rate per unit leaf area has decreased with increasing leaf size, however (81, 126). Apparently leaf area expansion has been more important than leaf photosynthetic rate as a determinant of wheat growth for higher yields. Keep in mind too, that the ear itself constitutes an important part of the photosynthetic system of wheat.

In soybean, high yielding ability has been reported to be closely associated with high photosynthesis rate (67, 117), although another report does not confirm such association (58).

Thus most evidence at present indicates that increase in yield potential of a variety is not associated with increase in photosynthetic rate, and it is difficult to find clear-cut evidence that a variety with high leaf photosynthetic rate of a given variety really has improved yielding potential. Probably leaf photosynthetic rate is just one of the parameters that determines total photosynthesis of a crop community. Other parameters such as LAI and leaf angle usually have been more important. But, if crop communities of a similar canopy structure are compared, leaf photosynthetic rate is responsible for differences in dry matter production (43, 135).

The causes of varietal differences in leaf photosynthetic rate have been studied by several people (47, 48, 104, 250). Heichell & Musgrave (104) found a large difference in CO_2 compensation point between corn varieties, and this difference is related to net photosynthetic rate.[2] On the other hand, out of 2458 soybean genotypes, Cannel et al (47) failed to find even a single variety with a low CO_2 compensation point.

The leaf photosynthetic rate of a given variety is subject to great variation due to changes in the environment under which the plant is grown, to age, and to demand by sink (79). Varietal differences in the leaf photosynthetic rate may be caused by variety-environment interactions since temperature and light regimes affect the morphological characters of a leaf (87) and since varieties differ in their response to changes in the environment.

EFFECTS OF TEMPERATURE, LIGHT INTENSITY, AND CARBON DIOXIDE ON GRAIN YIELD

TEMPERATURE

Temperature has a more complex relationship with spikelet formation, ripening, and grain yield than light intensity does since there is usually an optimum value for different processes. Therefore, the results of experiment depend on whether the range of temperature studied is above or below the optimum temperature.

Controlled environment studies have demonstrated that relatively low temperatures increase the size of inflorescence, number of spikelets, number of florets per spikelet, and grain yield of perennial ryegrass (204), wheat (183, 184, 243), and barley (246). Air temperature as low as 15 to 19°C at the meiotic stage of pollen mother cells (about 11 days before heading) causes very high sterility in rice (206). The mean optimum air temperature for ripening of rice in Japan has been reported to be about 20 to 22°C (3, 148, 151). This optimum temperature is in good agreement with the results of statistical analysis of the effects of climatic factors on rice yield (94, 158, 161).

There are optimum combinations of day and night temperatures for each stage of grain development of rice (148, 150, 151). It appears that low night temperature is favorable for ripening. This may be related to the effect of temperature on respiration (292). The optimum air temperatures decrease progressively from 21°C to 14°C with grain development (148, 150).

Low temperature itself just delays ripening of rice unless it is too low (148). Thorne et al (242) found that the ratio of grain weight to leaf area duration during the ripening period (G) was decreased when the temperature fell from 20.5°C–16°C (day-night) to 14.5°C–10°C. Since temperature coefficient for photosynthesis is close to unity, the temperature effect on G could be attributed to effect of temperature on translocation or on the capacity of the grain to accumulate carbohydrate (283). Many experiments have shown that translocation of carbohydrate and inorganic nutrients is dependent on temperature (2, 170, 217, 245, 284).

---

[2] Recently it was reported that only a very small variation exists in CO_2 compensation point between corn varieties, using 54 genotypes in 1969 and 114 genotypes in 1970 (Moss, D. N., Willmer, C. M., Crookston, R. K. *Plant Physiol.* 47:847–48).

Higher temperatures than the optimum impair ripening of rice (3, 147, 148, 150) and wheat (19). Asana & Williams (19) showed that the grain weight of wheat was decreased progressively with increasing air temperature in the field. The decrease in grain weight of wheat caused by high temperatures was confirmed by the use of controlled environment (19, 183, 184).

Why high night temperatures impair ripening and reduce grain yield is not well understood. It was thought that high night temperature increased respiration which accounted for the impaired ripening of rice (292). Moss et al (157) demonstrated, however, that a high night temperature increased respiration in the night, but this in turn resulted in increased photosynthesis in the daytime. For this reason, they concluded that increased respiration by high night temperature could not reduce grain yield.

In cereals, high temperatures increase growth rate but decrease growth duration. Since overall growth is the product of growth rate and growth duration, and since growth duration is more affected by temperature than growth rate, high temperatures usually result in decreased growth (64). Comparison of the ripening period of rice at different latitudes indicates that the lower the mean air temperature, the longer the ripening period (229). Therefore, high temperatures probably affect ripening more by shortening the period for kernel growth than by increasing loss by respiration. In this regard, Murata (161) attributed the detrimental effect of high temperatures on grain yield partly to loss by increased respiration and partly to decreased leaf surface (senescence of leaves) and decreased photosynthetic rate.

Light Intensity

The light-photosynthesis curve of a single rice leaf indicates that the light saturation point is around 50 K lux, which is lower than the maximum light intensity in sunny days (160). However, photosynthesis of a rice community increases with increasing light intensity until about 70 to 90 K lux (160, 220, 227, 233, 249). In corn, photosynthesis of both single leaf (107) and community (24) increases steadily with increasing light intensity up to 10,000 fc.

Thus light intensity tends to limit photosynthesis of a crop community under natural conditions. In his extensive studies on yield and yield components of rice, Matsushima (145) demonstrated that there are two stages at which low light intensity has a critical effect on grain yield. Low light intensity decreases grain yield either by increasing the number of degenerated spikelets at the reduction division stage or by decreasing ripening after flowering. Friend (86) showed that formation of large inflorescences with many spikelets in wheat was associated with large total dry weight at anthesis under high light intensity and low temperature. It was also demonstrated that low light intensity, when combined with high nitrogen supply, resulted in a high percentage of sterility in rice (247).

High light intensity combined with low temperature early in the development of grain increases grain set of wheat (270). On the other hand, low light intensity combined with high temperature markedly impairs the ripening of rice (145). Light intensity also determines the nitrogen response of a rice crop. At high light intensities, increasing application of nitrogen increased grain yield of rice whereas at low light intensities, grain yield did not increase in response to increase in nitrogen application (212, 228). Ample evidence indicates a close positive correlation of rice yield and the amount of solar radiation during the period from reproductive stage to maturity (62, 158, 161). Thus high light intensity is indispensable for high photosynthetic activity, formation of many spikelets, better grain filling, and greater nitrogen response.

CARBON DIOXIDE ENRICHMENT

Although CO_2 enrichment of greenhouse crops has been widely studied (289), few reports have been made on the effects of the CO_2 enrichment on grain crop yield. It is known that increasing CO_2 concentration in atmosphere above 300 ppm increases leaf photosynthetic rate of rice (291), soybean (42), wheat (198), and barley (83). It also increases the growth or grain yield of rice (202, 296), soybean (56, 202), barley (83), wheat (202), and sorghum (202).

The most spectacular results of the CO_2 enrichment on grain crops were reported by Riley & Hodges (202). All the crops tested responded positively to the CO_2 enrichment. The yield of rice was increased from 10 t/ha to 18.9 t/ha when the atmospheric CO_2 was increased from 300 ppm to 2400 ppm. This experiment was conducted in an enclosed inflated plastic greenhouse under solar radiation of about 550 to 600 cal cm^{-2} day^{-1}.

More recently, the effects of the CO_2 enrichment on grain yield of rice were studied by enclosing the plants in plastic film in the field under solar radiation of 400 to 500 cal cm^{-2} day^{-1} (296). Increasing the CO_2 concentration to about 900 ppm before heading increased grain yield by 29 percent. The increase in the grain yield was caused by increased number of grains and increased grain weight. The same treatment after heading also increased the grain yield by about 21 percent. The yield increase at this time was largely caused by increased grain weight and increased filled grain percentage.

These two experiments appear to have established that the concentration of CO_2 in the atmosphere really limits grain yield of rice under natural conditions, and that a sizable increase in grain yield can be expected by increasing CO_2 in the atmosphere. Under field conditions, however, it is extremely difficult or uneconomical to increase the concentration of atmospheric CO_2. There is much speculation on how significant the CO_2 from the soil is in photosynthesis of crops. Upland soils supply CO_2 at a rate of 0.13 to 2.20 g m^{-2} hr^{-1} (138), whereas the rate is less than 0.1 g m^{-2} hr^{-1} due to insulation effect of water for CO_2 diffusion from the soil into the atmosphere in submerged soils (164).

Moss et al (157) examined the possible contribution of CO_2 from the soil to photosynthesis of a corn crop and found that in cloudy weather conditions the CO_2 supply from the soil became quite significant because the total photosynthesis was low. Under bright sunny conditions, however, the CO_2 supply from the soil was only a fraction of the total CO_2 assimilation by the crop.

In a corn canopy the CO_2 concentration decreases in the daytime when photosynthesis is active. This decrease is more pronounced when the LAI is large and wind speed is low (256). Thus CO_2 may limit the photosynthesis of a good corn crop.

The natural increase in CO_2 in the atmosphere may affect crop yields, too. According to Bolin & Keeling (35), atmospheric CO_2 is increasing at a rate of 0.7 ppm per year. This increase would eventually affect photosynthesis and the yield of field crops in the future.

FACTORS LIMITING GRAIN YIELD

A crucial question for crop physiologists is whether yield capacity as determined before flowering is more limiting to yield than photosynthesis during the grain filling period. If so, how can yield capacity be increased?

In rice, Murata (162) gave three examples for the relative importance of yield capacity and assimilate supply for the grain yield: (a) yield capacity is limiting; (b) assimilate supply is limiting; and (c) yield capacity and assimilate supply are well balanced. Since the percentage of ripened grains in rice decreases with increasing number of grains per unit land area, and since partial removal of grains increases the percentage of the remaining grains that ripen (145), an optimum number of grains appears to exist for maximum grain yield under a given condition (263). These results were obtained in the temperate region. On the other hand, in the tropics, Yoshida et al (296) showed that the grain yield is closely correlated with grain number per unit land area; the ripened grain percentage is about the same for both dry and wet seasons. Thus whether the yield capacity or assimilate supply limits the grain yield of rice under field conditions is not clear-cut. Defoliation and shading experiments at or after heading clearly demonstrate, however, that photosynthesis during the ripening period can severely limit the grain yield of rice (145, 212).

The existence of the optimum number of grains for the maximum grain yield suggests that assimilate supply is limiting the grain yield under such conditions. If photosynthetic activity is limited by low solar radiation, or if translocation of assimilates into the grain decreases, a certain portion of the grains may remain unfilled.

In a good rice crop about 75 to 90 percent of the grain ripens (166), but sometimes as little as 50 percent ripens (162). Under such conditions, grain filling tends to determine the grain yield.

The causes of low ripened grain percentage vary. Under field conditions at high nitrogen levels, lodging is likely to be involved. Besides lodging, the ripening is possibly determined by assimilate supply, translocation of assimilates, and ability of the grain to accept assimilates. Recently Nakayama (169) demonstrated that the senescence of the grain starts with the conductive tissue of the rachilla, suggesting that translocation may limit grain filling.

For further increase in the yield potential of rice, however, the number of grains per unit land area is obviously the limiting factor because of the physical limitation on grain size.

In wheat and barley, grain size is more variable. Partial removal of grains from the ear increased the weight of the remaining grains (28, 174, 216), although the increased grain weight did not compensate for the reduced grain number (28).

Stoy (216) demonstrated that the grain weight of intact wheat ears decreased at a low light intensity. Partial removal of the grains produced larger grains at a low light intensity than those of intact ears at a high light intensity, however, suggesting that the assimilate supply limited grain size of the intact ears at the low light intensity, and increased photosynthetic activity relative to grain number increased grain weight. In fact, Thorne (241) showed that the increased grain yield of modern, barley varieties was closely correlated with the increased grain weight. Perhaps at the present level of grain yields and under climatic environment in England, yield capacity is not limiting the grain yield of these barley varieties.

On the other hand, Evans (80) gave several lines of evidence that the yield capacity constitutes a major limitation to grain yield in wheat. One of them was that sterilization of one or two florets in central spikelets before anthesis increased grain yield by 20 percent for one variety. The yield increase was associated with increased number of grains and with increased grain weight in other parts of the spikelet (200). The grain yield of wheat was apparently increased without changing the photosynthetic capacity. The relationship of grain yield to grain number also suggested that the photosynthetic system does not limit the grain yield of wheat (199, 243).

Thus it appears that not only feedback interactions between photosynthesis and yield capacity but their interactions with the climatic environment make it difficult to determine which usually limits grain yield. But, under favorable climate and with an abundant supply of nutrients, it is more likely that the yield capacity limits the grain yield of rice and wheat.

Ample evidence indicates that for rice changing the photosynthesis alters the yield capacity: (a) the degeneration of spikelet and grain size is affected by light intensity during reduction division stage (145), and the degeneration is negatively correlated with dry matter production per spikelet (263); (b) shading from panicle initiation to heading reduced the grain yield (212); (c) CO_2 enrichment before heading increased grain number and grain size, and increased grain yield without further CO_2 enrichment after heading (296). In wheat, mutual shading as a result of high plant density during the period from formation of double ridge to ear emergence decreased the number of fertile spikelets per plant (192). It appears that decreased photosynthesis per plant decreased the spikelet number per plant.

On the other hand, as Evans (80) emphasized, hormonal effects seem to be involved in determination of yield capacity, and hence they merit more attention. The partitioning of assimilates between developing panicles and leaves is probably under some hormonal control. Greater distribution of assimilates into developing panicles may produce larger panicles. Attempts to understand the mechanism of the partitioning and to find means of controlling it should receive more attention.

Thus attempts to increase yield capacity and to increase photosynthetic capacity appear to be equally important for further increases in grain yield.

LITERATURE CITED

1. Adams, M. W. 1967. *Crop Sci.* 7: 505–10
2. Aimi, R., Sawamura, H. 1959. *Proc. Crop Sci. Soc. Jap.* 28: 41–43
3. Aimi, R., Sawamura, H., Konno, S. 1959. *Proc. Crop Sci. Soc. Jap.* 27:405–7
4. Akimoto, S., Togari, Y. 1939. *Proc. Crop Sci. Soc. Jap.* 11: 168–84
5. Allison, J. C. S. 1964. *J. Agr. Sci.* 63:1–4
6. Allison, J. C. S., Watson, D. J. 1966. *Ann. Bot. N. S.* 30:365–81
7. Anderson, R. E., Musgrave, R. B. 1960. *Proc. 15th Ann. Hybrid Corn Ind. Res. Conf.* Publ. No. 15, 97–103
8. Apel, P. 1966. *Kulturpflanze* 14: 163–69
9. Archbold, H. K. 1938. *Ann. Bot. N.S.* 2:403–35
10. Ibid 1942. 6:487–531
11. Archbold, H. K., Mukerjee, B. N. 1942. *Ann. Bot. N.S.* 6:1–41
12. Asada, K., Konishi, S., Kawashima, Y., Kasai, Z. 1960. *Mem. Res. Inst. Food Sci. Kyoto Univ.* 22:1–11
13. Asakuma, S. 1958. *Proc. Crop Sci. Soc. Jap.* 27:61–66
14. Asana, R. D., Joseph, C. M. 1964. *Indian J. Plant Physiol.* 7:86–101
15. Asana, R. D., Mani, V. S. 1950. *Physiol. Plant.* 3:22–39
16. Ibid 1955. 8:8–19
17. Asana, R. D., Parvatikar, S. R., Saxena, N. P. 1969. *Physiol. Plant.* 22:915–24
18. Asana, R. D., Saini, A. D., Ray, D. 1958. *Physiol. Plant.* 11: 655–65
19. Asana, R. D., Williams, R. F. 1965. *Aust. J. Agr. Res.* 16:1–13
20. Asanuma, K., Naka, J., Tamaki, K. 1967. *Proc. Crop Sci. Soc. Jap.* 36:481–88
21. Athwal, D. S. 1971. *Quart. Rev. Biol.* 46:1–34
22. Baba, I. 1954. *Studies on Rice Breeding* (A separate volume of *Jap. J. Breed.*) 43:167–84
23. Baba, I. 1961. *IRC Newsletter* 10: 9–16
24. Baker, D. N., Musgrave, R. B. 1964. *Crop Sci.* 4:127–31
25. Barnell, H. R. 1936. *New Phytol.* 35:229–66
26. Beachell, H. M., Jennings, P. R. 1964. *The Mineral Nutrition of the Rice Plant*, 29–35. Baltimore: Johns Hopkins Press. 494 pp.
27. Bingham, J. 1967. *Ann. Appl. Biol.* 59:312–15
28. Bingham, J. 1967. *J. Agr. Sci. Cambridge* 68:411–22
29. Birecka, H., Dakick-Wlodkowska, L. 1963. *Acta Soc. Bot. Pol.* 32: 631–50
30. Ibid 1964. 33:407–26
31. Birecka, H., Skupinska, J. 1963. *Acta Soc. Bot. Pol.* 32:531–52
32. Birecka, H., Skupinska, J., Bernstein, J. 1964. *Acta Soc. Bot. Pol.* 33:601–18
33. Birecka, H., Skupinska, J., Wojcieska, U., Zinkiewiez, E. 1963. *Acta Soc. Bot. Pol.* 32:435–61
34. Black, J. N. 1963. *Aust. J. Agr. Res.* 14:20–38
35. Bolin, B., Keeling, C. D. 1963. *J. Geophys. Res.* 68:3899–3920
36. Bonnet, D. T. 1935. *J. Agr. Res.* 51:451–57
37. Ibid 1936. 53:445–51
38. Ibid 1937. 54:927–31
39. Boonstra, A. E. H. R. 1929. *Meded. Landbouwhogesch. Wageningen* 33:3–21
40. Borlaug, N. E. 1968. *Proc. 3rd Int. Wheat Genet. Symp. Canberra*, 1–36
41. Brougham, R. W. 1956. *Aust. J. Agr. Res.* 7:377–87
42. Brun, W. A., Cooper, R. L. 1967. *Crop Sci.* 7:451–54
43. Buttery, B. R. 1970. *Crop Sci.* 10: 9–13
44. Buttrose, M. S. 1962. *Aust. J. Biol. Sci.* 15:611–18
45. Buttrose, M. S., May, L. H. 1959. *Aust. J. Biol. Sci.* 12:40–52
46. Buttrose, M. S., May, L. H. 1965.

Ann. Bot. N. S. 29:79–81
47. Cannell, R. Q., Brun, W. A., Moss, D. N. 1969. *Crop Sci.* 9: 840–42
48. Carlson, G. E., Pearce, R. B., Lee, D. R., Hart, R. H. 1971. *Crop Sci.* 11:35–37
49. Chandler, R. F. Jr. 1968. *Science for Better Living.* U.S. Dep. Agr. Handbook
50. Chandler, R. F. Jr. 1969. *Physiological Aspects of Crop Yield,* ed. J. D. Eastin, 265–85. *Am. Soc. Agron., Crop Sci. Soc. Am., Madison, Wis.* 396 pp.
51. Chandler, R. F. Jr. 1969. Presented at *All-Congress Symp. World Food Supply.* 11th Int. Bot. Congr., Seattle
52. Chandler, R. F. Jr. 1970. Presented at *Symp. Role Fert. Intensification Agr. Prod.* Int. Potash Inst., Berne, Switzerland
53. Chang, T. T. 1967. *IRC Newsletter* special issue, 54–60
54. Cock, J. H. 1969. *The physiology of yield in semi-dwarf and standard winter wheats.* PhD thesis. Univ. Reading
55. Cock, J. H., Yoshida, S. 1972. *Proc. Crop Sci. Soc. Japan.* In press
56. Cooper, R. L., Brun, W. A. 1967. *Crop Sci.* 7:455–57
57. Cornelius, P. L., Russell, W. A., Woolley, D. G. 1961. *Agron. J.* 53:285–89
58. Curtis, P. E., Ogren, W. L., Hageman, R. H. 1969. *Crop Sci.* 9: 323–27
59. Davidson, J. L., Donald, C. M. 1958. *Aust. J. Agr. Res.* 9:53–72
60. Davidson, J. L., Philip, J. R. 1958. *Climatology and Microclimatology.* Proc. Canberra Symp. 1956 UNESCO, Paris, 181–87
61. Daynard, T. B., Tanner, J. W., Duncan, W. G. 1971. *Crop Sci.* 11:45–48
62. De Datta, S. K., Zarate, P. 1970. *Biometeorology* 4:71–89
63. De Wit, C. T. 1965. *Agr. Res. Rep.* 663. Cent. Agr. Publ. Doc. Wageningen. 57 pp.
64. Dobben, W. H. van. 1962. *Neth. J. Agr. Sci.* 10:377–89
65. Donald, C. M. 1962. *J. Aust. Inst. Agr. Sci.* 28:171–78
66. Donald, C. M. 1968. *Euphytica* 17:385–403

67. Dornhoff, G. M., Shibles, R. M. 1970. *Crop Sci.* 10:42–45
68. Dreger, R. H., Brun, W. A., Cooper, R. L. 1969. *Crop Sci.* 9:429–31
69. Duncan, W. G. See Ref. 50, 327–39
70. Duncan, W. G., Hatfield, A. L., Ragland, J. L. 1965. *Agron. J.* 57:221–23
71. Duncan, W. G., Hesketh, J. D. 1968. *Crop Sci.* 8:670–74
72. Duncan, W. G., Loomis, R. S., Williams, W. A., Hanau, R. 1967. *Hilgardia* 38:181–205
73. Duncan, W. G., Williams, W. A., Loomis, R. S. 1967. *Crop. Sci.* 7:37–39
74. Eastin, J. A. 1969. *Proc. 24th Ann. Corn Sorghum Res. Conf.,* 81–89
75. Eik, K., Hanway, J. J. 1966. *Agron. J.* 58:16–18
76. El-Sharkawy, M. A., Hesketh, J. D. 1964. *Crop Sci.* 4:514–18
77. Ibid 1965. 5:517–21
78. Enyi, B. A. C. 1962. *Ann. Bot.* 26: 529–31
79. Evans, L. T. 1970. Presented at *Symp. Plant Response Clim. Factors,* UNESCO, Uppsala, Sweden
80. Evans, L. T. 1971. Presented at *Rice Breeding Symp., Int. Rice Res. Inst.*
81. Evans, L. T., Dunstone, R. L. 1970. *Aust. J. Biol. Sci.* 23: 723–42
82. Evans, L. T., Rawson, H. M. 1970. *Aust. J. Biol. Sci.* 23:245–54
83. Ford, M. A., Thorne, G. N. 1967. *Ann. Bot.* 31:629–44
84. Frey, R. L., Janick, J. 1971. *Crop Sci.* 11:220–24
85. Frey-Wyssling, A., Buttrose, M. S. 1959. *Nature* 184:2031
86. Friend, D. J. C. 1965. *Can. J. Bot.* 43:345
87. Friend, D. J. C. 1966. *The Growth of Cereals and Grasses,* ed F. L. Milthorpe, J. D. Ivins, 181–99. London: Butterworths. 358 pp.
88. Friend, D. J. C., Helson, V. A., Fisher, J. E. 1962. *Can. J. Bot.* 40:1299–1311
89. Fujiwara, A., Ohira, K., Otsuki, M., Narita, S. 1951. *J. Sci. Soil Manure Jap.* 22:91–102
90. Fujiwara, A., Suzuki, M. 1957. *Tohoku J. Agr. Res.* 8:89–97
91. Garderner, C. J. 1966. *The physiological basis for yield differ-*

ences in three high and three low yielding varieties of barley. Thesis. Univ. Guelph, Ontario, Canada

92. Goldsworthy, P. R. 1970. *J. Agr. Sci. Cambridge* 74:523–31

93. Ibid. 75:109–22

94. Hanyu, J., Uchijima, T., Sugawara, S. 1966. *Bull. Nat. Tohoku Agr. Exp. Sta.* 34:27–36

95. Hatch, M. D., Slack, C. R. 1966. *Biochem. J.* 101:103–11

96. Hatch, M. D., Slack, C. R. 1970. *Ann. Rev. Plant Physiol.* 21: 141–62

97. Hatch, M. D., Slack, C. R., Johnson, H. S. 1967. *Biochem. J.* 102:417–22

98. Hayashi, K. 1966. *Proc. Crop Sci. Soc. Jap.* 35:205–11

99. Ibid 1967. 36:422–28

100. Ibid 1968. 37:528–33

101. Ibid 1969. 38:495–500

102. Hayashi, K., Itoh, H. 1962. *Proc. Crop Sci. Soc. Jap.* 30:327–33

103. Heichell, G. H., Musgrave, R. B. 1969. *Crop Sci.* 9:483–86

104. Heichell, G. H., Musgrave, R. B. 1969. *Plant Physiol.* 44:1724–28

105. Hesketh, J. D. 1963. *Crop Sci.* 3: 493–96

106. Hesketh, J. D., Moss, D. N. 1963. *Crop Sci.* 3:107–10

107. Hesketh, J. D., Musgrave, R. B. 1962. *Crop Sci.* 2:311–15

108. Hitaka, N. 1968. *Bull. Nat. Inst. Agr. Sci. Ser. A* 15:1–175

109. Hojo, Y., Kobayashi, H. 1969. *Bull. Nat. Inst. Agr. Sci. Ser. D* 20:35–77

110. Hoyt, P., Bradfield, R. 1962. *Agron. J.* 54:523–25

111. Hunter, R. B., Daynard, T. B., Hume, D. J., Tanner, J. W., Curtis, J. D., Karnbenberg, L. W. 1969. *Crop Sci.* 9:405–8

112. Irvine, J. E. 1967. *Crop Sci.* 2: 297–300

113. Ishizuka, Y., Tanaka, A. 1953. *J. Sci. Soil Manure Jap.* 23:159–65

114. Ishizuka, Y., Tanaka, A. 1969. *Nutrio-physiology of the Rice Plant.* Tokyo: Yokendo. Rev. ed. 364 pp.

115. Isobe, S. 1969. *Bull. Nat. Inst. Agr. Sci. Ser. A* 16:1–25

116. Ito, H., Hayashi, K. 1969. *Proc. Symp. Trop. Agr. Res., Tokyo*

117. Izhar, S., Wallace, D. H. 1967. *Crop Sci.* 7:457–60

118. Jennings, P. R. 1964. *Crop Sci.* 4: 13–15

119. Jennings, P. R., Beachell, H. M. See Ref. 26, 449–57

120. Jennings, V. M., Shibles, R. M. 1968. *Crop Sci.* 8:173–75

121. Johnson, H. S., Hatch, M. D. 1968. *Phytochemistry* 7:375–80

122. Johnson, V. A., Schmidt, J. W., Makasha, W. 1966. *Agron. J.* 58:438–40

123. Kanda, M., Sato, F. 1963. *Sci. Rep. Res. Inst. Tohoku Univ. Ser. D* 14:57–73

124. Kariya, K., Sakamoto, S. 1963. *Bull. Chugoku Agr. Exp. Sta.* A9:17–30

125. Kasanaga, H., Monsi, M. 1954. *Jap. J. Bot.* 14:304–24

126. Khan, M. A., Tsunoda, S. 1970. *Jap. J. Breed.* 20:133–40

127. Ibid, 305-14

128. Ibid, 344–50

129. Kiesselbach, T. A. 1948. *J. Am. Agron.* 40:216–36

130. King, R. W., Evans, L. T. 1967. *Aust. J. Biol. Sci.* 20:623–35

131. King, R. W., Wardlaw, I. F., Evans, L. T. 1967. *Planta* 77:261–76

132. Kiuchi, T., Yoshida, T., Kono, M. 1966. *Nogyo-gojitsu* 21:551–54

133. Kriedemann, P. 1966. *Ann. Bot. N. S.* 30:349–63

134. Lang, A. L., Pendleton, J. W., Gungan, G. H. 1958. *Agron. J.* 48:284–89

135. Loomis, R. S., Williams, W. A. See Ref. 50, 27–47

136. Loomis, R. S., Williams, W. A., Duncan, W. G., Dovrat, A., Nunez, F. A. 1968. *Crop Sci.* 8: 352–56

137. Ludwig, L. J., Saeki, T., Evans, L. T. 1965. *Aust. J. Biol. Sci.* 18: 1103–18

138. Lundegårdh, H. 1964. *Klima und Boden in Ihrer Wirkung aus des Pflanzenleven.* Transl. M. Monsi et al. Tokyo: Iwanami. 551 pp.

139. Lupton, F. G. H. 1961. *Ann. Appl. Biol.* 49:557–60

140. Ibid 1966. 57:355–64

141. Ibid 1968. 61:109–19

142. Lupton, F. G. H., Ali, M. A. M. 1966. *Ann. Appl. Biol.* 57:281–86

143. Lupton, F. G. H., Kirby, E. J. M. 1968. *Rep. Plant Breed. Inst. Cambridge 1966–1967,* 5–26

144. Matsumoto, T. 1968. *Jap. Agr. Res. Quart.* 4:22–26

145. Matsushima, S. 1957. *Bull. Nat. Inst. Agr. Sci. Jap. Ser. A* 5:1–271
146. Matsushima, S. 1970. *Crop Science in Rice.* Tokyo: Fuji. 379 pp.
147. Matsushima, S., Manaka, T. 1959. *Proc. Crop Sci. Soc. Jap.* 28:201–4
148. Matsushima, S., Manaka, T., Tsunoda, K. 1957. *Proc. Crop Sci. Soc. Jap.* 25:203–6
149. Matsushima, S., Tanaka, T., Hoshino, T. 1964. *Proc. Crop Sci. Soc. Jap.* 33:44–48
150. Ibid, 53–58
151. Matsushima, S., Tsunoda, K. 1958. *Proc. Crop Sci. Soc. Jap.* 26:243–44
152. McCree, K. J., Troughton, J. H. 1966. *Plant Physiol.* 41:1615–22
153. McNeal, F. H., Berg, M. A., Klages, M. G. 1960. *Agron. J.* 52:710–12
154. Monsi, M., Saeki, T. 1953. *Jap. J. Bot.* 14:22–52
155. Monteith, J. L. 1965. *Ann. Bot. N. S.* 29:17–37
156. Monteith, J. L. See Ref. 50, 89–111
157. Moss, D. N., Musgrave, R. B., Lemon, E. R. 1961. *Crop Sci.* 1:83–87
158. Munakata, K., Kawasaki, T., Kariya, K. 1967. *Bull. Chugoku Agr. Exp. Sta. Ser. A* 14:59–95
159. Murata, Y. 1957. *Nogyo-gijitsu (Tokyo)* 12:460–62
160. Murata, Y. 1961. *Bull. Nat. Inst. Agr. Sci. Jap.* D9:1–169
161. Murata, Y. 1964. *Proc. Crop Sci. Soc. Jap.* 35:59–63
162. Murata, Y. See Ref. 50, 235–59
163. Murata, Y., Iyama, J. 1963. *Proc. Crop Sci. Soc. Jap.* 31:315–22
164. Murata, Y., Osada, A., Iyama, J. 1957. *Agr. Hort. (Tokyo)* 32:11–14
165. Murayama, N. 1967. *Jap. Agr. Res. Quart.* 2:1–5
166. Murayama, N. 1971. *Agr. Hort. (Tokyo)* 46:145–49
167. Murayama, N., Oshima, M., Tsukahara, S. 1961. *J. Sci. Soil Manure Jap.* 32:261–65
168. Murayama, N., Yoshino, M., Oshima, M., Tsukahara, S., Kawarazaki, K. 1955. *Bull. Nat. Inst. Agr. Sci. Jap.* B4:123–66
169. Nakayama, H. 1969. *Proc. Crop Sci. Soc. Jap.* 38:338–41
170. Nelson, C. D. 1963. *Environmental Control of Plant Growth,* ed. L. T. Evans, 149–74. New York: Academic
171. Nichiporovich, A. A. 1954. *Photosynthesis and the theory of obtaining high crop yields.* 15th Timiryazev Lect. AN SSSR, Moscow, 1956. English Transl. Dep. Sci. Ind. Res. Great Britain, 1959
172. Nichiporovich, A. A. 1967. *Photosynthesis of Productive Systems.* Jerusalem: Israel Program Sci. Transl.
173. Nichiporovich, A. A., Malofeyer, V. 1965. *Fiziol. Rast.* 12:3–12
174. Nosberger, J., Thorne, G. N. 1965. *Ann. Bot. N. S.* 29:635–44
175. Ohta, Y., Yamada, N. 1965. *Nettai Nogyo* 9:76–79
176. Ojima, M., Kawashima, R. 1968. *Proc. Crop Sci. Soc. Jap.* 37:667–75
177. Ojima, M., Kawashima, R., Sakamoto, S. 1968. *Proc. Crop Sci. Soc. Jap.* 37:676–79
178. Okubo, J., Iwata, F. 1968. *Symp. Maize Prod. Southeast Asia.* Agr. Forest Fish Res. Counc. Jap. 131–38
179. Osada, A. 1971. *Bull. Nat. Inst. Agr. Sci.* D14:117–88
180. Oshima, M. 1966. *J. Sci. Soil Manure Jap.* 37:589–93
181. Owen, P. C. 1968. *Aust. J. Exp. Agr. Anim. Husb.* 8:587–93
182. Owen, P. C. 1969. *Exp. Agr.* 5:101–9
183. Ibid 1971. 7:33–41
184. Ibid, 43–47
185. Palmer, A. F. E. 1969. *Translocation pattern of ^{14}C-labelled photosynthate in the corn plant (Zea Mays, L.) during the earfilling stage.* PhD thesis. Cornell Univ., Ithaca, NY
186. Pearce, R. B., Brown, R. H., Blaser, R. E. 1965. *Crop Sci.* 5:553–56
187. Pearce, R. B., Carlson, G. E., Barnes, D. K., Hart, R. H., Hanson, C. H. 1969. *Crop Sci.* 9:423–26
188. Pendleton, J. W., Hammond, J. J. 1969. *Agron. J.* 61:911–13
189. Pendleton, J. W., Smith, G. E., Winter, S. R., Johnston, T. J. 1968. *Agron. J.* 60:422–24
190. Porter, H. K., Pal, N., Martin, R. V. 1950. *Ann. Bot. N.S.* 14:55–68
191. Porter, K. B., Atkins, I. M., Gilmore, E. C., Lahr, K. A. 1964.

Agron. J. 56:393–96
192. Puckridge, D. W. 1968. *Aust. J. Agr. Res.* 19:191–201
193. Ibid, 711–19
194. Ibid, 1969. 20:623–34
195. Puckridge, D. W., Donald, C. M. 1967. *Aust. J. Agr. Res.* 18: 193–211
196. Quinlan, J. D., Sagar, G. R. 1962. *Weed Res.* 2:264–73
197. Quinlan, J. D., Sagar, G. R. 1965. *Ann. Bot. N.S.* 29:683–97
198. Rabinowitch, E. I. 1951. *Photosynthesis and Related Processes*, vol. 2, part 1. New York: Interscience. 1208 pp.
199. Rawson, H. M. 1970. *Aust. J. Biol. Sci.* 23:1–16
200. Rawson, H. M., Evans, L. T. 1970. *Aust. J. Biol. Sci.* 23:753–64
201. Rawson, H. M., Hofstra, G. 1969. *Aust. J. Biol. Sci.* 22:321–31
202. Riley, J. J., Hodges, C. N. 1969. Presented at *Southwest. Rocky Mt. Div. Am. Assoc. Advan. Sci., Colorado Springs, Colo.*
203. Russell, W. A. 1968. *Crop Sci.* 8: 244–47
204. Ryle, G. J. A. 1965. *Ann. Appl. Biol.* 55:107–14
205. Saeki, T. 1960. *Bot. Mag. Tokyo* 73:55–63
206. Satake, T. 1969. *Jap. Agr. Res. Quart.* 4:5–10
207. Shibles, R. M., Weber, C. R. 1965. *Crop Sci.* 5:575–77
208. Shigemura, S. 1966. *Jap. Agr. Res. Quart.* 1:1–6
209. Shimizu, T. 1967. *Sakumotsuno Buttshitsu Seisan* 4:12–26
210. Smith, H. F. 1933. *J. Counc. Sci. Ind. Res. Aust.* 6:32–42
211. Soga, Y., Nozaki, M. 1957. *Proc. Crop Sci. Soc. Jap.* 26:105–8
212. Stansel, J. W., Bollich, C. N., Thysell, J. R., Hall, V. L. 1965. *Rice J.* 68:34–35
213. Stickler, F. C., Pauli, A. W. 1961. *Agron. J.* 53:99–102
214. Ibid, 352–53
215. Stoy, V. 1963. *Physiol. Plant.* 16: 851–66
216. Stoy, V. 1965. *Physiol. Plant. Suppl.* 4p. 1–125
217. Swanson, C. A., Whitney, J. B. 1953. *Am. J. Bot.* 40:816–23
218. Takahashi, J. et al 1955. *Bull. Nat. Inst. Agr. Sci. Jap.* B4:85–122
219. Takahashi, Y., Iwata, I., Baba, I. 1959. *Proc. Crop Sci. Soc. Jap.* 28:22–24
220. Takeda, T. 1961. *Jap. J. Bot.* 17: 403–37
221. Takeda, T., Kumura, A. 1959. *Proc. Crop Sci. Soc. Jap.* 28: 179–81
222. Takeda, T., Maruta, H. 1956. *Proc. Crop Sci. Soc. Jap.* 24: 181–84
223. Tanaka, A. 1958. *J. Sci. Soil Manure Jap.* 29:327–33
224. Tanaka, A., Ishizuka, Y. 1969. *J. Sci. Soil Manure Jap.* 40:113–20
225. Tanaka, A., Kawano, K. 1965. *Soil Sci. Plant Nutr.* 11:251–58
226. Tanaka, A., Kawano, K. 1966. *Plant Soil* 24:128–44
227. Tanaka, A., Kawano, K., Yamaguchi, J. 1966. *Int. Rice Res. Inst. Tech. Bull.* 7
228. Tanaka, A., Navasero, S. A., Garcia, C. V., Parao, F. T., Ramirez, E. 1964. *Int. Rice Res. Inst. Tech. Bull.* 3
229. Tanaka, A., Vergara, B. S. 1967. *IRC Newsletter* special issue, 26–42
230. Tanaka, A., Yamaguchi, J., Fujita, K. 1969. *J. Sci. Soil Manure Jap.* 40:498–503
231. Tanaka, A., Yamaguchi, J., Imai, M. 1971. *J. Sci. Soil Manure Jap.* 42:33–36
232. Tanaka, A., Yamaguchi, J., Shimazaki, Y., Shibata, K. 1968. *J. Sci. Soil Manure Jap.* 39:526–34
233. Tanaka, T., Matsushima, S., Kojo, S., Nitta, H. 1969. *Proc. Crop Sci. Soc. Jap.* 38:287–93
234. Tanner, J. W., Garderner, C. J., Stoskopf, N. C., Reinbergo, E. 1966. *Can. J. Plant Sci.* 46:690
235. Thorne, G. N. 1959. *Ann. Bot. N. S.* 23:365–70
236. Thorne, G. N. 1962. *J. Agr. Sci.* 58:89–96
237. Thorne, G. N. 1962. *Ann. Bot. N. S.* 26:37–54
238. Ibid 1963. 27:155–74
239. Ibid, 245–52
240. Ibid 1965. 29:317–29
241. Thorne, G. N. See Ref. 87, 88–105
242. Thorne, G. N., Ford, M. A., Watson, D. J. 1967. *Ann. Bot. N. S.* 31:71–101
243. Ibid 1968. 32:425–46
244. Thorne, G. N., Welbank, P. J., Blackwood, G. C. 1969. *Ann. Appl. Biol.* 63:241–51
245. Thrower, S. L. 1965. *Aust. J. Biol. Sci.* 18:449–61

246. Tingle, J. N., Paris, D. G., Ormrod, D. P. 1970. *Crop Sci.* 10: 26–28
247. Togari, Y., Kashiwakura, Y. 1958. *Proc. Crop Sci. Soc. Jap.* 27:3–5
248. Togari, Y., Okamoto, Y., Kumura, A. 1954. *Proc. Crop Sci. Soc. Jap.* 22:95–97
249. Togari, Y., Takeda, T., Maruta, H. 1956. *Proc. Crop Sci. Soc. Jap.* 24:254–59
250. Treharne, K. J., Eagles, C. F. 1969. *Progress in Photosynthesis Research,* ed. H. Metzner, 1: 377–82
251. Tsunoda, S. 1959. *Jap. J. Breed.* 9:161–68
252. Ibid, 237–44
253. Ibid 1960. 10:107–11
254. Ibid 1962. 12:49–56
255. Tsunoda, S. 1964. *A Developmental Analysis of Yielding Ability in Varieties of Field Crops.* Nihon-Gakujitsu-Shinkokai, Maruzen. 135 pp.
256. Uchijima, Z., Udagawa, T., Horie, T., Kobayashi, K. 1967. *J. Agr. Meteorol. Tokyo* 23:99–108
257. Vergara, B. S., Chang, T. T., Lilis, R. 1969. *Int. Rice Res. Inst. Tech. Bull. 8*
258. Verhagen, A. M. W., Wilson, J. H., Britten, E. J. 1963. *Ann. Bot. N. S.* 27:627–40
259. Vogel, O. A. 1964. *Crop Sci.* 4: 116–17
260. Vogel, O. A., Craddock, J. C. Jr., Muir, C. E., Everson, E. H., Rohda, C. R. 1956. *Agron. J.* 48:76–78
261. Wada, G. 1968. *Proc. Crop Sci. Soc. Jap.* 37:394–98
262. Ibid, 650–55
263. Wada, G. 1969. *Bull. Nat. Inst. Agr. Sci. Ser. A* 16:27–167
264. Wallace, D. H., Munger, H. M. 1965. *Crop Sci.* 5:343–48
265. Ibid 1966. 6:503–7
266. Wang, T. D., Wei, J. 1964. *Acta Bot. Sinica* 12:154–58
267. Wardlaw, I. F. 1965. *Aust. J. Biol. Sci.* 18:269–81
268. Ibid 1967. 20:25–39
269. Wardlaw, I. F. 1968. *Bot. Rev.* 34:79–105
270. Wardlaw, I. F. 1970. *Aust. J. Biol. Sci.* 23:765–74
271. Wardlaw, I. F., Porter, H. K. 1967. *Aust. J. Biol. Sci.* 20: 309–18
272. Wareing, P. F., Khalifa, M. M., Treharne, K. J. 1968. *Nature* 220:453–57
273. Watson, D. J. 1947. *Ann. Bot. N. S.* 11:41–76
274. Ibid, 375–437
275. Watson, D. J. 1951. *Advan. Agron.* 4:101–44
276. Watson, D. J. 1958. *Ann. Bot. N. S.* 22:37–54
277. Watson, D. J., French, S. A. W. 1958. *Rep. Rothamst. Exp. Sta. 1957.* 90 pp
278. Watson, D. J., French, S. A. W. 1962. *Ann. Appl. Biol.* 50:1–10
279. Watson, D. J., Norman, A. G. 1939. *J. Agr. Sci.* 29:321–46
280. Watson, D. J., Thorne, G. N., French, S. A. W. 1958. *Ann. Bot. N. S.* 22:321–52
281. Ibid 1963. 27:1–22
282. Welbank, P. J., French, S. A. W., Witts, K. J. 1966. *Ann. Bot. N. S.* 30:291–300
283. Welbank, P. J., Witts, K. J., Thorne, G. N. 1968. *Ann. Bot. N. S.* 32:79–95
284. Whittle, C. M. 1964. *Ann. Bot. N. S.* 28:339–44
285. Wilfong, R. T., Brown, R. H., Blaser, R. E. 1967. *Crop Sci.* 7: 27–30
286. Williams, W. A., Loomis, R. S., Duncan, W. G., Dovrat, A., Nunez, F. A. 1968. *Crop Sci.* 8: 303–8
287. Williams, W. A., Loomis, R. S., Lepley, C. R. 1965. *Crop Sci.* 5:211–15
288. Williams, W. A., Loomis, R. S., Reyes, C. R. 1965. *Crop Sci.* 5: 215–19
289. Wittwer, S. H., Robb, W. M. 1964. *Econ. Bot.* 18:34–56
290. Yamada, N. 1961. *Agr. Hort. (Tokyo)* 36:13–18
291. Yamada, N., Murata, Y., Osada, A., Iyama, J. 1955. *Proc. Crop Sci. Soc. Jap.* 24:112–18
292. Yamamoto, K. 1954. *Agr. Hort. (Tokyo)* 29:1161–63, 1303–4, 1425–27
293. Yin, H. Z., Wang, T. D., Li, Y. Z., Qiu, G. X., Yang, S. Y., Shen, G. M. 1960. *Sci. Sinica* 9:790–811
294. Yoshida, S., Ahn, S. B. 1968. *Soil Sci. Plant Nutr.* 14:153–61
295. Yoshida, S., Cock, J. H. 1971. Presented at *12th Pac. Sci. Congr. Canberra*
296. Yoshida, S., Cock, J. H., Parao, F. T. 1971. Presented at *Symp. Rice Breed., Int. Rice Res. Inst.*

AUTHOR INDEX

SUBJECT INDEX

A

Abscisic acid
  on auxin-induced root growth, 243
  on water transport, 167-68
Abscisin
  on enzyme induction, 143
Abscission
  ethylene role in, 260, 278-80
Acer
  transfer cells of, 185, 192
Acetabularia
  blue light effects on, 229
Acetate
  as ethylene precursor, 263, 265-66
Acetylcholine
  and phytochrome, 327-28
Acetyl coenzyme A
  stereospecificity of, 355-56, 358
Acid phosphatase
  ethylene effect on, 279
Aconitase hydratase
  stereospecificity of, 349-51
Aconitate
  cis- and trans-stereospecificity of, 350
Acrylate
  as ethylene precursor, 262
Acrylic acid
  as precursor of ethylene, 268
Actinomycin D
  on ethylene activity, 280
Adenine
  identification of, 7
  and nicotinic acid, 7-8
Adenosine diphosphate (ADP)
  in respiration, 430
Adenosine triphosphatase (ATPase)
  aging effects on, 55
  effect of ethylene on, 276, 279
  water transport across membranes, 163-64, 166-67
Adenosine triphosphate (ATP)
  on ion transport, 57-58, 69
  and salt transport, 378-82
Aging
  ethylene in, 279

and ion uptake, 55, 61
Agropyron
  salinity effects of 2 species of, 384
Agrostis
  salinity adaptation by, 385
Alanine
  blue light effects on formation of, 223
  as ethylene precursor, 262
Albizzia julibrissin
  ion transport in, 212
  phytochrome of, 325, 327
Alcaligenes faecalis
  symmetry studies of, 339
Aleurone transfer cells
  of grass seed, 184
Alfalfa
  basic substances in, 5-8
  yield studies of, 440
Algae
  blue light effects on, 219, 221
  chlorophyll of, 75, 77
  coenocytes of
    ion transport studies in, 51-53
  fluorescence in relation to photosynthesis in
    blue-green, 93-94, 97, 101, 104, 107
    brown, 94, 97-98, 101-2, 107
    green, 89-93, 96-97, 99-101, 103-7
    red, 93-94, 97, 101, 104, 107
  protoplasts of, 32
Alkaloids
  biosynthesis of, 349
Allium
  protoplasts of, 33
Alyssum
  protoplasts of, 29
Amaranthus
  phytochrome, 320
Amaryllidaceae
  transfer cells of, 189
Amino acids
  basic, 8-9
  of phytochrome, 303-4, 316
  revision of nomenclature of, 13-14

δ-Aminolevulinic acid (ALA)
  in photoregulation, 138-39
D-Amino-oxidase
  photoinactivation of, 226
Ammobium
  transfer cells of, 175
Ammonia lyases
  characterization of, 144-45
α-Amylase
  control of synthesis of, 141
  ethylene effect on, 276, 278-79
Anabaena
  chlorophyll of, 77, 80-81
  chlorophyll fluorescence of, 93
Anacystis
  chlorophyll of, 77
  chlorophyll fluorescence of, 93, 104
Anacystis nidulans
  blue light effects on, 221
Anesthetics
  and water permeability of membranes, 164-65
Angiospermae
  protoplasts of, 30
  transfer cells of, 183
Angiosperms
  photoregulation of enzyme synthesis in, 135
Ankistrodesmus braunii
  blue light effects on, 227
Antarafacial
  definition of, 345
Anthocyanin
  synthesis
    regulation of, 140-41
Antiperiplanar
  definition of, 345
Apium graveolens
  ion transport in, 208
Apple
  ethylene in, 265, 267, 273, 275, 280
Arabinofuranose
  in cell wall, 120
Arabinogalactans
  of larch cell wall, 122-24
  formula, 123
Arabino-(4-O-methylglucurono)
  xylans
    of cell wall, 120
Araceae

CUMULATIVE INDEXES

VOLUMES 14-23

INDEX OF CONTRIBUTING AUTHORS